The development of the supersymmetry technique has led to significant advances in the study of disordered metals and semiconductors. The technique has proved to be of great use in the analysis of modern mesoscopic quantum devices, but is also finding applications in a broad range of other topics, such as localization and quantum chaos. This book provides the first comprehensive treatment of the ideas and uses of supersymmetry.

The first four chapters set out the basic results and some straightforward applications of the technique. Thereafter, a range of topics is covered in detail, including random matrix theory, the physics of small metal particles, persistent currents in mesoscopic rings, transport in mesoscopic devices, localization in quantum wires and films, the Anderson metal-insulator transition, and the quantum Hall effect.

Each topic is covered in a self-contained manner, and the book will be of great interest to graduate students and researchers in condensed matter physics, quantum chaos, statistical mechanics, and quantum field theory.

Supersymmetry in disorder and chaos

Supersymmetry in disorder and chaos

KONSTANTIN EFETOV

Landau Institute for Theoretical Physics, Moscow

CAMBRIDGE
UNIVERSITY PRESS

PUBLISHED BY THE PRESS SYNDICATE OF THE UNIVERSITY OF CAMBRIDGE
The Pitt Building, Trumpington Street, Cambridge CB2 1RP, United Kingdom

CAMBRIDGE UNIVERSITY PRESS
The Edinburgh Building, Cambridge CB2 2RU, United Kingdom
40 West 20th Street, New York, NY 10011–4211, USA
10 Stamford Road, Oakleigh, Melbourne 3166, Australia

First published 1997

Printed in the United States of America

Typeset in Times Roman

Library of Congress Cataloging-in-Publication Data
Efetov, Konstantin
Supersymmetry in disorder and chaos / Konstantin Efetov.
p.   cm.
Includes bibliographical references.
ISBN 0–521–47097–8 (hc : alk. paper)
1. Supersymmetry.   2. Supersymmetry – Industrial applications.
3. Order-disorder in alloys.   4. Metals – Surfaces.
5. Semiconductors.   6. Quantum chaos.   I. Title.
QC174.17.S9E37   1996
530.4'13 – DC20   96-1137

*A catalog record of this book is available from
the British Library*

ISBN 0–521–47097–8 hardback

Contents

Preface

Recent progress in physics of disordered metals and semiconductors has led to the development of theoretical methods adequate for their description. Now, it is completely clear that such disciplines of theoretical physics as theories of disorder and quantum chaos are necessary to describe, for example, modern mesoscopic quantum devices. Moreover, these disciplines are converging toward each other, an exciting theoretical development. Although a lot of information can be obtained from numerical simulations, an analytical approach unifying disorder and chaos is definitely desirable. Besides, numerical simulation is often not conclusive and one has to have an analytical tool for calculations.

Currently the most efficient analytical method enabling us to achieve both goals is the supersymmetry technique, and many problems of disorder and chaos can be studied with a supermatrix nonlinear σ-model. The number of publications using the supersymmetry technique has been growing fast in the last 2–3 years. At the same time, many people still have a hesitation to start study of the method. The main reason is that they are afraid that manipulating the Grassmann anticommuting variables is something very difficult and, what is more important, that having spent a considerable time learning the technique, they would be able only to reproduce results that could be obtained by other more standard techniques. Such an attitude is to a great extent due to absence of a self-contained literature on the subject. Only two reviews have been written, and they were written more than 10 years ago.

This book is an attempt to present the subject in a self-contained way. In principle, the reader is supposed to be familiar only with general university courses on mathematics and physics (mathematical analysis, theory of functions of complex variables, electrodynamics, and quantum mechanics). Of course, knowledge of more advanced topics, such as theory of superconductivity and field theory, can make understanding the material easier because many ideas of the supersymmetry method are similar to those in the other fields. Everything else (including the definition of the Grassmann variables, Green functions, diagrammatic expansions, diffusion modes) is introduced in the book.

This includes not only formal mathematical objects and schemes but also new physics. Many different physical topics are considered by the supersymmetry technique (localization, mesoscopics, quantum chaos, quantum Hall effect, etc.), and each section begins with an extended introduction to the corresponding physics. I hope this book will be useful for readers with a different grounding. Those who want just to get an idea of mathematical aspects of supersymmetry may limit themselves to reading the chapter on supermathematics. Those who want to get a basic knowledge of what is going on in such topics as localization, persistent currents, transport through quantum dots, and small metal particles can read the first sections of the corresponding chapters. Everywhere I have tried to separate discussion of physical effects from presentation of mathematical details so that the reader can easily skip mathematical material that seems too complicated.

At the same time, I have tried to present all the most important achievements so that the present state of the art will be clear. For a more experienced reader, the book can serve as a review of what has been done recently. The book is conceived also to make it possible for a well-motivated beginner to start studying the supersymmetry technique, to learn where it can be useful, and, finally, to be abreast of today's research.

The book contains more "disorder" than "chaos." This is mainly because the super-symmetry technique was invented initially for disorder problems and only recently began intensively to penetrate quantum chaos, where study has been based mainly on numerical simulations and semiclassical formulae. Besides, many books on quantum chaos (without use of supersymmetry) have already been written, and there is no reason to repeat their contents here. Nevertheless, I hope that the present book can be quite useful for the "chaotic" community, too.

Stuttgart, August 1995

Acknowledgments

The most important results forming the basis for this book were obtained at the Landau Institute for Theoretical Physics at the beginning of the 1980s. It was a beautiful time when several generations of famous scientists worked together. Experts in condensed matter, field theory, and mathematics had joint seminars and discussed their problems. Now I feel that at that time new ideas were "hanging in the air." I am not sure that I would have been able to do the same work if I had been in another place. I am especially grateful to my teacher, Anatoly Larkin, for invaluable discussions. He drew my attention to several problems presented in this book and with him I completed several works on disorder.

During that period I learned a lot on disorder-related problems and field theory from discussions with Alexei Abrikosov, Alexander Finkelshtein, Lev Ioffe, David Khmelnitskii (with whom I also have a joint publication), Alexander Polyakov, and Paul Wiegmann. I thank also Tomas Bohr and Vladimir Marikhin for joint work on several problems.

I thank my co-workers in Stuttgart who worked with me on the subject presented in this book: Vladimir Falko, Shinji Iida, Stefan Kettemann, Peter Kopietz, Vladimir Prigodin, and Olaf Viehweger. In the last few years I have had intensive interactions with my colleagues and friends working on the problems of disorder and chaos using the supersymmetry technique. I am especially grateful to Alexander Altland, Boris Altshuler, Anton Andreev, Carlo Beenakker, Yan Fyodorov, Vladimir Kravtsov, Igor Lerner, Eduardo Mucciolo, Boris Muzykantskii, Ben Simons, Nabuhiko Taniguchi, Hans Weidenmüller, and Martin Zirnbauer. Useful discussions with Charlie Marcus that motivated a publication included in the book are appreciated.

I am grateful to Peter Fulde for his interest in this book and in the results of the research presented in it.

The book was written in Stuttgart during my stay as a guest at the Max-Planck-Institut für Festkörperforschung and the Max-Planck-Institut für Physik Komplexer Systeme. I thank these institutes for their hospitality.

1

Introduction

1.1 Historical remarks

The last 15–20 years has witnessed spectacular progress in the study of disordered metals and semiconductors. These systems are interesting not only from the point of view of different technical applications but also because they reveal new unusual physical properties that are very different from what one would expect in clean regular materials. Although very often thermodynamic characteristics are already quite influenced by disorder, the most remarkable effects are observed in kinetics. Of course, in many cases one may use the classical transport theory based on the Boltzmann equation for a description of electron motion. However, if the disorder is strong or temperature is low, quantum effects become important, and to construct a theory in this situation one has to start from the Schrödinger equation in a potential that is assumed to be random.

To get information about physical properties of the system one has to solve the Schrödinger equation for an arbitrary potential, calculate a physical quantity, and, at the end, average over the random potential. Sometimes it is important to have information not only about the average but also about fluctuations. In this case one has to calculate moments of the physical quantities and even an entire distribution function. Needless to say, generally speaking, this program cannot be carried out exactly even in the absence of electron–electron interaction and one should use different approximation schemes.

For quite a long time the only method of analytical study has been a perturbation theory in the random potential. This procedure can be performed in a convenient way using Green functions. Considerable progress has been achieved by applying this method to studying the phenomenon that is now known as "weak localization." A very important concept that has been developed is that of "diffusion modes." These modes describe multiple quantum interference of electron waves and are very sensitive to changing magnetic field, temperature, and other experimental parameters. This sensitivity leads to very strong dependence of physical quantities such as conductivity on the parameters, and this was confirmed in the beginning of the 1980s by numerous experiments. Somewhat later diagrammatic expansions proved to be very useful for investigation of mesoscopic effects, and such a famous phenomenon as "universal conductance fluctuations" was described in this way.

Although very powerful and useful, the diagrammatic expansions work well only when the quantum effects are weak and the properties of the system are almost classical. By increasing the disorder one can approach the Anderson metal–insulator transition and finally enter the insulating state. In this situation the perturbation theory is no longer valid. The diagrammatic expansion is also not very useful in weakly disordered mesoscopic conductors if they are completely isolated or weakly connected to leads. At the same time, localization is also possible in weakly disordered conductors of a one-dimensional geometry (wires), and this is one more example of instances in which a diagrammatic approach does not work.

One can continue the list of nontrivial problems that cannot be attacked using the perturbation theory. All this demonstrates that nonperturbative methods are in great demand. In fact, disordered systems can be treated by the so-called replica method. Within this approach one substitutes the initial disordered system by n identical systems. This makes it possible to average over the disorder in the very beginning. An average physical quantity can be obtained by calculating the partition function for all integers n and taking at the end the limit $n \to 0$. This method very often gives quite reasonable results for very nontrivial systems (e.g., spin glasses). Using the replica trick one can reduce the problem of localization (weak or strong) to matrix nonlinear σ-models containing $n \times n$ matrices.

At first glance, the replica method, being very general, should be very useful for studying electron motion in a random potential. However, up to now the replica nonlinear σ-models helped only to reproduce diagrammatic expansions and demonstrate renormalizability. Except for computation of the one-particle density of states in several situations, no reliable nonperturbative results have been obtained in localization theory as yet. The main difficulty of the replica method is that it is usually impossible to compute physical quantities exactly for an arbitrary n and one has to make approximations. However, the approximation may be good for all integers n but fail when taking the limit $n \to 0$. This is what happens with the replica nonlinear σ-models.

In contrast, the supersymmetry method based on using both commuting and anticommuting variables is free of these difficulties. Although this technique is not as general as the replica method and cannot be applied, for example, to interacting particles, it is perfectly suitable for studying localization, mesoscopic effects, integer quantum Hall effect, and so on. Recently it was discovered that it can even help in computing correlation functions for one-dimensional models of particles with some special interparticle interaction.

Using the supersymmetry approach one can also derive a nonlinear σ-model. This model contains 8×8 supermatrices (matrices consisting of both commuting and anticommuting elements) Q with the constraint $Q^2 = 1$. The matrix size is not very large and this makes direct calculations feasible. Of course, using the supermatrix σ-model one can reproduce diagrammatic expansions and demonstrate renormalizability, but this is the least of its merits. The supermatrix σ-model has been solved in many more interesting situations and in most cases this was the first solution of the corresponding problems.

The first nontrivial situation where the σ-model was solved is a small disordered metal particle. To find, for example, a level–level correlation function one has to solve a zero-dimensional version of the σ-model. Technically this means that one should calculate a definite integral over the elements of the supermatrix Q. The computation of the corresponding integrals proved not to be very difficult, although attempts to do the same with the replica trick failed completely. The computation of the level–level correlation function of a disordered metal particle turned out to be extremely important not only in itself. Quite unexpectedly the correlation function exactly coincided with the corresponding function known from random matrix theory. The latter theory was initially proposed on a purely phenomenological basis for a description of complex nuclei. The solution of the zero-dimensional σ-model was the first analytical confirmation of the random matrix theory. Moreover, somewhat later the zero-dimensional σ-model was derived from random matrix theory, thereby demonstrating the equivalence of the two theories.

At approximately the same time (beginning of the 1980s) problems of quantum chaos started to attract attention. The phenomenon of quantum chaos proved to be quite general. It was discovered that the level statistics in, for example, a ballistic billiard was strongly dependent on whether the classical motion in this system was regular or chaotic. It was demonstrated in many numerical simulations that correlation functions of chaotic systems agreed very well with the predictions of random matrix theory, whereas the level statistics of regular (integrable) billiards obeyed the Poisson law.

Thus, one comes to the conclusion that the problems of disorder are equivalent to the problems of quantum chaos and even to those of physics of complex nuclei. The supersymmetry technique plays a very important role in this unification. All these theories are very popular now also because of new developments in the physics of mesoscopic objects. Recent success in nanotechnology has made it possible to make very small devices and such theoretical models as quantum billiards can be directly applied now to some experimentally available objects (quantum dots). The zero-dimensional σ-model is thus a very convenient and adequate tool for a theoretical description of these mesoscopic objects.

Recently, it was found that the one-dimensional supermatrix σ-model describing disordered wires is equivalent to random band matrices. The latter were used to describe such quantum chaos problems as the kicked rotator problem. So, the similarity between problems of disorder and problems of chaos is really deep and fully deserves further investigation.

However, it makes sense to stress again that the supermatrix σ-model not only is a way to establish a connection between the different disciplines but, primarily, is a powerful method of calculation. Even if a problem can be reduced to the random matrix theory, the zero-dimensional σ-model is more convenient for computations and very often it is the only way to get explicit results. As concerns higher dimensions, in most cases the supermatrix σ-model is the only tool now available for an analytical study of nonperturbative problems.

1.2 What is this book about?

The main idea of this book is to present the supersymmetry method and to show how it can help to solve very important physical problems. The reader has no need to know any special mathematics. All necessary mathematics that is beyond the scope of standard university courses in physics is given in Chapter 2, which contains the definition of Grassmann anticommuting variables, supervectors, supermatrices, and operations with these objects. It is shown how one can introduce integrals over the Grassmann variables and how one can change variables. Some important integrals are computed. The reader who wants to be able to follow the details of the calculations in the rest of the book should look through Chapter 2. Everywhere I have tried to avoid complicated mathematical notations that very often frighten physicists. In fact, except for Chapter 2 the book does not contain any new mathematics, and this can be an attractive feature for a broad physics audience.

The number of problems solved now by the supersymmetry technique is quite large and this method can be considered as unifying different fields. This gives a unique opportunity to expound on such different topics as localization, mesoscopics, quantum chaos, integer quantum Hall effect, and even some one-dimensional models of interacting particles using just the supermatrix σ-model. The book is an attempt to bring about the idea. Each chapter is devoted to a new physical topic. Therefore, to make the book self-contained, each chapter starts with an extended introduction to physical problems considered there. This can make it useful even for those readers who do not want to follow the calculations but would like to get a general idea of the subject of the chapter. Everywhere in the text I have tried to separate formulation of problems and results from computations to make reading without going into mathematical details possible.

Symmetries of the σ-model reflect some hidden symmetries existing in disordered and chaotic systems. This means that the basic physics of the completely different physical objects considered in the book is dependent solely on some underlying symmetries. Possibly, this can give a second meaning to the word *supersymmetry* used in the title.

Chapter 3 has an introductory character and is devoted to a discussion of diffusion modes. The reader can learn that it is the quantum interference that is responsible for the well-known effect of weak localization and understand from simple qualitative arguments why conductivity can be very sensitive to an external magnetic field and inelastic processes. It is shown how one should use Green functions to make expansions in a random potential and how one can obtain the diffusion modes summing a certain class of diagrams. It is argued that these diffusion modes are a manifestation of quantum interference.

Chapter 4 is most important because in it the supermatrix σ-model is derived. To simplify understanding the scheme of the derivation, some kind of a "mean field" approximation is used in the beginning. This approximation is very similar to the BCS theory of superconductivity and the analogy with superconductivity is widely used in the book. Then, a more formal derivation based on a Hubbard–Stratonovich transformation is given. The mean field approximation corresponds to a saddle point of an effective Lagrangian. The saddle-point approximation leads to the constraint $Q^2 = 1$ for the supermatrices Q, which corresponds to the well-known phenomenon of spontaneous breaking of symmetry. Goldstone modes appearing as a consequence of the breaking are just diffusion modes, and they are described by the σ-model. It is shown how one can include magnetic and spin-orbit interactions.

In Chapter 5 it is demonstrated how all weak localization effects can be derived using a perturbation theory for the σ-model. This perturbation theory works provided the frequency or inverse time of inelastic scattering is large. It is also shown that for the supermatrix σ-model one can use a renormalization group approach that has been popular for studying conventional σ-models. Using a standard scheme, renormalization group equations are obtained, and a discussion of results for physical quantities is presented. It is shown how by using the scheme in $2 + \epsilon$ dimensions one can get information about the metal–insulator transition in three dimensions.

Chapter 6 begins with an introduction of the Wigner–Dyson random matrix theory. Then it is shown that the problem of level statistics in small disordered metal particles is described by the zero-dimensional σ-model. It is demonstrated in detail how one can calculate integrals using a parametrization for the matrices Q. At the end of the chapter, it is shown that the zero-dimensional σ-model can be derived directly from random matrix theory. Reading this chapter is important because it contains all technical details of the computation of integrals over the supermatrices Q, and the scheme of the computation will be used everywhere in subsequent chapters. This chapter also bridges disorder problems with random matrix theory and, hence, with quantum chaos.

Chapter 7 is devoted to a study of physical properties of small metal particles. After a general introduction to the physics of these experimentally available objects some early phenomenological description is presented. Then, using the zero-dimensional σ-model, electric susceptibility as a function of frequency is calculated. The static electric susceptibility is shown to be dependent on magnetic and spin-orbit interactions, a relationship that is reminiscent of the effects of weak localization. The second problem considered in the chapter is nuclear magnetic resonance. To obtain the resonance line shape one has to calculate the distribution function of local density of states. This is done explicitly and it is shown that at low temperatures or small particle sizes the resonance line becomes very broad and asymmetric.

In Chapter 8 a popular problem of persistent currents in mesoscopic rings is considered. Several theoretical approaches developed recently are presented, including calculations within a canonical ensemble and a dynamic approach. It is demonstrated why the diffusion modes become important in this purely thermodynamic problem. It is shown that at low temperatures or in the use of the dynamic approach one inevitably comes to the σ-model. Using the zero-dimensional version the persistent current as a function of the magnetic flux through the ring is calculated and the results are compared with results of numerical simulations. Some interesting effects of spin-orbit interactions are considered.

Chapter 9 is devoted to studying transport through mesoscopic devices (quantum dots). This is the "hottest" topic now because different experiments are being carried out. The chapter starts by presenting the theory of universal conductance fluctuations. The derivation can be performed by using the perturbation theory in the diffusion modes. However, by making the coupling to the leads weaker one gets into a regime where the perturbation theory is no longer valid and it is necessary to use the σ-model. Now the σ-model must contain terms describing the coupling to the leads. Treating the leads is presented in detail. After the σ-model is derived, different physical quantities are calculated. In particular, the average conductance and conductance variance as functions of the number of channels in the leads are presented. To make a comparison with existing experiments temperature effects are considered. Finally, the entire distribution function of conductances is obtained.

In the rest of the chapter a statistical theory of Coulomb blockade oscillations is developed. By simple arguments it is shown that the distribution function of conductance peaks is related to a distribution function of wave functions. The latter function is calculated exactly at an arbitrary magnetic field.

In Chapter 10 a new universality in spectra of chaotic systems is discussed. This universality shows up in correlation functions of physical quantities taken at different values of a parameter, for example, an external field. After a proper rescaling the functions prove to be universal, or, in other words, they do not depend on any parameters. A Brownian motion model is discussed and the relation between the correlation functions at different parameters computed with the zero-dimensional σ-model and corresponding functions for the Brownian motion model is established. The Brownian motion models are related to one-dimensional models of interacting particles with the interaction $1/r^2$, where r is the distance between the particles. Thus, by using the zero-dimensional σ-model it became possible to compute some correlation functions for the one-dimensional models for the first time. Moreover, these models can be equivalent to continuous matrix models analogous to those used in string theories and theories of quantum gravity. The chapter ends with a discussion of connections among all these different theories.

Chapter 11 begins with a discussion of problems that can be mapped onto the one-dimensional σ-model. As concerns disordered systems, the one-dimensional σ-model can be derived from a model of a conductor with one-dimensional geometry (wire). However, it can also be derived from band matrix models used to describe some systems with chaotic motion. Moreover, provided some assumptions are made, a kicked rotator model well known to exhibit crossover from quantum chaos to dynamic localization can be described by a model with a disorder that, in its turn, can be mapped onto the one-dimensional σ-model.

Thus, the discussion demonstrates that the one-dimensional σ-model is relevant for very important problems of both disorder and chaos.

As the next step, it is shown how to solve the one-dimensional model. A transfer matrix technique is developed. In the limit of weak disorder the problem of computation of correlation functions is reduced to solving partial differential equations that become rather simple in the most interesting limit of low frequencies. It is demonstrated that an infinite wire is an insulator for any weak disorder. Parameters characterizing the insulator are calculated. The chapter continues with computation of an inverse participation ratio and conductance of finite wires. The result obtained for the former problem makes it possible to check some scaling relations derived from numerical simulations for the kicked rotator model.

At the end of the chapter a model of a chain of coupled metallic grains that is assumed to mimic a strong disorder is considered. If the coupling constant is not very large, one obtains integral equations instead of the differential equation. The physical quantities are calculated explicitly in this case, too. An interesting phenomenon, namely, dynamic diamagnetism due to localization in a chain of metal rings, is discussed.

Chapter 12 is devoted to investigation of the Anderson metal–insulator transition. A lattice version of the σ-model that corresponds to a system of coupled metallic grains is proposed. Changing the coupling constant between the grains one can have the transition. The chapter starts with a discussion of different general schemes for studying phase transitions. It is emphasized that the renormalization group scheme in $2 + \epsilon$ dimensions may miss very important features.

After that, an effective medium approximation is suggested and it is argued to be quite a natural scheme of calculations. By following this scheme a nonlinear integral equation that is solved near the transition is derived. Because of the noncompactness of the symmetry group of the supermatrices Q very unusual critical behavior, which is very different from that predicted by the renormalization group calculations in $2 + \epsilon$ dimensions, is obtained. A physical picture explaining the results is suggested.

The chapter then considers the metal–insulator transition on a Bethe lattice. It is shown that the effective medium approximation becomes exact on such a lattice. It is argued then that this model may be quite relevant for a description of transport in highly conducting polymers. The chapter ends with a discussion of models of random sparse matrices. Calculation of physical quantities in these models can be reduced to solving nonlinear integral equations analogous to those obtained on the Bethe lattice.

Chapter 13 contains a discussion of some nonperturbative problems of two-dimensional disordered metals. The first problem considered is that of electron motion in a strong magnetic field. It is shown how one can compute the density of states using the supersymmetry technique and prove localization of eigenstates that are not located very close to the centers of Landau levels. This results in the existence of Hall plateaus, thus demonstrating that the integer quantum Hall effect can be obtained from the model under consideration. To describe transitions between the plateaus, a σ-model with an additional topological term is derived. It is argued that this model may be sufficient to describe the critical behavior near the transitions.

The second part of the chapter deals with wave functions of an isolated sample that is smaller than the localization length. A magnetic field is assumed to be weak or absent. The distribution function of local amplitudes is calculated. Although the same distribution function is considered in Chapter 9, here the computation is not restricted by the use of the zero-dimensional σ-model. It is demonstrated that the asymptotics of the distribution function at large amplitudes is determined by nontrivial saddle points of a reduced σ-model and obeys a log-normal law. By calculating moments of the distribution it is demonstrated that the eigenstates exhibit multifractal behavior.

The Appendices contain some of the technical details of the calculations presented in the book.

Thus, many very different physical phenomena are considered. At the same time, I want to emphasize that the book is not a collection of complete reviews on localization, mesoscopics, quantum chaos, and so forth. Its main purpose is to demonstrate that these fields can be described in an efficient way by using only one general scheme. Therefore, discussion of some interesting physical effects may be absent only because it would not require supersymmetry. During the process of writing many publications containing calculations with the supersymmetry technique have appeared. It was not easy to update the book after every publication and therefore some recent works are not mentioned. I hope nevertheless that the reader can get an impression of what is currently being done in the field.

2

Supermathematics

2.1 What is supermathematics?

All supersymmetric theories are based on the use of anticommuting classical variables first introduced by Grassmann in the last century. At first glance, these objects look very artificial and seem to have no relation to the real world. There is a certain threshold for physicists to start using the Grassmann anticommuting variables for calculations because one expects the game to have very unusual rules. Surprisingly, it is not true, and provided proper definitions are given, one can simply generalize conventional mathematical constructions so that it is possible to treat both commuting and anticommuting variables on an equal footing. Sometimes the corresponding branch of mathematics is called supermathematics.

Of course, the main purpose of this book is to consider different physical results obtained with the use of the Grassmann variables, and therefore one could try to demonstrate how these variables work while making some concrete calculations. However, it seems to be more reasonable to present the basic formulae of supermathematics in one place, first, because it may be the best way to get used to the anticommuting variables, and, second, because one can see that practically all the rules of operating with "superobjects" are quite standard.

Today the mathematical analysis and algebra of functions of both commuting and anticommuting variables are very well developed. Historically, the use of anticommuting variables by Martin (1959) and the introduction of integrals over them by Berezin (1961, 1965) can be considered as the beginning of supermathematics. Shortly afterward, many mathematical aspects were developed in a number of works (Pakhomov (1974), Leites (1975), Berezin (1983)).

Simultaneously with the purely mathematical activity applications by Gol'fand and Lichtmann (1972), by Volkov and Akulov (1973), and by Volkov and Soroka (1973) in quantum field theory appeared. After the famous work by Wess and Zumino (1974) interest in supermathematics became general. In the first works the supersymmetry between commuting and anticommuting variables played an auxiliary role, helping to write Ward identities in complex field theories. According to the modern point of view the supersymmetries in field theories reflect the existence of a fundamental symmetry between coordinates and fields.

The word *supersymmetry* has appeared in condensed matter physics in a publication by Parisi and Sourlas (1979), who discovered a complex symmetry in a model describing ferromagnets in a random magnetic field. They used a concept of superspace including both commuting and anticommuting variables. By averaging over the random field they managed to reduce calculations to studying a regular model in superspace.

Later it was found that field theoretical n-component models with ϕ^4 interactions are equivalent in the artificial limit $n = 0$ to those containing both commuting and anticommuting fields (Mc Kane (1980), Parisi and Sourlas (1981)). The latter models have symmetry under mixing of these two types of the fields without mixing the fields and coordinates. This symmetry can also be called supersymmetry, although it differs from the supersymmetry under mixing anticommuting fields and coordinates. The models with ϕ^4 interaction arise naturally in the problem of a particle moving in a random potential, in the physics of polymers, in the theory of random matrices, and so on.

In this class of problems the supersymmetry approach turned out to be an extremely powerful method of calculations. Quite a large number of different physical problems have been solved. Some of them can be found in the reviews by Efetov (1983, 1990) and by Verbaarschot, Weidenmüller, and Zirnbauer (1985).

In several reviews and books properties of mathematical objects containing both commuting and anticommuting variables and operations using them are described in great detail (see, for example, Berezin (1979, 1983, 1987), Corwin, Neemann, and Sternberg (1975), Kostant (1977), DeWitt (1984)). Readers who are interested in mathematical aspects of the superobjects and in rigorous proofs should read these references. However, the main idea of the present book is to present the supersymmetry method for the audience not familiar with complicated mathematics and not much interested in rigorous proofs. Besides, many formulae that can be useful for calculations are often presented in the mathematically oriented articles in an implicit form and sometimes are absent altogether. Therefore, I will present in this chapter the basic concepts and formulae of the supermathematics in a simple and self-contained form. Reading the mathematical introduction does not require special knowledge but is necessary for understanding the material presented in the book.

2.2 Grassmann variables

The Grassmann anticommuting variables χ_i, $i = 1, 2, \ldots, n$ (the elements of a Grassmann algebra) are introduced in a completely formal way. Construction of new objects is usual for mathematicians and they are quite satisfied by formal definitions. To the contrary, physicists usually try to find some physical correspondence to mathematical symbols and one can hear very often questions about the physical meaning of the Grassmann variables. Unfortunately, in many applications it is not easy to answer this question. I can say only that these mathematical objects help a lot in calculations. We know that in many cases abstract mathematical constructions drastically influence the development of physics. For example, nobody can dispute the usefulness of complex numbers for physics, but what is the physical meaning of $\sqrt{-1}$?

So, I will not discuss how one should imagine the Grassmann variables χ_i intuitively and start with a formal definition. These variables are some mathematical objects obeying the following anticommutation rules (Berezin (1965));

$$\{\chi_i, \chi_j\} \stackrel{\text{def}}{=} \chi_i \chi_j + \chi_j \chi_i = 0 \tag{2.1}$$

for any $1 \le i, j \le n$.

The anticommutation rules Eq. (2.1) hold in particular for $i = j$ and we see that the square of an arbitrary variable χ_i is zero

$$\chi_i^2 = 0 \tag{2.2}$$

I want to emphasize again that the Eqs. (2.1, 2.2) formally describe the rules of a game that involves some completely abstract objects. One does not need to try to imagine these objects as being like matrices or other more familiar mathematical quantities.

Because of the very important property Eq. (2.2), any function K of the anticommuting variables is a finite polynomial

$$K(\chi_1, \chi_2, \ldots, \chi_n) = \sum_{\alpha_i = 0, 1} a(\alpha_1, \alpha_2, \ldots, \alpha_n)\chi_1^{\alpha_1}\chi_2^{\alpha_2} \cdots \chi_n^{\alpha_n} \qquad (2.3)$$

It is convenient to define some important operations analogous to those used for conventional numbers. Let us assume that for any element χ one can put into correspondence another element χ^* and let us call the element χ^* the complex conjugate to χ. The variables χ_i^* anticommute with each other and with all the variables χ_i. By the complex conjugate of a product of variables $\chi_1, \ldots, \chi_n$ we mean the product of the complex conjugates of these variables

$$(\chi_1\chi_2 \cdots \chi_n)^* = \chi_1^*\chi_2^* \cdots \chi_n^* \qquad (2.4)$$

Now we have to define what the complex conjugate of the complex conjugate $(\chi_i^*)^*$ is. In principle, one has different options and it is a matter of taste to choose the proper one. Let us do it as follows:

$$(\chi_i^*)^* = -\chi_i \qquad (2.5)$$

Such a definition has been used in different books and articles (Berezin (1983), Rittenberg and Scheunert (1978), Efetov (1982a, 1983)).

At first glance, the definition Eq. (2.5) is very strange because it differs in the sign of the right-hand side from the corresponding one for conventional numbers. However, it is quite convenient for the anticommuting variables; for example, for the quantity $\chi_i^*\chi_i$ we have

$$(\chi_i^*\chi_i)^* = -\chi_i\chi_i^* = \chi_i^*\chi_i \qquad (2.6)$$

We see from Eq. (2.6) that the quantity $\chi^*\chi$ does not change under complex conjugation and therefore can be considered as "real." It is an analogue of the square of the modulus for conventional complex numbers.

One can introduce linear transformations of a set of anticommuting variables ρ_i, $i = 1, 2, \ldots, n$, to the set of variables χ_i

$$\chi_i = \sum_{k=1}^{n} a_{ik}\rho_k \qquad (2.7)$$

Calculating the product of different χ_i we can check the following identity:

$$\chi_1\chi_2 \cdots \chi_n = \rho_1\rho_2 \cdots \rho_n \det a \qquad (2.8)$$

Using the same procedure as for functions of the conventional variables one can introduce derivatives of a function Ω of the anticommuting variables, distinguishing, however, between left and right derivatives. As usual, one changes a variable $\chi_i : \chi_i \rightarrow \chi_i + \Delta\chi_i$ and calculates the increment $\Delta\Omega = \Omega(\chi_1, \chi_2, \ldots, \chi_i + \Delta\chi_i, \ldots, \chi_n) - \Omega(\chi_1, \chi_2, \ldots, \chi_i, \ldots, \chi_n)$. Because of the property Eq. (2.2) the function Ω can contain only the first power of χ_i for an arbitrary i or does not contain this variable at all. Therefore, in order to calculate $\Delta\Omega$ one should substitute the variable χ_i in Ω by $\Delta\chi_i$. Then, using the anticommutation relations Eq. (2.1), one can transfer $\Delta\chi_i$ either to the right or to the left. The left (right) derivative of the function Ω with respect to χ_i is defined as the coefficient in the dependence of the increment $\Delta\Omega$ on $\Delta\chi_i$ standing after (before) $\Delta\chi_i$.

Using the notation $\overrightarrow{\partial}/\partial\chi_i$ for the left derivative and $\overleftarrow{\partial}/\partial\chi_i$ for the right one we have by the definition

$$\Delta\Omega = \Delta\chi_i(\frac{\overrightarrow{\partial}}{\partial\chi_i}\Omega) = (\Omega\frac{\overleftarrow{\partial}}{\partial\chi_i})\Delta\chi_i \tag{2.9}$$

The left derivative of a homogeneous polynomial of nth degree coincides with the right derivative for odd n and differs by the sign from the right derivative for even n.

The anticommutation of the variables χ_i leads to the anticommutation of the operations of differentiation over several variables, for example,

$$\frac{\partial^2\Omega}{\partial\chi_1\partial\chi_2} = -\frac{\partial^2\Omega}{\partial\chi_2\partial\chi_1} \tag{2.10}$$

As we will see later the definition of the derivatives coincides with the definition of an integration over the Grassmann variables.

The formulae written previously define operations with the anticommuting variables and enable us to introduce more complex objects containing both commuting and anticommuting variables.

2.3 Supervectors and supermatrices

2.3.1 Definitions and basic properties

As the reader has possibly realized, throughout this book I use the word *super* to denote everything containing both commuting and anticommuting variables. I am aware of the fact that for some more mathematically oriented readers such notation may seem to be "super" simplified. Other notations are used by many authors of papers on field theory and statistical mechanics. For example, the mathematical objects that I call supervectors and supermatrices are called graded vectors and graded matrices. A space of the commuting and anticommuting elements is a graded space.

However, for a condensed matter physicist all these notions are not very familiar and writing correct mathematical words would require additional definitions or references to mathematical literature. As a result it would lead to additional labor and, possibly, to a decision not to read such a sophisticated book at all. At the same time the supersymmetry method presented is only a calculational tool, and one can see how the mechanism works without knowing what its details are called in scientific language. Those who are interested in the correct notation can find it in the review by Verbaarschot, Weidenmüller and Zirnbauer (1985). The following notations are the same as in the review by Efetov (1983). In many cases they coincide with those introduced in the book by Berezin (1983).

A supervector Φ is defined as a sequence of anticommuting χ and commuting S variables

$$\Phi = \begin{pmatrix} \chi \\ S \end{pmatrix}, \qquad \chi = \begin{pmatrix} \chi_1 \\ \chi_2 \\ \vdots \\ \chi_n \end{pmatrix}, \qquad S = \begin{pmatrix} S_1 \\ S_2 \\ \vdots \\ S_m \end{pmatrix} \tag{2.11}$$

The transposed supervector Φ^T is given by the expression

$$\Phi^T = (\; \chi^T \quad S^T \;), \tag{2.12}$$

$$\chi^T = (\; \chi_1 \quad \chi_2 \quad \cdots \quad \chi_n \;), \qquad S^T = (\; S_1 \quad S_2 \quad \cdots \quad S_m \;)$$

Addition of the supervectors is performed as usual by the addition of the corresponding components. The supervectors can be multiplied by a number by multiplying all the components by this number.

Let us introduce the Hermitian conjugate Φ^+ of the supervector Φ by the following relation:

$$\Phi^+ = (\Phi^T)^* \tag{2.13}$$

The operation of the complex conjugation $*$ on the anticommuting variables is specified by Eqs. (2.4, 2.5). Its action on conventional complex numbers is given by the conventional rules.

Now, by analogy with the corresponding definition for conventional vectors we can introduce a scalar product of two supervectors Φ^i and Φ^j

$$\Phi^{i+}\Phi^j = \sum_{\alpha=1}^{n} \chi_\alpha^{i*}\chi_\alpha^j + \sum_{\alpha=1}^{m} S_\alpha^{i*}S_\alpha^j \tag{2.14}$$

The expression $\Phi^{i+}\Phi^j$ in this equation represents a conventional matrix product of the supervectors Φ^{i+} and Φ^j, Eqs. (2.11–2.13). The scalar product, Eq. (2.14), for $i = j$ gives the square of the "modulus" or, in other words, of the "length" of the supervector Φ^i. It is easy to see that this quantity is real in the sense of the complex conjugation Eq. (2.5)

$$(\Phi^{i+}\Phi^i)^* = \Phi^{i+}\Phi^i \tag{2.15}$$

For $i \neq j$ one obtains from the definition Eq. (2.14) the following property for the scalar product

$$(\Phi^{i+}\Phi^j)^* = \Phi^{j+}\Phi^i \tag{2.16}$$

A linear transformation F in the space of the supervectors converts a supervector Φ into another supervector $\tilde{\Phi}$

$$\tilde{\Phi} = F\Phi \tag{2.17}$$

Of course, the supervector $\tilde{\Phi}$ must have the same structure Eq. (2.11) as the supervector Φ. This imposes a restriction on the structure of the supermatrix F corresponding to the linear transformation F: it has to be of the form

$$F = \begin{pmatrix} a & \sigma \\ \rho & b \end{pmatrix} \tag{2.18}$$

In Eq. (2.18) a and b are $n \times n$ and $m \times m$ matrices containing only commuting variables, σ and ρ are $n \times m$ and $m \times n$ matrices consisting of anticommuting ones. Matrices having the structure Eq. (2.18) can be called supermatrices. In what follows commuting elements of supermatrices and supervectors are designated by Latin letters and anticommuting by Greek letters.

Two supermatrices F and G of the rank $(m + n) \times (m + n)$ are assumed to multiply according to the conventional rules

$$(FG)_{ik} = \sum_{l=1}^{m+n} F_{il} G_{lk} \tag{2.19}$$

One can see from Eq. (2.19) that the product of supematrices of the form, Eq. (2.18), is also a supermatrix of the same form. In order to define the supertranspose F^T of the supermatrix F one should use the notion of the scalar product of two supervectors Eq. (2.14). Again, by analogy with the conventional definition the supermatrix F^T is introduced as

$$\Phi_1^T F^T \Phi_2 = (F\Phi_1)^T \Phi_2 \tag{2.20}$$

The transpose of a conventional matrix is obtained by transposing its indexes. This is not as simple for the supermatrices. Writing out the scalar products on both sides of the Eq. (2.20) explicitly and using the anticommutation relation Eq. (2.1) one can see that the supermatrix F^T is equal to

$$F^T = \begin{pmatrix} a^T & -\rho^T \\ \sigma^T & b^T \end{pmatrix} \tag{2.21}$$

where

$$(a^T)_{\alpha\beta} = a_{\beta\alpha}, \qquad (\sigma^T)_{\alpha\beta} = \sigma_{\beta\alpha}, \qquad (b^T)_{\alpha\beta} = b_{\beta\alpha}, \qquad (\rho^T)_{\alpha\beta} = \rho_{\beta\alpha}$$

Eq. (2.21) shows that supertransposition differs from conventional transposition by the sign of the upper right block. Later the conventional transposition is not used for the supermatrices and therefore the notation T cannot lead to misunderstanding. One should be careful only in distinguishing in the text between conventional matrices and supermatrices. To simplify this task all elements of the conventional matrices are written as either Greek or Latin letters. In the supermatrices commuting variables are designated by Latin letters and anticommuting by Greek ones, so they contain both.

Using Eq. (2.20) and writing the scalar products properly one can find immediately the transpose of the product of two arbitrary supermatrices F_1 and F_2.

$$(F_1 F_2)^T = F_2^T F_1^T \tag{2.22}$$

It is clear from Eq. (2.22) that the supertransposition Eqs. (2.20, 2.21) is better than the conventional transposition because it enables us to use the rules of conventional matrix algebra when operating with supermatrices.

The Hermitian conjugate F^+ of the matrix F can be defined in a standard way

$$F^+ = (F^T)^* \tag{2.23}$$

Combining Eqs. (2.4, 2.22, 2.23) one can obtain

$$(F_1 F_2)^+ = F_2^+ F_1^+ \quad \text{and} \quad (F^+)^+ = F \tag{2.24}$$

The latter equality in Eq. (2.24) shows that the operation of the Hermitian conjugation is inverse to itself. Let us observe that this is not true for supertransposition. Generally speaking

$$(F^T)^T \neq F$$

as can easily be seen from Eq. (2.21).

Having defined the operation of the Hermitian conjugation one can introduce Hermitian and unitary supermatrices. The Hermitian matrices H satisfy the condition

$$H = H^+ \tag{2.25}$$

whereas the unitary supermatrices U have the following properties:

$$UU^+ = U^+U = 1 \tag{2.26}$$

As one can see from Eqs. (2.13, 2.20, 2.23, 2.26) the unitary transformations acting on supervectors Φ conserve the length of the supervectors

$$(U\Phi)^+U\Phi = \Phi^+U^+U\Phi = \Phi^+\Phi \tag{2.27}$$

Just like conventional matrices, supermatrices can be diagonalized; this property is very important for different applications. However, the corresponding procedure needs some explanation and therefore in the following discussion I present the basic steps of the diagonalization of a Hermitian supermatrix H.

As usual, one should seek supervector solutions Φ of the equation

$$H\Phi = \lambda\Phi \tag{2.28}$$

where

$$H = \begin{pmatrix} a & \sigma \\ \sigma^+ & b \end{pmatrix}, \qquad \Phi = \begin{pmatrix} \chi \\ S \end{pmatrix},$$

$$a = a^+, \qquad b = b^+$$

Eq. (2.28) can be rewritten in the following explicit form:

$$\begin{cases} (a - \lambda)\chi + \sigma S = 0 \\ \sigma^+\chi + (b - \lambda)S = 0 \end{cases} \tag{2.29}$$

Now we can exclude from Eqs. (2.29) either S or χ and obtain linear equations for χ or S correspondingly:

$$\left(a - \lambda - \sigma(b - \lambda)^{-1}\sigma^+\right)\chi = 0 \tag{2.30}$$

$$\left(b - \lambda - \sigma^+(a - \lambda)^{-1}\sigma\right)S = 0 \tag{2.31}$$

The expressions in the parentheses in Eqs. (2.30, 2.31) contain products of only even numbers of the anticommuting elements. The fact that the elements of the vector χ are anticommuting does not lead to any difference with respect to conventional linear equations. Therefore, nonzero solutions of Eqs. (2.30, 2.31) exist, provided the following conditions are fulfilled:

$$\det\left(a - \lambda - \sigma(b - \lambda)^{-1}\sigma^+\right) = 0 \tag{2.32}$$

$$\det\left(b - \lambda - \sigma^+(a - \lambda)^{-1}\sigma\right) = 0 \tag{2.33}$$

Solutions of Eqs. (2.32, 2.33) are finite polynomials in the anticommuting variables, and therefore they can be found by iterations using the eigenvalues of the matrices a and b as the zero approximation.

With the definition of the scalar product, Eq. (2.14), one can prove that that eigenvectors corresponding to different eigenvalues are orthogonal to each other. If several eigenvalues coincide, the orthogonalization process is performed in the usual way. Using the definition, Eq. (2.5), of the complex conjugate for the Grassmann variables one can show that all eigenvalues λ_i of any Hermitian matrix H are real. Therefore, any Hermitian matrix H can be represented in the form

$$H = U \hat{\lambda} U^+ \tag{2.34}$$

where $\hat{\lambda}$ is a diagonal real matrix and U is a unitary matrix transforming the original basis to the orthogonal basis of the eigenvectors of the matrix H.

An arbitrary non-Hermitian matrix F can be diagonalized by using two different unitary matrices U and V

$$F = U \hat{\lambda} V^+ \tag{2.35}$$

where $\hat{\lambda}$ is diagonal and real.

Correspondingly, the Hermitian matrices $F F^+$ and $F^+ F$ have the form

$$F F^+ = U \hat{\lambda}^2 U^+, \qquad F^+ F = V \hat{\lambda}^2 V^+ \tag{2.36}$$

Eqs. (2.36) help to illustrate the meaning of the diagonal matrix $\hat{\lambda}$ in Eq. (2.35).

2.3.2 Supertrace and superdeterminant

A very important operation in the theory of conventional matrices is taking the trace of a matrix. However, the usual definition of the trace as the sum of all diagonal elements is not useful for the supermatrices because the most important property of the invariance under cyclic permutations of the matrices is lost. Nevertheless, an operation conserving this invariance can be introduced for the supermatrices, too. By analogy with the trace in the theory of conventional matrices it can be called supertrace. For a supermatrix F of the form Eq. (2.18) the supertrace str F is defined as

$$\operatorname{str} F = \operatorname{tr} a - \operatorname{tr} b \tag{2.37}$$

where the symbol tr stands for the conventional trace.

Although somewhat strange, the definition Eq. (2.37) is very useful because it is this operation that provides the invariance under the cyclic permutations. Taking two supermatrices F_1 and F_2 of the form Eq. (2.18) we have

$$\operatorname{str} F_1 F_2 = \operatorname{str} \begin{pmatrix} a_1 & \sigma_1 \\ \rho_1 & b_1 \end{pmatrix} \begin{pmatrix} a_2 & \sigma_2 \\ \rho_2 & b_2 \end{pmatrix} \tag{2.38}$$

$$= \operatorname{tr} (a_1 a_2 + \sigma_1 \rho_2 - \rho_1 \sigma_2 - b_1 b_2)$$

At the same time

$$\operatorname{str} F_2 F_1 = \operatorname{str} \begin{pmatrix} a_2 & \sigma_2 \\ \rho_2 & b_2 \end{pmatrix} \begin{pmatrix} a_1 & \sigma_1 \\ \rho_1 & b_1 \end{pmatrix} \tag{2.39}$$

$$= \operatorname{tr} (a_2 a_1 + \sigma_2 \rho_1 - \rho_2 \sigma_1 - b_2 b_1)$$

Now we can use the properties of the conventional trace and permute the matrices in all the products under the trace in Eq. (2.39). The first and the fourth terms do not change, but the second and the third change sign because all elements of the matrices $\sigma_{1,2}$ and $\rho_{1,2}$ are anticommuting. Comparing the result of the permutations with Eq. (2.38) we obtain

$$\text{str } F_1 F_2 = \text{str } F_2 F_1 \tag{2.40}$$

for arbitrary supermatrices F_1 and F_2.

From Eq. (2.40) the invariance under cyclic permutations follows immediately

$$\text{str } (F_1 F_2 \ldots F_n) = \text{str } (F_n F_1 \ldots F_{n-1}) \tag{2.41}$$

Furthemore the following relation holds:

$$\text{str } F = \text{str } F^T \tag{2.42}$$

The definition of the supertrace can be found in Berezin (1979, 1983). In these publications one can also find the definition of the superdeterminant of a supermatrix F of the form Eq. (2.18). This superdeterminant sdet is defined as

$$\text{sdet } F = \det \left(a - \sigma b^{-1} \rho \right) \det b^{-1} \tag{2.43}$$

Again the definition Eq. (2.43) looks very unusual. However, one can see with this definition that the condition of the existence of nonzero solutions of a system of linear equations

$$F\Phi = 0 \tag{2.44}$$

where Φ is a supervector of the form Eq. (2.28) can be written as

$$\text{sdet } F = 0 \tag{2.45}$$

This can be shown easily by eliminating the commuting components S of the supervector Φ in Eq. (2.29) analogously to the derivation of Eqs. (2.32, 2.33). It follows from the definitions Eqs. (2.21, 2.43) that the superdeterminant does not change under the supertransposition of F

$$\text{sdet } F = \text{sdet } F^T \tag{2.46}$$

The supertrace and the superdeterminant are related to each other analogously to the conventional trace and determinant

$$\ln \text{sdet } F = \text{str } \ln F \tag{2.47}$$

To prove Eq. (2.47) it is convenient to use the identity

$$F = \begin{pmatrix} a & \sigma \\ \rho & b \end{pmatrix} = \begin{pmatrix} a & 0 \\ 0 & b \end{pmatrix} \begin{pmatrix} 1 & a^{-1}\sigma \\ b^{-1}\rho & 1 \end{pmatrix} \tag{2.48}$$

With the property Eqs. (2.40, 2.41) we can use all the conventional rules of matrix algebra and therefore write

$$\text{str } \ln F = \text{str } \ln \begin{pmatrix} a & 0 \\ 0 & b \end{pmatrix} + \text{str } \ln \left[1 + \begin{pmatrix} 0 & a^{-1}\sigma \\ b^{-1}\rho & 0 \end{pmatrix} \right]$$

$$= \text{tr } \ln a - \text{tr } \ln b + \frac{1}{2} \text{str } \ln \left[1 - \begin{pmatrix} a^{-1}\sigma b^{-1}\rho & 0 \\ 0 & b^{-1}\rho a^{-1}\sigma \end{pmatrix} \right] \tag{2.49}$$

Calculating the supertrace of the last logarithm in Eq. (2.49) we have

$$\text{str} \ln F = \text{tr} \ln a - \text{tr} \ln b + \text{tr} \ln \left(1 - a^{-1} \sigma b^{-1} \rho \right)$$

$$= \text{tr} \left(\ln a \left(1 - a^{-1} \sigma b^{-1} \rho \right) + \ln b^{-1} \right) \tag{2.50}$$

Using the definition of the superdeterminant Eq. (2.43) and the relation between the trace and the determinant of conventional matrices we reduce Eq. (2.50) to the form Eq. (2.47), thus proving it.

The connection, Eq. (2.47), between the supertrace and the superdeterminant enables us to prove immediately the multiplicativity of the superdeterminant

$$\text{sdet} \, F_1 F_2 = \text{sdet} \, F_1 \, \text{sdet} \, F_2 \tag{2.51}$$

The rules of the operations with supervectors and supermatrices introduced in this section are very convenient because they are similar to those of conventional linear algebra. In fact, one can manipulate superobjects in exactly the same way as conventional objects. It simplifies calculations with quantities containing both types of variables considerably.

2.4 Integrals

2.4.1 Integrals over anticommuting variables

Now let us consider integrals over anticommuting variables first introduced by Berezin (1961). They are defined formally as follows:

$$\int d\chi_i = \int d\chi^* = 0, \qquad \int \chi_i d\chi_i = \int \chi_i^* d\chi_i^* = 1 \tag{2.52}$$

I want to emphasize that the definition Eq. (2.52) is completely formal. The notation $\int$ is only a symbol and one should not try to imagine this integral as a sum of something. It is implied that the "differentials" $d\chi_i, d\chi_i^*$ anticommute with each other and with the variables χ_i, χ_i^*:

$$\left\{ d\chi_i, d\chi_j \right\} = \{ d\chi_i, d\chi_j^* \} = \left\{ d\chi_i^*, d\chi_j^* \right\} = 0$$

$$\{ d\chi_i, \chi_j \} = \{ d\chi_i, \chi_j^* \} = \{ d\chi_i^*, \chi_j \} = \{ d\chi_i^*, \chi_j^* \} = 0 \tag{2.53}$$

The definition Eq. (2.52) is sufficient for introducing integrals of an arbitrary function. If such a function depends only on one variable χ_i it must be linear in χ_i because already $\chi_i^2 = 0$. Assuming that the integral of a sum of two functions equals the sum of the integrals we calculate the integral of the function with Eq. (2.52). The repeated integrals are implied by integrals over several variables. This enables us to calculate the integral of a function of an arbitrary number of variables.

The definitions Eqs. (2.52) are the only ones compatible with the conditions Eqs. (2.53). The choice of unity in the right-hand side of Eqs. (2.52) is completely arbitrary; one might write any other number.

The accepted definition leads to the following formula for the integral of the function $K(\chi)$, Eq. (2.3):

$$\int K(\chi) d\chi = a(1, 1, \ldots, 1) \tag{2.54}$$

where $\chi = (\chi_1, \ldots, \chi_n)$, $d\chi = d\chi_n d\chi_{n-1} \ldots d\chi_1$. One can see from Eq. (2.54) that, in fact, integration over the anticommuting variables is equivalent to the differentiation introduced in Section 2.2.

For the integrals over the anticommuting variables one can introduce formulae for linear change of the variables. Changing the variables from χ to ρ according to Eq. (2.7) and using the definition of the integrals, Eq. (2.52), we obtain the following relation between the differentials

$$d\chi_i = \sum_{k=1}^{n} (a^{-1})_{ik} d\rho_k \tag{2.55}$$

where the matrix a^{-1} is inverse with respect to a. Correspondingly the transformation of the elementary volume $d\chi$ is given by

$$d\chi = (\det a)^{-1} d\rho \tag{2.56}$$

Changing the variables according to Eq. (2.7) one can transform integrals as

$$\int K(\chi) d\chi = (\det a)^{-1} \int K(a\rho) d\rho \tag{2.57}$$

Analogous expressions can be written for the complex conjugates.

In integrals over anticommuting variables one can shift the variables of the integration

$$\int K(\chi + \rho) d\chi = \int K(\chi) d\chi \tag{2.58}$$

for an arbitrary vector ρ with anticommuting components.

One of the most important formulae for further calculations is the formula for integration of the Gaussian exponential

$$I = \int \exp(-\chi^+ A\chi) \prod_{i=1}^{n} d\chi_i^* d\chi_i = \det A \tag{2.59}$$

where A is an $n \times n$ Hermitian matrix, $\chi^+ = (\chi^*)^T$.

Eq. (2.59) can be proved by diagonalizing the matrix $A = u^+ \hat{\lambda} u$, where $\hat{\lambda}$ is a diagonal real matrix and u is a unitary one. Changing the variables $\rho = u\chi$, $\rho^+ = \chi^+ u^+$ and using Eqs. (2.55, 2.56), we come to the integral

$$I = \int \exp\left(- \sum_{i=1}^{n} \rho_i^* \lambda_i \rho_i\right) \prod_{i=1}^{n} d\rho_i^* d\rho_i \tag{2.60}$$

where λ_i are the elements of the matrix $\hat{\lambda}$. The integral I in Eq. (2.60) can be calculated easily by writing the exponential of the sum as the product of the exponentials and integrating each multiplier separately

$$I = \prod_{i=1}^{n} \int \exp(-\rho_i^* \lambda_i \rho_i) d\rho_i^* d\rho_i = \prod_{i=1}^{n} \int (1 - \rho_i^* \lambda_i \rho_i) d\rho_i^* d\rho_i = \prod_{i=1}^{n} \lambda_i \tag{2.61}$$

From Eq. (2.61) one immediately obtains Eq. (2.59). Eq. (2.59) differs from the corresponding equation for the commuting variables by giving $\det A$ instead of $(\det A)^{-1}$. This remarkable difference is the basis of the supersymmetry method presented in this book.

Shifting the variables $\chi' = \chi + A^{-1}\eta$, $\chi^{+'} = \chi^+ + \eta^+ A^{-1}$ one can extend Eq. (2.59) to

$$\int \exp(-\chi^+ A\chi - \chi^+\eta - \eta^+\chi) \prod_{i=1}^{n} d\chi_i^* d\chi_i = \det A \exp(\eta^+ A^{-1}\eta) \tag{2.62}$$

In addition to Eqs. (2.59, 2.62) one can write one more useful integral I_2

$$I_2 = \frac{\int \chi_i \chi_k^* \exp(-\chi^+ A\chi) \prod_{l=1}^{n} d\chi_l^* d\chi_l}{\int \exp(-\chi^+ A\chi) \prod_{l=1}^{n} d\chi_l^* d\chi_l} = (A^{-1})_{ik} \tag{2.63}$$

Eq. (2.63) is completely similar to the corresponding integral over conventional numbers. It is the basis of the use of the Grassmann variables in fermion field theories. The Grassmann variables in such theories correspond to classical fermion fields (Bogoliubov and Shirkov (1976)). Eq. (2.63) can be derived from Eq. (2.59) by using the following relation between the integrals I and I_2:

$$I_2 = \frac{\partial \ln I}{\partial A_{ki}} \tag{2.64}$$

Differentiating $\det A$ in the right-hand side of Eq. (2.59) we prove Eq. (2.63). Of course, Eq. (2.63) can also be proved directly by diagonalizing the matrix A and changing to the variables ρ as in deriving Eq. (2.59). In the same way one can perform Gaussian averaging of products of an arbitrary number of variables. Proceeding in this way one can prove the Wick theorem, which is the most convenient way to calculate averages of the products of many variables.

2.4.2 Superintegrals

Now let us consider integrals containing both commuting and anticommuting variables (superintegrals). Let the variables $S_1, S_2, \ldots, S_m$ be coordinates in a domain B in the space of conventional real numbers and $\chi_1, \chi_2, \ldots, \chi_n$ be anticommuting Grassmann variables. For any function $f(\chi, S)$ the following integral is well defined:

$$I_{nm} = \int f(\chi, S) d\chi \, dS, \tag{2.65}$$

$$dS = \pi^{-m/2} dS_1 dS_2 \ldots dS_m, \quad d\chi = d\chi_n d\chi_{n-1} \ldots d\chi_1$$

In the integral Eq. (2.65) the integration over S is performed over the domain B. It is assumed that the integral is convergent when the integration over S is performed. For integration over χ such a question does not arise.

In many applications when calculating integrals over both commuting and anticommuting variables it is convenient to change the variables, thus mixing them. It can happen that new commuting variables are not conventional numbers but general even elements of the Grassmann algebra (e.g., they are equal to a sum of a conventional number and a sum of products of an even number of the elements χ_i). Then it is necessary to determine what integration over the general even elements of the Grassmann algebra means and what the limits of the integration are.

Let $S_i = S_i(\chi, X)$ now be even elements, with X_i being coordinates in a domain of the conventional numbers $\tilde{B}$. Let us introduce formally the following integral $\tilde{I}_{nm}$:

$$\tilde{I}_{nm} = \int f(\chi, S(\chi, X)) \det \left(\frac{\partial S_i}{\partial X_k} \right) d\chi dX, \tag{2.66}$$

$$dX = \pi^{-m/2} dX_1 dX_2 \dots dX_m, \qquad X \in \tilde{B}$$

Changing from the variables X to S we can formally reduce the integral $\tilde{I}_{nm}$ Eq. (2.66) to the form Eq. (2.65). The integral Eq. (2.66), although after this change of variables very similar to the integral I_{nm}, Eq. (2.65), is, in principle, different from it because of the different domains of the integration. The variables X change in the domain $\tilde{B}$ and this generally speaking means that the variable S changes in a domain of even elements of Grassmann algebra that cannot be reduced to a domain of conventional numbers. However, provided certain restrictions are imposed on the function f these two integrals can coincide. We will discuss this question in the next section.

Many properties of conventional integrals remain valid for integrals of the types Eq. (2.65, 2.66). Provided functions f and g are equal to zero on the boundary of a domain B, formulae for partial integration can be written as

$$\int f \frac{\partial g}{\partial S_i} d\chi dS = -\int \frac{\partial f}{\partial S_i} g d\chi dS \tag{2.67}$$

$$\int f \frac{\overrightarrow{\partial} g}{\partial \chi_i} d\chi dS = \int f \frac{\overleftarrow{\partial}}{\partial \chi_i} g d\chi dS \tag{2.68}$$

Changing variables in a conventional integral leads to the appearance of a Jacobian that is equal to the determinant of the partial derivatives matrix. Analogous formulae exist in supermathematics, too. For a function f that is equal to zero on the boundary of B we have

$$\int f(\chi, S) d\chi dS = \int f((\chi(X, \eta), S(X, \eta)) J(\chi, S/\eta, X) d\eta dX \tag{2.69}$$

The function $J(\chi, S/\eta, X)$ in Eq. (2.69) is the superdeterminant of the partial derivatives supermatrix.

$$J(\chi, S/\eta, X) = \mathrm{sdet}\, R, \qquad R = \begin{pmatrix} p & \alpha \\ \beta & q \end{pmatrix} \tag{2.70}$$

$$p_{ik} = \frac{\partial S_i}{\partial X_k}, \qquad \alpha_{ik} = S_i \frac{\overleftarrow{\partial}}{\partial \eta_k} \tag{2.71}$$

$$\beta_{ik} = \frac{\partial \chi_i}{\partial X_k}, \qquad q_{ik} = \chi_i \frac{\overleftarrow{\partial}}{\partial \eta_k} \tag{2.72}$$

Another representation for J is

$$J(\chi, S/\eta, X) = \left(\mathrm{sdet}\, \tilde{R} \right)^{-1} \tag{2.73}$$

where

$$\tilde{R} = \begin{pmatrix} q & \beta \\ \alpha & p \end{pmatrix}$$

The function J is often called Berezinian and this term will be used here. The formulae Eqs. (2.67–2.72) have been obtained in works by Berezin (1965, 1967, 1979, 1983, 1987) and Pakhomov (1974).

For integration over supervectors an analogue of the formulae Eq. (2.59, 2.63) exists. Using the notations for the supervectors and the supermatrices introduced in the preceding section and integrating first over the commuting variables and then over the anticommuting ones we obtain

$$\int \exp(-\Phi^+ F \Phi) d\Phi^* d\Phi = \text{sdet } F, \tag{2.74}$$

$$d\Phi^* d\Phi = \pi^{-m} \prod_{i=1}^{n} d\chi_i^* d\chi_i \prod_{k=1}^{m} dS_i^* dS_i$$

Shifting the variables of the integration $\Phi \to \Phi - F^{-1} \Phi_0$, $\Phi^+ \to \Phi^+ - \Phi_0^+ F^{-1}$ one can derive the formulae for integration of the Gaussian exponential with linear terms

$$\int \exp\left(-\Phi^+ F \Phi - \Phi_0^+ \Phi - \Phi^+ \Phi_0\right) d\Phi^* d\Phi = \text{sdet } F \exp\left(\Phi_0^+ F^{-1} \Phi_0\right) \tag{2.75}$$

Finally, taking the derivative of Eq. (2.74) with respect to F_{ki} we find the analogue of Eq. (2.63)

$$\frac{\int \Phi_i \Phi_k^+ \exp(-\Phi^+ F \Phi) d\Phi^* d\Phi}{\int \exp(-\Phi^+ F \Phi) d\Phi^* d\Phi} = \left(F^{-1}\right)_{ik} \tag{2.76}$$

For averaging a product of a large number of Φ_i one can again use the Wick theorem, again keeping in mind that permutation of two anticommuting elements changes the sign.

The formulae written in this section give complete information about integrals over the Grassmann variables, supervectors, and supermatrices. However, one more important procedure has to be worked out, namely, changing the variables. Of course, this problem is trivial when both the conventional numbers and the anticommuting elements transform into themselves separately. Formulae for the conventional numbers are well known, whereas those for the Grassmann variables are given by Eqs. (2.55, 2.58). The case when a transformation mixes both types of variables is less trivial. In this case a set of the conventional numbers over which the integration is performed is transformed into a set of general even elements of Grassmann algebra. For example, very often instead of integrating over all elements of a supermatrix F it is more convenient to integrate over the eigenvalues λ_i of the supermatrix and unitary supermatrices U and V diagonalizing F (see Eqs. (2.34–2.36)). In this situation the eigenvalues λ_i are not conventional numbers but general even elements of Grassmann algebra. It is not clear in advance how to integrate over such variables and one should be careful in doing so. In the next section some examples of nontrivial changes of the variables of integration will be considered.

2.5 Changing the variables

2.5.1 General formulae

Let us calculate the integral, Eq. (2.66), changing from the variables X to $S(\chi, X)$. Of course, the calculations can be done for arbitrary n and m, but to avoid complicated formulae let us consider the case $n = 2$. Then we have two Grassmann variables $\chi = \chi_1$ and $\chi^* = \chi_2$. If the variable X changes in a domain $\tilde{B}$ the variable $S(0, X)$ changes in another domain of conventional numbers B. Denoting $S(0, X)$ as $S(0, X) = Y$, $Y = (Y_1, Y_2 \ldots Y_m)$ we can represent $S(\chi, X)$ in the following most general form:

$$S(\chi, X) = Y + Z(Y)\chi^*\chi, \quad Z = (Z_1, Z_2, \ldots, Z_m) \tag{2.77}$$

Then the function $f(\chi, S(\chi, X))$ is written as

$$f(\chi, S(\chi, X)) = f(\chi, Y) + \sum_{k=1}^{m} \frac{\partial f(\chi, Y)}{\partial Y_k} Z_k(Y)\chi^*\chi \tag{2.78}$$

The determinant in Eq. (2.66) can be rewritten in the form

$$\det\left(\frac{\partial S_i}{\partial X_k}\right) = \det\left(\frac{\partial S_i}{\partial Y_k}\right) \det\left(\frac{\partial Y_i}{\partial X_k}\right) \tag{2.79}$$

Substituting Eq. (2.77) into Eq. (2.79) we obtain

$$\det\left(\frac{\partial S_i}{\partial X_k}\right) = \det\left(\delta_{ik} + \frac{\partial Z_i(Y)}{\partial Y_k}\chi^*\chi\right) \det\left(\frac{\partial Y_i}{\partial X_k}\right)$$

$$= \left(1 + \sum_{k=1}^{m} \frac{\partial Z_k(Y)}{\partial Y_k}\chi^*\chi\right) \det\left(\frac{\partial Y_i}{\partial X_k}\right) \tag{2.80}$$

Substituting Eqs. (2.78, 2.80) into Eq. (2.66) and changing integration over X to integration over Y we find

$$\tilde{I}_{2m} = \int \left[f(\chi, Y) + \sum_{k=1}^{m} \frac{\partial (Z_k(Y) f(\chi, Y))}{\partial Y_k} \chi^*\chi \right] dY d\chi^* d\chi \tag{2.81}$$

The integral of the second term in Eq. (2.81) reduces immediately to the integral over the surface S_B of the region B. If the function $f(0, Y)$ vanishes at the surface, the second term in the integrand in Eq. (2.81) vanishes and we obtain

$$\tilde{I}_{nm} = I_{nm} \tag{2.82}$$

where the integral I_{nm} is given by Eq. (2.65) and $n = 2$.

So, we have proved a remarkable property: If the integrand is zero on the boundary of integration over conventional numbers, changing the variables one need not worry that new variables can become general even elements of the Grassmann algebra instead of conventional numbers. All integrals should be calculated as if these new variables were conventional numbers. Besides, the presence of the Grassmann variables does not influence the domain of integration over the new variables. In the example considered previously the domain B was determined by variation of the variable $S(0, X)$.

Nontrivial contributions can arise if the function $f(0, Y) \neq 0$ when Y is on the surface S_B of B. In this case we can derive from Eq. (2.81) that

$$\tilde{I}_{2m} = I_{2m} - \oint_{Y \in S_B} f(0, Y) \left[\vec{Z}(Y)\right]_t dS_B \tag{2.83}$$

where the notation $[]_t$ stands for the vector component perpendicular to the surface and the integration in Eq. (2.83) is performed over the surface S_B.

In many applications the surface contribution Eq. (2.83) is quite important and should be taken into account. The results presented previously can be extended to $n > 2$ and one can prove Eq. (2.82) for an arbitrary n provided not only the function f itself but also a certain number of its derivatives turn to zero on the surface. The way to prove this general statement seems to be clear from the considerations presented. In order to avoid long expressions the general proof is not written. Instead, I present several examples that are useful for applications.

2.5.2 Integral over supervector

First, let us consider the following integral:

$$I = \int g((\Phi^+\Phi)^{1/2})d\Phi^*d\Phi, \tag{2.84}$$

$$\Phi = \begin{pmatrix} \chi \\ Y \end{pmatrix}, \quad d\Phi^*d\Phi = \pi^{-1}d\chi^*d\chi \, dY^*dY$$

where g is a function decreasing fast enough at infinity. The integrand in Eq. (2.69) contains only the scalar product $\Phi^+\Phi = \chi^*\chi + |Y|^2$. Expanding the function g in the anticommuting variables and integrating over them we have

$$I = \int 2|Y| \left[g(|Y|) + \frac{1}{2|Y|}\chi^*\chi g'(|Y|) \right] d\chi^*d\chi \, d|Y| \tag{2.85}$$

$$= -\int\limits_0^\infty g'(|Y|) \, d|Y| = g(0)$$

Now, let us calculate the same integral I by using "polar" coordinates r, ρ, ρ^*

$$|Y| = r(1 - \frac{1}{2}\chi^*\chi), \quad \chi = r\rho, \quad \chi^* = r\rho^* \tag{2.86}$$

It follows from Eq. (2.86) that

$$\Phi^+\Phi = r^2 \tag{2.87}$$

Using the variables r, ρ, ρ^* and calculating the Berezinian we rewrite the integral Eq. (2.84) in the form

$$I = 2 \int\limits_0^\infty \int \frac{g(r)}{r} dr d\rho^* d\rho = ? \tag{2.88}$$

We see from Eq. (2.88) that with the parametrization Eq. (2.86) we obtain a singular integral. At first glance, it must be zero because the integrand does not contain the Grassmann variables, which would contradict Eq. (2.85). But, at the same time, the integral over r in Eq. (2.88) diverges at $r = 0$ and therefore one cannot conclude that this integral is really zero. The simplest reaction to such a discrepancy might be just that the coordinates Eq. (2.86) are bad and lead to incorrect results. As we will see later the same problem arises when integration over all elements of a supermatrix is substituted by integration over the eigenvalues and diagonalizing unitary matrices. However, in many problems integrands contain functions that are invariant with respect to rotations in the superspace and generalized polarlike coordinates would considerably simplify the calculations. Therefore, one should try to circumvent the discrepancy in order to use coordinates like those in Eq. (2.86).

We can do so by first restricting the integration over Y and Y^* in Eq. (2.84) to the region $|Y| > \delta$ and taking the limit $\delta \to 0$ only at the end of the calculations. Now, let us repeat the procedure that enabled us to obtain Eq. (2.83). The function g is not equal to zero at the boundary $|Y|$ of the domain of the integration over $|Y|$; Eq. (2.83) can be directly used and we obtain

$$I = 2 \int_{\delta}^{\infty} \int \frac{g\,(r)}{r} dr d\rho^* d\rho + \frac{1}{2\pi} \int_{|Y|=\delta} \frac{g\,(|Y|)}{|Y|} dY^* dY \qquad (2.89)$$

The first integral in Eq. (2.89) is now definitely zero for any finite δ because integration over r is not singular, but the integrand does not contain the anticommuting variables. The second integral can easily be calculated in the limit $\delta \to 0$ and we come finally to Eq. (2.85). A practical lesson that one can learn from this example is that each time after changing variables of integration one obtains an integrand that does not contain the anticommuting variables, one should check whether the remaining integral is singular or not. If the singularity is of $1/r$ type, one should substitute the integral by the value of the integrand at $r = 0$. If the singularity is weaker than $1/r$ the integral is really zero. The latter statement gives rise to one more method of deriving Eq. (2.85).

Instead of the integral I, Eq. (2.84), let us calculate a more general integral $I\,(h)$ (Efetov (1983))

$$I\,(h) = \int g\left(\left(\Phi^+\Phi\right)^{1/2}\right) \exp\left(-h\Phi^+\Phi\right) d\Phi^* d\Phi \qquad (2.90)$$

The integral I can be obtained from $I\,(h)$ putting $h = 0$. Differentiating the integral $I\,(h)$ we have

$$\frac{dI\,(h)}{dh} = -\int \Phi^+\Phi g\left(\left(\Phi^+\Phi\right)^{1/2}\right) \exp\left(-h\Phi^+\Phi\right) d\Phi^* d\Phi \qquad (2.91)$$

Using the variables r, ρ, ρ^* we reduce Eq. (2.91) to

$$\frac{dI\,(h)}{dh} = -2 \int \int_0^{\infty} rg\,(r) \exp\left(-hr^2\right) dr d\rho^* d\rho \qquad (2.92)$$

The integrand in Eq. (2.92) does not contain the Grassmann variables as in Eq. (2.88) but now the integral over r converges. Therefore we have

$$\frac{dI\,(h)}{dh} = 0, \quad I\,(h) = \text{const} \qquad (2.93)$$

To determine const in Eq. (2.93) it is sufficient to know the integral $I\,(h)$ for an arbitrary fixed h. The most convenient is the limit $h \to \infty$. In this limit one can substitute the quantity $\Phi^+\Phi$ in the function $g\left(\Phi^+\Phi\right)$ in Eq. (2.90) by 0 and then the integration becomes trivial

$$I\,(h) = I\,(h \to \infty) = g\,(0) \int \exp\left(-h\Phi^+\Phi\right) d\Phi^* d\Phi = g\,(0) \qquad (2.94)$$

which agrees with Eq. (2.85).

The last trick is most convenient for practical calculations. Of course, nontrivial expressions can be obtained only if the integrand of a calculated integral is not completely invariant under rotations in the superspace. For example, we can take the following integral

$$\tilde{I} = \int \chi \chi^* g\left(\left(\Phi^+\Phi\right)^{1/2}\right) d\Phi^* d\Phi \qquad (2.95)$$

with the supervector Φ specified as before by Eq. (2.84). Integrating over χ and χ^* and over the phase of the complex variable Y we have

$$\tilde{I} = 2 \int_0^\infty g(|Y|) |Y| \, d|Y| \tag{2.96}$$

The integral $\tilde{I}$ already depends on the entire function $g(|Y|)$ and not only on its value at $|Y| = 0$.

2.5.3 Integral over supermatrix

Similar formulae can be obtained for integration over supermatrices. If the integrand of an integral contains invariant combinations of the supermatrices it is often convenient to use the eigenvalues and the corresponding diagonalizing unitary matrices as the variables of integration. Let us consider the following example:

$$K = \int f((\mathrm{str}\, H^2)^{1/2}) dH \tag{2.97}$$

where

$$H = \begin{pmatrix} a & \sigma^* \\ \sigma & ib \end{pmatrix}, \quad dH = \pi^{-1} d\sigma^* d\sigma \, da \, db$$

with a and b real numbers, σ, σ^* Grassmann variables.

Writing $\mathrm{str}\, H^2 = a^2 + b^2 + 2\sigma^* \sigma$ and substituting this expression into Eq. (2.97) we obtain practically the same integral as I in Eq. (2.84) and the result of the integration is

$$K = 2f(0) \tag{2.98}$$

Now, diagonalizing the supermatrix H we have

$$H = u\hat{\lambda}u^+, \quad \hat{\lambda} = \begin{pmatrix} p & 0 \\ 0 & iq \end{pmatrix} \tag{2.99}$$

$$u = \begin{pmatrix} 1 - 2\eta\eta^* & 2\eta \\ -2\eta^* & 1 + 2\eta\eta^* \end{pmatrix}$$

where

$$p = a - \frac{\sigma\sigma^*}{a - ib}, \quad iq = ib - \frac{\sigma\sigma^*}{a - ib} \tag{2.100}$$

$$\eta = -\frac{\sigma^*}{2(a - ib)}, \quad \eta^* = -\frac{\sigma}{2(a - ib)}$$

To calculate the Berezinian J of the transformation one can use Eqs. (2.70–2.72). The procedure is straightforward and we obtain

$$J = -\frac{1}{4(p - iq)^2} \tag{2.101}$$

The integral K, Eq. (2.97), in this new parametrization takes the form

$$K = (4\pi)^{-1} \int f\left((p^2 + q^2)^{1/2}\right) \frac{d\eta d\eta^* dp dq}{(p - iq)^2} \qquad (2.102)$$

Again we encounter the same problem that arises when calculating the integral I, Eq. (2.88). The integrand in Eq. (2.102) does not depend on the Grassmann variables, but the integral over the commuting variables p and q has a singularity at $p = q = 0$. Therefore, one should take into account surface contributions using Eq. (2.83). A simpler trick that will often be used in this book is to consider, as in Eqs. (2.90–2.94), the integral $K(h)$

$$K(h) = \int f\left((\operatorname{str} H^2)^{1/2}\right) \exp(-h \operatorname{str} H^2) dH \qquad (2.103)$$

Differentiating the integral $K(h)$ and changing to the variables p, q, η, η^* we have

$$\frac{dK(h)}{dh} = -(4\pi)^{-1} \int f(r) r^2 \exp(-hr^2) \frac{d\eta d\eta^* dp dq}{(p - iq)^2}, \qquad (2.104)$$

$$r = (p^2 + q^2)^{1/2}$$

Now the singularity at $p = q = 0$ has been eliminated and clearly

$$\frac{dK(h)}{dh} = 0, \quad K(h) = \text{const} \qquad (2.105)$$

To determine the const in Eq. (2.105) we can take the limit $h \to \infty$ and calculate the integral in Eq. (2.103) using the variables a, b, σ, σ^*. In this limit one can immediately substitute the argument in the function f by 0 and integrate only the exponential. As the result we obtain

$$K(h) = 2f(0) \qquad (2.106)$$

which is in agreement with Eq. (2.98).

To obtain a nontrivial result that does not depend only on the integrand at a single point one should violate the invariance of the integrand with respect to rotations in the space of the supermatrices. As an example we can consider the following integral:

$$\tilde{K} = \int \operatorname{str}(kH)^2 f\left((\operatorname{str} H^2)^{1/2}\right) dH, \quad k = \begin{pmatrix} 1 & 0 \\ 0 & -1 \end{pmatrix} \qquad (2.107)$$

The invariance of the integrand in the integral $\tilde{K}$ is violated by the presence of the matrix k. The integrand is not singular in the sense that it turns to zero at $H = 0$. Using the parametrization Eq. (2.99) it is convenient to integrate first over the unitary supermatrix u. The argument of the function f does not depend on u and so one should integrate over this variable only the first multiplier $\operatorname{str}(kH)^2$ in the integrand

$$\int \operatorname{str}(kH)^2 d\eta d\eta^* = \int \operatorname{str}\left(u^+ ku\hat{\lambda}\right)^2 d\eta d\eta^*$$

$$= \int \operatorname{str} \begin{pmatrix} (1 - 8\eta\eta^*) p & 4\eta iq \\ 4\eta^* p & -(1 + 8\eta\eta^*) iq \end{pmatrix}^2 d\eta d\eta^* \qquad (2.108)$$

$$= \int \left[p^2 + q^2 + 16\eta^*\eta (p - iq)^2 \right] d\eta d\eta^* = 16(p - iq)^2$$

Substituting the result of the integration in Eq. (2.108) into Eq. (2.107) we obtain with the use of Eq. (2.101)

$$\tilde{K} = 4\pi^{-1} \int f\left((a^2 + b^2)^{1/2}\right) da db \tag{2.109}$$

The examples presented previously give the basic idea of how integration over supervectors and supermatrices can be performed. All the manipulations with the superobjects are very similar to those in conventional mathematical analysis. The only important new peculiarity is that one should take care of surface terms like the second term in Eq. (2.83). We have seen that such a contribution leads to the nonzero results in Eqs. (2.94, 2.98). One can also find a detailed discussion of the singular contributions in Zirnbauer (1986).

2.5.4 Jacobian (Berezinian) and length in superspace

Often when changing variables in an integral it is convenient instead of calculating the superdeterminant of the partial derivatives Eqs. (2.70–2.72) to derive the Jacobian writing the elementary length $(\Delta l)^2$ in the superspace. Suppose that we calculate the integral I

$$I = \int f(\Phi) d\Phi^* d\Phi \tag{2.110}$$

and that we want to write this integral in terms of the variables $\tilde{\Phi}$ determined by the following relations:

$$\Phi = \Phi\left(\tilde{\Phi}\right), \quad \Phi^+ = \Phi^+\left(\tilde{\Phi}\right) \tag{2.111}$$

Let us write the square of the elementary length $(dl)^2$

$$(dl)^2 = d\Phi^+ d\Phi = d\tilde{\Phi}^+ F d\tilde{\Phi} \tag{2.112}$$

where

$$F = g^+ g, \quad g_{ik} = \Phi_i \frac{\overleftarrow{\partial}}{\partial \tilde{\Phi}_k}$$

In Eq. (2.112) the right derivative is used. This remark is important when the derivative is taken with respect to anticommuting variables. Of course, when differentiating over commuting variables the difference between the left and the right derivatives does not exists.

On the other hand, we know that changing the variables Eq. (2.111) yields the Berezinian J that is equal to

$$J = (\text{sdet } F)^{-1} \tag{2.113}$$

(cf. Eqs. (2.70–2.72)).

It follows from Eqs. (2.112, 2.113) that knowing the square of the elementary length $(dl)^2$ in the superspace written in terms of the increments $d\tilde{\Phi}$ and $d\tilde{\Phi}^+$ in the form Eq. (2.112) one can obtain the Berezinian of the transformation by calculation of the inverse superdeterminant of the metric supertensor F.

Let us demonstrate how it can be done for changing the variables given by Eq. (2.99). Writing the elementary length in the space of the elements of the supermatrix H and using the invariance of the supertrace under cyclic permutations we have

$$(dl)^2 = \text{str}\,(dH)^2 = \text{str}\,\left(du\hat{\lambda}u^+ + ud\hat{\lambda}u^+ + u\hat{\lambda}du^+\right)^2 \qquad (2.114)$$

$$= \text{str}\,\left(\left[\delta u, \hat{\lambda}\right]^2 + \left(d\hat{\lambda}\right)^2 + 2\delta u\left[\hat{\lambda}, d\hat{\lambda}\right]\right)$$

where $\delta u = u^+ du$ and $[a,\ b]$ represents the commutator of a and b. Note that $(\delta u)^+ = -\delta u$.

The last commutator in Eq. (2.114) vanishes and we see that the quadratic form is diagonal in the variables $\delta u,\ d\hat{\lambda}$. Now we have to express it in terms of the variables $d\eta, d\eta^*, dp, dq$. The matrix δu has the form

$$\delta u = 2\begin{pmatrix} \eta d\eta^* + \eta^* d\eta & d\eta \\ -d\eta^* & \eta^* d\eta + \eta d\eta^* \end{pmatrix} \qquad (2.115)$$

Substituting Eq. (2.115) into Eq. (2.114) we have finally

$$(dl)^2 = (dp)^2 + (dq)^2 + 8\,(p - iq)^2\,d\eta d\eta^* \qquad (2.116)$$

Reconstructing the Berezinian from Eq. (2.116) with the use of Eqs. (2.112, 2.113) we come to Eq. (2.101).

In the next chapters different physical phenomena will be discussed using a nonlinear supermatrix σ-model. The formulae and calculational schemes presented in this chapter are sufficient for the derivation and use of this model. I want to emphasize that all the necessary mathematics has been developed in this chapter and after looking through it the reader should have no difficulty in reading the rest of the book. Some other ideas and notions, such as spontaneous breaking of symmetry, Goldstone modes, and order parameter, that will be used later are quite common in condensed matter physics and quantum field theory.

Before starting the formal derivation of the σ-model I want to present recent new quantum phenomena and discuss the difficulties in describing them by conventional methods that motivated the creation of the supersymmetry technique. The central concept of these phenomena is that of diffusion modes, which are the subject of the next chapter.

3

Diffusion modes

3.1 Quantum interference on defects

3.1.1 Localization

Periodicity of crystalline materials permits their description in terms of Bloch waves. In reality different types of defects are always present and the impurity concentration can vary in a broad region such that one can have the whole spectrum of materials ranging from a very good metal to an insulator. The weak disorder limit is usually described by the scattering of Bloch waves by impurities. In the classical limit it leads to a Boltzmann kinetic equation that gives, for example, for the conductivity σ_0 a very well known formula

$$\sigma_0 = \frac{ne^2\tau}{m} \tag{3.1}$$

where n is the electron density, e and m are the electron charge and mass, τ is the mean free time determined by the impurity concentration and the amplitude of the scattering on a single impurity. The classical conductivity Eq. (3.1) is not sensitive to a magnetic field or inelastic scattering as soon as $\omega_c \tau, \tau/\tau_{\text{in}} \ll 1$, where $\omega_c = eH/mc$ is the cyclotron frequency, τ_{in} is the mean free time of the inelastic scattering.

The situation changes if the disorder is sufficiently strong. Then, according to Anderson (1958), the system is no longer metal and must exhibit insulating properties. This means that the classical formula Eq. (3.1) is not valid in this limit and the proper description must be based on quantum mechanics. In a metal electron wave functions corresponding to eigenenergies near the Fermi energy are extended over the sample but in the insulating regime the electron wave function can become localized and decay exponentially from some point r_0 in the space

$$|\psi(r)| \sim \exp\left(-\frac{|r - r_0|}{L_c}\right) \tag{3.2}$$

where L_c is the localization length.

The localization in the limit of strong disorder $\tau\epsilon_0 \sim 1$, where ϵ_0 is the Fermi energy, can be understood in terms of bound states in deep quantum wells. As soon as the disorder becomes weaker the depth of the wells decreases, leading to an increase in the localization length. The wave functions of different wells overlap substantially, and in the region of weak disorder one can expect always to have an extended wave function.

Surprisingly, it is not quite so. It was argued later (Mott and Twose (1961)) that in a one-dimensional chain the electrons can be localized by an arbitrarily weak disorder. Qualitatively such behavior is due to multiple interference of the electron waves. If the potential of an impurity is small the probability of backward scattering is also small. But after each scattering the electron wave can return to the fixed point and complex amplitudes (not intensities) of these waves are added to each other. Although each term in the sum is small, the number of such terms is large and finally the electron gets localized in some region of the space. Of course, the localization length can be very large in the limit of weak disorder. Different inelastic processes destroy the coherence and conduction can become possible. The result that all states are localized in a chain has also been proved by Berezinsky (1973) by diagrammatic technique and since then the problem has been discussed in many works (see, e.g., Berezinsky and Gorkov (1979), Gogolin (1982), Abrikosov and Ryzhkin (1978)).

Now it is clear that the class of weakly disordered systems in which electrons are localized is considerably wider. Using scaling arguments Thouless (1977) predicted that in long wires the resistivity increases exponentially with the length of the sample L. In this case the system is one-dimensional in geometry only. The Fermi surface can be three-dimensional and at short distances electrons diffuse classically. The crossover from short wires, where the diffusion is classical and the dependence of the resistance on L is linear, to long ones with exponentially large resistivities occurs at some localization length $L_c \sim p_0^2 S l$, where p_0 is the Fermi momentum, S is the cross section of the wire, l is the mean free path. The resistance of a sample of the length L_c does not depend on disorder and the size of the sample and is approximately equal to 10 kΩ. In Thouless's reasoning the only quantity determining the behavior of the system is its total resistivity (or its inverse, g, the conductance).

Using this conjecture Abrahams et al. (1979) considered the behavior of a disordered metal in an arbitrary dimension d. They described relations for a cube of a size bL in d dimensions by combining b^d cubes of size L into blocks of the size bL. If, according to Thouless's conjecture, the conductance g is the only parameter determining the behavior of the system, then one can write the following relation

$$g(bL) = f(b, g(L)) \tag{3.3}$$

where f is an unknown function. The scaling relation Eq. (3.3) can be rewritten in a differential form. Taking the limit $b \to 1$ we have

$$\frac{d \ln g(L)}{d \ln L} = \beta(g(L)) \tag{3.4}$$

The scaling arguments presented here cannot help to calculate the function β. However, its asymptotes can be determined quite easily. For large g the system is a good metal and one can use classical formulae. In this range we have

$$g(L) = \sigma_0 L^{d-2} \tag{3.5}$$

where σ_0 is the classical conductivity Eq. (3.1).

Substituting Eq. (3.5) into Eq. (3.4) we find for large g

$$\beta(g) = d - 2 \tag{3.6}$$

At small g, the system must be in the regime of localization in an arbitrary dimension and the conductance falls off exponentially with the size of the system

$$g = g_0 \exp(-\alpha L) \quad \text{and} \quad \beta(g) = \ln\left(\frac{g}{g_0}\right) \tag{3.7}$$

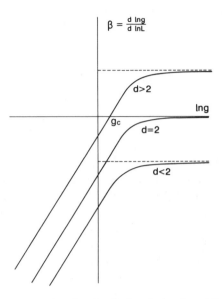

Fig. 3.1. Plot of $\beta\,(g)$ versus $\ln g$ for $d > 2, d = 2, d < 2$; $g\,(L)$ is the conductance.

where g_0 is of the order of $e^2/\hbar \sim 0, 1(k\Omega)^{-1}$. Using the asymptotes Eqs. (3.6, 3.7) and assuming that the function $\beta\,(g)$ is continuous and monotonous, one can plot the function $\beta\,(g)$ (Fig. 3.1).

It is seen from Fig. 3.1 that the behavior of the function β strongly depends on the dimensionality of the space. If $d = 3$, an unstable fixed point g_c exists where β turns to zero. At this point the conductance g does not change with change in the size of the sample and this corresponds to the point of the Anderson metal–insulator transition. If $d = 1$, then $\beta < 0$ for any disorder and the conductance decays with increasing size. Such behavior corresponds to localization and agrees with the predictions made for chains and wires. It comes as a surprise that in two dimensions β is also everywhere negative, indicating localization. Of course, this result holds only if the function β is smooth and monotonous. One should also remember that the main assumption of this approach is that everything is determined by the single parameter g.

Later it was argued that instead of this one-parameter picture one had to scale a distribution function of conductances (Altshuler, Kravtsov and Lerner (1985, 1986), Shapiro (1986, 1987), Efetov (1988)). The localization in two dimensions in weak disorder has not been proved analytically yet and only numerical evidence supports this statement (Lee and Fisher (1981), MacKinnon and Kramer (1983)). Although more work had to be done and remains to be done now to understand all aspects of the localization and the metal–insulator transition, the article by Abrahams et al. (1979) was a milestone in the study of disordered systems and stimulated a lot of works in the field. Now, it is difficult to list all the articles and reviews written about localization, so I am providing references only for several reviews and conference proceedings: Altshuler et al. (1983), Altshuler and Aronov (1985), Bergmann (1984), Kramer, Bergmann and Bruynseraede (1985), Lee and Ramakrishnan (1985), Nagaoka (1985), Chakravarty and Schmid (1986), Ando and Fukuyama (1988), Benedict and Chalker (1991), Vollhardt and Wölfle (1992),

Kramer and MacKinnon (1993), Belitz and Kirkpatrick (1994). They are quite complete
and describe the research activity very well.

3.1.2 Multiple interference

The main question that has arisen since the work of Abrahams et al. (1979) is how to
explain the behavior of the β function in the limit of weak disorder, for example, in two
dimensions (2D). It was commonly accepted that 2D transport could be well described by
the Bolzmann equation and therefore the system had to be a metal. The only possibility
could be that there were some diverging quantum corrections and soon the mechanism
of such divergencies was discovered (Gorkov, Larkin and Khmelnitskii (1979), Anderson,
Abrahams and Ramakrishnan (1979), Abrahams and Ramakrishnan (1980)). Some of the
formal derivations will be presented in the next section, but now let us discuss a qualitative
picture suggested by Khmelnitskii, which can also be found in reviews by Altshuler et al.
(1983) and Lee and Ramakrishnan (1985) and in a book by Abrikosov (1988). Bergmann
(1982) used similar arguments and incorporated the possibility of spin-orbit scattering.

   In the limit of weak disorder $p_0 l \gg 1$ we may speak of an electron path from a point
r_1 to another point r_2 (Fig. 3.2a). According to basic principles of quantum mechanics in
order to obtain the probability of getting from the point r_1 to the point r_2 one should add
not the probabilities W_i of the paths but the corresponding complex amplitudes A_i. As a
result of scattering by impurities very complicated paths can contribute to the process and
the total probability W can be written as

$$W = |\sum_i A_i|^2 = \sum_i |A_i|^2 + \sum_{i \neq j} A_i A_j^* \tag{3.8}$$

The first term in Eq. (3.8) describes the sum of the probabilities for each path, and the
second corresponds to the interference of different amplitudes. For most of the trajectories
the interference is not important because the lengths of different paths can be very different.
This leads to large differences $\Delta \varphi = \varphi_1 - \varphi_2$ of the phases of the wave function at the
points r_1 and r_2

$$\Delta \varphi = \hbar^{-1} \int_1^2 \mathbf{p} d\mathbf{r} \tag{3.9}$$

When performing summation over all possible trajectories the second term in Eq. (3.8)
oscillates very fast and its contribution is negligible.

   However, for some trajectories one cannot neglect interference. These are self-intersecting
paths, Fig. 3.2b. Each trajectory of the type represented by Fig. 3.2b corresponds to two
amplitudes that differ by the direction of travel around the loop. Changing the direction of
motion around the loop one should substitute $\mathbf{p}$ with $-\mathbf{p}$ and $d\mathbf{r}$ with $-d\mathbf{r}$, which does not
change $\Delta \varphi$. Then both amplitudes have the same phases and hence the total probability is
equal to

$$|A_1 + A_2|^2 = |A_1|^2 + |A_2|^2 + A_1 A_2^* + A_2 A_1^* = 4 |A_1|^2$$

and is two times larger than the classical probability calculated without the interference.
We see that for self-intersecting paths the quantum interference leads to an increase of the
scattering and, hence, to a larger resistivity.

   Until now all arguments have not depended on dimensionality. What is special about
1D and 2D? Why can the quantum corrections be so important in low dimensionality? To

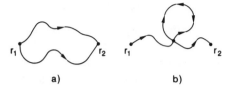

Fig. 3.2. a) Paths without intersections from the point r_1 to the point r_2; b) self-intersecting path.

answer these questions we should estimate the probability of the self-intersecting paths. The difference between 1D and 2D arises because the diffusion process in 1D and 2D is qualitatively different from that in 3D. The probability of returning to the point where the walk started is smaller than unity in 3D, whereas in 2D and 1D this probability is unity. In other words, making a random walk one cannot lose his or her way in 1D and 2D and the probability of returning home is unity although the travel can take quite a long time.

Let us explain this point in more detail. The probability of returning would be zero for an infinitely narrow trajectory. However, in the quantum description the electron path should be considered as a tube with thickness of the order of the electron wavelength $\lambda \sim \hbar/p_0$. Let the electron move diffusively during a time $t \gg \tau$, where τ is the mean free path. In this time it can reach any point located at the distance $r \sim (D_0 t)^{1/2}$ from the initial point, where $D_0 = v_0^2 \tau/d$ is the classical diffusion coefficient, v_0 is the Fermi velocity. Therefore, after that time the electron can be at any point in the volume $(D_0 t)^{d/2}$. The probability that moving during a time dt it can intersect its own trace at the initial point is proportional to the ratio of the volume $v\lambda^{d-1} dt$ (the volume of the tube of the length vdt) to the volume $(D_0 t)^{d/2}$. We are interested in the total probability that determines the relative correction $\Delta\sigma/\sigma_0$ to the classical conductivity, and to find it one should integrate over the time. As a result we have

$$\frac{\Delta\sigma}{\sigma_0} \sim -\int_\tau^{\tau_\varphi} v\lambda^{d-1} (D_0 t)^{-d/2} dt \qquad (3.10)$$

The sign of the integral is due to the fact that interference increases the probability of scattering and, hence, decreases conductivity. The lower limit of the integration τ in Eq. (3.10) is written because the previous picture is valid for times larger than the elastic mean free time τ. As the upper limit an inelastic mean free time τ_φ is chosen. This is because inelastic interactions such as electron–electron interaction destroy the phase coherence and the interference effects at larger times are no longer important. The time τ_φ must increase with lowering of the temperature and turn to infinity at $T = 0$.

We see that the integral in Eq. (3.10) at $T = 0$ drastically depends on the dimensionality d. It is finite for $d = 3$ and the correction to the conductivity is small as soon as the mean free path l much exceeds the electron wavelength λ but it diverges at $d = 2$ and $d = 1$. This means that in 3D the classical description is justified in the limit $l \gg \lambda$ but it is no longer valid at low temperatures in 1D and 2D. The result, Eq. (3.10), was first obtained by Gorkov, Larkin, and Khmelnitskii (1979) and by Anderson, Abrahams, and Ramakrishnan (1979).

3.1.3 Magnetic field effects

The quantum correction to the conductivity $\Delta\sigma$ is very sensitive to magnetic fields. As we have seen this correction is due to the large probability of self-intersections of the electron paths, Fig. 3.2b. However, the presence of the magnetic field $\mathbf{H}$ makes the amplitudes A_1 and A_2 corresponding to motion in the opposite directions different. To estimate the difference $\Delta\varphi_H$ of the phases of these amplitudes one should substitute $\mathbf{p} \to \mathbf{p} - (e/c)\mathbf{A}$. Changing the direction of motion along the path one should change the sign of the momentum $\mathbf{p}$, but the sign of the vector potential $\mathbf{A}$ remains the same and the phase difference $\Delta\varphi_H$ takes the form

$$\Delta\varphi_H = \frac{2e}{c\hbar} \oint \mathbf{A}d\mathbf{l} = \frac{2e}{c\hbar} \int \mathrm{curl}\,\mathbf{A}d\mathbf{S}$$

$$= \frac{2e}{c\hbar}\phi = 2\pi\frac{2\phi}{\phi_0} \tag{3.11}$$

where ϕ is the magnetic flux through the loop and $\phi_0 = ch/e$ is the flux quantum.

For different trajectories the phase difference varies considerably, leading to destruction of the quantum interference (Altshuler et al. (1980), Altshuler et al. (1983)). Therefore, applying the magnetic field leads to a decrease of the resistivity. One can estimate the characteristic magnetic field leading to the destruction at $T = 0$ substituting the time τ_φ in Eq. (3.10) by another time τ_H. The latter can be estimated as follows: We again assume diffusive motion, which gives the characteristic size $(Dt)^{1/2}$ of the loop. We see from Eq. (3.11) that the interference is destroyed when the magnetic flux through the loop is of the order of the flux quantum ϕ_0. This immediately gives the value of τ_H

$$t_H \sim \frac{\phi_0}{HD_0} \sim \frac{l_H^2}{D_0} \tag{3.12}$$

where $l_H = (c\hbar/2eH)^{1/2}$ is the magnetic length.

At finite temperature the magnetic field is essential when the corresponding time τ_H becomes shorter than τ_φ and so its characteristic value can be estimated as $H \sim \phi_0/(D_0\tau_\varphi)$. This field is much smaller than the classical field H_c determined by the relation $\omega_c\tau \sim 1$.

At first glance, both the magnetic field and the inelastic processes act on the interference in a similar way, destroying it, but this is true only for the first quantum correction. In the limit of weak disorder the quantum corrections are small and the one-loop approximation leading to Eq. (3.10) is sufficient for describing many experiments. However, when increasing the disorder or decreasing the temperature one has to include many-loop trajectories in the description of the system. The inelastic processes really lead to the destruction of all quantum corrections, but the magnetic field destroys only part of them. As a result one can have localization and insulating properties of the system in the magnetic field but inelastic scattering always leads to finite conductivity.

A very interesting effect based on the influence of the magnetic field on quantum interference was predicted by Altshuler, Aronov, and Spivak (1981). They considered a hollow metallic cylinder with thin walls and a magnetic field concentrated inside. It was assumed that the magnetic field did not penetrate the metallic walls and that the mean free path l was much smaller than the circumference L. In such a situation the electron motion is diffusive and the electron feels the vector potential A only. We can repeat the arguments leading to Eq. (3.11) for the cylinder. The phase difference $\Delta\varphi_H$ of the wave functions of the electrons moving around the cylinder in the opposite directions is given as before by Eq. (3.11), but now the magnetic flux ϕ is the same for all trajectories rounding the cylinder one time. Therefore, the quantum interference must lead to oscillations of the resistivity in magnetic flux ϕ.

One more striking point is that the period of the oscillations is equal, as in superconductors, to $\phi_0/2$, but the effect considered has nothing in common with superconductivity and should be observed in normal metals. Although the effect is analogous to the Aharonov–Bohm (1959) effect, it differs from the latter by the period of the oscillations ($\phi_0/2$ instead of ϕ_0). Of course, such periodicity does not contradict the more general argument of Aharonov and Bohm (1959). Simply, all odd harmonics in ϕ_0 in the expansion of the resistivity in a Fourier series are destroyed by the disorder. The oscillations of the resistivity in flux with the period $\phi_0/2$ were observed shortly afterward by Sharvin and Sharvin (1981) and Sharvin (1984). For a review see also Aronov and Sharvin (1987).

The first quantum correction is destroyed not only by a magnetic field that acts on the orbital motion but by magnetic impurities because of their interaction with the electron spin. All the magnetic interactions violate the time reversal symmetry and so, one may say that the main quantum correction is zero in the absence of this symmetry. Another type of symmetry, namely, central symmetry, can be destroyed by spin-orbit impurities, which at the same time preserve time reversal symmetry. Surprisingly, the presence of the spin-orbit impurities results in changing of the sign of the main quantum correction Eq. (3.10), thus increasing the conductivity (Hikami, Larkin, and Nagaoka (1980)). One can find a simple explanation of this effect in the book by Abrikosov (1988). The opposite sign of the interference correction leads to decreasing of the resistivity with lowering of the temperature and to a positive magnetoresistance. Both effects have been observed experimentally (Bergmann (1984)).

Thus, we see that simple qualitative arguments can help to explain different interesting phenomena. The effects of the decrease of the conductivity due to the quantum interference are often called "weak localization" because they can be considered as precursors of the localization. However, one should be careful when trying to draw conclusions about the localization using the first correction only. In many cases more powerful methods of calculation are necessary. In fact, historically the quantum correction Eq. (3.10) and effects of different interactions on it were first calculated by using diagrammatic technique (Gorkov, Larkin, and Khmelnitskii (1979), Altshuler et al. (1980), Hikami, Larkin, and Nagaoka (1980)) and the simple physical arguments were developed later. In the next sections I present a diagrammatic technique that makes it possible to calculate the effects of the weak localization in a quite straightforward manner. The direct calculations will show in many important cases the necessity for a more general treatment than just calculating diagrams, thus giving a motivation for constructing more sophisticated field theory techniques.

3.2 Choice of the model and correlation functions

In order to present a formal treatment of the effects discussed in the previous section it is necessary to choose the model and physical quantities to be calculated. Electron–electron interaction is not important to these effects and they are mainly due to disorder. So, the behavior of a noninteracting electron gas in a random potential will now be studied. This problem is equivalent to that of one-particle motion in the random potential but in what

follows I will speak in terms of the electron gas with the Fermi energy determined by the
density of the electrons. Properties of the electron gas can be described by solutions of the
Schrödinger equation

$$\mathcal{H}\varphi_k = \varepsilon_k\varphi_k, \qquad \mathcal{H} = \mathcal{H}_0 + \mathcal{H}_1, \qquad \langle\mathcal{H}_1\rangle = 0 \tag{3.13}$$

In Eq. (3.13), ε_k and φ_k are eigenvalues and eigenfunctions of the Hamiltonian $\mathcal{H}$. The term
$\mathcal{H}_0$ stands for the kinetic energy $\varepsilon(\hat{\mathbf{p}})$ of a free electron where $\hat{\mathbf{p}}$ is the momentum operator.
In the simplest case this energy is

$$\varepsilon(\hat{\mathbf{p}}) = \frac{\hat{\mathbf{p}}^2}{2m} \tag{3.14}$$

where m is the electron mass.

The term $\mathcal{H}_1$ describes the interaction with the random potential that can originate from
impurities in the system. The angular brackets in Eq. (3.13) stand for the averaging over
random realizations of the potential. The interaction with an external magnetic field $\mathbf{H}$ can
be included by the usual substitution $\hat{\mathbf{p}} \rightarrow \hat{\mathbf{p}} - (e/c)\mathbf{A}$, where $\mathbf{A}$ is the vector potential
corresponding to the magnetic field $\mathbf{H}$. Eq. (3.13) can include interactions with magnetic or
spin-orbit impurities and therefore one should remember that the Hamiltonian acts on spin
variables, too.

There is no possibility of solving Eq. (3.13) analytically for an arbitrary potential. However, such a solution would be unnecessary because usually one has to deal with statistical
properties only and thus has to calculate different averages of quantities containing the
solutions of the equation. For example, in order to calculate thermodynamic quantities it is
sufficient to know the average density of states $\rho(\varepsilon, \mathbf{r})$ given by the following expression:

$$\langle\rho(\varepsilon, \mathbf{r})\rangle = \left\langle \sum_{k,\sigma} \varphi_k^*(\mathbf{r}, \sigma)\, \varphi_k(\mathbf{r}, \sigma)\, \delta(\varepsilon - \varepsilon_k) \right\rangle \tag{3.15}$$

In Eq. (3.15), the summation is performed over all eigenstates k of the system, and $\sigma = \pm 1$
is the spin variable. The local density of states $\rho(\varepsilon, \mathbf{r})$ can be expressed in terms of the
advanced G^A and retarded G^R Green functions of Eq. (3.13)

$$\rho(\varepsilon, \mathbf{r}) = \frac{1}{2\pi i} \sum_\sigma \left(G_\varepsilon^A(y, y) - G_\varepsilon^R(y, y) \right) = \frac{1}{\pi} \sum_\sigma \operatorname{Im} G_\varepsilon^A(y, y) \tag{3.16}$$

where

$$G_\varepsilon^{R,A}(y, y') = \sum_k \frac{\varphi_k(y)\, \varphi_k^*(y')}{\varepsilon - \varepsilon_k \pm i\delta} = \sum_k G_{\varepsilon k}^{R,A} \varphi_k(y)\, \varphi_k^*(y') \tag{3.17}$$

In Eqs. (3.16, 3.17) $y = (\mathbf{r}, \sigma)$ and the limit $\delta \rightarrow 0$ is implied. In the absence of spin
interactions summation over the spin variable is easily performed by counting each state
twice. Expressing physical quantities in terms of the Green functions is necessary for most
analytical calculations. Let us mention one important property of the Green functions Eq.
(3.17)

$$G_\varepsilon^R(y, y') = \left(G_\varepsilon^A(y', y) \right)^* \tag{3.18}$$

Calculation of the average density of states, Eq. (3.15), is usually not difficult and can be
done for a metal that is not very dirty by perturbation theory. Problems (and new effects)
arise when one has to calculate quantities containing products of several Green functions.

For example, calculating the average square of the density of states $\langle \rho^2 (\varepsilon, \mathbf{r}) \rangle$ is already a much more difficult problem because the quantum interference effects discussed in the previous section become very important. In the perturbation theory this interference leads to divergencies. In some cases one has to calculate the entire distribution function of a physical quantity. As we will see in the next chapters the distribution function of the local density of states $P(x)$

$$P(x) = \langle \delta (x - \rho (\varepsilon, \mathbf{r})) \rangle \tag{3.19}$$

can describe the nuclear magnetic resonance (NMR) line.

Most kinetic quantities can be written in terms of products of two Green functions. For such quantities the calculation of the average is already not trivial and difficulties increase when calculating higher moments. Usually, all interesting information can be extracted from the density–density X^{00} or current–current $X^{\alpha\beta}$ ($\alpha, \beta = 1, 2, 3$) correlation functions, which determine the dielectric permeability or the conductivity of the system, respectively. These functions are introduced as follows:

$$X^{ab} (\mathbf{r}, \mathbf{r}', t) = i\theta (-t) \left\langle \left[\hat{j}^a (\mathbf{r}, t), \hat{j}^b (\mathbf{r}', 0) \right] \right\rangle_T \tag{3.20}$$

where $\theta(t)$ is the step function

$$\theta (t) = \begin{cases} 0, & t > 0 \\ 1, & t < 0 \end{cases}$$

the angular brackets $\langle \ldots \rangle_T$ stand for the thermodynamic averaging, and $[.., ..]$ is the commutator. The superscripts a and b can take the values 0, 1, 2, 3. By definition $\hat{j}^0 (\mathbf{r}, t)$ is the local charge density operator and $\hat{j}^\alpha (\mathbf{r}, t)$ ($\alpha = 1, 2, 3$) are components of the local current density operator. Later by writing the superscripts in Greek letters, the current related quantities are implied. The correlation function X^{ab} Eq. (3.20) is a retarded density–density or current–current Green function. Averaging over a random potential is not implied in Eq. (3.20) and should be performed additionally when calculating physical quantities.

The local charge density and current operators can be written in a standard second quantization form

$$\hat{j}^0 (\mathbf{r}, t) = e \sum_{k, k', \sigma} \varphi_k^* (y) \varphi_{k'} (y) a_{k\sigma}^+ a_{k'\sigma} \tag{3.21}$$

$$\hat{j}^\alpha (\mathbf{r}, t) = \sum_{k, k', \sigma} \left[\frac{ie}{2m} \left(\nabla^\alpha \varphi_k^* (y) \varphi_{k'} (y) - \varphi_k^*(y) \nabla^\alpha \varphi_{k'} (y) \right) \right.$$
$$\left. - \frac{e^2}{mc} \varphi_k^* (y) \varphi_{k'} (y) \mathbf{A} (\mathbf{r}) \right] a_{k\sigma}^+ a_{k'\sigma}$$

where a_k^+ and a_k are the electron creation and destruction operators for state k. The second equation (3.21) is written for the simplest spectrum of the form Eq. (3.14). If the spectrum is more complicated, corresponding changes can be made without difficulties. By substituting Eqs. (3.21) into Eq. (3.20) and performing thermodynamic averaging one can reduce the correlation functions X^{00} to the form

$$X^{00} (\mathbf{r}, \mathbf{r}', t) = ie^2 \theta(-t) \sum_{k, k', \sigma, \sigma'} \exp(i (\varepsilon_k - \varepsilon_{k'}) t) \tag{3.22}$$

$$(n (\varepsilon_k) - n (\varepsilon_{k'})) \varphi_k^* (y) \varphi_{k'} (y) \varphi_{k'}^* (y') \varphi_k (y')$$

where $n(\varepsilon)$ is the Fermi function $n(\varepsilon) = (1 + \exp((\varepsilon - \mu)/T))^{-1}$ and μ is the chemical potential.

Using the definition of the Green function, Eq. (3.17), and the relation Eq. (3.18), we can represent the density–density correlation function X^{00} in the form

$$X^{00}(\mathbf{r}, \mathbf{r}', t) = \frac{e^2}{(2\pi)^2 i} \sum_{\sigma, \sigma'} \iint_{-\infty}^{\infty} \exp(-i\omega t) n(\varepsilon) \left[Y^{00}(y, y', \varepsilon, \omega) \right.$$

$$\left. - (Y^{00}(y, y', \varepsilon, -\omega))^* \right] d\varepsilon d\omega \qquad (3.23)$$

The function $Y^{00}(y, y')$ in Eq. (3.23) is expressed in terms of the Green functions as

$$Y^{00}(y, y', \varepsilon, \omega) = G^A_{\varepsilon-\omega}(y', y) \left(G^R_\varepsilon(y, y') - G^A_\varepsilon(y, y') \right) \qquad (3.24)$$

Eq. (3.23) is derived for the noninteracting fermions but is in fact more general. Another way to derive it is to write the correlation function X^{00} at Matsubara frequencies, represent it as a sum over the frequencies of two temperature Green functions, and, finally, make the analytical continuation to the axis of real frequencies. This procedure is due to Eliashberg (1961) and can also be found in the book by Abrikosov, Gorkov, and Dzyaloshinskii (1963).

For the calculation of kinetic coefficients one should average the correlation functions X^{ab} and take the momentum and frequency representation. Let us define the corresponding averaged representation K^{ab} as

$$K^{ab}(\mathbf{k}, \omega) = \int \langle X^{ab}(\mathbf{r}, \mathbf{r}', t) \rangle \exp(i\omega t - i\mathbf{k}(\mathbf{r} - \mathbf{r}')) d\mathbf{r}' dt \qquad (3.25)$$

Then, such a physical quantity as conductivity $\sigma^{\alpha\beta}(\omega)$ determining the current in a homogeneous electric field is equal to (see, e.g., Isihara (1971))

$$\sigma^{\alpha\beta}(\omega) = \frac{1}{i\omega} \left(K^{\alpha\beta}(0, \omega) - K^{\alpha\beta}(0, 0) \right) \qquad (3.26)$$

Notice that Eq. (3.26) is valid only if the current induced by the electric field vanishes after switching off of the electric field. This equation cannot be used for studying such problems as superconductivity or persistent currents. In these cases one should directly use formulae for the response to vector potential $\mathbf{A}$, and the correlation function $K^{\alpha\beta}$ is the paramagnetic part of this response. In the absence of a magnetic field the matrix $\sigma^{\alpha\beta}$ is diagonal. Applying the magnetic field one also obtains a nondiagonal component, which is called the Hall conductivity.

Eq. (3.25) can be rewritten in a form that is very useful for explicit calculations. Using Eq. (3.23) and the corresponding equation for the current–current correlation function we obtain

$$K^{ab}(\mathbf{k}, \omega) = K^{ab}_1(\mathbf{k}, \omega) + K^{ab}_2(\mathbf{k}, \omega), \qquad (3.27)$$

$$K^{ab}_1(\mathbf{k}, \omega) = \frac{1}{2\pi i} \int (n(\varepsilon) - n(\varepsilon - \omega)) R^{ab}(\mathbf{k}, \varepsilon, \omega) d\varepsilon \qquad (3.28)$$

$$K^{ab}_2(\mathbf{k}, \omega) = \frac{1}{2\pi i} \sum_{\sigma, \sigma'} \int n(\varepsilon) < \left[\hat{\pi}^a_r G^R_{\varepsilon+\omega}(y, y') \hat{\pi}^b_{r'} G^R_\varepsilon(y', y) \right.$$

$$\left. - \hat{\pi}^a_r G^A_\varepsilon(y, y') \hat{\pi}^b_{r'} G^A_{\varepsilon-\omega}(y', y) \right] > \exp(i\mathbf{k}(\mathbf{r}' - \mathbf{r})) d\varepsilon d\mathbf{r}' \qquad (3.29)$$

In Eqs. (3.27–3.29)

$$\hat{\pi}_r^0 = e, \text{ and } \hat{\pi}_r^\alpha = \frac{e}{m}\left[-i\nabla_r^\alpha - (e/c)A^\alpha\right] \tag{3.30}$$

are the charge and the current operator. The function $R^{ab}(\mathbf{k}, \varepsilon, \omega)$ in Eq. (3.28) has the form

$$R^{ab}(\mathbf{k}, \varepsilon, \omega) = \sum_{\sigma,\sigma'}\int \langle\hat{\pi}_r^a G_\varepsilon^R(y, y')\hat{\pi}_r^b G_{\varepsilon-\omega}^A(y', y)\rangle$$

$$\times \exp\left(i\mathbf{k}(\mathbf{r}' - \mathbf{r})\right)d\mathbf{r}' \tag{3.31}$$

The function K_1^{ab} is most important when calculating kinetic quantities and contains, in particular, all the effects of the quantum interference discussed in the previous section. As concerns the function K_2^{ab}, it is weakly dependent on frequency for $\omega \ll \mu$ and can be obtained by calculating second derivatives of the free energy over the scalar or vector potentials. It has purely thermodynamic origin. In the limit of weak disorder this function is very small.

The function $Y^{00}(y, y', \varepsilon, \omega)$, Eq. (3.24), plays an important role in a description of wave functions in situations when the energy spectrum is discrete, as in finite size systems or in insulators. Using Eq. (3.18) and taking the limit $\omega \to 0$ we obtain

$$\sum_{\sigma,\sigma'} Y^{00}(y, y', \varepsilon, \omega) = \frac{2\pi}{-i\omega}p_\infty(\mathbf{r}, \mathbf{r}', \varepsilon) \tag{3.32}$$

where

$$p_\infty(\mathbf{r}, \mathbf{r}', \varepsilon) = \sum_{k,\sigma,\sigma'} |\varphi_k(y)|^2 |\varphi_k(y')|^2 \delta(\varepsilon - \varepsilon_k) \tag{3.33}$$

The function $p_\infty(\mathbf{r}, \mathbf{r}', \varepsilon)$ was introduced by Anderson (1958) to distinguish between the cases when the spin diffusion is possible in a disordered system and when it is not. In a metal the electron wave functions are extended and the function $p_\infty(\mathbf{r}, \mathbf{r}', \varepsilon)$ vanishes in the limit of large volume. Integrating in Eq. (3.33) over $\mathbf{r}'$ we obtain the following relation:

$$\int p_\infty(\mathbf{r}, \mathbf{r}', \varepsilon)d\mathbf{r}' = \rho(\mathbf{r}, \varepsilon) \tag{3.34}$$

where $\rho(\mathbf{r}, \varepsilon)$ is the local density of states introduced in Eqs. (3.15, 3.16).

Averaging the function $p_\infty(\mathbf{r}, \mathbf{r}', \varepsilon)$ over disorder one can describe the decay of the wave functions of the localized states.

Eq. (3.34) can be written in a more general form applicable not only in the localized regime but also in the metallic one

$$\lim_{\omega \to 0}\sum_{\sigma'}\int Y^{00}(y, y', \varepsilon, \omega)d\mathbf{r}' = \frac{2\pi}{-i\omega}\rho(\mathbf{r}, \varepsilon) \tag{3.35}$$

This equation is also correct for nonaveraged quantities. In fact, Eq. (3.35) describes the particle conservation law.

Although the correlation functions K^{ab} and $\langle Y^{00}\rangle$ describe averaged kinetic coefficients very well, it is often more convenient for technical reasons to calculate some other correlation functions. In particular, the energy level–level correlation function $R(\omega)$ is of interest in many applications. This function is defined as

$$R(\omega) = \left\langle \frac{1}{4\omega\nu^2 V^2} \sum_{k,m} (n(\varepsilon_k) - n(\varepsilon_m)) \delta(\omega - \varepsilon_m + \varepsilon_k) \right\rangle \tag{3.36}$$

where $\nu = mp_0/2\pi^2$ is the density of states per one spin degree of freedom of the electron gas at the Fermi surface, and V is the volume of the metal particle.

The function $R(\omega)$ is equal to the probability of finding two energy levels at the distance ω. It is clear that at large energy distances levels are not correlated and the probability is unity. This can be seen from Eq. (3.36), substituting the sums over the states with corresponding integrals. In the absence of spin interactions each energy level is double degenerate and one should use the following substitution

$$\frac{1}{V} \sum_k \rightarrow 2 \int d^3\mathbf{p}$$

Calculating integrals arising in Eq. (3.36) one gets unity; this shows that the function $R(\omega)$ is properly normalized. Eq. (3.36) can also be rewritten in terms of the retarded G^R and advanced G^A Green functions

$$R(\omega) = \frac{1}{8\pi^2\omega\nu^2 V^2} < \mathrm{Re} \int_{-\infty}^{\infty} \sum_{k,m} (n(\varepsilon - \omega) - n(\varepsilon))$$

$$\times G_{k,\varepsilon-\omega}^A \left(G_{m,\varepsilon}^R - G_{m,\varepsilon}^A \right) > d\varepsilon \tag{3.37}$$

which allows the possibility of calculating the level–level correlation function using the same methods as those for calculation of the functions X^{ab}.

In the next section some calculations using the standard "cross" technique (Abrikosov, Gorkov, Dzyaloshinskii (1963)) are presented

3.3 Averaging over impurities: Classical formulae

In order to carry out explicit calculation it is necessary to define correlation properties of the random part $\mathcal{H}_1$ of the Hamiltonian Eq. (3.13). Let us consider the simplest case when impurities do not act on the electron spin. Assume also that the potential of a single impurity falls off fast enough with distance and the scattering amplitude a_0 is small. We can describe such a situation by writing $\mathcal{H}_1 = u(\mathbf{r})$ and assuming the impurity potential $u(\mathbf{r})$ to be random and distributed according to the Gaussian δ-correlated law

$$\langle u(\mathbf{r}) \rangle = 0, \qquad \langle u(\mathbf{r}) u(\mathbf{r}') \rangle = \frac{1}{2\pi\nu\tau} \delta(\mathbf{r} - \mathbf{r}') \tag{3.38}$$

where τ is the mean free time.

Strictly speaking, the Gaussian distribution corresponds to the limit $a_0 \rightarrow 0$ as $a_0^2 n$ remains finite, where n is the impurity concentration. The value $a_0^2 n$ is inversely proportional to the mean free time in Eq. (3.38). In the general case of a non-Gaussian distribution the results obtained later remain correct provided an appropriate renormalization of constants is made. Before starting more complicated calculations it is useful to demonstrate how the

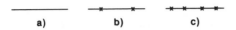

Fig. 3.3. Perturbation theory in the impurity potential: a) zeroth order, b) second order, c) fourth order.

perturbation theory works for the average Green functions $\langle G_\varepsilon^{R,A} \rangle$. The Green functions $G_\varepsilon^{R,A}$ satisfy the following equation:

$$(\varepsilon - \mathcal{H}_0 - u(\mathbf{r}))G_\varepsilon^{R,A}(\mathbf{r}, \mathbf{r}') = \delta(\mathbf{r} - \mathbf{r}') \qquad (3.39)$$

Eq. (3.39) can be rewritten in integral form

$$(\varepsilon - \mathcal{H}_0)\, G_\varepsilon^{R,A}(\mathbf{r}, \mathbf{r}') = \delta(\mathbf{r} - \mathbf{r}') + \int G_{0\varepsilon}^{R,A}(\mathbf{r}, \mathbf{r}'')\, u(\mathbf{r}'')\, G_\varepsilon^{R,A}(\mathbf{r}'', \mathbf{r})\, d\mathbf{r}'' \quad (3.40)$$

which is most convenient for writing perturbation series. In Eq. (3.40) the function $G_{0\varepsilon}^{R,A}(\mathbf{r}, \mathbf{r}')$ is the Green function of the operator $(\varepsilon - \mathcal{H}_0)$. In the absence of external fields additional to the field of the impurities $u(\mathbf{r})$ the function $G_{0\varepsilon}^{R,A}(\mathbf{r}, \mathbf{r}')$ depends on $\mathbf{r} - \mathbf{r}'$ only and in the momentum representation takes the form

$$G_{0\varepsilon}^{R,A}(\mathbf{p}) = (\varepsilon - \varepsilon(\mathbf{p}) \pm i\delta)^{-1} \qquad (3.41)$$

In the zero-order approximation one neglects the second term in the right-hand side of Eq. (3.40). The first-order approximation is obtained by substituting the function $G_\varepsilon^{R,A}(\mathbf{r}, \mathbf{r}')$ in the second term by $G_{0\varepsilon}^{R,A}(\mathbf{r}, \mathbf{r}')$. Each term of the perturbation theory is represented by a line with a corresponding number of crosses. The zeroth, second, and fourth orders of perturbation theory are represented in Fig. 3.3a, b, c.

Each line in Fig. 3.3 corresponds to the Green function $G_{0\varepsilon}^{R,A}(\mathbf{r}, \mathbf{r}')$, each cross to the potential $u(\mathbf{r})$, and one should integrate over the coordinates of the crosses.

Averaging with the Gaussian distribution Eq. (3.38) can be performed graphically, connecting each two crosses by dashed lines (Edwards (1958), Abrikosov, Gorkov, and Dzyaloshinskii (1963)). It is assumed that two crosses connected by the impurity line are located at the same point in space. Each impurity line gives the factor $(2\pi\nu\tau)^{-1}$. After averaging, the Green function $\langle G_\varepsilon^{R,A} \rangle$ depends on $\mathbf{r} - \mathbf{r}'$ and can be Fourier transformed. The result can be written by introducing the self-energy $\Sigma_\varepsilon^{R,A}(\mathbf{p})$ in the form

$$(G_\varepsilon^{R,A}(\mathbf{p}))^{-1} = (G_{0\varepsilon}^{R,A}(\mathbf{p}))^{-1} - \Sigma_\varepsilon^{R,A}(\mathbf{p}) \qquad (3.42)$$

The second- and fourth-order diagrams for the self-energy $\Sigma_\varepsilon^{R,A}(\mathbf{p})$ are represented in Fig. 3.4. In the limit $\varepsilon\tau \gg 1$ summation of all diagrams for $\langle G_\varepsilon^{R,A} \rangle$ is considerably simplified. It is well known (Abrikosov, Gorkov, and Dzyaloshinskii (1963)) that diagrams without intersections of the impurity lines in this limit in 2D and 3D provide the main contribution. For example, the diagram in Fig. 3.4c is generally speaking smaller than the diagram in Fig. 3.4b (for the δ-correlated potential, Eq. (3.38), the graph in Fig. 3.4b is accidentally small, too). In one dimension this selection rule does not work and one should use a more sophisticated technique (Berezinsky (1973), Abrikosov and Ryzhkin (1978), Berezinsky and Gorkov (1979)).

Neglecting diagrams with the intersections of the impurity lines one obtains the following equation for the self-energy $\Sigma_\varepsilon^{R,A}$; the latter is independent in this approximation of the

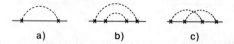

Fig. 3.4. a) Second- and b, c) fourth-order diagrams for self-energy.

momentum $\mathbf{p}$

$$\Sigma_\varepsilon^{R,A} = \frac{1}{2\pi \nu \tau} \int \frac{1}{\varepsilon - \varepsilon(\mathbf{p}) - \Sigma_\varepsilon^{R,A}} \frac{d^3\mathbf{p}}{(2\pi)^3} \tag{3.43}$$

The solution of Eq. (3.43) is a complex number. However, the real part of $\Sigma_\varepsilon^{R,A}$ is not interesting because it simply renormalizes the energy ε. As concerns the imaginary part, calculation of this quantity is in the limit $\varepsilon\tau \gg 1$ rather simple because the main contribution is from the region of the momenta $\mathbf{p}$ such that $\varepsilon(\mathbf{p})$ is close to ε. Renormalizing the energy $\varepsilon \rightarrow \varepsilon - \mathrm{Re}\,\Sigma_\varepsilon^{R,A}$ and substituting the integration over $\mathbf{p}$ by integration over $\xi = \varepsilon(\mathbf{p}) - \varepsilon$ such that $d\xi = \nu d^3\mathbf{p}$ (we assume that the energy ε is close to the Fermi energy ε_0) we obtain

$$\mathrm{Im}\,\Sigma_\varepsilon^{R,A} = \frac{1}{2\pi\tau} \int_{-\infty}^{\infty} \frac{d\xi}{\xi + i\,\mathrm{Im}\,\Sigma_\varepsilon^{R,A}} \tag{3.44}$$

At first glance, the solution of Eq. (3.44) should be the same for both $\mathrm{Im}\,\Sigma_\varepsilon^R$ and $\mathrm{Im}\,\Sigma_\varepsilon^A$. However, one should remember that in the limit $\tau \rightarrow \infty$ the Green functions have the form Eq. (3.41). This determines the solutions uniquely and we have

$$\mathrm{Im}\,\Sigma_\varepsilon^{R,A} = \mp 1/2\tau, \qquad \langle G_\varepsilon^{R,A}(\mathbf{p})\rangle = (\varepsilon - \varepsilon(\mathbf{p}) \pm i/2\tau)^{-1} \tag{3.45}$$

Knowing the average Green function Eq. (3.45) one can calculate all thermodynamic quantities. However, the fact that the mean free time τ is finite does not lead in the limit $\varepsilon\tau \gg 1$ to considerable changes in results obtained for the ideal electron gas. Averaging, for example, the density of states $\rho(\varepsilon, \mathbf{r})$ Eq. (3.16) we find

$$\langle \rho(\varepsilon, \mathbf{r})\rangle = 2\pi^{-1}\,\mathrm{Im} \int \langle G_\varepsilon^A(\mathbf{p})\rangle \frac{d^3\mathbf{p}}{(2\pi)^3}$$

$$= -2\pi^{-1}\nu\,\mathrm{Im} \int_{-\infty}^{\infty} \frac{d\xi}{\xi + i/2\tau} = 2\nu \tag{3.46}$$

The finiteness of the mean free time is very important for kinetic quantities because it is disorder that makes the conductivity finite. In order to calculate the average conductivity σ it is convenient to use Eqs. (3.26–3.31). The correlation functions $K_1^{\alpha\beta}(\omega)$ and $K_2^{\alpha\beta}(\omega)$ are represented by the loop in Fig. 3.5a with the current operators $\hat{\pi}_r^\alpha$ at the vertices and all possible crosses representing the impurity potential. In principle, when averaging one should connect not only the crosses on the same line in Fig. 3.5b but also the crosses belonging to different lines representing different Green functions. This reflects the fact that the average of the product of two Green functions is not necessarily equal to the product of the two corresponding averages. However, in the particular case of the δ-correlated Gaussian distribution for the random potential Eq. (3.38) the simplest "ladder" diagrams of the type in Fig. 3.5c give zero after the integration over the momenta because they contain the current operator at the vertices. A nonzero contribution can arise provided one substitutes the δ-function in Eq. (3.38) with a function with a finite range. In this case it would be necessary

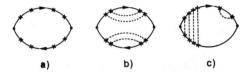

Fig. 3.5. Diagrams for conductivity.

to sum up all the ladder diagrams in Fig. 3.5c. Diagrams with intersections of the impurity lines are again small as soon as $\varepsilon\tau \gg 1$.

Taking into account the diagrams in Fig. 3.5b for the function $R^{\alpha\beta}(\omega)$, Eq. (3.31), we obtain in 3D in the white-noise potential, Eq. (3.38),

$$R^{\alpha\beta}(0, \varepsilon, \omega) = 2\left(\frac{e}{m}\right)^2 \int p^\alpha p^\beta \langle G^R_\varepsilon(\mathbf{p})\rangle\langle G^A_{\varepsilon-\omega}(\mathbf{p})\rangle \frac{d^3\mathbf{p}}{(2\pi)}$$

$$= \left(\frac{e}{m}\right)^2 \frac{2\nu p_0^2 \delta_{\alpha\beta}}{3} \int_{-\infty}^{\infty} \frac{d\xi}{(\xi - i/2\tau)(\xi + \omega + i/2\tau)} \tag{3.47}$$

The integral over ξ in Eq. (3.47) is not equal to zero because the integrand has poles both below and above the axis of real numbers. This is due to the presence of both retarded G^R and advanced G^A Green functions. The situation is different when one calculates the correlation function $K_2^{\alpha\beta}$, Eq. (3.28). This function is expressed in terms of integrals of the products of two retarded or two advanced Green functions. The integrands in the corresponding integrals over the variable ξ have poles both above or both below the real axis and therefore the integrals vanish. (In fact, they do not vanish exactly because substitution of the integrals over $\mathbf{p}$ with the integrals over ξ from $-\infty$ to ∞ is approximate but the result is small in the limit $\varepsilon\tau \gg 1$.)

Calculating the integral in Eq. (3.47) and using Eqs. (3.26–3.31) we find the frequency-dependent conductivity $\sigma(\omega)$

$$\sigma(\omega) = \frac{\sigma_0}{1 - i\omega\tau}, \qquad \sigma_0 = 2e^2\nu D_0 \tag{3.48}$$

where $D_0 = v_0^2\tau/3$ is the classical diffusion coefficient. This is the same coefficient used in Section 3.1. The expression for the classical conductivity σ_0, Eq. (3.48), agrees with Eq. (3.1). Eq. (3.48) can be obtained from the classical kinetic equation too.

The correlation function

$$Y^{00}(\mathbf{k}, \varepsilon, \omega) = \left\langle \sum_{\sigma,\sigma'} \int Y^{00}(y, y', \varepsilon, \omega) \exp\left(i\mathbf{k}(\mathbf{r}' - \mathbf{r})\right) d\mathbf{r} d\mathbf{r}'\right\rangle \tag{3.49}$$

with the function $Y^{00}(y, y', \varepsilon, \omega)$ defined in Eq. (3.24) is also an important quantity describing the kinetics. Calculation of this function is similar to that of $R^{\alpha\beta}$ but now one should take into account all the ladder diagrams in Fig. 3.5c and keep $\mathbf{k}$ finite. Summing the ladder diagrams we have

$$Y^{00}(\mathbf{k}, \varepsilon, \omega) = \frac{2}{\Pi^{-1}(\mathbf{k}, \omega) - (2\pi\nu\tau)^{-1}} \tag{3.50}$$

where

$$\Pi\left(\mathbf{k}, \omega\right) = \int \langle G_\varepsilon^R\left(\mathbf{p}\right)\rangle\langle G_{\varepsilon-\omega}^A\left(\mathbf{p} - \mathbf{k}\right)\rangle \frac{d^3\mathbf{p}}{(2\pi)^3} \tag{3.51}$$

The function $Y^{00}\left(\mathbf{k}, \varepsilon, \omega\right)$ has a simple form in the limit ω, $kv_0 \ll \tau^{-1}$. Making an expansion in ω and k in the function $\Pi^{-1}\left(\mathbf{k}, \omega\right)$ we obtain in this limit

$$Y^{00}\left(\mathbf{k}, \varepsilon, \omega\right) = \frac{4\pi\nu}{D_0\mathbf{k}^2 - i\omega} \tag{3.52}$$

We see that the correlation function $Y^{00}\left(\mathbf{k}, \varepsilon, \omega\right)$ is the propagator of the equation of classical diffusion. If the dimensionality d is different from 3 one should substitute d for the number 3 in the expression for the diffusion coefficient. In the weakly disordered limit when contributions of products of two retarded or two advanced Green functions can be neglected the function $R^{00}\left(\mathbf{k}, \varepsilon, \omega\right)$, Eq. (3.31), coincides with $Y^{00}\left(\mathbf{k}, \varepsilon, \omega\right)$, Eq. (3.49), after multiplication by e^2.

Thus, we have learned from this section that in the language of the diagrams the classical motion in a random potential is described by diagrams without intersections of impurity lines. The interference effects discussed in Section 3.1 are not seen in this class of diagrams. At the same time diagrams with the intersections of the impurity lines are small in the limit $\varepsilon\tau \gg 1$ (Abrikosov, Gorkov, and Dzyaloshinskii (1963)) in dimensionality larger than 1 and at first glance cannot give large contributions. The resolution of this paradox lies in the existence of a certain class of diagrams with intersections. Although each diagram is small, their sum can be large because of the large number of diagrams. These special sequences of diagrams give the diverging quantum corrections described in Section 3.1 and are discussed in the next section.

3.4 Quantum corrections

Apparently, the first attempt to go beyond the classical formulae for motion in a weak random potential was made by Langer and Neal (1966), who considered maximally crossed diagrams for kinetic coefficients and pointed out that anomalous contributions can arise when calculating kinetic quantities. As concerns thermodynamics, the classical description continued to be accurate. Although the idea that maximally crossed diagrams were important turned out to be correct, appropriate treatment of the phenomenon was done much later by Gorkov, Larkin, and Khmelnitskii (1979) and Abrahams, Anderson, and Ramakrishnan (1980), who showed for the first time that summation of these diagrams gave divergencies for the space dimensionality $d \leq 2$. This was an explicit confirmation of the scaling ideas of Abrahams et al. (1979).

Let us demonstrate how the calculations are carried out for the conductivity σ. As in the previous section one should calculate the correlation function $K_1^{\alpha\beta}\left(\omega\right)$. The maximally crossed diagrams for this quantity are represented in Fig. 3.6a. It becomes more clear why maximally crossed diagrams play a special role when one substitutes for the diagram in Fig. 3.6a an interaction of the electron with an effective mode, represented by a wavy line in Fig. 3.6b. This effective mode is in fact the sum of all possible "ladder" diagrams represented in Fig. 3.6c. The ladder in Fig. 3.6c is very similar to the sequence of diagrams leading to superconducting instability (see, e.g., Abrikosov, Gorkov, and Dzyaloshinskii (1963)). The only difference is that now one should draw impurity lines

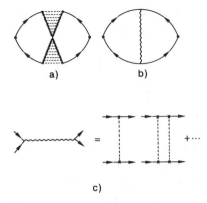

Fig. 3.6. a) The maximally crossed diagram, b) representation in terms of interaction of the electron with an effective mode, c) effective diffusion mode (cooperon).

instead of interaction lines. Both electron lines have the same direction. As a result of this similarity the effective mode in Fig. 3.6c is often called a "cooperon." It is very important that one electron line relates to the retarded Green function and the other to the advanced one. Denoting the cooperon by $C(\mathbf{k}, \omega)$ and summing up all the ladder diagrams we have

$$C(\mathbf{k}, \omega) = 2\pi v\tau \left[1 - \frac{1}{2\pi v\tau} \int \langle G_\varepsilon^R(\mathbf{p})\rangle \langle G_{\varepsilon-\omega}^A(\mathbf{p} - \mathbf{k})\rangle \frac{d^d\mathbf{p}}{(2\pi)^d}\right]^{-1} \qquad (3.53)$$

The product of the averaged Green functions in the brackets in Eq. (3.53) is the same as that in Eq. (3.51) for $\Pi(\mathbf{k}, \omega)$ although the direction of the arrows of the Green functions in these cases is different. This occurs if the system possesses time reversal invariance. At the same time, the two cases can differ if the time reversal symmetry is violated by magnetic interactions. The calculations can be performed in an arbitrary dimension d. Integrating in Eq. (3.53) one obtains for ω, $v_0 k \ll \tau^{-1}$

$$C(\mathbf{k}, \omega) = \frac{2\pi v}{D_0 \mathbf{k}^2 - i\omega} \qquad (3.54)$$

where D_0 is again the classical diffusion coefficient. The normalization coefficient in Eq. (3.53) is chosen quite arbitrarily. Using the result of the summation of the ladder diagrams, Eq. (3.53), we can write the integral corresponding to Fig. 3.6b as

$$-\frac{1}{(2\pi v\tau)^2} \int \langle G_\varepsilon^R(\mathbf{p})\rangle^2 \langle G_\varepsilon^A(\mathbf{p})\rangle^2 v^\alpha v^\beta C(\mathbf{k}, \omega) \frac{d^d\mathbf{p}}{(2\pi)^3} \qquad (3.55)$$

where v^α are components of the velocity. In order to obtain the quantum correction to the conductivity one should integrate the expression in Eq. (3.55) over the momenta $\mathbf{k}$ assuming that the main contribution comes from the region of small $k \ll (\tau v_0)^{-1}$. At the same time, the main contribution when integrating over $\mathbf{p}$ is from the region $|\varepsilon(\mathbf{p}) - \varepsilon| \sim \tau^{-1}$. This means that the region of integration over momenta for the electron lines is completely different from that for the effective diffusion mode, Eq. (3.54). Integrating first over the momenta of the electron lines we integrate out "fast degrees of freedom" and are left with some "quasi particles" with the propagator given by Eq. (3.54).

Substituting the propagator Eq. (3.54) into the diagram in Eq. (3.55) and calculating the integrals over $\mathbf{p}$ the conductivity $\sigma(\omega)$ can be written in the form

$$\sigma(\omega) = \sigma_0 \left(1 - \frac{1}{\pi \nu} \int \frac{1}{D_0 \mathbf{k}^2 - i\omega} \frac{d^d \mathbf{k}}{(2\pi)^d} \right) \tag{3.56}$$

The assumption that the main contribution is from the region of small k is well justified in 1D and 2D in the low-frequency region and where one can really speak of a contribution of a mode with low-lying excitations with the propagator $C(\omega)$, Eq. (3.54). In 3D the contribution of small ω and k in Eq. (3.56) is small for weak disorder but is very sensitive to a magnetic field and inelastic processes and can be observed experimentally (Altshuler et al. (1983), Bergmann (1984), Lee and Ramakrishnan (1985)).

It is important to note that the dimensionality in Eq. (3.56) is determined by the geometry of the sample rather than by the dimensionality of the Fermi surface. One can apply this equation to describe kinetics of thick wires or films. Microscopically, such systems are 3D metals but as soon as $\omega \ll D_0/L_{\min}^2$, where $L_{\min}$ is the minimal size of the sample (the thickness in films, the diameter of the cross section in wires), one can neglect all nonzero space harmonics in this direction, and thus the dimensionality in the integral in Eq. (3.56) is reduced. If all the sizes of the sample are small and $\omega \ll D_0/L_{\max}^2$, where $L_{\max}$ is the maximal size, one can speak of a zero-dimensional situation, which means that only the harmonics with $\mathbf{k} = 0$ is taken into account in all diffusion modes.

At finite temperatures, when inelastic processes are important, Eq. (3.56) is applicable in the region $\omega \gg \tau_\varphi^{-1}$, where τ_φ is an inelastic mean free time. One can see that Eq. (3.56) is equivalent to Eq. (3.10). This shows that the quantum interference effects discussed in Section 3.1 are formally due to the existence of the effective diffusion modes that arise after summation of certain classes of diagrams.

It is not difficult to write an expression for the cooperon in the presence of a magnetic field $\mathbf{H}$. One should again sum the ladder diagrams in Fig. 3.6c but write the Green functions $\tilde{G}_\varepsilon^{R,A}(\mathbf{r}, \mathbf{r}')$ in the presence of the field in the form

$$\tilde{G}_\varepsilon^{R,A}(\mathbf{r}, \mathbf{r}') = G_\varepsilon^{R,A}(\mathbf{r}, \mathbf{r}') \exp \left(\frac{ie}{c} \int_{\mathbf{r}}^{\mathbf{r}'} \mathbf{A} d\mathbf{r} \right) \tag{3.57}$$

where $\mathbf{A}$ is the vector potential corresponding to the field $\mathbf{H}$ and the integral in the exponential should be calculated along the straight line between the points $\mathbf{r}$ and $\mathbf{r}'$. The average Green functions decay at the distances $|\mathbf{r} - \mathbf{r}'| \sim l$ and, therefore, if the magnetic flux through a square with the size l is smaller than the flux quantum $\phi_0 = ch/e$, one can expand the exponential in Eq. (3.57). Because the Green functions of the two electron lines in the ladder have the same direction the magnetic field contributions in each pair of the Green functions add together. As a result the cooperon $C(\mathbf{r}, \mathbf{r}')$ can be written in the form of the solution of the following differential equation

$$\left(D_0 \left(-i\nabla - \frac{2e}{c} \mathbf{A} \right)^2 - i\omega \right) C(\mathbf{r}, \mathbf{r}') = 2\pi \nu \delta(\mathbf{r} - \mathbf{r}') \tag{3.58}$$

Eq. (3.58) is gauge-invariant and contains charge $2e$, which is reminiscent of the corresponding gauge-invariant term in the Ginzburg–Landau equation for superconductors. However, the effects considered here have nothing in common with superconductivity. The solution

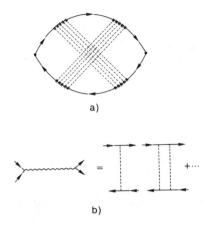

a)

b)

Fig. 3.7. a) A second-order quantum correction to the conductivity; b) diffuson.

of Eq. (3.58) can be found in the form of an expansion in eigenfunctions of the Hamiltonian describing electron motion in the magnetic field. The corresponding coefficients $C_E(p)$ of the expansion equal

$$C_E(p) = 2\pi\nu \left(\frac{4eHD_0}{c} \left(n + \frac{1}{2} \right) + D_0 p^2 - i\omega \right)^{-1} \tag{3.59}$$

where p is the component of the momentum parallel to the magnetic field. We see that the magnetic field cuts off the cooperon.

In the presence of the magnetic field one should substitute the cooperon from Eq. (3.59) for the diffusion propagator in Eq. (3.56). Eq. (3.56) is the first correction to classical conductivity. One can continue the procedure, drawing more complicated diagrams. An example of a more complicated sequence of diagrams is represented in Fig. 3.7a. The diagram contains two ladders, and the effective modes corresponding to the ladders are similar to the one in Figs. 3.6, but now the directions of the arrows on two different electron lines are opposite (Fig. 3.7b). The diagram of Fig. 3.7a can be represented schematically by Fig. 3.8a. Summing the ladder diagrams in Fig. 3.7b we find for this diffusion mode $D(\mathbf{k}, \omega)$, which is often called a "diffuson"

$$D(\mathbf{k}, \omega) = \frac{1}{2} Y^{00}(\mathbf{k}, \varepsilon, \omega) \tag{3.60}$$

where $Y^{00}(\mathbf{k}, \varepsilon, \omega)$ is the density–density correlation function, Eqs. (3.24, 3.49, 3.52). In Eq. (3.60) the normalization for diffuson $D(\mathbf{k}, \omega)$ is the same as for the cooperon, Eq. (3.54). The correlation function $Y^{00}(\mathbf{k}, \varepsilon, \omega)$ differs by a factor of 2 as a result of the summation over the spin degree of freedom. Again, it is convenient to integrate first over the momenta of the electron lines, reducing the problem to study of interacting diffusion modes.

The diagram in Figs. 3.7a and 3.8a is not the only one in this order of the perturbation theory. One can draw, for example, a diagram with two cooperons, Fig. 3.8b. The third diagram that gives a contribution of the same order as those in Figs. 3.8a and 3.8b is represented in Fig. 3.8c. It contains three diffusion modes. Calculations in the second order of perturbation theory in the diffusion modes have been done by Gorkov, Larkin, and Khmelnitskii (1979) and Hikami (1981). The results of the calculations are not given here because perturbation theory will be discussed in Chapter 4.

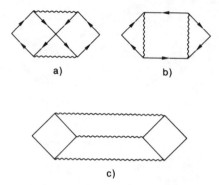

Fig. 3.8. Diagrams representing the second order of the perturbation theory in the diffusion modes.

Because the Green functions in the ladder, Fig. 3.7b, have opposite directions the magnetic field included via substitution, Eq. (3.57), does not change the form of the diffuson because the contributions of the two Green functions cancel each other. This is the main difference between the diffusons and cooperons. A strong magnetic field can destroy the first cooperon correction to the conductivity, but that does not mean that quantum interference is destroyed, because higher-order corrections can still be important.

Although taking into account the first quantum correction, Eqs. (3.10, 3.56), is sufficient to explain many experiments, at low temperature the quantum corrections to the conductivity in 0D, 1D, and 2D diverge in the limit $\omega \to 0$ and the kinetic coefficients cannot be calculated by considering any finite number of the diagrams. The same occurs in 3D in the limit of strong disorder. Therefore, many effects in disordered and mesoscopic systems cannot be described at low temperatures by summing the diagrams because their number grows dramatically when considering higher orders of perturbation theory. Furthermore, even if it became possible to sum up all diagrams, that would not necessarily solve the problem in the region of low frequencies, where terms that are nonanalytic in the expansion parameter can become important. In particular, the results of Chapter 6 show that the level–level correlation function for metallic particles cannot be obtained by summing perturbation theory graphs only. All this means that one really needs a new nonperturbative method. The possibility of constructing a more powerful scheme of calculations is indicated by the way the quantum correction to the conductivity, Eq. (3.56), was calculated. One had to integrate out the fast "electron degrees of freedom" and then investigate the contribution of the low-lying diffusion modes (cooperons and diffusons). A more general approach is to try to integrate out the "electron degrees of freedom" at the beginning of the calculations without diagram expansion. As a result of such manipulations, a Lagrangian describing the diffusion modes only would be obtained.

Such an approach is often used in condensed matter physics and field theory. For example, starting from a system of strongly interacting electrons one can obtain the Heisenberg model describing spin excitations. Spin waves and their interactions can be studied using the Heisenberg model but this would be much more difficult to do on the basis of the initial electron Hamiltonian. A more sophisticated example is the superfluid ^{3}He. The theory of the superfluidity in this system starts with a Hamiltonian of interacting fermions. The interaction of the fermions results in spontaneous breaking of symmetry. The order parameter is a matrix and its fluctuations lead to a variety of different gapless excitations (Salomaa and Volovik

(1987), Wölfle and Vollhardt (1987)). According to the Goldstone theorem the existence of the gapless excitations is a general consequence of the spontaneous breaking of symmetry and they are often called Goldstone modes.

It will be shown in the next chapter that the Grassmann anticommuting variables make it possible to develop a method of integrating out the "electron degrees of freedom" and deriving an effective Lagrangian containing diffusion modes only. Moreover, it will be seen that these modes are also due to a spontaneous breaking of symmetry and they are, in fact, Goldstone modes. The use of the Lagrangian allows one to solve many problems of disorder and chaos for the first time.

In concluding this chapter let us emphasize that the diffusion modes arise only when the average of the products of the retarded G^R and advanced G^A Green functions, $\langle G^R_\varepsilon G^A_\varepsilon \rangle$, are calculated. Calculation of averages such as $\langle G^R_\varepsilon G^R_{\varepsilon-\omega} \rangle$ and $\langle G^A_\varepsilon G^A_{\varepsilon-\omega} \rangle$ in the limit of weak disorder can be carried out in the framework of standard perturbation theory. The same is valid for calculation of the one-particle Green functions $\langle G^R_\varepsilon \rangle$ and $\langle G^A_\varepsilon \rangle$. The direct calculation of first-order terms of perturbation theory shows that the correlation functions $\langle G^A_\varepsilon (0, \mathbf{r}) G^A_{\varepsilon-\omega} (\mathbf{r},0) \rangle$ and $\langle G^R_\varepsilon (0, \mathbf{r}) G^R_{\varepsilon-\omega} (\mathbf{r},0) \rangle$ fall off rapidly at atomic distances. Therefore, in calculating the density–density correlation function and other kinetic quantities at large distances these averages need not be considered.

4

Nonlinear supermatrix σ-model

4.1 Reduction to a regular model, mean field theory

4.1.1 Regular model with "interaction"

The main task of this section is to represent the physical quantities discussed in the preceding chapter in such a form that one could average over the random potential at the beginning of the calculations. One way to do so is the replica trick suggested by Edwards and Anderson (1975) for a study of spin glasses. The first works on the application of field theoretical models to disorder problems were based on this trick (Wegner (1979), Schäfer and Wegner (1980), Efetov, Larkin, and Khmelnitskii (1980)). In the works of Wegner (1979) and Schäfer and Wegner (1980) kinetic quantities were written in terms of functional integrals over conventional numbers, whereas in the work of Efetov, Larkin, and Khmelnitskii (1980) integration over the anticommuting Grassmann variables was used. Then, in both approaches nonlinear σ-models that contained $n \times n$ matrices were derived. At the end of calculations one had to take the limit $n \to 0$. The formalism of Efetov, Larkin, and Khmelnitskii (1980) was extended later to include electron–electron interactions (Finkelstein (1983, 1984)) and strong magnetic fields (Levine, Libby, and Pruisken (1984)).

The nonlinear supermatrix σ-model derived later has many common features with the replica σ-model, and it may seem that the calculations within all these models are equivalent. However, they are far from equivalent and everything depends on the kind of calculations being carried out. In Chapter 5 it is demonstrated how perturbation theory in diffusion modes can be obtained from the supermatrix σ-model and how renormalization group calculations can be carried out. This type of calculation can be done in the same way using the replica σ-models, and, in fact, such calculations were done in that way before the supersymmetry technique appeared. However, all other results presented in the rest of the book can be obtained only by using the supermatrix σ-model. It is not clear how to reproduce them with the replica σ-model even now. For example, there have been attempts to study the level statistics in a limited volume with replica models (Verbaarschot and Zirnbauer (1984)), but the conclusion was that the limit $n \to 0$ led to nonphysical results. Another example is localization, for which nobody has managed to get any result using the replica technique.

So the supersymmetry technique is not just a new way to obtain old results but is for some problems the only method of calculation available. This chapter started with an advertisement of the advantages of the supersymmetry method to give to the reader additional motivation to make the effort necessary to follow the derivation of the σ-model. Although rather simple in principle the calculations presented are sometimes cumbersome and one should have some patience to go through them.

The main idea of the supersymmetry technique has been explained in Chapter 2 and is contained in Eq. (2.76). Using this formula one can write the retarded $G_{\varepsilon k}^{R}$ and advanced $G_{\varepsilon k}^{A}$ Green functions Eq. (3.17) in the form

$$G_{\varepsilon k}^{R,A} = \mp i \int \Phi_{\alpha k} \Phi_{\alpha k}^{+} \exp\left(i\Phi_{k}^{+}\left(\pm\left(\varepsilon - \varepsilon_{k}\right) + i\delta\right)\Phi_{k}\right) d\Phi_{k}^{*} d\Phi_{k} \qquad (4.1)$$

where

$$\Phi_{k} = \begin{pmatrix} \chi_{k} \\ S_{k} \end{pmatrix}$$

In Eq. (4.1), in contrast to Eqs. (2.74–2.76), χ_{k} and S_{k} are anticommuting and commuting variables but not the sets of variables. So α can take two values, 1 and 2. At first glance, it is somewhat inconvenient to write the Gaussian integral in Eq. (4.1) not with a real number but with $i = \sqrt{-1}$ in the exponential. However, this form is the only one providing the convergence of the integral for arbitrary $\varepsilon - \varepsilon_{k}$. Furthermore, the difference between the retarded and advanced Green functions is clearly seen from the integral, Eq. (4.1).

The formal absence of the weight denominator in Eq. (4.1) is very important. This absence is due to the fact that now the supermatrix F from Eqs. (2.74–2.76) is proportional to the unit matrix I. The superdeterminant of a matrix of the form bI, where b is a number, is equal to unity. The possibility of writing the Green functions in the form of Eq. (4.1) is central in constructing the supersymmetry method. For those to whom integration over the supervectors is unfamiliar, the absence of the weight denominator in Eq. (4.1) can be explained by simpler arguments. Integration of the Gaussian exponential over anticommuting variables, Eq. (2.59), gives $\det A$ instead of $(\det A)^{-1}$ on the right-hand side. Therefore, when integrating over both types of variables with the same matrix A for both integrals, these determinants in the denominator cancel each other. If the subscript α in Eq. (4.1) stands for the commuting variables, the integral over the anticommuting variables plays the role of the weight denominator in Gaussian averaging, and vice versa, if α stands for the Grassmann elements, the integral over the conventional numbers plays this role.

Introducing the field variables Φ, Φ^{+} in a standard way

$$\Phi(y) = \sum_{k} \Phi_{k}\varphi_{k}(y), \qquad \Phi^{+}(y) = \sum_{k} \Phi_{k}^{+}\varphi_{k}^{*}(y) \qquad (4.2)$$

the Green functions $G_{\varepsilon}^{R,A}(y, y')$ can be rewritten in the form of functional integrals

$$G_{\varepsilon}^{R,A}(y, y') = \mp \int \Phi_{\alpha}(y)\Phi_{\alpha}^{+}(y') \exp\left[i \int \Phi^{+}(x)(\pm(\varepsilon - \mathcal{H})\right.$$
$$\left. + i\delta)\Phi(x)dx\right]D\Phi^{*}D\Phi \qquad (4.3)$$

In Eqs. (4.1, 4.2) x and y stand for both the space and spin variables. Because of the absence of the weight denominator Eq. (4.3) enables us to average the Green functions over the impurities immediately and to calculate exactly the average density of states in some cases. However, let us concentrate now on studying kinetic quantities containing products of two Green functions. As we have seen in the preceding chapter calculation of the average of the product $G_{\varepsilon}^{R}G_{\varepsilon-\omega}^{A}$ is the main difficulty. It is this average that will be considered. Calculation of the averages of the products $G_{\varepsilon}^{A}G_{\varepsilon-\omega}^{A}$ and $G_{\varepsilon}^{R}G_{\varepsilon-\omega}^{R}$ in the cases under consideration is not difficult and can be done in the framework of conventional perturbation theory.

In order to write the average of a product of two Green functions at different energies ε and $\varepsilon - \omega$ one has to introduce a double set of the variables with Φ^1 related to the advanced Green function and Φ^2 to the retarded one. For simplicity let us consider first a system without spin interactions; the latter is included in Section 4.3. Without spin interactions one can make calculations for the states with the spin "up" and "down" independently and write all intermediate formulae as if the electrons were spinless. Only at the end should one multiply, for example, conductivity by a factor of 2. The appropriate functional integral for the product $G_\varepsilon^R G_{\varepsilon-\omega}^A$ has the form

$$
\left\langle G_{\varepsilon-\omega}^A (\mathbf{r}, 0) \, G_\varepsilon^R (0, \mathbf{r}) \right\rangle
$$

$$
= \left\langle \int \Phi_\alpha^1 (\mathbf{r}) \, \Phi_\alpha^{1+} (0) \, \Phi_\beta^2 (0) \, \Phi_\beta^{2+} (\mathbf{r}) \exp (-\mathcal{L}) \, D\Phi^{1*} D\Phi^2 D\Phi^{2*} D\Phi^2 \right\rangle, \qquad (4.4)
$$

$$
\mathcal{L} = i \int \left[\sum_{m=1}^{2} \Phi^{m+} \left(\left(\varepsilon - \frac{\omega}{2} - \mathcal{H} \right) (-1)^{m-1} - \frac{\omega}{2} - i\delta \right) \Phi^m \right] dy \qquad (4.5)
$$

In Eqs. (4.4, 4.5) the subscripts α and β stand for arbitrary components of the supervector Φ. All calculations and formulae can be made more compact if we introduce the eight-component supervectors ψ consisting of the four-component supervectors ψ^1 and ψ^2 such that

$$
\psi^m = \begin{pmatrix} \vartheta^m \\ v^m \end{pmatrix}, \, \vartheta^m = \frac{1}{\sqrt{2}} \begin{pmatrix} \chi^{m*} \\ \chi^m \end{pmatrix}, \, v^m = \frac{1}{\sqrt{2}} \begin{pmatrix} S^{m*} \\ S^m \end{pmatrix}, \, m = 1, 2 \qquad (4.6)
$$

For the supervectors ψ of the form of Eq. (4.6) one can define, in addition to transposition and Hermitian conjugation, the operation of "charge conjugation"

$$
\bar{\psi} = (C\psi)^T, \, \bar{\psi}^m = \begin{pmatrix} \bar{\vartheta}^m & \bar{v}^m \end{pmatrix} \qquad (4.7)
$$

In Eq. (4.7), T stands for transposition, and C is the supermatrix of the form

$$
C^{mn} = \Lambda^{mn} \begin{pmatrix} c_1 & 0 \\ 0 & c_2 \end{pmatrix} \equiv \Lambda^{mn} C_0 \qquad (4.8)
$$

where Λ is the diagonal supermatrix

$$
\Lambda = \begin{pmatrix} \mathbf{1} & 0 \\ 0 & -\mathbf{1} \end{pmatrix} \qquad (4.9)
$$

with $\mathbf{1}$ the 4×4 unity matrix.

The 2×2 matrices c_1 and c_2 have the form

$$
c_1 = \begin{pmatrix} 0 & -1 \\ 1 & 0 \end{pmatrix}, \, c_2 = \begin{pmatrix} 0 & 1 \\ 1 & 0 \end{pmatrix} \qquad (4.10)
$$

Here and in what follows the superscripts relate to the advanced and retarded Green functions. The charge conjugate field $\bar{\psi}$ is related to the Hermitian conjugate field ψ^+ as

$$
\bar{\psi} = \psi^+ \Lambda \qquad (4.11)
$$

In terms of the supervectors ψ the Lagrangian $\mathcal{L}$ in Eqs. (4.4, 4.5) takes the form

$$\mathcal{L} = i \int \bar{\psi}(\mathbf{r}) \left(-\tilde{\mathcal{H}}_0 - u(\mathbf{r}) - \frac{1}{2}(\omega + i\delta) \Lambda \right) \psi(\mathbf{r}) \, d\mathbf{r} \qquad (4.12)$$

where $u(\mathbf{r})$ is the impurity potential. The operator $\tilde{\mathcal{H}}_0$ equals

$$\tilde{\mathcal{H}}_0 = \varepsilon(-i\nabla_{\mathbf{r}}) - \varepsilon + \frac{\omega}{2} \qquad (4.13)$$

At first glance, there is no need to introduce eight-component vectors because already four-component ones would enable us to write the integral, Eq. (4.4), in a more compact form. The convenience of the eight-component supervectors will become clear somewhat later.

As in the preceding chapter, the impurities are supposed to be modeled by a random potential $u(\mathbf{r})$ with the Gaussian distribution, Eq. (3.38). The use of the Gaussian distribution lets us easily average Eq. (4.4) with the Lagrangian Eq. (4.12) over the impurities. This possibility is a consequence of the absence of the weight denominator. After averaging, Eq. (4.4) keeps its form but now the Lagrangian $\mathcal{L}$ is written as

$$\mathcal{L} = \int \left[-i\bar{\psi}\tilde{\mathcal{H}}_0\psi + \frac{1}{4\pi\nu\tau}(\bar{\psi}\psi)^2 - \frac{i(\omega + i\delta)}{2}\bar{\psi}\Lambda\psi \right] d\mathbf{r} \qquad (4.14)$$

The Lagrangian $\mathcal{L}$ in Eq. (4.14) is similar to those studied in field theory and belongs to the class of models with the interaction ψ^4. Let us remark that at $\omega = 0$ the Lagrangian is invariant under rotations of the supervectors in the superspace because it depends on the square of "length" $\bar{\psi}\psi$ only. The frequency ω violates this symmetry and, if one uses an analogy with spin models, plays the role of an "external field." The violation of supersymmetry by the frequency is due to the fact that Eq. (4.14) is written for the product $G^R G^A$ (the presence of the matrix Λ is a direct consequence of this). Symmetry would not be violated in the corresponding integrals for $\langle G^R G^R \rangle$, $\langle G^A G^A \rangle$. The averaging of the simpler Lagrangian, Eq. (4.3), for the one-particle Green function results in a model with the interaction ψ^4, but the supersymmetry in this case is also not violated. As we will see the diffusion modes discussed in the preceding chapter exist only as a result of the violation of symmetry.

4.1.2 Mean field theory

It is clearly not possible to calculate any correlation functions with the Lagrangian, Eq. (4.14), exactly (except for the case of the one-dimensional chain or the Bethe lattice, where one can write recurrence equations). Further calculations will be performed in the limit of large mean free times τ, which corresponds to a weak interaction in the Lagrangian $\mathcal{L}$, Eq. (4.14). However, the use of the standard perturbation theory as we have seen is impossible even in this limit because of the existence of the diffusion modes and so one should try to use nonperturbative approaches.

The Lagrangian $\mathcal{L}$, Eq. (4.14), is formally very similar to that used in the theory of superconductivity and it is natural to try to use methods developed for studying this phenomenon. Most superconductors are well described by the theory of Bardeen, Cooper, and Schrieffer (1957), which is a type of mean field theory. This theory is applicable in the limit of weak electron–electron interaction and this corresponds to the weak coupling limit ($\varepsilon\tau \gg 1$) of the Lagrangian $\mathcal{L}$, Eq. (4.14). So let us try to construct a mean field approximation by following the line of a popular book (De Gennes (1966)). In the mean field scheme one has to write a quadratic term in the Lagrangian instead of the quartic term, substituting different pairs of the fields by averages. For interaction ψ^4 there can be six different pairings, which can be written as follows:

$$\mathcal{L}_{\text{int}} \equiv \frac{1}{4\pi\nu\tau} \int (\bar{\psi}\psi)^2 d\mathbf{r} \rightarrow \mathcal{L}_1 + \mathcal{L}_2 + \mathcal{L}_3, \tag{4.15}$$

$$\mathcal{L}_1 = \frac{1}{4\pi\nu\tau} \sum_{\alpha,\beta} \int 2\langle\bar{\psi}_\alpha\psi_\alpha\rangle_{\text{eff}} \bar{\psi}_\beta\psi_\beta d\mathbf{r} \tag{4.16}$$

$$\mathcal{L}_2 = \frac{1}{4\pi\nu\tau} \sum_{\alpha,\beta} \int 2\bar{\psi}_\alpha \langle\psi_\alpha\bar{\psi}_\beta\rangle_{\text{eff}} \psi_\beta d\mathbf{r} \tag{4.17}$$

$$\mathcal{L}_3 = \frac{1}{4\pi\nu\tau} \sum_{\alpha,\beta} \int \left(\langle\bar{\psi}_\alpha\bar{\psi}_\beta\rangle_{\text{eff}} \psi_\beta\psi_\alpha + \bar{\psi}_\alpha\bar{\psi}_\beta \langle\psi_\beta\psi_\alpha\rangle_{\text{eff}} \right) d\mathbf{r} \tag{4.18}$$

In Eqs. (4.16–4.18), the symbol

$$\langle\ldots\rangle_{\text{eff}} = \int (\ldots) \exp(-\mathcal{L}_{\text{eff}}) D\psi$$

stands for the functional averaging with the effective Lagrangian $\mathcal{L}_{\text{eff}} = \mathcal{L}_0 + \mathcal{L}_1 + \mathcal{L}_2 + \mathcal{L}_3$, where $\mathcal{L}_0$ is the quadratic part of the Lagrangian $\mathcal{L}$ in Eq. (4.14), and α and β stand for the components of the supervectors $\bar{\psi}$ and ψ.

In fact, all the terms in Eqs. (4.17, 4.18) are equal to each other and this can be seen directly from the definitions Eqs. (4.6–4.10) of the supervectors $\bar{\psi}$ and ψ. The average $\langle\bar{\psi}\psi\rangle$ in $\mathcal{L}_1$, Eq. (4.16), is invariant with respect to rotations in the superspace and, therefore, according to the property Eqs. (2.84, 2.85), is equal to zero. The averages in Eqs. (4.17, 4.18) can be both commuting and anticommuting, depending on the subscripts α and β. The final Lagrangian $\mathcal{L}_{\text{eff}}$ takes the form

$$\mathcal{L}_{\text{eff}} = \int \left[-i\bar{\psi} \left(\tilde{\mathcal{H}}_0 + \frac{i}{2}(\omega + i\delta)\Lambda + \frac{i}{2\tau}Q \right) \psi \right] d\mathbf{r} \tag{4.19}$$

with the 8×8 supermatrix Q satisfying the following self-consistency equation:

$$Q = \frac{2}{\pi\nu} \langle\psi\bar{\psi}\rangle_{\text{eff}} \tag{4.20}$$

The effective Lagrangian $\mathcal{L}_{\text{eff}}$, Eq. (4.19), has a compact form that justifies the introduction of the eight-component supervectors ψ. As mentioned, one could write the Lagrangian $\mathcal{L}$, Eq. (4.14), using four-component supervectors Φ. However, making the mean field theory in this case one would have to include averages of the type $\langle\Phi_\alpha^*\Phi_\beta^*\rangle$ and $\langle\Phi_\alpha\Phi_\beta\rangle$. As in the theory of superconductivity, this would lead anyway to doubling the number of components of the supervectors. So, eight is really the minimal number of components of supervectors that is necessary for calculations of the products $\langle G^R G^A \rangle$ using the supersymmetry method.

Calculating the Gaussian integral in Eq. (4.20) with the help of Eq. (2.76) we obtain

$$Q = \frac{1}{\pi\nu} \int g_0(\mathbf{p}) \frac{d^d\mathbf{p}}{(2\pi)^d}, \tag{4.21}$$

$$g_0(\mathbf{p}) = i \left(\varepsilon(\mathbf{p}) + \frac{\omega + i\delta}{2} - \varepsilon + \frac{1}{2}(\omega + i\delta)\Lambda + \frac{iQ}{2\tau} \right)^{-1} \tag{4.22}$$

The integral over the momenta $\mathbf{p}$ in Eq. (4.21) has both a real and an imaginary part. As concerns the imaginary part the main contribution comes from the region $|\varepsilon(\mathbf{p}) - \varepsilon|$ $\sim \varepsilon \gg \tau^{-1}$, ω and, therefore, is proportional to the unit matrix $\mathbf{1}$. Using Eqs. (4.13, 4.19) we see that this contribution leads to a small renormalization of the energy ε. Assuming that this energy has already been renormalized we can forget about the imaginary part of Q and concentrate on the real part. The main contribution to the real part of Eq. (4.21) comes from the region $|\varepsilon(\mathbf{p}) - \varepsilon| \sim \tau^{-1} \ll \varepsilon$. Introducing, as in Section 3.3 of the preceding chapter, the variable $\xi = \varepsilon(\mathbf{p}) - \varepsilon$ we can rewrite Eq. (4.21) in the form

$$Q = \frac{i}{\pi} \int\limits_{-\infty}^{\infty} \left(\xi + \frac{1}{2}(\omega + i\delta)\Lambda + \frac{iQ}{2\tau} \right)^{-1} d\xi \qquad (4.23)$$

Eq. (4.23) determines the contribution to the real part of Q only and has at $\omega \neq 0$ one evident solution

$$Q = \Lambda, \qquad \omega \neq 0 \qquad (4.24)$$

Formally, Eq. (4.23) also has the solutions $Q = \pm \mathbf{1}$, but these solutions should be discarded; otherwise the analyticity properties of the retarded G_ε^R or advanced G_ε^A Green functions in the half-planes of the complex variable ε would be violated.

Substituting the solution Eq. (4.24) for Q into the effective Lagrangian Eq. (4.19) and calculating the corresponding integrals over the supervectors ψ we come to Eq. (3.45) for the retarded G^R and advanced G^A Green functions. Thus, as concerns the one-particle Green functions, the mean field approximation developed in this section is equivalent to summation of the diagrams without the intersections of the impurity lines. However, such a simple mean field theory is not sufficient for calculation of two-particle correlation functions such as the function R^{ab}, Eq. (3.31). Substituting in Eq. (4.5) the Lagrangian $\mathcal{L}$ by $\mathcal{L}_{\text{eff}}$, Eq. (4.19), with the supermatrix Q determined by Eq. (4.24) we obtain, for example, for R^{00} the result $R^{00} = 4\pi\nu\tau$, which is in strong disagreement with Eq. (3.52). In order to obtain the result Eq. (3.52) one should go beyond the mean field theory and take into account fluctuations of the supermatrix Q. It turns out that at low frequencies ω there are low-lying excitations that change the mean field results drastically. They are due to a degeneracy of the solution of the Eq. (4.23) at $\omega = 0$. One can see easily that in this case not only the supermatrices Q specified by Eq. (4.24) but any Q of the form

$$Q = V\Lambda\bar{V} \qquad (4.25)$$

where V is an arbitrary unitary supermatrix, $V\bar{V} = 1$, satisfies Eq. (4.23). This degeneracy leads to the existence of the low-lying Goldstone modes, and their contribution to physical quantities should be taken into account properly. It will be seen that these Goldstone modes are just the diffusion modes considered in detail in the preceding chapter. The degeneracy, Eq. (4.25), is similar to the degeneracy of the ground state in superconductors where the phase of the superconducting order parameter can be arbitrary. In the language of spin models the degeneracy of the solution, Eq. (4.25), is equivalent to the degeneracy due to an arbitrary spin direction.

To consider low-lying excitations one has to be able to derive a functional describing the energy of the fluctuations of the supermatrix Q. For spin systems the corresponding fluctuations are described by the Heisenberg model. The free energy functional for the fluctuations of the supermatrix Q proves to be very similar to that for the Heisenberg model and will be derived in the next section.

4.2 Derivation of the σ-model

4.2.1 Hubbard–Stratonovich transformation

The discussion of the mean field scheme presented in the preceding section helps in understanding the main idea of calculations, but for explicit calculations a more sophisticated approach is needed. As has been mentioned, for a calculation of physical quantities one should take into account fluctuations of the supermatrix Q and therefore one should be able to describe them, and that is beyond the mean field scheme. Besides, a natural question arises: If supermatrix Q determines the physical quantities, then what should one do with "unphysical" Grassmann elements of the supermatrix? No observable quantity can contain the anticommuting variables and so one should know how to eliminate these variables when obtaining physical results.

To start the consideration we recall the main lesson of Section 4.1, namely, that averages of the type $\langle \psi \bar{\psi} \rangle$ are very important. Now we want to consider their slow fluctuations. Let us make the following approximation for the Lagrangian $\mathcal{L}_{\text{int}}$ Eq. (4.15)

$$
\mathcal{L}_{\text{int}} = \frac{1}{4\pi\nu\tau} \int (\bar{\psi}\psi)^2 d\mathbf{r} = \frac{1}{4\pi\nu\tau} \sum_{\mathbf{p}_1+\mathbf{p}_2+\mathbf{p}_3+\mathbf{p}_4=0} (\bar{\psi}_{\mathbf{p}_1}\psi_{\mathbf{p}_2})(\bar{\psi}_{\mathbf{p}_3}\psi_{\mathbf{p}_4})
$$

$$
\approx \frac{1}{4\pi\nu\tau} \sum_{\mathbf{p}_1,\mathbf{p}_2,\mathbf{q}<q_0} \Big[(\bar{\psi}_{\mathbf{p}_1}\psi_{-\mathbf{p}_1+\mathbf{q}})(\bar{\psi}_{\mathbf{p}_2}\psi_{-\mathbf{p}_2-\mathbf{q}})
$$

$$
+ (\bar{\psi}_{\mathbf{p}_1}\psi_{\mathbf{p}_2})(\bar{\psi}_{-\mathbf{p}_2-\mathbf{q}}\psi_{-\mathbf{p}_1+\mathbf{q}}) + (\bar{\psi}_{\mathbf{p}_1}\psi_{\mathbf{p}_2})(\bar{\psi}_{-\mathbf{p}_1+\mathbf{q}}\psi_{-\mathbf{p}_2-\mathbf{q}}) \Big] \tag{4.26}
$$

In Eq. (4.26), the domains with small $q < q_0 \ll l^{-1}$ are explicitly singled out. The three terms in square brackets are reminiscent of the terms $\mathcal{L}_1$, $\mathcal{L}_2$ and $\mathcal{L}_3$, Eqs. (4.16–4.18). They would coincide with the latter terms if one kept only the term with $\mathbf{q} = 0$ in the sum over the momenta in Eq. (4.26) and substituted one of the brackets in each of the three terms by the corresponding average. This means that now we do not try to construct a completely new scheme but extend the mean field scheme. Domains with large momenta q in many problems give a small contribution, and they are not included later. Using the charge-conjugation operation, Eqs. (4.7, 4.8), and the commutation rules for the components of the supervectors ψ one can see that the second term in the square brackets in Eq. (4.26) is equal to the third.

With the help of a Gaussian integration over auxiliary fields the interaction in Eq. (4.26) can be reduced to quadratic form. This procedure is often called the Hubbard–Stratonovich transformation. The reduction of the first term in Eq. (4.26) is very simple

$$
\exp\left[-\frac{1}{4\pi\nu\tau} \sum_{\mathbf{p}_1,\mathbf{p}_2,\mathbf{q}} (\bar{\psi}_{\mathbf{p}_1}\psi_{-\mathbf{p}_1+\mathbf{q}})(\bar{\psi}_{\mathbf{p}_2}\psi_{-\mathbf{p}_2-\mathbf{q}}) \right]
$$

$$
= \int \exp\left[-\frac{i}{4\tau} \int \mathcal{E}(\mathbf{r})\bar{\psi}(\mathbf{r})\psi(\mathbf{r}) d\mathbf{r} - \frac{\pi\nu}{16\tau} \int \mathcal{E}^2(\mathbf{r}) d\mathbf{r} \right] D\mathcal{E}
$$

$$
\times \left(\int \exp\left[-\frac{\pi\nu}{16\tau} \int \mathcal{E}^2(\mathbf{r}) d\mathbf{r} \right] D\mathcal{E} \right)^{-1} \tag{4.27}
$$

In Eq. (4.27), functional integration over slow-varying real functions $\mathcal{E}(\mathbf{r})$ is performed. Comparing this equation with Eqs. (4.13, 4.14) we can see that the presence of the function $\mathcal{E}(\mathbf{r})$ adds slow fluctuations of the electron energy. These fluctuations in the limit $\varepsilon\tau \gg 1$ do not lead to any nontrivial phenomena and will not be considered.

All the other terms in Eq. (4.26) can only be decoupled into slow-varying parts by integration over a supermatrix. Taking into account the equality of the last two terms in Eq. (4.26) to each other and neglecting the first one the interaction can be written as

$$\mathcal{L}_{\text{int}} = \frac{1}{2\pi\nu\tau} \sum_{p_1,p_2,q<q_0} \left[(\bar{\vartheta}_{\mathbf{p}_1}\vartheta_{\mathbf{p}_2})(\bar{\vartheta}_{-\mathbf{p}_2-\mathbf{q}}\vartheta_{-\mathbf{p}_1+\mathbf{q}}) \right.$$

$$\left. + (\bar{v}_{\mathbf{p}_1}v_{\mathbf{p}_2})(\bar{v}_{-\mathbf{p}_2-\mathbf{q}}v_{-\mathbf{p}_1+\mathbf{q}}) + 2(\bar{\vartheta}_{\mathbf{p}_1}\vartheta_{\mathbf{p}_2})(\bar{v}_{-\mathbf{p}_2-\mathbf{q}}v_{-\mathbf{p}_1+\mathbf{q}}) \right] \tag{4.28}$$

In Eq. (4.28) the fermionic ϑ^a and bosonic v^a components of the supervector ψ are written out explicitly. The decoupling of the first two terms in Eq. (4.28) can be carried out by integration over Bose fields. However, for the decoupling of the third term it is necessary to integrate over anticommuting fields. Using the formula for integration over the anticommuting variables of the exponential with linear terms, Eq. (2.62), and that for integration over the conventional ones we find

$$\exp(-\mathcal{L}_{\text{int}}) = P_1 P_2 P_3, \tag{4.29}$$

$$P_1 = \int \exp\left[-\frac{1}{2\tau} \sum_{\mathbf{q}} \sum_{m,n=1}^{2} \left(\frac{\pi\nu}{4} \operatorname{tr}\left((a_{\mathbf{q}}^{mn})^+ a_{-\mathbf{q}}^{mn}\right) \right.\right.$$

$$\left.\left. + \sum_{\mathbf{p}} \bar{\vartheta}_{\mathbf{p}}^m a_{\mathbf{q}}^{mn} \vartheta_{-\mathbf{p}-\mathbf{q}}^n i^{n-m} \right) \right] Da$$

$$P_2 = \int \exp\left[-\frac{1}{2\tau} \sum_{\mathbf{q}} \sum_{m,n=1}^{2} \left(\frac{\pi\nu}{4} \operatorname{tr}\left((b_{\mathbf{q}}^{mn})^+ b_{\mathbf{q}}^{mn}\right) \right.\right.$$

$$\left.\left. + \sum_{\mathbf{p}} \bar{v}_{\mathbf{p}}^m b_{\mathbf{q}}^{mn} v_{-\mathbf{p}-\mathbf{q}}^n i^{n-m+1} \right) \right] Db$$

$$P_3 = \int \exp\left[-\frac{1}{2\tau} \sum_{\mathbf{q}} \sum_{m,n=1}^{2} \left(\frac{\pi\nu}{2} (-1)^n \operatorname{tr}\left((\sigma_{\mathbf{q}}^{mn})^+ \sigma_{-\mathbf{q}}^{mn}\right) \right.\right.$$

$$\left.\left. + \sum_{\mathbf{p}} \left(\bar{\vartheta}_{\mathbf{p}}^m \sigma_{\mathbf{q}}^{mn} v_{-\mathbf{p}-\mathbf{q}}^n i^{2n-m-1} + \bar{v}_{\mathbf{p}}(\sigma_{\mathbf{q}}^{mn})^+ \vartheta_{-\mathbf{p}-\mathbf{q}}^m i^{m-1} \right) \right) \right] D\sigma$$

In Eqs. (4.29) the integration is performed over 2×2 matrices a^{mn}, b^{mn}, and σ^{mn}. The elements of the matrices a^{mn} and b^{mn} are conventional numbers and those of the matrices σ^{mn} are anticommuting variables. All these matrices satisfy the conditions

$$(a^{mn})^+ = a^{mn} = c_1 (a^{mn})^T c_1^T,$$

$$(b^{mn})^+ = b^{mn} = c_2 (b^{mn})^T c_2^T, \tag{4.30}$$

$$(\sigma^{mn})^+ = c_2 (\sigma^{mn})^T c_1^T$$

where the matrices c_1 and c_2 are defined in Eq. (4.10).

The matrices a^{mn}, b^{mn}, and σ^{mn} satisfying the relations Eq. (4.30) can be written out explicitly

$$a^{mn} = \begin{pmatrix} a_1^{mn} & a_2^{mn} \\ -a_2^{mn*} & a_1^{mn*} \end{pmatrix}, \qquad b^{mn} = \begin{pmatrix} b_1^{mn} & b_2^{mn} \\ b_2^{mn*} & b_1^{mn*} \end{pmatrix}$$

$$\sigma^{mn} = \begin{pmatrix} \sigma_1^{mn} & \sigma_2^{mn} \\ -\sigma_2^{mn*} & -\sigma_1^{mn*} \end{pmatrix}, \tag{4.31}$$

$$a_1^{mn} = a_1^{nm*}, \qquad a_2^{mn} = -a_2^{nm}, \qquad b_1^{mn} = b_1^{nm*}, \qquad b_2^{mn} = b_2^{nm}$$

The matrices a, b, and σ contain sixteen independent complex variables, eight of the variables bosonic and eight fermionic. The equality of the numbers of the commuting and anticommuting variables enables us to omit the weight denominators because their product is equal to unity. Let us emphasize that the integrals in Eqs. (4.29) are convergent.

A similar decoupling of the interaction ψ^4 when considering the problem involved by the replica method was first proposed by Wegner (1979), who studied a model with a random amplitude of hops from one site to another and a large number of orbitals per site. Wegner's model differs from the model considered previously but this difference is not important provided we consider the limit $\varepsilon\tau \gg 1$ in the model with the random potential and the limit of a large number of the orbitals in the model with the random hops. After the separation of the slowly varying parts Eq. (4.26), which makes sense if $\varepsilon\tau \gg 1$, both the models coincide. A careful decoupling within the replica method has been done by Schäfer and Wegner (1980) and Houghton et al. (1980) using functional integration over commuting variables, and by Efetov, Larkin, and Khmelnitskii (1980) using Green functions written in terms of functional integrals over anticommuting variables.

The symmetry of the matrix a in Eqs. (4.31) is the same as the symmetry of the collective variable Q from the work of Efetov, Larkin, and Khmelnitskii (1980). The symmetry of the matrix b coincides with the symmetry of the corresponding matrix Q from the work of Schäfer and Wegner (1980). Of course, in the replica method one has to deal with $n \times n$ matrices for an arbitrary n. In the supersymmetry method the matrices have finite sizes that are not very large.

Using the supermatrices introduced in Chapter 2 a somewhat cumbersome expression, Eqs. (4.31), can be rewritten in a more compact form. Returning to the supervectors ψ, $\bar{\psi}$, Eqs. (4.6, 4.7), we obtain instead of Eq. (4.31)

$$\exp(-\mathcal{L}_{\text{int}}) = \int \left[\exp\left(-\frac{1}{2\tau} \int \left(\bar{\psi} Q \psi + \frac{\pi \nu}{4} \operatorname{str} Q^2 \right) d\mathbf{r} \right) \right] DQ \tag{4.32}$$

where Q consists of four supermatrices Q^{11}, Q^{12}, Q^{21}, and Q^{22}

$$Q = \begin{pmatrix} Q^{11} & Q^{12} \\ Q^{21} & Q^{22} \end{pmatrix} \tag{4.33}$$

$$Q^{11} = \begin{pmatrix} a^{11} & \sigma^{11} \\ \sigma^{11+} & ib^{11} \end{pmatrix}, \quad Q^{12} = i \begin{pmatrix} a^{12} & i\sigma^{12} \\ \sigma^{21+} & ib^{12} \end{pmatrix}$$

$$Q^{22} = \begin{pmatrix} a^{22} & i\sigma^{22} \\ i\sigma^{22+} & ib^{22} \end{pmatrix}, \quad Q^{21} = -i \begin{pmatrix} a^{12+} & \sigma^{21} \\ i\sigma^{12+} & ib^{12+} \end{pmatrix}$$

Strictly speaking, the form of the matrix Q differs from that of the supermatrices Eq. (2.18) because of a somewhat different arrangement of the commuting and anticommuting elements. However, the matrix Q can be reduced to the canonical supermatrix form, Eq. (2.18), by permutations of elements. The same can be said about the supervector ψ. So in this book Q and ψ are called the *supermatrix* and the *supervector*. For the matrices Q of the form Eq. (4.33) the supertrace is written as

$$\text{str}\, Q = \text{str}\, Q^{11} + \text{str}\, Q^{22} \tag{4.34}$$

Analogously, one can write all other operation rules for the supermatrices of the form Eq. (4.33). Of course, the supermatrix Q and the supervector ψ might also be written in the canonical form, Eqs. (2.11, 2.12, 2.18). However, the representations Eqs. (4.6, 4.7, 4.33) are more convenient because they enable one explicitly to separate the blocks (superelements) that relate to the retarded and advanced Green functions. Calculating the averaged one-particle Green function it would be sufficient to use only the superelements Q^{11} or Q^{22} that are true supermatrices in the sense of the definition, Eq. (2.18). According to their role in the decoupling Eqs. (4.29) the matrices a^{mn} can be called fermion–fermion, b^{mn}–boson–boson, and σ^{mn} fermion–boson blocks.

The supermatrix Q in Eq. (4.32) has a certain symmetry. To write down this symmetry let us first define the charge conjugate $\bar{Q}$ of a supermatrix Q as

$$\bar{Q} = CQ^T C^T \tag{4.35}$$

where the matrix C is specified by Eqs. (4.8, 4.10) and Q^T is

$$Q^T = \begin{pmatrix} Q^{11T} & Q^{21T} \\ Q^{12T} & Q^{22T} \end{pmatrix} \tag{4.36}$$

Using Eqs. (2.20, 2.22) the following equations for arbitrary supermatrices Q_1 and Q_2 and supervectors ψ_1 and ψ_2 can be checked easily:

$$\overline{Q_1 Q_2} = \bar{Q}_2 \bar{Q}_1 \quad \text{and} \quad \bar{\psi}_1 Q \psi_2 = \bar{\psi}_2 \bar{Q} \psi_1 \tag{4.37}$$

The supermatrix Q of the form Eq. (4.33) has a very important symmetry

$$Q = \bar{Q} \tag{4.38}$$

which can be directly checked.

Substituting the interaction term $\mathcal{L}_{\text{int}}$ of the Lagrangian Eq. (4.32) into Eq. (4.14) we come to the following expression:

$$\exp(-\mathcal{L}) = \int \exp(-\mathcal{L}_{\text{eff}}[\psi, \bar{\psi}, Q]) \exp\left(-\frac{\pi \nu}{8\tau} \text{str}\, Q^2\right) DQ \tag{4.39}$$

The effective Lagrangian $\mathcal{L}_{\text{eff}}[\psi, \bar{\psi}, Q]$ in Eq. (4.39) has the form determined by Eq. (4.19), but now the supermatrix Q can vary in space and the functional integration in Eq. (4.39) should be performed over all such variations. So we see that the decoupling scheme presented here is very closely connected with the mean field theory, but now we are able to take into account all fluctuations of the collective variable Q. This variable looks like an order parameter analogous, for example, to the order parameter Δ in superconductors. However, the supermatrix Q is not an order parameter because it is not related to any phase transition. This question will be discussed in more detail in Chapter 12.

The Lagrangian $\mathcal{L}_{\text{eff}}$ is quadratic in ψ and, therefore, in order to obtain physical quantities, one should calculate Gaussian integrals using, for example, the Wick theorem. These integrals cannot be calculated exactly for an arbitrary dependence of the supermatrix Q on the coordinate. At first glance, after all the manipulations we have a problem that is no more simple than the initial problem because integrals with the Lagrangian $\mathcal{L}$ in Eq. (4.12) were also Gaussian.

Of course, it is not so. In the initial problem one had to consider all possible variations of the random potential $u(\mathbf{r})$ but now we may take into account only slow variations of Q. In many problems the slow variations determine the physics completely. Naturally, the method does not work in situations where fast variations of Q are important. The fact that we need to consider the slow variations only simplifies calculations because, in the first approximation, one can neglect the fluctuations of Q in space, enabling us to get explicit results, and then make a gradient expansion.

Let us follow this procedure and calculate the correlation function $Y^{00}(\mathbf{k}, \varepsilon, \omega)$, Eqs. (3.24, 3.49). We start by rewriting Eqs. (4.4, 4.5) as

$$Y^{00}(\mathbf{r}, \varepsilon, \omega) = 2\langle G^A_{\varepsilon-\omega}(\mathbf{r}, 0)\, G^R_\varepsilon(0, \mathbf{r})\rangle \tag{4.40}$$

$$= -8 \int \psi^1_\alpha(\mathbf{r})\, \bar{\psi}^1_\alpha(0)\, \psi^2_\beta(0)\, \bar{\psi}^2_\beta(\mathbf{r}) \exp(-\mathcal{L}_{\text{eff}}[\psi, \bar{\psi}, Q])$$

$$\times \exp\left(-\frac{\pi \nu}{8\tau} \operatorname{str} Q^2\right) D\psi\, DQ$$

In Eq. (4.40) the subscripts α, β stand for the components of the supervectors ψ^m, $\bar{\psi}^m$ and no summation over the subscripts is implied. (The product $\langle G^A_{\varepsilon-\omega}(\mathbf{r}, 0)\, G^A_\varepsilon(0, \mathbf{r})\rangle$ decays very fast with distance and therefore is neglected.)

The first task now is to integrate over the supervectors ψ. Using the Wick theorem one can express the result of the integration in terms of an effective supermatrix Green function $g_{\alpha\beta}(\mathbf{r}, \mathbf{r}', Q)$ introduced as

$$g_{\alpha\beta}(\mathbf{r}, \mathbf{r}', Q) = 2 \int \psi_\alpha \bar{\psi}_\beta \exp(-\mathcal{L}_{\text{eff}}[\psi, \bar{\psi}, Q]) D\psi \tag{4.41}$$

Calculation of the Gaussian integral in Eq. (4.41) can be reduced to solving the following differential equation:

$$\left(\tilde{\mathcal{H}}_0 + \frac{1}{2}(\omega + i\delta)\Lambda + \frac{iQ(\mathbf{r})}{2\tau}\right) g(\mathbf{r}, \mathbf{r}', Q) = i\delta(\mathbf{r} - \mathbf{r}') \tag{4.42}$$

In Eq. (4.42), the energy operator $\tilde{\mathcal{H}}_0$, Eq. (4.13), acts on the coordinate $\mathbf{r}$. If Q did not depend on the coordinate the function $g(\mathbf{r}, \mathbf{r}', Q)$ would coincide with the function $g_0(\mathbf{r} - \mathbf{r}')$ that can be obtained from Eq. (4.22). The function $g(\mathbf{r}, \mathbf{r}', Q)$ falls off at the distances $|\mathbf{r} - \mathbf{r}'|$ of the order of the mean free path l. If we calculate the correlation function $\langle G^A_{\varepsilon-\omega} G^R_\varepsilon\rangle$ in Eq. (4.40) at large distances $|\mathbf{r} - \mathbf{r}'| \gg l$ using the Wick theorem, only one type of pairing can contribute and we obtain for the function Y^{00} in the coordinate representation

$$Y^{00}(\mathbf{r}, \varepsilon, \omega) = 2 \int g^{12}_{\alpha\beta}(\mathbf{r}, \mathbf{r})\, k_{\beta\beta} g^{21}_{\beta\alpha}(0, 0) \exp(-F[Q])\, DQ \tag{4.43}$$

where k is the 4×4 superelement of the form

$$k = \begin{pmatrix} 1 & 0 \\ 0 & -1 \end{pmatrix} \tag{4.44}$$

with **1** the 2×2 unity matrix. The superelement k is in the same space as the supermatrices Q^{mn} and should not be confused with the supermatrix Λ in Eq. (4.9).

The free energy functional $F[Q]$ in Eq. (4.43) can be written as

$$F[Q] = \int \left[-\frac{1}{2} \operatorname{str} \ln \left(-i\tilde{\mathcal{H}}_0 - \frac{i(\omega + i\delta)}{2} \Lambda + \frac{Q(\mathbf{r})}{2\tau} \right) + \frac{\pi \nu}{8\tau} \operatorname{str} Q^2 \right] d\mathbf{r} \quad (4.45)$$

The Hamiltonian $\tilde{\mathcal{H}}_0$ in Eq. (4.45) acts on the coordinate $\mathbf{r}$. Let us emphasize that the integral over Q, Eq. (4.43), with the energy functional Eq. (4.45) converges as before.

4.2.2 Saddle point

As has been mentioned, the singling out of the averages slowly varying in space makes sense only for large mean free times $\tau \varepsilon \gg 1$, because only in this case are the omitted terms small. In this limit, calculation of the integral, Eq. (4.43), is more simple because the main contribution comes from the minima of the free energy functional $F[Q]$ and one can use the saddle point method. Varying the free energy functional $F[Q]$, Eq. (4.45); putting the first variation to zero; and using the property of the invariance of the supertrace under cyclic permutations, Eq. (2.41), we obtain the following equation for $Q(\mathbf{r})$:

$$Q(\mathbf{r}) = \frac{1}{\pi \nu} g(\mathbf{r}, \mathbf{r}) \quad (4.46)$$

Assuming that the solution of this equation does not depend on coordinates the function $g(\mathbf{r}, \mathbf{r}')$ reduces to the function $g_0(\mathbf{p})$, Eq. (4.22), and one comes immediately to Eqs. (4.23–4.25). We see that the saddle point approximation in the scheme involved is completely equivalent to the mean field approximation considered in the preceding section.

However, one should be careful with the saddle point because the solution $Q = \Lambda$, Eq. (4.24), does not belong to the set of the matrices a, b, and σ, Eqs. (4.33), because the diagonal elements of the matrices ib^{11} and ib^{22} are always imaginary. The analogous discrepancy arises when calculating by the replica method (Schäfer and Wegner (1980)). In order to reach the saddle point it is necessary to move the contour of the integration over the matrices b^{mm} into the complex plane. Carrying out this procedure one should be sure that convergence of the integrals over these matrices is not violated and that, when moving the contours, singular points of the integrand in the complex plane are not crossed. The problem with the singular points arises when integrating over the boson–boson blocks b because the Gaussian integration over the boson components of the supervectors ψ gives expressions like $(\det B)^{-1}$ with B determined by

$$B = -i\tilde{\mathcal{H}}_0 - \frac{i(\omega + i\delta)}{2} \Lambda + \frac{b}{2\tau}$$

Zeros of B are singular points of the integrand. As concerns the fermion–fermion blocks, after integration over the fermion components of the supervectors ψ one obtains the result $\det A$, where A is the corresponding expression containing the matrices a^{mm}. Zeros of A are zeros of the integrand and are of no importance. The fermion–boson blocks σ^{mn} do not influence convergence integrals and do not determine singular points. Therefore, they need not be considered when discussing these problems.

Shifting the contours of the integration over the matrices b^{11} and b^{22} one can see that the point $Q = \Lambda$ can really be reached such that the new contour that can be chosen to be parallel to the real axis crosses it. This is the only saddle point at $\omega \neq 0$ that can be reached without crossing singularities. For example, the points $-\Lambda$ and ± 1 cannot be reached in this way and should not be taken into account although they are formal solutions of Eq. (4.23). Using an analogy with Fermi and Bose gases one can say that the existence of nonzero values of Q is the effect of the spontaneous breaking of symmetry. What is considered here corresponds to the gases with an attraction. The Fermi gas with an attraction is a very well-defined quantity, so no problems arise when integrating over the matrices a. The necessity to shift contours in the complex plane when integrating over the matrices b is the consequence of the fact that a Bose gas with an attraction is badly defined.

At $\omega = 0$ the saddle point is degenerate and one should use the general solution $Q = V\Lambda\bar{V}$, Eq. (4.25). In addition to the condition $V\bar{V} = 1$ the supermatrix V must obey another constraint to provide equal numbers of independent elements in the matrices Q and V. This constraint will be presented later, but now I want to say a few words about the procedure of shifting the contours of integration in order to reach the general solution. The point $Q = \Lambda$ was a rather simple saddle point because it corresponded to just zero in integrals over all nondiagonal elements, so one had to shift contours in the integrals over the diagonal elements of the matrices a^{11}, a^{22}, b^{11}, and b^{22} only. The procedure of shifting the contours for reaching the general point, Eq. (4.25), is more difficult because the saddle point is a nontrivial point in a multiple integral over many variables.

The simplest way to carry out the procedure is the following: Let us move the contours in the integrals over Q^{11} and Q^{22} keeping Q^{12} and Q^{21} fixed. One has to careful about the correct shifts in the integrals over b^{11} and b^{22} only. This is because matrices a and b are connected by the anticommuting variables only and any function of these variables is a finite polynomial. After the expansion of the exponential $\exp(-F[Q])$ in Eq. (4.43) in a finite series in the anticommuting variables, the matrices a and b are the only ones remaining in the exponent in the integrand. Then, one should move the contours in the integrals over the matrices b^{11} and b^{22} only because the matrices a^{11} and a^{22} have the correct form. Shifting in the integrals over b^{11} and b^{22} can be carried out by diagonalizing these matrices and by making shifts in the integrals over the eigenvalues of these matrices. Shifting in the integrals over the anticommuting variables can be done with the use of Eq. (2.58). As a result, the general saddle point solution that can be obtained from the saddle point $Q = \Lambda$ by a continuous transformation is written in the form

$$Q = W + \Lambda\left(1 - W^2\right)^{1/2}, \qquad W = \begin{pmatrix} 0 & Q^{12} \\ Q^{21} & 0 \end{pmatrix} \qquad (4.47)$$

The square root in Eq. (4.47) is defined by Taylor expansion in W^2; the supermatrices Q^{12} and Q^{21} are specified by Eq. (4.33). The supermatrix W has eight independent complex variables.

For the solutions Q of the form Eq. (4.47) the following relation holds in addition to the condition Eq. (4.38):

$$\bar{Q} = KQ^+K, \qquad K = \begin{pmatrix} 1 & 0 \\ 0 & k \end{pmatrix} \qquad (4.48)$$

where k is the superelement, Eq. (4.44).

Eq. (4.48) can be checked using Eqs. (4.30). The somewhat strange relation between the Hermitian and charge conjugation, Eq. (4.48) (the presence of the supermatrix K), is due to the multiplication of the matrices b^{12}, b^{12+}, σ^{12}, and σ^{12+} by i in Eqs. (4.33). Comparing

Eqs. (4.25) and (4.48) one can write the additional constraint for the matrix V

$$\bar{V} = KV^+K \qquad (4.49)$$

Thus, the saddle point solution at $\omega = 0$, Eq. (4.25), is degenerate. This degeneracy is a consequence of the invariance of the original Lagrangian, Eq. (4.14), with respect to rotations in the space of the supervectors ψ. It is essential that the average of the product $\langle G^R G^A \rangle$ of the retarded and advanced Green functions be considered. If the product $\langle G^R G^R \rangle$ or $\langle G^A G^A \rangle$ is considered, the matrix Λ in Eq. (4.25) must be replaced by the unit matrix. In this case the degeneracy of the extremum solution disappears.

4.2.3 σ-Model

Because of the degeneracy of the saddle point solution, it is not correct just to take the free energy functional at the extremum; one should integrate over all the saddle point solutions described by Eq. (4.25). In fact, the free energy functional $F[Q]$, Eq. (4.45), at the saddle point is equal to zero. At $\omega \neq 0$ the saddle point solution is not degenerate. However, for nonzero frequencies integration over all Q is necessary anyway; otherwise the anticommuting variables would enter physical quantities. Besides, at low frequencies ω, all Q of the form Eq. (4.25), being substituted to the free energy functional, give low energies and therefore should be taken into account.

The degeneracy of the ground state at $\omega = 0$ leads to the existence of Goldstone modes that can be obtained by expansion of the free energy functional near the extremum. For small deviations δQ from the equilibrium value, the free energy at $\omega = 0$ is quadratic in these deviations. At a low frequency ω one can also make an expansion in ω and it is sufficient to retain a linear term only. Making the expansion one obtains

$$F[Q] = \frac{1}{16\tau^2} \operatorname{str} \int \left[g(\mathbf{r}, \mathbf{r}', Q) \delta Q(\mathbf{r}') g(\mathbf{r}', \mathbf{r}, Q) \delta Q(\mathbf{r}) \right.$$

$$\left. + 2\pi\nu\tau\delta(\mathbf{r} - \mathbf{r}') \delta Q(\mathbf{r}) \delta Q(\mathbf{r}') \right] d\mathbf{r} d\mathbf{r}' - \frac{i\omega\pi\nu}{4} \operatorname{str} \int \Lambda Q(\mathbf{r}) d\mathbf{r} \qquad (4.50)$$

where the function $g(\mathbf{r}, \mathbf{r}')$ has been introduced in Eq. (4.42). In order to transform Eq. (4.50) to a simpler form one should distinguish between longitudinal fluctuations $\delta_l Q$ changing the eigenvalues of Q and transverse fluctuations $\delta_t Q$ described by fluctuations of the matrix V in Eq. (4.25). The deviations $\delta_l Q$ and $\delta_t Q$ satisfy the following commutation relations:

$$Q\delta_l Q - \delta_l QQ = 0, \qquad Q\delta_t Q + \delta_t QQ = 0 \qquad (4.51)$$

In the quadratic approximation, these two types of the fluctuations decouple and can be considered independently. The first term F_1 in the square brackets in Eq. (4.50) can be written using Eq. (4.51) as

$$F_1 = F_{1l} + F_{1t},$$

$$F_{1l} = \frac{1}{16\tau^2} \operatorname{str} \int g(\mathbf{r}, \mathbf{r}', Q)g(\mathbf{r}', \mathbf{r}, Q)\delta_l Q(\mathbf{r}')\delta_l Q(\mathbf{r})d\mathbf{r} d\mathbf{r}'$$

$$F_{1t} = \frac{1}{16\tau^2} \operatorname{str} \int g(\mathbf{r}, \mathbf{r}', Q)g(\mathbf{r}', \mathbf{r}, -Q)\delta_t Q(\mathbf{r}')\delta_t Q(\mathbf{r})d\mathbf{r} d\mathbf{r}' \qquad (4.52)$$

As mentioned, the function $g(\mathbf{r}, \mathbf{r}', Q)$ falls off at distances $|\mathbf{r} - \mathbf{r}'| \sim l$, and now we want to consider fluctuations with Q slowly varying at these distances. Therefore, we can make a gradient expansion. In the zero approximation, we take both δQ in Eqs. (4.52) at the same point $\mathbf{r}$ and integrate the products of the functions g over $\mathbf{r}'$. Using the Fourier representation and changing to the variable $\xi = \varepsilon\,(\mathbf{p}) - \varepsilon$ we obtain in this approximation

$$F_{1l}^{(0)} = -\frac{\nu}{16\tau^2}\,\mathrm{str}\left[(\delta_l Q\,(\mathbf{r}))^2 \int\limits_{-\infty}^{\infty}\left(\xi + \frac{iQ}{2\tau}\right)^{-2} d\xi\right] = 0 \qquad (4.53)$$

$$F_{1t}^{(0)} = -\frac{\nu}{16\tau^2}\,\mathrm{str}\left[(\delta_t Q\,(\mathbf{r}))^2 \int\limits_{-\infty}^{\infty}\left(\xi^2 + \frac{1}{4\tau^2}\right)^{-2} d\xi\right]$$

$$= -\frac{\pi\nu}{8\tau}\,\mathrm{str}\int (\delta_t Q\,(\mathbf{r}))^2\, d\mathbf{r}$$

We see that in the quadratic approximation the energy of the the longitudinal fluctuations is described by the second term in square brackets in Eq. (4.50). In the limit $\varepsilon\tau \gg 1$ these fluctuations give a small contribution to physical quantities and it is sufficient to calculate the Gaussian integral over the variables $\delta_l Q$; this gives

$$\int \exp\left(-\frac{\pi\nu}{8\tau}\int \mathrm{str}\,(\delta_l Q)^2\, d\mathbf{r}\right) D\delta_l Q = 1 \qquad (4.54)$$

Eq. (4.54) follows immediately from the equal number of commuting and anticommuting variables of the integration.

The term $F_{1t}^{(0)}$ compensates exactly for the second term in the square brackets in Eq. (4.50) and one should take into account gradients of $\delta_t Q$. The term linear in the gradients vanishes after integration over momenta and one should expand $\delta_t Q\,(\mathbf{r}')$ near point $\mathbf{r}$ in Eq. (4.52) up to the second order. The calculation is straightforward in the momentum representation and finally we obtain

$$F\,[Q] = \frac{\pi\nu}{8}\,\mathrm{str}\int\left[D_0(\nabla Q)^2 + 2i\,(\omega + i\delta)\,\Lambda Q\right] d\mathbf{r} \qquad (4.55)$$

where $D_0 = v_0^2\tau/d$ is the classical diffusion coefficient and the supermatrix Q is described by Eq. (4.25).

The free energy functional, Eq. (4.55), describes the Goldstone modes, which are just the diffusion modes discussed in the preceding chapter. Their existence is a consequence of the spontaneous breaking of supersymmetry. The system with the free energy Eq. (4.55) and the constraint Eq. (4.25) belongs to the class of nonlinear σ-models. The functional $F\,[Q]$ is similar to the Heisenberg model of a ferromagnet. The frequency ω plays the role of an external field. In Eq. (4.55) a small imaginary part $i\delta$ is added to the frequency. In many cases this is necessary to provide convergence of integrals. At $\omega = 0$ the free energy $F\,[Q]$ is invariant with respect to the transformation

$$Q \to V Q \bar{V} \qquad (4.56)$$

where V is an arbitrary unitary supermatrix such that $V\bar{V} = 1$. The σ-model Eq. (4.55) describes a large variety of different physical phenomena, which will be considered in the next chapters.

Using the same approximation as when deriving Eq. (4.55) one can further simplify Eq. (4.43) for the density–density correlation function $Y^{00}(\mathbf{r}, \varepsilon, \omega)$. At the saddle point, Eq. (4.46) is fulfilled; that leads immediately to

$$Y^{00}(\mathbf{r}, \varepsilon, \omega) = 2(\pi v)^2 \int Q^{12}_{\alpha\beta}(\mathbf{r}) k_{\beta\beta} Q^{21}_{\beta\alpha}(0) \exp(-F[Q]) DQ \qquad (4.57)$$

$$= \frac{(\pi v)^2}{32} \int \mathrm{str}\,[k(1+\Lambda)(1-\tau_3) Q(\mathbf{r})(1-\Lambda)(1-\tau_3) kQ(0)] \qquad (4.58)$$

$$\times \exp(-F[Q]) DQ$$

where

$$\tau_3 = \begin{pmatrix} 1 & 0 \\ 0 & -1 \end{pmatrix} \qquad (4.59)$$

is the 2×2 matrix in the space of the matrices a, b, σ, Eq. (4.31); α and β are, as before, arbitrary subscripts (no summation is implied). Let us emphasize the differences among the matrices Λ, Eq. (4.9); k, Eq. (4.44); and τ_3, Eq. (4.59).

The correlation function $Y^{00}(\mathbf{r}, \varepsilon, \omega)$ determined by Eqs. (4.57, 4.55) does not depend on ε, and in the following discussion this argument will not be written in this function. Of course, this is correct in the limit of weakly disordered metal. The function Y^{00} is similar to a two-spin correlation function in the Heisenberg model.

Using the formal procedure developed for the solution one can rewrite also the average density of states $\rho(\varepsilon)$, Eq. (3.16), as

$$\langle \rho(\varepsilon) \rangle = \frac{v}{4} \int \mathrm{str}\,(k\Lambda Q) \exp(-F[Q]) DQ \qquad (4.60)$$

As discussed earlier, the average density of states in the limit $\varepsilon\tau \gg 1$ is not an interesting quantity and the result Eq. (3.46) can be obtained without using the σ-model. Nevertheless, Eq. (4.60) is useful because it relates the density of states to the average value of the "order parameter" Q in the theory involved. In particular, the assertion that the density of states does not have singularities and does not turn to zero at finite impurity concentrations (Wegner (1982)) leads to the conclusion that the average $\langle Q \rangle_F$ with the functional F, Eq. (4.55), does not have singularities and cannot equal to zero. So the σ-model under consideration drastically differs from models of ferromagnets, where the order parameter is zero in the disordered phase. This means also that the average of Q cannot play the role of an order parameter for the Anderson metal–insulator transition.

Now let us show that the supermatrix σ-model really describes the diffusion modes discussed in the preceding chapter. One can do this by calculating the density–density correlation function $Y^{00}(\mathbf{r}, \varepsilon, \omega)$, Eqs. (4.24, 4.49), again. This function has been computed in the approximation of the nonintersecting impurity lines, Eq. (3.52), and has the form of a free diffusion mode propagator. To reproduce this form one should expand the expression for Q, Eq. (4.47), in W and substitute the expansion into Eqs. (4.55, 4.57). In the lowest order we obtain for the free energy

$$F_0 = \frac{\pi v}{8} \mathrm{str} \int \left[D_0(\nabla W)^2 - i\omega W^2 \right] d\mathbf{r} \qquad (4.61)$$

and for the density–density correlation function

$$Y^{00}(\mathbf{r}, \omega) = -2(\pi\nu)^2 \int W_{\alpha\beta}^{12}(\mathbf{r}) k_{\beta\beta} W_{\beta\alpha}^{21}(0) \exp(-F_0) DW \qquad (4.62)$$

Calculating the Gaussian integrals in Eq. (4.62) we come to Eq. (3.52).

Performing the same expansion in the integral for the density of states Eq. (4.61) one can see that the diffusion modes do not contribute to this quantity. The only thing one should do to calculate $\langle \rho(\varepsilon) \rangle$ is to replace the supermatrix Q by Λ in the preexponential in Eq. (4.60); then the remaining integral is equal to unity.

All the formulae in this chapter have been written under the assumption that electron spins do not interact with impurities, so one can consider "up" and "down" spins separately, adding the factor 2 in the final equations. Magnetic and spin-orbit interactions will be considered in the next sections.

The supermatrix σ-model, Eq. (4.55), and the density–density correlation function, Eq. (4.57), were first derived by Efetov (1982a).

4.3 Magnetic and spin-orbit interactions

4.3.1 Magnetic field

When deriving the supermatrix nonlinear σ-model, Eq. (4.55), only potential impurity scattering has been considered so far. The system has been assumed to be invariant under the time reversal and rotations in spin space. The high symmetry of the free energy functional, Eq. (4.55), under the transformations V, Eq. (4.56), at $\omega = 0$ is related to the invariance of the system under the time reversal and spin rotations.

An external magnetic field and magnetic impurities break the time reversal symmetry. The spin rotation symmetry is violated by magnetic impurities and spin-orbit interactions. It is shown later that the magnetic and spin-orbit interactions result in the appearance of "external fields" of different symmetries in the free energy functional Eq. (4.55). These fields partially break the invariance Eq. (4.56).

The effect of the external magnetic field on the orbital motion of the electrons can be considered by the simple replacement $\hat{\mathbf{p}} \to \hat{\mathbf{p}} - (e/c)\mathbf{A}$ in Eqs. (3.14, 4.3, 4.5). However, writing a proper Lagrangian in terms of the supervectors ψ, Eq. (4.6, 4.7), is somewhat more delicate. In order to obtain the corresponding equations one has to use the formulae of partial integration

$$\int \chi^* \varepsilon\left(-i\nabla - \frac{e}{c}\mathbf{A}\right)\chi \, d\mathbf{r} = \int \left(\varepsilon\left(-i\nabla + \frac{e}{c}\mathbf{A}\right)\chi^*\right)\chi \, d\mathbf{r} \qquad (4.63)$$

$$\int S^* \varepsilon\left(-i\nabla - \frac{e}{c}\mathbf{A}\right) S \, d\mathbf{r} = \int \left(\varepsilon\left(-i\nabla + \frac{e}{c}\mathbf{A}\right) S^*\right) S \, d\mathbf{r}$$

As a result, instead of Eq. (4.13) for $\tilde{\mathcal{H}}_0$ one obtains the following expression:

$$\tilde{\mathcal{H}}_0 = \varepsilon\left(\hat{\mathbf{p}} - \frac{e}{c}\tau_3\mathbf{A}\right) - \varepsilon + \frac{\omega}{2} \qquad (4.64)$$

where the matrix τ_3 is defined by Eq. (4.59). This matrix appears to be due to changing the sign of $\mathbf{A}$ in the partial integration. If the magnetic field is weak, its effect can be considered as a small perturbation. When deriving the σ-model all energies have to be small with respect to τ^{-1} and writing the word *small* implies the smallness of the corresponding energies with respect to τ^{-1}. To derive the free energy functional one can use again Eq. (4.50) with the function $g(\mathbf{r}, \mathbf{r}')$ determined by Eq. (4.42). Now $\tilde{\mathcal{H}}_0$ in Eq. (4.42) is given by Eq. (4.64). By making an expansion in both ∇Q and $\mathbf{A}$ and calculating the remaining integrals we find

$$F = \frac{\pi \nu}{8} \, \text{str} \int \left[D_0 \left(\nabla Q - \frac{ie}{c} \mathbf{A} [Q, \tau_3] \right)^2 + 2i(\omega + i\delta) \Lambda Q \right] d\mathbf{r} \qquad (4.65)$$

The derivation of the free energy functional Eq. (4.65) is completely analogous to the derivation of the Ginzburg–Landau free energy functional in the theory of superconductivity (see, e.g., De Gennes (1966)). The vector-potential $\mathbf{A}$ enters the functional in a gauge-invariant form. The matrix τ_3 in the commutator $[Q, \tau_3]$, Eq. (4.65), breaks the symmetry, Eq. (4.56), and results in a gap in the spectrum of some of the diffusion modes. This effect has been discussed in the preceding chapter in terms of perturbation theory.

Substituting Eqs. (4.31, 4.33) into Eq. (4.65) one can see that only the excitations corresponding to the fields a_1, b_1, and σ_1 do not acquire any gap related to the magnetic field. They correspond to the so-called diffusons considered in the preceding chapter. Fluctuations of a_2, b_2, and σ_2 in the long-wavelength limit are suppressed by the magnetic field and the related modes were called *cooperons*.

Eq. (4.65) can be used for both spinless particles and particles with spin provided magnetic impurities and spin-orbit interaction are absent. The influence of the external magnetic field on the particle spin is small and is neglected. For calculation of the density–density correlation function Y^{00} one can still use Eqs. (4.58). As concerns Eq. (4.57), now the subscripts α and β can correspond only to the components a_1, b_1, and σ_1 in the matrices Eqs. (4.31, 4.33).

To study spin interactions one should use a double set of the components of the supervectors Φ in the equations for the Green functions Eqs. (4.1, 4.3, 4.5). The Hamiltonian $\mathcal{H}$ in Eq. (4.5) including these interactions is

$$\mathcal{H} = \mathcal{H}_0 + \mathcal{H}_1, \qquad \mathcal{H}_1 = u(\mathbf{r}) + \sigma \mathbf{u}_s(\mathbf{r}) + \sigma [\mathbf{u}_{so} \times \hat{\mathbf{p}}] \qquad (4.66)$$

where $\mathcal{H}_0$ stands for the kinetic energy and the interaction with the external magnetic field. As before the potential $u(\mathbf{r})$ describes the interaction with potential impurities, $\mathbf{u}_s$ is proportional to the random magnetic field of the magnetic impurities, and $\mathbf{u}_{so}$ stands for the spin-orbit interactions. The components of the vector σ in Eq. (4.66) are the Pauli matrices σ_x, σ_y, and σ_z. (Note the difference in the notations for the Pauli matrices from those for the elements σ_1 and σ_2 in Eqs. (4.31).) The second and third terms in $\mathcal{H}_1$ are essentially different from each other because the second term violates the time reversal symmetry but the third does not.

4.3.2 Magnetic impurities

To include spin interactions in the supersymmetric scheme it is convenient to define the anticommuting and commuting components of the supervector ψ as

$$\vartheta^m = \frac{1}{\sqrt{2}} \begin{pmatrix} \chi^{m*} \\ i\sigma_y \chi^m \end{pmatrix}, \qquad v^m = \frac{1}{\sqrt{2}} \begin{pmatrix} S^{m*} \\ i\sigma_y S^m \end{pmatrix} \qquad (4.67)$$

$$\bar{\psi} = (C\psi)^T, \qquad c_1 = \begin{pmatrix} 0 & i\sigma_y \\ i\sigma_y & 0 \end{pmatrix}, \qquad c_2 = \begin{pmatrix} 0 & -i\sigma_y \\ i\sigma_y & 0 \end{pmatrix} \qquad (4.68)$$

In Eqs. (4.67, 4.68) χ^m and S^m are two component vectors

$$\chi^m = \begin{pmatrix} \chi_1^m \\ \chi_2^m \end{pmatrix}, \qquad S^m = \begin{pmatrix} S_1^m \\ S_2^m \end{pmatrix}$$

in accordance with the two possible spin directions. It is supposed that the matrix C is related to the matrices c_1 and c_2 according to Eqs. (4.7, 4.8). Let us note that the matrices c_1 and c_2 in Eq. (4.68) are antisymmetric and symmetric correspondingly, like those in Eq. (4.10). When a system without spin interactions is considered, all the formulae of the preceding sections remain unchanged provided the matrix C is specified now by Eqs. (4.7, 4.8, 4.68). In particular the symmetry of the 16×16 supermatrix Q is determined by Eqs. (4.38, 4.48) as before.

We assume, as in the consideration of the effect of the magnetic field, that the interaction with the magnetic and spin-orbit impurities is weak; that means that the mean free times of the magnetic τ_s and spin-orbit scattering τ_{so} much exceed the mean free time τ. Then the effect of the spin interactions can be considered as a small perturbation and can therefore be calculated for each type of interaction independently. Averaging for spin interactions can be performed in the same way as for the potential scattering. Assuming that the distribution of the random magnetic fields U_s is Gaussian with the pair correlation

$$\langle U_s^i(\mathbf{r}) U_s^j(\mathbf{r}') \rangle = \frac{\delta_{ij} \delta(\mathbf{r} - \mathbf{r}')}{6\pi \nu \tau_s}$$

and averaging over the magnetic impurities we find that the second term in $\mathcal{H}_1$, Eq. (4.66), leads to the additional term $\mathcal{L}_s$ in the Lagrangian $\mathcal{L}$, Eq. (4.14)

$$\mathcal{L}_s = \frac{1}{12\pi \nu \tau_s} \int (\bar{\psi} \Sigma \psi)^2 \, d\mathbf{r}, \qquad \Sigma = \sigma \otimes \tau_3 \equiv \begin{pmatrix} \sigma & 0 \\ 0 & -\sigma \end{pmatrix} \qquad (4.69)$$

The symbol $\otimes$ in Eq. (4.69) stands for the direct product of matrices. Let us notice that

$$\bar{\Sigma} = \Sigma \qquad (4.70)$$

where the bar indicates the "charge conjugation" introduced in Eq. (4.35) with the matrices C from Eqs. (4.8, 4.68). Integration over the supervectors ψ can be performed by substituting $\mathcal{L}_s$ by $\langle \mathcal{L}_s \rangle_{\text{eff}}$, where the symbol $\langle \ldots \rangle_{\text{eff}}$ stands for the functional integral with the Lagrangian $\mathcal{L}_{\text{eff}}$, Eq. (4.19). Using the Wick theorem one should calculate the averages of all possible pairs, and, using Eq. (4.20), each average can be expressed in terms of the supermatrix Q. The average $\langle \bar{\psi} \Sigma \psi \rangle_{\text{eff}}$ is proportional to $\text{str}(Q\Sigma)$. The remaining two pairings are equal to each other. A simple calculation of these averages with the use of Eq. (4.20) or Eqs. (4.41, 4.42) shows that the magnetic impurities can properly be taken into account by adding the following additional term $F_s[Q]$ in the free energy $F[Q]$ Eq. (4.65)

$$F_s[Q] = \frac{\pi \nu}{48\tau_s} \int \left(-\text{str}[Q, \Sigma]^2 + (\text{str}\, Q\Sigma)^2 \right) d\mathbf{r} \qquad (4.71)$$

The density of states per one spin degree of freedom $\nu = mp_0/2\pi^2$ used in this chapter is the same as before and $[.., ..]$ is the commutator. We see that the magnetic impurities enter the free energy as an "external field" and in the long-wavelength limit some of the diffusion modes are cut. To see explicitly which modes are cut it is convenient to use spin quaternions representing the supermatrix Q in the form $Q = Q_0 + \sigma \mathbf{Q}$, where Q_0 and Q^i

are 8×8 supermatrices having numbers instead of spin blocks. This representation can be considered as a decomposition into singlet and triplet parts. The density–density correlation function Eq. (4.58) preserves its form provided the supermatrix Q is substituted by Q_0. In other words, supermatrix Q_0 describes variation of the number of particles, whereas $\mathbf{Q}$ corresponds to spin diffusion. In the variables Q_0 and $\mathbf{Q}$ Eq. (4.71) can be rewritten in the form

$$F_s = \frac{\pi \nu}{24 \tau_s} \int \left\{ \mathrm{str} \left(-\frac{3}{2} [Q_0, \tau_3]^2 + \frac{1}{2} [\mathbf{Q}, \tau_3]^2 + 4\mathbf{Q}^2 \right) \right. \tag{4.72}$$
$$\left. + (\mathrm{str}\, \tau_3 \mathbf{Q})^2 \right\} d\mathbf{r}$$

At the same time the "bare part" of the free energy Eq. (4.55) decouples into two parts, one of them containing only the variables Q_0; the other, the variables $\mathbf{Q}$. This equation with Eq. (4.72) shows that the excitations corresponding to fluctuations of $\mathbf{Q}$ have a gap. A gap also appears in the excitations corresponding to the fluctuations of a_2, b_2, and σ_2 in Q_0. Therefore the symmetry of the supermatrix Q in the presence of the magnetic impurities acting on the electron spin is the same as in the presence of the magnetic field acting on the orbital motion. However, the size of supermatrix Q is different in the two cases (8×8 for the system without spin interactions and 16×16 for the system with magnetic impurities). Of course, one could start consideration of the system without spin interactions using 16×16 supermatrices Q, but in this case all the elements of the spin blocks would correspond to the gapless excitations and these blocks would not reduce to conventional numbers as occurs in a system with magnetic impurities. The difference between the effects of the magnetic field and magnetic impurities can lead to different numbers in physical quantities such as localization length (see, e.g., Chapter 11).

In the long-wavelength limit the excitations with the gaps can be neglected and corresponding elements of the supermatrices can be put to zero. The supermatrix Q describing the remaining Goldstone modes is specified as before by Eqs. (4.33) but now instead of Eqs. (4.30, 4.31) one should use the following equations:

$$a^{mn} = \begin{pmatrix} a_1^{mn} & 0 \\ 0 & a_1^{mn*} \end{pmatrix}, \qquad b^{mn} = \begin{pmatrix} b_1^{mn} & 0 \\ 0 & b_1^{mn*} \end{pmatrix} \tag{4.73}$$

$$\sigma^{mn} = \begin{pmatrix} \sigma_1^{mn} & 0 \\ 0 & -\sigma_1^{mn*} \end{pmatrix}, \qquad a_1^{mn} = a_1^{nm*}, \qquad b_1^{mn} = b_1^{nm*}$$

Eqs. (4.73) are written for the case of an external magnetic field. In the presence of magnetic impurities one should multiply the elements a_1^{mn}, b_1^{mn}, and σ_1^{mn} by the unit matrix, thus obtaining the 16×16 supermatrix Q. All the other formulae of the preceding section, for example, Eqs. (4.55, 4.57), remain unchanged. Using formulae like Eq. (4.25) one should remember that, in the supermatrix V, only the part commuting with Σ has to be kept.

4.3.3 Spin-orbit impurities

The weak spin-orbit interaction described by the third term in $\mathcal{H}_1$ in Eq. (4.66) yields the term $\mathcal{L}_{\mathrm{so}}$ in the Lagrangian $\mathcal{L}$, Eq. (4.14)

$$\mathcal{L}_{\mathrm{so}} = \frac{1}{2} \left\langle \left(\int \bar{\psi}(\mathbf{r}) \sigma \left[\nabla u_{\mathrm{so}} \times \hat{\mathbf{p}} \right] \psi(\mathbf{r}) \, d\mathbf{r} \right)^2 \right\rangle \tag{4.74}$$

In contrast to the case of magnetic impurities, instead of Σ the matrix σ enters Eq. (4.74). This is due to the presence of the momentum operator $\hat{\mathbf{p}}$ in the formulae for the spin-orbit interaction, Eq. (4.66). Changing from the supervectors Φ to ψ one has to perform partial integration, as when deriving Eq. (4.65). The partial integration leads to additional multiplication by matrix τ_3, thus substituting Σ with σ. Note that $\bar{\sigma} = -\sigma$, which is different from Eq. (4.70) for Σ. Assuming the distribution of the electric fields ∇u_{so} is Gaussian and taking into account space anisotropy we obtain the corresponding additional term F_{so} in the free energy functional

$$F_{so}[Q] = -\frac{\pi \nu}{48} \sum_{i=x,y,z} \frac{1}{\tau_{so}^{(i)}} \operatorname{str} \int [Q, \sigma_i]^2 \, d\mathbf{r} \tag{4.75}$$

where $\tau_{so}^{(x)}$, $\tau_{so}^{(y)}$, and $\tau_{so}^{(z)}$ are mean free times of the spin-orbit scattering for the corresponding spin directions related to the fluctuations of the potential u_{so} by the relation

$$(\tau_{so}^{(i)})^{-1} = 6\pi \nu \left\langle \left\{ [\mathbf{p}' \times \mathbf{p}]^i u_{so} (\mathbf{p} - \mathbf{p}') \right\}^2 \right\rangle$$

where the angular brackets stand for both averaging over the impurities and the momenta of the Fermi surface. The total inverse mean free time τ_{so}^{-1} of spin-orbit scattering is equal to

$$\tau_{so}^{-1} = \sum_{i=x,y,z} (\tau_{so}^{(i)})^{-1}$$

The mean free times τ_s and τ_{so} in Eqs. (4.69, 4.75) are defined in such a way that to include them in the average one-particle Green function Eq. (3.45) one should make a simple substitution $\tau^{-1} \to \tau^{-1} + \tau_s^{-1} + \tau_{so}^{-1}$. A term analogous to the last term in Eq. (4.71) does not appear in Eq. (4.75) because, as a result of the relations $\sigma = -\bar{\sigma}$ and $Q = \bar{Q}$, one gets $\operatorname{str} \sigma Q = 0$. Writing the supermatrix Q as the sum of the singlet Q_0 and triplet parts again we can rewrite the free energy $F_{so}[Q]$ as

$$F_{so}[Q] = \frac{\pi \nu}{12} \operatorname{str} \int \left(\frac{1}{\tau_{so}} \mathbf{Q}^2 - \sum_{i=x,y,z} \frac{1}{\tau_{so}^{(i)}} (Q^i)^2 \right) d\mathbf{r} \tag{4.76}$$

We see from Eq. (4.76) that the excitations corresponding to fluctuations of $\mathbf{Q}$ have a gap, but, unlike in the magnetic case, all fluctuations of Q_0 remain gapless. The symmetry of the singlet part Q_0 of the supermatrix Q is conserved because the spin-orbit interactions do not break the time reversal symmetry. In the long-wavelength limit the excitations corresponding to the elements of the matrix $\mathbf{Q}$ are frozen. The matrix elements a, b, and σ of the supermatrix Q_0 that describe all the remaining diffusion modes can be written as

$$a^{mn} = \begin{pmatrix} a_1^{mn} & a_2^{mn} \\ a_2^{mn*} & a_1^{mn*} \end{pmatrix}, \qquad b^{mn} = \begin{pmatrix} b_1^{mn} & -b_2^{mn} \\ b_2^{mn*} & b_1^{mn*} \end{pmatrix} \tag{4.77}$$

$$\sigma^{mn} = \begin{pmatrix} \sigma_1^{mn} & \sigma_2^{mn} \\ -\sigma_2^{mn*} & \sigma_1^{mn*} \end{pmatrix}$$

$$a_1^{mn} = a_1^{nm*}, \qquad a_2^{mn} = a_2^{nm}, \qquad b_1^{mn} = b_1^{nm*}, \qquad b_2^{mn} = -b_2^{nm}$$

In Eqs. (4.77) a_i, b_i, and σ_i are 2×2 matrices proportional to the unit matrix. This structure can be obtained in a rather simple way. The elements of the supermatrix must obey the relations (4.30, 4.33, 4.38) with the matrices c_1 and c_2 from Eq. (4.68). If the matrices a_i, b_i, and σ_i are just unity matrices the presence of the matrices $i\sigma_y$ in the spin blocks of the matrices c_1 and c_2 is of no importance and, hence, the matrices $\tilde{c}_1$ and $\tilde{c}_2$ can be substituted for c_1 and c_2 in Eq. (4.68)

$$\tilde{c}_1 = \begin{pmatrix} 0 & 1 \\ 1 & 0 \end{pmatrix}, \qquad \tilde{c}_2 = \begin{pmatrix} 0 & -1 \\ 1 & 0 \end{pmatrix} \tag{4.78}$$

Substituting matrices c_1 and c_2 in Eqs. (4.30, 4.33, 4.38) with matrices $\tilde{c}_1$ and $\tilde{c}_2$, Eq. (4.78), we come to Eqs. (4.77).

Thus, depending on the presence of spin-orbit or magnetic interactions one can have in the long-wavelength limit three possible symmetries of the supermatrices Q, Eqs. (4.31, 4.73, 4.77), in the nonlinear σ-model with the free energy functional $F[Q]$, Eq. (4.55). Combinations of magnetic and spin-orbit interactions do not lead to new classes of symmetries. For example, adding the magnetic field to the system with spin-orbit impurities leads to reducing the degrees of freedom related to fluctuations of the elements a_2, b_2, and σ_2. Putting these elements to zero in Eqs. (4.77) we come to Eqs. (4.73). (The difference of the sign of the element σ_1^* is of no importance.)

The three different classes of symmetries have been used in the random matrix theory proposed long ago for the description of complex nuclei by Wigner (1951) (see also Mehta (1991)). Following the classification proposed by Wigner it is reasonable to call a system without magnetic or spin-orbit interactions an *orthogonal ensemble*, a system with magnetic interactions a *unitary ensemble*, and a system without magnetic but with spin-orbit interactions a *symplectic ensemble*. In other words the orthogonal ensemble corresponds to a system with time reversal and central symmetries, the unitary ensemble corresponds to systems with broken time reversal invariance, and, in the symplectic ensemble, central symmetry is broken although time reversal symmetry is preserved.

A very important question can be asked about the sensitivity of the σ-model to details of the initial Hamiltonian. The derivation presented previously was based on a model with white noise potential Eq. (3.38). However, the nonlinear σ-model can be derived as well, starting from the N-orbital model of Wegner (1979), which is a model with random hopping from site to site, each site having a large number N of states. In fact, the form of the Lagrangian and the supermatrices Q is more sensitive to the symmetries of the superspace than to the details of the impurity correlations. Using impurities with different correlations instead of white noise potential, after averaging over the impurities one would obtain not the term $\left(\bar{\psi}\psi\right)^2$ in the Lagrangian, Eq. (4.14), but another type of interaction.

Nevertheless, one can first try the mean field theory, which would give supermatrix Q, and then derive a Lagrangian describing the energy of the slow variations of this supermatrix. Of course, it is very important that fluctuations of the eigenvalues of this supermatrix can be neglected. Then, expanding in the frequency and gradients one comes to the standard form, Eq. (4.55), with a diffusion coefficient D_0 depending on the impurity correlations. In principle, the dependence can be very complicated, but in such a case D_0 in Eq. (4.55) can be considered a phenomenological parameter.

Recently the σ-model was derived for a model with a δ-correlated magnetic field (Aronov, Mirlin, and Wölfle (1994)). This corresponds to very-long-range correlations of the vector potential **A** entering the Hamiltonian. The long-range correlations lead to divergencies in the perturbation theory for the one-particle Green function and finally change its form (Altshuler and Ioffe (1992)). At the same time, transport properties do not seem to be affected by the unusual correlations. As soon as the classical diffusion coefficient is finite (Khveshchenko and Meshkov (1993)) one comes to the σ-model, Eq. (4.55), with symmetry corresponding to the unitary ensemble.

The problem of under what conditions the σ-model can be derived is analogous to the corresponding problem in the theory of superconductivity. It is rather difficult to calculate the transition temperature and other parameters characterizing superconductivity, but one can start with the phenomenological Ginzburg–Landau energy functional and describe many physical effects using this approach. The only thing one needs to know is the symmetry of the superconducting order parameter.

In spite of the difficulty of formulating mathematical criteria for the validity of the σ-model the physical situations when it can be used are quite clear. Each time electrons move diffusively at short distances or times one may use the σ-model and this makes it possible to describe the electron motion at large distances and times.

Technically, the unitary ensemble is the simplest. Within this ensemble, in effect 4×4 supermatrices Q are used; the reduction of size is due to the absence of nondiagonal elements in the matrices a, b, and σ in Eqs. (4.73). Calculations for the orthogonal and symplectic ensembles are more difficult because one should perform integrations over what are in effect 8×8 supermatrices Q. Note a certain symmetry between the orthogonal and symplectic cases. One can transform formulae for the former ensemble to the corresponding formulae for the latter simply by making the substitution $c_1 \rightarrow \tilde{c}_1 = c_2$, $c_2 \rightarrow \tilde{c}_2 = c_1$ (see Eqs. (4.10, 4.78)).

To calculate the density–density correlation function in the crossover regime between the ensembles, one can use Eq. (4.57) as before, provided each of the supermatrices $Q(\mathbf{r})$ and $Q(0)$ is replaced by the following expression:

$$Q \rightarrow Q\left(1 - \frac{\tau}{6}\left[Q, \left(\frac{1}{\tau_s}\left(\Sigma Q \Sigma - \frac{1}{2}\Sigma \operatorname{str} \Sigma Q\right) + \sum_{i=x,y,z} \frac{1}{\tau_{so}^{(i)}} \sigma_i Q \sigma_i\right)\right]\right) \quad (4.79)$$

The free energy must now contain the terms Eqs. (4.71, 4.75). Eq. (4.79), as well as Eqs. (4.71, 4.75), are written in the principal approximation with respect to the magnetic and spin-orbit interactions, $\tau \ll \tau_s$, τ_{so}. These formulae make it possible to describe all possible effects of the interactions on the quantum corrections, thus reproducing the known formulae of weak localization. The corresponding calculations complemented by calculations within a renormalization group scheme are presented in the next chapter.

5

Perturbation theory and renormalization group

5.1 Perturbation theory

5.1.1 First quantum correction

In the preceding chapter the σ-model describing the diffusion modes was derived. This model is rather complicated, so before studying it for nontrivial cases let us consider the high-frequency limit. In this limit one can make an expansion assuming that characteristic deviations of the supermatrix Q from Λ are small. In the zero-order approximation the diffusion form Eq. (3.52) of the density–density correlation function Y^{00} has been reproduced using the expansion Eqs. (4.47, 4.61, 4.62). It is interesting to calculate corrections to the classical result because in this limit one can use the diagrammatic approach described in Chapter 3, which allows one to check the σ-model. Besides, one can get a feeling of how the σ-model works and describe many known results in a unified manner.

In the Gaussian approximation, the density–density correlation function describes the free diffusion mode, Eq. (3.52). For calculation of corrections to the correlation function that are due to mode interactions one should consider higher-order terms in W in the parametrization Eq. (4.47). In principle, one might expand the supermatrix Q in W by using this parametrization and consider the resulting nonlinear terms. However, another parametrization of the supermatrix Q is more convenient for perturbation theory

$$Q = \Lambda \left(1 + iP\right)\left(1 - iP\right)^{-1}, \qquad P = \begin{pmatrix} 0 & B \\ \bar{B} & 0 \end{pmatrix}, \qquad \bar{B} = C_0 B^T C_0^T \qquad (5.1)$$

where the matrix C_0 is specified by Eqs. (4.8, 4.10).

In Eq. (5.1), B is a supermatrix that has the same symmetry as the supermatrix $-iQ^{12}$, Eq. (4.33). In particular, the supermatrix P satisfies the following relation analogous to Eq. (4.48):

$$P = KP^+K \qquad (5.2)$$

Of course, Eq. (4.48) follows from Eq. (5.2). The matrix P anticommutes with Λ and therefore the constraint $Q^2 = 1$ is satisfied.

By substituting Eq. (5.1) into Eq. (4.65) for free energy and adding the terms $F_s\left[Q\right]$, Eq. (4.71), and $F_{so}\left[Q\right]$, Eq. (4.75), describing all the magnetic and spin-orbit interactions, one can transform the total free energy functional $F[Q]$ to the form

$$F = \pi \nu \int \Bigg\{ \mathrm{str} \Big\{ D_0 \left(1 + B\bar{B}\right)^{-1} \partial B \left(1 + \bar{B}B\right)^{-1} \partial \bar{B} + i\omega \left(1 + B\bar{B}\right)^{-1} \tag{5.3}$$

$$- \frac{1}{6} \sum_{i=x,y,z} \frac{1}{\tau_{so}^{(i)}} \left(1 + B\bar{B}\right)^{-1} [B, \sigma_i] \left(1 + \bar{B}B\right)^{-1} [\bar{B}, \sigma_i]$$

$$- \frac{1}{6\tau_s} \left(1 + B\bar{B}\right)^{-1} [B, \Sigma] \left(1 + \bar{B}B\right)^{-1} [\bar{B}, \Sigma] \Big\}$$

$$+ \frac{1}{12\tau_s} \left(\mathrm{str} \left([\bar{B}, \Sigma] B \left(1 + \bar{B}B\right)^{-1} \right) \right)^2 \Bigg\} d\mathbf{r}$$

where $\partial B = \nabla B - (ie/c) \mathbf{A}[B, \tau_3]$.

The density–density correlation function Y^{00}, Eqs. (4.57, 4.79), can also be rewritten in terms of the supermatrices B. For explicit calculations it is necessary to know the Jacobian (Berezinian) that arises when changing integration over supermatrices Q to integration over supermatrices B, Eq. (5.1). The question about the Jacobian is common in studying σ-models in a certain parametrization. The parametrization Eq. (5.1) is most convenient because the corresponding Jacobian equals unity. The simplest way to prove this property is to write an appropriate quadratic form (cf. Eqs. (2.111–2.113)). The elementary length in the space of the supermatrix elements dQ is

$$\mathrm{str}(dQ)^2 = 8 \, \mathrm{str} \left(\left(1 + B\bar{B}\right)^{-1} dB \left(1 + \bar{B}B\right)^{-1} d\bar{B} \right) \tag{5.4}$$

In order to calculate the Berezinian corresponding to the quadratic form Eq. (5.4) it is convenient to consider a more general form

$$\mathrm{str}(F \, dB \, G \, d\bar{B}), \qquad F = \bar{F}, \qquad G = \bar{G} \tag{5.5}$$

If supermatrices F and G are diagonal, the quadratic form, Eq. (5.5), can be rewritten explicitly in terms of all elements of supermatrices B and $\bar{B}$. The commuting elements of the supermatrices give the product $\prod_\alpha F_\alpha G_\alpha$ in the numerator of the Berezinian. The anticommuting elements yield the same product in the denominator, thus giving unity for the Berezinian. If the self-conjugate supermatrices F and G are not diagonal they can be diagonalized

$$F = z_1 f \bar{z}_1, \qquad G = z_2 g \bar{z}_2, \qquad z_1 \bar{z}_1 = 1, \qquad z_2 \bar{z}_2 = 1$$

Changing the variables dB and $d\bar{B}$ to variables dA and $d\bar{A}$ with the same structure

$$dB = z_1 \, dA \, \bar{z}_2, \qquad d\bar{B} = z_2 \, d\bar{A} \, \bar{z}_1$$

instead of Eq. (5.5) we obtain the quadratic form $\mathrm{str}(f \, dA \, g \, d\bar{A})$, which shows that the Berezinian corresponding to the quadratic form, Eq. (5.5), and, hence, to Eq. (5.4) is really equal to unity. The equality of the Berezinian to unity makes the parametrization, Eq. (5.1), especially useful for perturbation theory at high frequencies.

Before making calculations in the presence of a magnetic field and magnetic and spin-orbit impurities let us show how to obtain the first quantum correction in their absence. This can help us to understand the computational scheme without writing cumbersome formulae. In this simple case we may use the 8×8 supermatrices Q derived in Section 4.2. Then, the matrices B in Eq. (5.1) are 4×4 supermatrices. By expanding Eq. (5.3) up to quartic terms, as is necessary to get the first quantum correction, one obtains

$$F = F_2 + F_4, \tag{5.6}$$

$$F_2 = \pi \nu \int \mathrm{str} \left(D_0 \nabla B \nabla \bar{B} - i\omega B\bar{B} \right) d\mathbf{r} \tag{5.7}$$

$$F_4 = \pi \nu \int \mathrm{str} \left(-D_0 \left(\nabla \bar{B} \nabla B \bar{B} B + \nabla B \nabla \bar{B} B \bar{B} \right) + i\omega \left(B\bar{B} \right)^2 \right) d\mathbf{r} \tag{5.8}$$

A corresponding expansion should also be done in Eq. (4.58) for the correlation function $Y^{00} (\mathbf{r}, \omega)$, and this function takes the form

$$Y^{00} (\mathbf{r}, \omega) = \frac{(\pi \nu)^2}{2} \int \mathrm{str}[k \left(1 - \tau_3 \right) \left(B (\mathbf{r}) - B (\mathbf{r}) \bar{B} (\mathbf{r}) B (\mathbf{r}) \right) k \left(1 - \tau_3 \right)$$

$$\times \left(\bar{B} (0) - \bar{B} (0) B (0) \bar{B} (0) \right) \exp \left(-F_4 \right) \exp \left(-F_2 \right) DBD\bar{B} \tag{5.9}$$

As the next step one has to make an expansion in F_4 considering this term as a perturbation and calculate Gaussian integrals with free energy F_2. In the zero order for the function $Y^{00} (\mathbf{k}, \omega)$ in the momentum representation one has

$$Y^{00} (\mathbf{k}, \omega) = \frac{4\pi \nu}{D\mathbf{k}^2 - i\omega} \tag{5.10}$$

with $D = D_0$.

Terms arising from the expansion in F_4 give, after solving the corresponding Dyson equation, a contribution to the self-energy of the propagator $Y^{00} (\mathbf{k}, \omega)$, whereas the nonlinear terms in the preexponential renormalize the residue. The self-energy can also be found by substituting the quartic terms in F_4 with a quadratic term multiplied by the averages of the remaining pairs of B and $\bar{B}$. Because of the equal numbers of commuting and anticommuting elements in supermatrices B one can check that $\langle B\bar{B} \rangle_2 = \langle \bar{B}B \rangle_2 = \langle \nabla B \nabla \bar{B} \rangle_2 = \langle \nabla \bar{B} \nabla B \rangle_2 = 0$. Making pairs of B, $\bar{B}$ with ∇B, $\nabla \bar{B}$ also gives zero because these averages do not depend on coordinates but must change sign after inversion of the coordinates. So the first term in F_4, Eq. (5.8), does not contribute at all.

As concerns the second term in Eq. (5.8), one can get nonzero averages by pairing B with B and $\bar{B}$ with $\bar{B}$. Calculating the Gaussian integrals one obtains

$$\langle \bar{B} M \bar{B} \rangle_2 = -\bar{M} I_1, \qquad \langle B \bar{M} B \rangle_2 = -M I_1, \tag{5.11}$$

$$I_1 = \frac{1}{2\pi \nu} \int \frac{1}{D_0 \mathbf{k}^2 - i\omega} \frac{d^d \mathbf{k}}{(2\pi)^d}$$

where M is a supermatrix with the same structure as B. After such decoupling instead of F_4 one has the following expression:

$$F_4 \rightarrow -\pi \nu 2 i\omega I_1 \int \mathrm{str} \left(B\bar{B} \right) d\mathbf{r} \tag{5.12}$$

which can be included in F_2, resulting in a change in the coefficient in the second term.

Then one should make two possible pairings in the preexponential in Eq. (5.9), and this gives

$$Y^{00} (\mathbf{k}, \omega) = \frac{4\pi \nu \left(1 + 2I_1 \right)}{D_0 \mathbf{k}^2 - i\omega \left(1 + 2I_1 \right)} \tag{5.13}$$

We see that the residue and the term with ω are renormalized in the same way and we come to Eq. (5.10) with the diffusion coefficient D given by the expression

$$D = D_0 \left(1 - \frac{1}{\pi \nu} \int \frac{1}{D_0 \mathbf{k}^2 - i\omega} \frac{d^d \mathbf{k}}{(2\pi)^d} \right) \tag{5.14}$$

The diffusion coefficient given by Eq. (5.14) is in agreement with Eq. (3.56) obtained by the conventional perturbation theory.

5.1.2 Magnetic field and magnetic and spin-orbit impurities

Having understood the scheme of calculation for the simplest example let us consider the general case. In order to be able to perform computations with magnetic and spin-orbit interactions let us represent the supermatrix B as

$$B = \sum_{a,\alpha} B_{a\alpha} s_\alpha \tag{5.15}$$

where $s_0 = 1$, $s_{1,2,3} = i\sigma_{x,y,z}$, $a = \|, \perp$. In Eq. (5.15), the supermatrix $B_\|$ commutes with τ_3, and $B_\perp$ anticommutes with τ_3. The condition Eq. (5.2) leads to the following relations for the supermatrices $B_{\|\alpha}$, $B_{\perp\alpha}$:

$$\tilde{C}_0 B_{a\alpha}^T \tilde{C}_0^T = k B_{a\alpha}^+, \qquad \tilde{C}_0 = \begin{pmatrix} \tilde{c}_1 & 0 \\ 0 & \tilde{c}_2 \end{pmatrix} \tag{5.16}$$

where matrices $\tilde{c}_1$ and $\tilde{c}_2$ are determined by Eq. (4.78).

Using the representation Eq. (5.15) and expanding $B(\mathbf{r})$ in a series in the eigenfunctions of the operator $\nabla - i(e/c)\mathbf{A}$ one can rewrite the quadratic part F_2 of the functional F, Eq. (5.3), as

$$F_2 = 2\pi \nu \operatorname{str} \sum_{E,a,\alpha} (-i\omega + \mathcal{E}_a + X_{a\alpha}) B_{a\alpha}^E \bar{B}_{a\alpha}^E \tag{5.17}$$

In Eq. (5.17), $B_{a\alpha}^E$ are coefficients in the expansion of the function $B(\mathbf{r})$, and $\mathcal{E}_a$ are eigenenergies

$$\mathcal{E}_\| = D_0 \mathbf{k}^2, \qquad \mathcal{E}_\perp = \frac{4 D_0 |e| H}{c} \left(n + \frac{1}{2} \right) + D_0 k_H^2 \tag{5.18}$$

where H is the magnetic field and k_H is the component of the momentum parallel to the magnetic field.

The energies $\mathcal{E}_a$ do not depend on α because the interaction of the magnetic field with the electron spin is neglected. The interaction with the magnetic and spin-orbit impurities is described by the coefficients $X_{a\alpha}$ that can be obtained by substituting Eq. (5.15) into Eq. (5.3) and calculating the commutators. The result can be written as

$$X_{\|0} = 0, \qquad X_{\|i} = \frac{2}{3} \left(\frac{2}{\tau_s} + \frac{1}{\tau_{so}} - \frac{1}{\tau_{so}^{(i)}} \right) \tag{5.19}$$

$$X_{\perp 0} = \frac{2}{\tau_s}, \qquad X_{\perp i} = \frac{2}{3} \left(\frac{1}{\tau_s} + \frac{1}{\tau_{so}} - \frac{1}{\tau_{so}^{(i)}} \right)$$

where $i = x, y, z$ and $1/\tau_{so} = \sum_{i=x,y,z} 1/\tau_{so}^{(i)}$.

Notice that in the principal approximation the last term in Eq. (5.3) does not contribute. The correlation functions in this approximation are determined by Gaussian integrals. Using the relations Eq. (5.16) we obtain

$$\left\langle \left(B_{a\alpha}^{E} \right)_{ik} \left(\bar{B}_{b\beta}^{E} \right)_{lm} \right\rangle_2 = \frac{1}{4\pi \nu} \frac{\delta_{ab}\delta_{\alpha\beta}\delta_{im}k_{kl}}{-i\omega + \mathcal{E}_{\alpha} + X_{a\alpha}} \tag{5.20}$$

where $\langle \ldots \rangle_2$ stands for averaging with the free energy F_2, Eq. (5.17). Eqs. (5.18, 5.20) explicitly show which diffusion modes acquire gaps and determine these gaps. The gap is absent only in the correlations of the elements of the supermatrix $B_{\parallel 0}$. The density–density correlation function Y^{00}, as mentioned in Section 4.3, can be calculated by using Eq. (4.58) provided supermatrix Q is replaced by the singlet part Q_0. This means that in the Gaussian approximation this correlation function reduces to the average $\langle B_{\parallel 0} B_{\parallel 0} \rangle$, which does not have a gap. The fact that the gap does not appear in the density–density correlation function is natural because the break of the time reversal and spin rotation symmetries does not influence the particle conservation law. This law is expressed by Eq. (3.35), which shows that the density–density correlation function integrated over the space (taken at $\mathbf{k} = 0$ in the momentum representation) and summed over the spin variable diverges as $(-i\omega)^{-1}$ in the limit $\omega \to 0$. (In the limit of small ω a difference between the averaged functions Y^{00} and R^{00} is not essential.)

In the presence of magnetic interactions only excitations corresponding to supermatrix $B_{\parallel}$ remain gapless. It follows from Eq. (5.16) that the structure of supermatrix B is the same as that of supermatrix Q^{12} determined by Eqs. (4.33, 4.73). Without magnetic interactions but in the presence of spin-orbit interactions, only the excitations corresponding to super-matrices $B_{\parallel 0}$ and $B_{\perp 0}$ are gapless. In this case the structure is determined by Eqs. (4.33, 4.77).

Now we can calculate the first correction arising from the interaction of the diffusion modes. Writing the interaction we can neglect the magnetic field and both the magnetic and spin-orbit impurities in the corresponding terms. Using the expansion Eq. (5.15) and calculating the trace of products of the matrices s_{α} we can rewrite the interaction energy F_{int} in the lowest order in B in the form

$$F_{\text{int}} = 2\pi \nu \sum_{\substack{a,b,c,d \\ \alpha,\beta,\gamma,\delta}} \left(e_{\alpha\beta\gamma\delta} + \delta_{\alpha\beta}\delta_{\gamma\delta} + \delta_{\alpha\delta}\delta_{\beta\gamma} - \delta_{\alpha\gamma}\delta_{\beta\delta} \right) \tag{5.21}$$

$$\times \int \text{str}\left[D_0 \left(\nabla B_{a\alpha} \nabla \bar{B}_{b\beta} B_{c\gamma} \bar{B}_{d\delta} + \nabla \bar{B}_{a\alpha} \nabla B_{b\beta} \bar{B}_{c\gamma} B_{d\delta} \right) \right.$$
$$\left. - i\omega B_{a\alpha} \bar{B}_{b\beta} B_{c\gamma} \bar{B}_{d\delta} \right] d\mathbf{r}$$

where $e_{\alpha\beta\gamma\delta}$ is the completely antisymmetric tensor ($e_{0123} = 1$). Calculating the density–density correlation function Y^{00} with Eq. (4.57, 4.79) one should also make an expansion in B in the preexponential. In the main order of the expansion the substitution Eq. (4.79) does not yield anything new with respect to Eq. (4.57) and one obtains

$$Y^{00}(\mathbf{r}, \omega) = (\pi \nu)^2 \langle \text{str}[(B_{\parallel 0}(\mathbf{r}) - \sum_{\substack{a,b,c \\ \alpha,\beta,\gamma}} T_{\alpha\beta\gamma} B_{a\alpha}(\mathbf{r}) \bar{B}_{b\beta}(\mathbf{r}) B_{c\gamma}(\mathbf{r})) \tag{5.22}$$

$$\times k(1 - \tau_3)(\bar{B}_{\parallel 0}(0) - \sum_{\substack{k,l,m \\ \sigma,\rho,\zeta}} T_{\sigma\rho\zeta}^* \bar{B}_{k\sigma}(0) B_{l\rho}(0) \bar{B}_{m\zeta}(0))k(1 - \tau_3)]\rangle$$

where $T_{\alpha\beta\gamma} = 1/2 \, \text{tr}(s_{\alpha} s_{\beta}^{+} s_{\gamma})$.

In Eq. (5.22) the angular brackets stand for averaging with the functional $F_2 + F_{int}$ from Eqs. (5.17, 5.21). Now, one should make the expansion considering the term F_{int} as a perturbation and calculate arising Gaussian integrals. Solving the corresponding Dyson equation we come to Eq. (5.10) with the diffusion coefficient D given by the following expression:

$$D = D_0 \left(1 - \frac{1}{2\pi\nu} \sum_{E_\perp,\alpha} \frac{1 - 2\delta_{\alpha 0}}{\mathcal{E}_\perp + X_{\perp\alpha} - i\omega} \right) \tag{5.23}$$

Thus, diffusion mode interaction does not change the diffusion form of the density–density correlation function but it does change the diffusion coefficient. The form of the density–density correlation function Eq. (5.10) is characteristic of metals and is valid for an arbitrary impurity concentration. It changes at the point of the metal–insulator transition, where the diffusion coefficient D turns to zero. Of course, the average density of states ν entering the diffusion propagator, Eq. (5.10), can change with impurity concentration, but in the model of the noninteracting particles it is always positive and cannot have singularities as a function of the concentration (Wegner (1982)). Let us emphasize that there are no corrections to the frequency ω in Eq. (5.10), and the interaction of the diffusion modes contributes to the diffusion coefficient only. This is a consequence of the general relation Eq. (3.35). In the formal expansion in B presented previously this property is due to the fact that the contribution from the last term in Eq. (5.21) cancels the contribution to the numerator from the cubic terms in Eq. (5.22).

Eq. (5.23) contains all known effects of the first order in the diffusion mode interaction. In the absence of magnetic and spin-orbit impurities the sum over α in Eq. (5.23) is trivial and reduces to a factor of 2. In the absence of a magnetic field we immediately reproduce Eq. (3.56). Calculation of the integral over the momenta shows that in 2D the quantum correction to the conductivity is negative and proportional to $|\ln \omega\tau|$ (Abrahams et al. (1979), Gorkov, Larkin, and Khmelnitskii (1979), Anderson, Abrahams, and Ramakrishnan (1979)). In long thin wires, where Eq. (5.23) is also valid, the correction is proportional to $(\omega\tau)^{-1/2}$.

If the magnetic field is not zero, using the direct diagram summation one had to replace the diffusion propagator in Eq. (3.56) by the cooperon in the magnetic field, Eq. (3.59), and this is just what one obtains from Eq. (5.23). The magnetic field cuts the quantum correction thus leading to negative magnetoresistance (Altshuler et al. (1980)). If the system does not contain magnetic impurities but central symmetry is violated by spin-orbit interactions, the term with $\alpha = 0$ in the sum in Eq. (5.23) is most singular because $X_{\perp 0} = 0$, Eq. (5.19). But this quantum correction is positive and therefore quantum interference increases conductivity (Hikami, Larkin, and Nagaoka (1980)). The correction in the presence of the spin-orbit interaction, having the opposite sign, is two times smaller than that for the system without both magnetic and spin-orbit impurities. Applying the magnetic field one cuts the quantum correction in the spin-orbit case too, but this leads to decreasing the conductivity (positive magnetoresistance). One can find a detailed discussion of experiments where all these effects were observed in the review by Bergmann (1984).

Eq. (5.23) was also derived by using the σ-model based on the replica method by Efetov, Larkin, and Khmelnitskii (1980); the derivation is completely similar throughout all steps. So the supersymmetry technique reproduces the known formulae of perturbation theory in diffusion modes. I have presented the derivation to demonstrate that the σ-model does not contradict to perturbation theory and to give a feeling with simple examples of how one can manipulate the model. The perturbation theory works provided the frequency ω is high enough. For lower frequencies more complicated methods are necessary. The next section investigates transport properties using a renormalization group method in $2 + \epsilon$ dimensions with small ϵ.

5.2 Renormalization group in $2 + \epsilon$ dimensions

5.2.1 Renormalization group procedure

Another example of calculations using the supermatrix σ-model, Eq. (4.55), is the renormalization group study in $(2 + \epsilon)$ dimensions. This is the simplest generalization of the perturbation theory calculations and as soon as the existence of a renormalization group is established both approaches coincide. Renormalization group calculations for conventional vector and matrix σ-models, for many types of symmetry of matrices, have been very popular in the last 20 years (Polyakov (1975), Brezin and Zinn-Justin (1976), Brezin, Zinn-Justin, and Le Guillou (1976), Brezin, Hikami, and Zinn-Justin (1980), Nelson and Pelcovits (1977), Hikami (1983, 1990), Zinn-Justin (1993), Wegner (1987, 1989), Kravtsov, Lerner, and Yudson (1989)). The limit $n \to 0$ for $n \times n$ matrices corresponds to the localization problem (Wegner (1979), Efetov, Larkin, and Khmelnitskii (1980), Schäfer and Wegner (1980)). Renormalization group treatment of supermatrix σ-models was first carried out by Efetov (1982a).

Following the latter work let us show how the renormalizability of the supermatrix σ-model, Eq. (4.55), can be proved for all three types of symmetry of supermatrix Q determined by Eqs. (4.31, 4.33, 4.73, 4.77). Supermatrices Q are 8×8 supermatrices for systems without spin interactions and 16×16 for systems with magnetic or spin-orbit impurities. In principle, one could write 16×16 supermatrices for all cases but that would make calculations more complicated. The ability to write 8×8 supermatrices for systems without spin interactions is a consequence of the decoupling of subsystems with "up" and "down" spin. In this case the spin degree of freedom results in an additional factor of 2 in physical quantities. However, in the presence of magnetic and spin-orbit interactions the long-wavelength limit of the corresponding σ-models is also described by what are effectively 8×8 supermatrices because some excitations acquire gaps and, thus, are frozen out. Then, each spin block in the initial 16×16 supermatrices consists of the unity matrix and this is equivalent to the 8×8 supermatrix. For the same reason the long-wavelength limit in systems with magnetic interactions (unitary ensemble) is described in fact by 4×4 supermatrices, and therefore the unitary ensemble is the simplest for explicit nonperturbative calculations.

Assuming that the long-wavelength limit is the only one we are interested in, the 8×8 supermatrices are used in the following discussion. To write formulae in a general form it is convenient to describe the unitary ensemble by the supermatrices of this size, too, remembering, however, that all elements in this case commute with the matrix τ_3, Eq. (4.59). The free energy functional can be written in a unified form

$$F = \frac{1}{t} \int \mathrm{str} \left[(\nabla Q)^2 + 2i\tilde{\omega}\Lambda Q \right] d\mathbf{r} \tag{5.24}$$

with the 8×8 supermatrix Q specified by Eqs. (4.31, 4.33, 4.73, or 4.77).

Let the model in which only potential scattering on normal impurities is possible be called Model I. This model corresponds to the orthogonal ensemble (Wigner (1951), Mehta (1991)). The system with potential scattering in an external magnetic field will be called Model IIa and addition of magnetic impurities to Model I or Model IIa leads to Model IIb.

Models IIa and IIb correspond to the unitary ensemble and the difference between them is minimal. The system in which only potential and spin-orbit scattering are possible will be called Model III, which is equivalent to the symplectic ensemble. Model I is invariant under time reversal and spin rotations. In Model IIa time reversal invariance is violated but the spin rotation invariance is conserved, whereas in Model IIb both invariances are broken. And in Model III time reversal symmetry is conserved but spin rotation invariance is broken. The system with potential and spin orbit scattering only but placed in a magnetic field relates to Model IIb. In Model I the matrices a, b, and σ are specified by Eqs. (4.31), in the Models IIa, IIb by Eqs. (4.73), and in Model III by Eqs. (4.77). The coefficient t in Eqs. (5.24) for this models is

$$
t = \begin{cases} 8\,(\pi \nu D_0)^{-1}, & \text{Models I, IIa} \\ 4\,(\pi \nu D_0)^{-1}, & \text{Models IIb, III} \end{cases} \tag{5.25}
$$

and in all the cases $\tilde{\omega} = \omega/D_0$.

The quasi-diagonal structure, Eq. (4.73), of supermatrix Q in Models II enables one, as mentioned, to reduce the problem to the σ-model with the 4×4 supermatrix.

To complete the definition of the σ-model, Eq. (5.24), one should introduce a minimal length that will serve as a cutoff at low distances. Recalling the derivation of the σ-model presented in the preceding chapter it is natural to set this length equal to the mean free path l. The main idea of renormalization group calculations is that the form of the free energy functional Eq. (5.24) does not depend on the cutoff in the low frequency and momentum limit. This is because the free energy functional $F[Q]$, Eq. (5.24), with the constraint $Q^2 = 1$ is the only possible functional in this limit. Only if some symmetries are violated can one write additional terms. Such terms can appear in a strong magnetic field (Levine, Libby and Pruisken (1984), Pruisken (1985)), and they are thought to describe the quantum Hall effect.

The invariance of the form of the free energy functional with respect to the cutoff k_0 makes it possible to integrate out the fields with the momenta within the interval (λ, k_0), where $\lambda < k_0$, and to obtain the same form, Eq. (5.24), of the free energy functional. Of course, the coefficients t and $\tilde{\omega}$ can change as a result of this procedure and, thus, they are functions of λ. However, changing the scale of all momenta $k \to k(\lambda/k_0)$ one must obtain the initial free energy and this makes it possible to write equations for the effective "charges" t and $\tilde{\omega}$.

For explicit calculations one can use a procedure analogous to that suggested by Polyakov (1975) for studying a vector σ-model. Representing supermatrix Q in the form

$$
Q(\mathbf{r}) = V(\mathbf{r}) \Lambda \bar{V}(\mathbf{r}) \tag{5.26}
$$

with $V\bar{V} = 1$, let us represent the unitary supermatrix $V(\mathbf{r})$ in Eq. (5.26) as the product of fast $V_0(\mathbf{r})$ and slow $\tilde{V}(\mathbf{r})$ parts

$$
V(\mathbf{r}) = \tilde{V}(\mathbf{r}) V_0(\mathbf{r}) \tag{5.27}
$$

The matrices $V_0(\mathbf{r})$ and $\tilde{V}(\mathbf{r})$ are assumed to satisfy the condition Eq. (4.49). Substituting Eqs. (5.26, 5.27) into Eq. (5.24) we obtain

$$
F = \frac{1}{t} \int \operatorname{str}[(\nabla Q_0)^2 + 2[Q_0, \nabla Q_0]\Phi + [Q_0, \Phi]^2 + 2i\tilde{\omega}\overline{\tilde{V}} \Lambda \tilde{V} Q_0]d\mathbf{r}, \tag{5.28}
$$

$$
Q_0 = V_0 \Lambda \bar{V}_0, \qquad \Phi = \overline{\tilde{V}} \nabla \tilde{V} = -\bar{\Phi}
$$

Calculating, for example, the density–density correlation function Y^{00} we should integrate in Eq. (4.57) over the fast variable $Q_0(\mathbf{r})$ using the free energy functional, Eq. (5.28). After such integration the free energy F in Eqs. (5.24, 5.28) is replaced by energy $\tilde{F}$ describing the slow fluctuations

$$\tilde{F} = -\ln \int \exp\left(-F\right) DQ_0 \tag{5.29}$$

To simplify the calculations it is convenient to choose a certain gauge of supermatrices $\tilde{V}$ and Φ. We can assume that supermatrix V in Eqs. (5.27, 5.28) is close to unity in the vicinity of a coordinate $\mathbf{r}$. The slowness of the variations of this matrix enables us to integrate over Q_0 in different domains independently. In the gauge chosen in the vicinity of the coordinate $\mathbf{r}$ the superelements Φ^{12} and Φ^{21} are the only ones in the supermatrix Φ that are not equal to zero. These superelements are related to each other by the anti-self-conjugacy condition of supermatrix Φ.

The question of separating the fast and the slow variables and cutting divergent integrals in an invariant manner is extremely important. It is assumed that the momenta of the space variations of the supermatrix Q_0 are in the interval $\lambda < k < k_0$. For invariant "infrared" cutting in the integral Eq. (5.29) over Q_0 in Eq. (5.28) one may add the following term:

$$F_{\text{reg}} = -\frac{2\lambda^2}{t} \int \text{str}\left(\Lambda Q\right) d\mathbf{r} \tag{5.30}$$

Integrals divergent at large momenta can be cut by dimensional regularization. According to this approach an analytical continuation is made from $d = 2 - \delta$ dimensions, where the "ultraviolet" divergencies are absent, to $d = 2 + \epsilon$ dimensions. In principle, other possibilities for regularization exist, but dimensional regularization seems to be most simple. For calculation of the integral over Q_0 it is convenient to parametrize supermatrix Q_0 with Eq. (5.1).

Now, after exclusion of spin structure, matrices c_1 and c_2 entering C and C_0 are specified for Models I and II by Eqs. (4.8, 4.10) again. The permutation $c_1 \rightarrow c_2, c_2 \rightarrow c_1$ of matrices c_1 and c_2 gives, as discussed in Section 4.3, the matrices C and C_0 in Model III.

It has been mentioned that integration over supermatrices B is performed with the unit Berezinian. Using Eq. (5.1), expanding Q_0 over P, and retaining, in the functional F, Eqs. (5.28, 5.30), the number of terms necessary for obtaining the first two orders, we find

$$F = F_0 + F_0' + F_1 + F_2' + \tilde{F}_0,$$

$$F_0 = \frac{4}{t} \, \text{str} \int \left[(\nabla P)^2 + \lambda^2 P^2\right] d\mathbf{r}$$

$$F_0' = -\frac{4}{t} \, \text{str} \int \left[2P^2 (\nabla P)^2 - \lambda^2 P^4\right] d\mathbf{r} \tag{5.31}$$

$$F_1 = -\frac{16i}{t} \, \text{str} \int P\nabla P P \Phi d\mathbf{r}$$

$$F_2 = \frac{8}{t} \, \text{str} \int \left[P^2 \Phi^2 - (P\Phi)^2\right] d\mathbf{r}$$

$$F'_2 = -\frac{8}{t} \operatorname{str} \int \left[P^4 \Phi^2 + \left(P^2 \Phi \right)^2 - 2P\Phi P^3 \Phi \right.$$

$$\left. + \frac{i\tilde{\omega}}{2} \Lambda \tilde{V} \Lambda \left(P^2 - P^4 \right) \overline{\tilde{V}} \right] d\mathbf{r}$$

$$\tilde{F}_0 = \operatorname{str} \int \left[\left(\nabla \tilde{Q} \right)^2 + 2i\tilde{\omega} \Lambda \tilde{Q} \right] d\mathbf{r}$$

In Eqs. (5.31), supermatrix $\tilde{Q}$ is defined as $\tilde{Q} = \tilde{V} \Lambda \overline{\tilde{V}}$.

Terms linear in P in Eq. (5.31) are absent because P varies rapidly and integration over the volume gives zero. It is assumed that $\lambda^2 \gg \tilde{\omega}$. Keeping the first-order terms in t and the quadratic terms in Φ, the free energy functional $\tilde{F}$ in Eq. (5.29) can be reduced to the form

$$\tilde{F} = \tilde{F}_0 + \langle F_2 \rangle_0 + \langle F'_2 \rangle_0 - \frac{1}{2} \langle F_1^2 \rangle_0 - \langle F'_0 F_2 \rangle_0 \tag{5.32}$$

In Eq. (5.32) the brackets $\langle \ldots \rangle_0$ stand for averaging with the functional F_0 from Eqs. (5.31). The Gaussian integrals can be calculated using the Wick theorem. It is convenient to write down all arising averages in matrix form. The following formulae can be checked by direct calculation:

$$\langle PMP \rangle_0 = \frac{\alpha t}{16 \left(k^2 + \lambda^2 \right)} \bar{M}', \quad \text{where } \bar{M}' = C_0 M^T C_0^T \tag{5.33}$$

Eq. (5.33) is valid for supermatrices M of the form

$$M = \begin{pmatrix} 0 & M^{12} \\ M^{21} & 0 \end{pmatrix} \tag{5.34}$$

For block diagonal supermatrices N of the form

$$N = \begin{pmatrix} N^{11} & 0 \\ 0 & N^{22} \end{pmatrix} \tag{5.35}$$

the following equation holds:

$$\langle PNP \rangle_0 = \frac{t}{64 \left(k^2 + \lambda^2 \right)} \left[\left(1 + \alpha^2 \right) \left(\operatorname{str} N - \Lambda \operatorname{str} \Lambda N \right) \right. \tag{5.36}$$

$$\left. + \left(1 - \alpha^2 \right) \tau_3 \left(\operatorname{str} \left(\tau_3 N \right) - \Lambda \operatorname{str} \left(\Lambda \tau_3 N \right) \right) \right]$$

And, at last, for arbitrary supermatrices M_1 and M_2 of the form Eq. (5.34) satisfying the conditions $M_1 = \bar{M}'_1$, $M_2 = \bar{M}'_2$ one can obtain the formula

$$\langle \operatorname{str} \left(PM_1 \right) \operatorname{str} \left(PM_2 \right) \rangle_0 = \frac{t}{8 \left(k^2 + \lambda^2 \right)} \operatorname{str} \left(M_1 M_2 \right) \tag{5.37}$$

Supermatrices M_i^{12} and M_i^{21} in Eqs. (5.34, 5.37) are assumed to have the same structure as supermatrices B and $\bar{B}$ correspondingly. As concerns supermatrices N^{11} and N^{22} they should have the structure of $B\bar{B}$ and $\bar{B}B$. Eqs. (5.33, 5.36, 5.37) are valid for all three ensembles. Only coefficient α in these equations depends on the model and is equal to

$$
\alpha = \begin{cases} -1, & \text{Model I} \\ 0, & \text{Model II} \\ 1, & \text{Model III} \end{cases} \tag{5.38}
$$

Using Eqs. (5.33–5.38) and calculating the averages in Eq. (5.32) after lengthy but simple calculations we obtain

$$
F = \frac{1}{t} \int \left\{ \text{str}(\nabla \tilde{Q})^2 \left[1 + \frac{\alpha t}{8} \int \frac{d^d k}{(2\pi)^d (k^2 + \lambda^2)} \right. \right.
$$

$$
\left. + \frac{t^2}{64} \left(1 - \alpha^2\right) \left(\frac{1}{2} - \frac{1}{d}\right) \int \frac{1}{\left(k_1^2 + \lambda^2\right)\left(k_2^2 + \lambda^2\right)} \frac{d^d k_1 d^d k_2}{(2\pi)^{2d}} \right] \tag{5.39}
$$

$$
\left. + 2i\tilde{\omega} \, \text{str} \, \Lambda Q \right\} d\mathbf{r}
$$

In Eq. (5.39) only those terms that are logarithmic in 2D are written down and averaging over angles of the integrals over the momenta $\mathbf{k}$ has been carried out. The term with coefficient $1/d$ in the second line of Eq. (5.39) is a result of such averaging in the term with $\langle F_1^2 \rangle$ in Eq. (5.32).

In the supermatrix σ-model, quadratic divergencies, usual in conventional σ-models (Brezin, Zinn-Justin, and Le Guillou (1976), Brezin and Zinn-Justin (1976)), are absent. Furthermore, divergencies arising after expansion of the Jacobian are also absent. Let us mention an interesting peculiarity of the model under consideration. In this model only the effective "temperature" t (in spin models the coefficient t is the real temperature and one can use this analogy) is renormalized. The quantity $\tilde{\omega}/t$, which is proportional to frequency ω, does not change under renormalization. This property holds in all orders of renormalization group calculations and is a consequence of the general equation (3.35) reflecting the particle conservation law.

5.2.2 Gell-Mann–Low function and physical quantities

The first-order correction in t in Eq. (5.39) gives a logarithmic contribution in 2D and can be easily calculated. However, when evaluating the second-order correction one should be careful. Naively substituting $d = 2$ would give zero, which would imply that the second-order correction is zero in 2D. But such a conclusion is wrong because the integral over the momenta k_1 and k_2 in the corresponding integral gives in 2D the square of the logarithm. That is why one should go below two dimensions, calculate the corrections there, and then make an analytical continuation into 2 and $2 + \epsilon$ dimensions. Carrying out this procedure one can see that both the first- and second-order corrections in Eq. (5.39) are logarithmic and the Gell-Mann–Low function $\tilde{\beta}$ can be written for small $\epsilon = d - 2$ as

$$
\tilde{\beta}\left(\tilde{t}\right) \equiv \frac{d\tilde{t}}{d \ln \lambda} = (d - 2)\,\tilde{t} + \alpha \tilde{t}^2 - \frac{1}{2}\tilde{t}^3 \left(1 - \alpha^2\right) + \beta_4\left(\tilde{t}\right), \tag{5.40}
$$

$$
\tilde{t} = \frac{t}{2^{d+1}\pi d \Gamma\left(d/2\right)}
$$

The contribution to the function $\tilde{\beta}(\tilde{t})$ from the first two orders of the perturbation theory is written explicitly in Eq. (5.40). The contribution of higher orders is denoted by $\tilde{\beta}_4(\tilde{t})$. It is seen from Eqs. (5.38, 5.40) that the t^3 term is zero in Models I and III. In strictly two dimensions the solution of Eq. (5.40) for these two models can be written as

$$\tilde{t}(\omega) = \frac{\tilde{t}_0}{1 + \alpha \tilde{t}_0 \ln(1/\omega\tau)} \tag{5.41}$$

For sufficiently high frequencies ω the "temperature" t and hence the diffusion coefficient D coincide with their bare values.

When decreasing the frequency in Model I, coefficient t, which is proportional to resistivity, grows and diffusion coefficient D decreases. Unfortunately, it is impossible to reach the low-frequency region because at $t \sim 1$ all higher-order terms in the Gell-Mann–Low function are of the same order and Eq. (5.41) is no longer valid. The renormalization group treatment based on the non-linear σ-model agrees very well with the scaling hypothesis proposed by Abrahams et al. (1979). The $\tilde{\beta}$ function for Model I in 2D is negative for small t (large conductance g), and this corresponds to the right tail of the β function drawn for two dimensions in Fig. 3.1. (Note that the definition of the function $\tilde{\beta}$, Eq. (5.40), is somewhat different from that of the function β, Eq. (3.4).) Assuming that the continuation of the curve in Fig. 3.1 can describe the further growth of coefficient t one concludes that the statement about the localization in 2D at any weak disorder is justified.

In Models II the t^2 term in the Gell-Mann–Low function, Eq. (5.40), vanishes but the t^3 term is not equal to zero. The sign of this term also corresponds to the growth of resistivity as ω decreases. The solution of Eq. (5.40) in 2D can be written for Models II as

$$\tilde{t} = \frac{\tilde{t}_0}{(1 - \tilde{t}_0^2 \ln(1/\omega\tau))^{1/2}} \tag{5.42}$$

As in Model I coefficient $\tilde{t}$ grows when decreasing the frequency and the solution, Eq. (5.42), is not valid after $\tilde{t}$ becomes of the order of 1. This occurs at a much lower frequency than in Model I as a result of the higher power of $\tilde{t}_0$ in front of the logarithm in the denominator, which shows that the tendency to localization is weaker in systems with broken time reversal invariance.

A very unusual result follows from Eq. (5.41) for Model III (a system with spin-orbit impurities). In this case $\alpha = 1$ and we see that the resistivity that is proportional to $\tilde{t}$ decreases when frequency ω decreases. This result is in agreement with the corresponding result of perturbation theory discussed in Section 5.1. However, renormalization group treatment leads to a much stronger conclusion, namely, that resistivity goes to zero when $\omega \rightarrow 0$. (Of course, one should start with $\tilde{t}_0 \ll 1$. For $\tilde{t}_0 \geq 1$ localization behavior is possible at low frequencies.) This is because when $\tilde{t} \rightarrow 0$ the $\tilde{t}^2$ term is sufficient for describing the frequency dependence of resistivity at *all frequencies*. In this limit all higher-order terms fall off faster and can be neglected.

Then the solution, Eq. (5.41), is really the exact solution of the problem. The vanishing of resistivity in the limit $\omega \rightarrow 0$ is not a well-understood phenomenon and is somehow at odds with common sense. However, I want to emphasize that if one really believes that everything is fine with the one-parameter renormalization group discussed previously one should reconcile oneself to the possibility of zero resistance in a disordered system. It is relevant to mention that Eq. (5.40) (without the term β_4) was first obtained for Models I and II by Wegner (1979) by using the replica method. The diminishing of resistivity in

Model III in the first order was first obtained by Hikami, Larkin, and Nagaoka (1980), and the appropriate renormalization group equation was written by using the replica method by Efetov, Larkin, and Khmelnitskii (1980).

In $2 + \epsilon$ dimensions the parameter $\tilde{t}$ plays the role of dimensionless resistance of the system. In Models I and II a fixed point t_c, where the Gell-Mann–Low function $\beta(t)$ turns to zero, exists:

$$\beta(t_c) = 0 \tag{5.43}$$

At this point the total resistance of the sample does not depend on the sample size. The scaling approach is assumed to be applicable at distances larger than a minimal length ξ_0, which is of the order of inter-atomic distances at strong disorder or of the order of the mean free path at weak disorder. Accordingly one can introduce resistance t_0 of the sample with size ξ_0. All the dependence on the microscopic structure of the sample is contained in the value of t_0. Linearizing the Gell-Mann–Low function near the fixed point t_c, Eq. (5.43), we can solve this equation assuming that the resistance of the sample $t(\xi)$ is small when the size of the sample ξ is large. As a result one can find how the characteristic length (correlation length) ξ varies near the fixed point

$$\xi \sim \xi_0 \left(\frac{t_c - t_0}{t_c} \right)^{-1/y}, \qquad y = -\beta'(t_c) \tag{5.44}$$

Using Eqs. (5.44) for correlation length ξ, assuming that this length is the only characteristic length in the system and that the conductivity σ is proportional to $t_c^{-1}\xi^{2-d}$, one can write the equation for conductivity in the following form (Wegner 1976):

$$\sigma = A \frac{e^2}{\hbar \xi_0^{d-2}} \frac{1}{t_c} \left(\frac{t_c - t_0}{t_c} \right)^s, \qquad s = \frac{d-2}{y} \tag{5.45}$$

where A is a number of the order of unity, and the factor $e^2/\hbar \xi_0^{d-2}$ is the Mott conductivity. Resistance t_0 is a smooth function of microscopic disorder and Eq. (5.45) can be rewritten in terms of the impurity concentration. At $t_0 = t_c$ conductivity σ, Eq. (5.45), vanishes and this is the point of the Anderson metal–insulator transition. Eqs. (5.43–5.45) can be used for any dimensionality $d = (2 + \epsilon) > 2$. If ϵ is small, one may use only the first terms in the Gell-Mann–Low function $\tilde{\beta}(\tilde{t})$ written in Eq. (5.40). In this case the fixed points t_c for the different models are equal to

$$\tilde{t}_c = \begin{cases} d - 2, & \text{Model I} \\ (2(d-2))^{1/2}, & \text{Model II} \end{cases} \tag{5.46}$$

It is not difficult in this approximation to calculate the critical exponents y and s

$$y = \begin{cases} d - 2 \\ 2(d-2) \end{cases}, \qquad s = \begin{cases} 1 \\ 1/2 \end{cases} \tag{5.47}$$

In Model III the fixed point is absent provided the disorder is weak and $d - 2$ is small. Nevertheless, one can hardly imagine that the mobility edge does not exist in three dimensions.

The Gell-Mann–Low function $\tilde{\beta}$ was later calculated up to higher orders in t by Wegner (1989) and confirmed by Hikami (1990) using string theory methods. The corresponding derivation is too complicated to be presented here and I write the final result only. The function $\tilde{\beta}(\tilde{t})$ can be written for all three models as

$$
\tilde{\beta}(\tilde{t}) = \begin{cases} \epsilon\tilde{t} - \tilde{t}^2 - \frac{3}{4}\zeta(3)\tilde{t}^5 + O\left(\tilde{t}^6\right), & \text{orthogonal ensemble} \\[2mm] \epsilon\tilde{t} - \frac{1}{2}\tilde{t}^3 - \frac{3}{8}\tilde{t}^5 + O\left(\tilde{t}^7\right), & \text{unitary ensemble} \\[2mm] \epsilon\tilde{t} + \tilde{t}^2 - \frac{3}{4}\zeta(3)\tilde{t}^5 + O\left(\tilde{t}^6\right), & \text{symplectic ensemble} \end{cases} \qquad (5.48)
$$

where $\zeta(x)$ is the Riemann ζ-function, and $\zeta(3) \simeq 1,202$. (Recently Hikami (1992) calculated terms of the order of $\tilde{t}^6$ for the orthogonal and symplectic ensembles.)

Eqs. (5.48) help in estimating the accuracy of the results Eqs. (5.46, 5.47) in 3D. Using the first equation in Eqs. (5.48) one obtains for small ϵ the value of the critical resistance $\tilde{t}_c$ in the orthogonal case

$$
\tilde{t}_c = \epsilon - \frac{3}{4}\zeta(3)\epsilon^4 + O\left(\epsilon^5\right) \qquad (5.49)
$$

At $\epsilon = 1$ the value of $\tilde{t}_c$ given by the first two terms in Eq. (5.49) is ten times smaller than that given in the main approximation, Eq. (5.46). As concerns the exponent s introduced in Eq. (5.45) one easily finds for the orthogonal ensemble

$$
s = 1 - \frac{9}{4}\zeta(3)\epsilon^3 + O\left(\epsilon^4\right) \qquad (5.50)
$$

One can see that adding the second term in Eq. (5.50) even changes the sign of the exponent s at $\epsilon = 1$. The situation for the other two ensembles is no better. We see that by using the renormalization group scheme in $2 + \epsilon$ dimensions and then extrapolating to $\epsilon = 1$ one can hardly obtain a good description of the Anderson metal–insulator transition in 3D and therefore other approaches are necessary. One should remember also that the $2 + \epsilon$ treatment of models of statistical physics always gives second-order phase transitions. So if this treatment were always correct, first-order transitions would not exist in nature.

Recently, studying the nonlinear σ-models appropriate for disorder problems, Kravtsov, Lerner, and Yudson (1988, 1989) discovered that in the procedure of $(2 + \epsilon)$-expansion higher-order gradient operators neglected in the previous discussion can become very important. Usually such operators are not considered because their "naive" dimension $d - 2n$ is negative at $n > 1$ and they seem to be irrelevant. However, a more sophisticated analysis shows that in the one-loop order the dominant scalar operator with $2n$ gradients in the orthogonal ensemble has a positive correction proportional to $n(n-1)$. If n is large enough, the dimensionality of the corresponding gradient operators becomes positive and these high gradients should be taken into account. The same phenomenon was found later for the σ-model describing the unitary ensemble (Lerner and Wegner (1990)), and even more, Wegner (1990) found the same effect in the standard N-vector model. Now, the fact that the higher gradients can be relevant seems to be general for σ-models and has also been confirmed in the two-loop approximation by Castilla and Chakravarty (1993). All this means that the conventional results with the one-parameter renormalization group can drastically change when higher-order gradients are taken into account. It is not clear what the correct results are, but one should be very careful when trying to draw a physical picture from a continuous σ-model using the renormalization group technique. My personal opinion is that one should not trust results obtained from the $(2 + \epsilon)$-expansion unless they are confirmed by other methods. An alternative consideration based on some kind of effective medium approximation will be presented in Chapter 12 and the results obtained there are in very strong disagreement with the results presented in this section.

The calculations carried out in this chapter show that by using the supermatrix σ-model one can reproduce all the results obtained by diagram expansions or replica methods in a straightforward and rather simple manner. The difficulties related to analytical continuations over the number of fields met in the replica method do not exist here. Of course, such an improvement is not very important for problems that can be treated by perturbation theory or renormalization group calculations, and the present chapter was written to "warm up" before beginning essentially nonperturbative calculations. In the rest of the book nontrivial problems for which the supersymmetry technique is really necessary will be considered. Most of the results presented in the next chapters were obtained first by the supersymmetry method, and for many of the problems this technique is the only one now available that makes it possible to get analytical results.

6

Energy level statistics

6.1 Random matrix theory

6.1.1 General formulation

According to the basic principles of quantum mechanics the energy spectrum of a particle in a limited volume is discrete. The positions of the energy levels and the spacings between them depend on the boundary conditions and interactions in the system. In the simplest cases these quantities can be calculated exactly or approximately. However, often the inter-actions are so complicated that calculations for the levels become impossible. On the other hand, the complexity and variety of interactions lead to the idea of a statistical description in which information about separate levels is neglected and only averaged quantities are studied. Density of states, energy level and wave function correlations, and the like, can be so considered. The analogous approach is used in statistical physics, where information about separate particles is neglected and only averages over large number of particles are calculated.

The idea of statistical description of the energy levels was first proposed by Wigner (1951, 1958) for study of highly excited nuclear levels in complex nuclei. In such nuclei a large number of particles interact in an unknown way and it is plausible to assume that all the interactions are equally probable. Of course, the first question one can ask is what the characterization "equally probable" means and therefore one should introduce a measure for averaging. Using the analogy with statistical physics, one should formulate a statistical hypothesis in terms of a probability distribution that would play the role of the Gibbs distribution.

Any energy spectrum can be characterized by mean level spacing and fluctuations. When considering high energy levels the mean level spacing can be the same for different parts of the spectrum and only its fluctuations are of interest. In many cases the mean level spacing slowly varies as a function of the energy, but if the scale of the variation is larger than the scale of correlations between the levels it can be removed from the description after a proper rescaling. The level fluctuations are more universal than the dependence of the mean level spacing on energy and are determined by the symmetries of the Hamiltonian describing the system.

Choosing a complete set of eigenfunctions as a basis, one represents the Hamiltonian $\mathcal{H}$ as a matrix. Within the statistical description one can assume that all the matrix elements $\mathcal{H}_{mn}$ of a system having N relevant energy levels are random variables that can take any values with equal probabilities. However, this assumption leads to divergent integrals in calculated averages at large $\mathcal{H}_{mn}$. Large matrix elements would correspond to a finite probability of having arbitrarily strong interactions. To exclude this nonphysical possibility one should introduce a cutoff. It is very important that this cutoff be introduced in an invariant manner that does not depend on the basis chosen. In the language of the random matrices $\mathcal{H}_{mn}$ the corresponding weight function for integration over the matrix elements can contain only tr $f(\mathcal{H})$, where f is a function.

The first statistical theory (Wigner 1951) was based on a Gaussian distribution. According to the Gaussian statistical hypothesis a physical system having N quantum states has the statistical weight $\mathcal{D}(\mathcal{H})$:

$$\mathcal{D}(\mathcal{H}) = A \exp\left[-\sum_{m,n=1}^{N} \frac{|\mathcal{H}_{mn}|^2}{2a^2}\right] = A \exp\left[-\frac{\mathrm{tr}\,\mathcal{H}^2}{2a^2}\right] \tag{6.1}$$

In Eq. (6.1) the parameter a is the cutoff excluding the strong interactions, $\mathcal{H}$ is a random $N \times N$ matrix, and A is the normalization coefficient.

The weight $\mathcal{D}(\mathcal{H})$ is rather arbitrary and one can invent many other invariant forms of the distribution. Which one is better? The answer given by the random matrix theory (RMT) is that, although the dependence of the mean level spacing $\Delta(\varepsilon)$

$$\Delta^{-1}(\varepsilon) = \langle \mathrm{tr}\,\delta(\varepsilon - \mathcal{H})\rangle_{\mathcal{D}} \tag{6.2}$$

on the energy ε is different for different $\mathcal{D}(\mathcal{H})$, level fluctuations described, for example, by the level–level correlation function $R(\omega)$

$$R(\omega) = \Delta^2(\varepsilon) \langle \mathrm{tr}\,\delta(\varepsilon - \mathcal{H})\,\mathrm{tr}\,\delta(\varepsilon - \omega - \mathcal{H})\rangle_{\mathcal{D}} \tag{6.3}$$

are universal in the limit $N \to \infty$ provided the energy ω is much smaller than the characteristic scale of the variation of $\Delta(\varepsilon)$. In Eqs. (6.2, 6.3) the symbol $\langle\ldots\rangle_{\mathcal{D}}$ stands for the averaging with a distribution $\mathcal{D}(\mathcal{H})$. For example, the distribution function $\mathcal{D}(\mathcal{H})$, Eq. (6.1), gives a nontrivial function $\Delta^{-1}(\varepsilon)$ (semicircle Wigner law). This function is nonzero on a finite interval $(-\varepsilon_0, \varepsilon_0)$ only, but provided ε is not close to the ends of the interval one gets universal formulae for the function $R(\omega)$.

According to Wigner one has to distinguish among three classes of universality. A system without magnetic or spin-orbit interactions possesses time-reversal and spin-rotation invariance and can be described by symmetric random matrices $\mathcal{H}$. Hence, one should integrate with $\mathcal{D}(\mathcal{H})$ remembering that $\mathcal{H}_{mn} = \mathcal{H}_{nm}$. The ensemble of the symmetric matrices with the Gaussian distribution is often called the *Gaussian orthogonal ensemble* (GOE).

In the presence of magnetic interactions the time-reversal and spin-rotation symmetry is violated and the wave functions are no longer real. This means that one should deal with general Hermitian matrices without any additional symmetry and integrate over the matrix elements $\mathcal{H}_{mn}$ using only the constraint $\mathcal{H}_{mn} = (\mathcal{H}_{nm})^*$. This system is called the *Gaussian unitary ensemble* (GUE).

The third possible type of symmetry arises when the system is time-reversal-invariant but does not have central symmetry. In this case it is also impossible to make all the matrix elements real. Nevertheless an additional symmetry exists in this case, too. According to the Kramers theorem all levels in the system remain doubly degenerate. Hence, every eigenvalue of the matrix $\mathcal{H}$ must appear twice. Matrices consisting of real quaternions $\mathcal{H}_{mn}$ of the form

$$\begin{pmatrix} p_{mn} & q_{mn} \\ -q_{mn}^* & p_{mn}^* \end{pmatrix} \tag{6.4}$$

and satisfying the condition $\mathcal{H}_{mn} = (\mathcal{H}_{nm})^+$ have this property. The corresponding ensemble is called the *Gaussian symplectic ensemble* (GSE).

In the Wigner statistical theory based on the Gaussian ensembles the interactions corresponding to different matrix elements of the Hamiltonian are not equally weighted and, in addition, one has to use a phenomenological cutoff a. Because of these unsatisfactory features another approach was developed by Dyson (1962a), who suggested characterizing the system not by its Hamiltonian but by a unitary $N \times N$ matrix S whose elements give the transition probabilities between the different states, where, again, N is the number of the levels. This matrix is related to the Hamiltonian $\mathcal{H}$ of the system in a complicated way that is not specified in the theory. As an example one can write two possible relations

$$S = (1 + i\mathcal{H})(1 - i\mathcal{H})^{-1}, \qquad S = \exp(i\mathcal{H}) \tag{6.5}$$

An arbitrary unitary matrix S can be diagonalized

$$S = Z S_0 Z^+ \tag{6.6}$$

where $(S_0)_{ab} = \delta_{ab} \exp(i\theta_a)$, Z is a unitary matrix

$$Z^+ Z = 1 \tag{6.7}$$

According to the Dyson hypothesis the correlation properties of n successive energy levels of the system ($n \ll N$) are statistically equivalent to those of n successive angles θ_a. All the unitary matrices S are assumed to have equal probabilities. Again, depending on the symmetry, one can distinguish among the three different ensembles. In the absence of magnetic and spin-orbit interactions all elements of matrix S can be chosen to be real; this means that the matrix S is orthogonal. The corresponding ensemble is called the *circular orthogonal ensemble* (COE). In the presence of magnetic interactions no additional symmetry is imposed on the unitary matrix S and the ensemble is called the *circular unitary ensemble* (CUE). A time-reversal-invariant system with broken central symmetry is described by unitary matrices containing as elements the real quaternions, Eq. (6.4). Such matrices are called *symplectic* and the corresponding ensemble is called the *circular symplectic ensemble* (CSE).

In the Wigner theory, when averaging with the distribution $\mathcal{D}(\mathcal{H})$, Eq. (6.1), one should integrate over all the matrix elements $\mathcal{H}_{mn}$ as if they were independent variables. Let us discuss now how to integrate over the elements of the unitary matrix S in the Dyson circular ensembles. Of course, introducing an integration measure is one more hypothesis. However, one can obtain the measure in a quite natural way assuming that all the unitary matrices S are equally probable. One may consider an $N \times N$ complex matrix as a point in the $(2N)^2$-dimensional space of real numbers. Then, all possible unitary $N \times N$ matrices constitute a hypersurface $\{S\}$ in this space. The equal probability of all matrices S in calculating average quantities implies equal contributions of all elementary volumes on the hypersurface $\{S\}$. In other words, when calculating averages one should integrate over the hypersurface $\{S\}$ choosing an appropriate coordinate basis.

In many cases one calculates correlation functions of energy levels that can be described by the angles θ_a. For such calculations the coordinates θ_a and Z_{ab}, Eq. (6.6), are convenient. The elementary volume on the hypersurface $\{S\}$ can be written easily in these coordinates. As discussed in Section 2.5 it is convenient first to write the square of the elementary length dl

$$dl^2 = \text{tr}\left(d\,S d\,S^+\right) = 2\,\text{tr}\left(\delta Z \delta Z^+ - \delta Z S_0^+ \delta Z^+ S_0\right) + \sum_a (d\theta_a)^2 \qquad (6.8)$$

where $\delta Z = Z^+ dZ = -(\delta Z)^+$, meaning that the matrix δZ is anti-Hermitian. Using this property one can rewrite Eq. (6.8) in the more convenient form

$$(dl)^2 = \sum_{a,b} \left(\delta Z_{ab}(\delta Z_{ab})^+\right)|e^{i\theta_a} - e^{i\theta_b}|^2 + \sum_a (d\theta_a)^2 \qquad (6.9)$$

The quadratic form $(dl)^2$, Eq. (6.9), is written for all ensembles. In the COE and CUE δZ_{ab} are real and complex numbers, respectively; in the CSE these elements are real quaternions. Because the quadratic form, Eq. (6.9), is diagonal with respect to the elements δZ_{ab} and $d\theta_a$, we easily write the corresponding elementary volume. Taking into account the structure of the elements δZ_{ab} for each ensemble one can write the elementary volume ΔS in the form

$$\Delta S = \prod_{a<b} |e^{i\theta_a} - e^{i\theta_b}|^\beta \prod_a d\theta_a \Delta Z \qquad (6.10)$$

where $\beta = 1$ for the orthogonal ensemble, $\beta = 2$ for the unitary ensemble, and $\beta = 4$ for the symplectic ensemble. The vectors δZ_{ab} constitute the volume ΔZ. Calculating averages of arbitrary functions $f(\theta_1, \theta_2, \ldots, \theta_N)$ one can integrate separately over θ_a and δZ_{ab}. Therefore, the volume of the group of the matrices Z is not part of the result of such calculations. Let us notice that the elementary volume ΔS, Eq. (6.10), is invariant with respect to rotations of the form $S \to Z_1 S Z_2^+$, where Z_1 and Z_2 are, depending on the ensemble, arbitrary orthogonal, unitary, or symplectic matrices.

The distribution Eq. (6.10) makes it possible to calculate any correlation function of the angles θ_a and hence those of energy levels. The average density of states is the simplest correlation function. Eq. (6.10) contains only differences $\theta_a - \theta_b$, indicating that the density of states is a constant. Intuitively, one expects that in the limit $N \to \infty$ the level fluctuations of the circular and Gaussian ensembles must be the same provided one considers a small energy interval in the Gaussian ensembles where the average density of states does not change much. Formally, one can choose as independent variables of the integration over the matrices $\mathcal{H}$ their eigenvalues r_a and the unitary matrices U diagonalizing $\mathcal{H}$. Repeating the calculations that lead to Eq. (6.10), for the elementary volume $\Delta\mathcal{H}$ of the integration over these variables one obtains

$$\Delta\mathcal{H} = \prod_{a<b} |r_a - r_b|^\beta \prod_a dr_a \Delta U \qquad (6.11)$$

where ΔU is the elementary volume analogous to ΔZ in Eq. (6.10). In terms of the eigenvalues r_a the weight $\mathcal{D}(\mathcal{H})$, Eq. (6.1), takes the form

$$\mathcal{D}(\mathcal{H}) = A \exp\left[-\frac{1}{2a^2}\sum_{b=1}^{N} r_b^2\right] \qquad (6.12)$$

Eqs. (6.10) and (6.11, 6.12) are very similar. In the circular ensembles the variables $\exp(i\theta_a)$ change on the unit circumference; in the Gaussian ensemble large values of the variables r_a do not contribute because of the weight $\mathcal{D}(\mathcal{H})$. The fluctuations are determined mainly by the products of $|e^{i\theta_a} - e^{i\theta_b}|^\beta$ and $|r_a - r_b|^\beta$ in Eqs. (6.10, 6.11). One can study the level fluctuations by calculating the level–level correlation functions $R(\omega)$, Eq. (6.3). Integrations with the measures given by Eqs. (6.10, 6.11) are very difficult and many efforts have been made to calculate the function $R(\omega)$ (Dyson (1962a), Mehta and Dyson (1963), Porter (1965)). In fact, calculations for the orthogonal and the symplectic ensembles require a great deal of mathematical art. Only the unitary ensemble is simpler and the corresponding calculations are much easier. I do not want to present here the whole story because a very complete book by Mehta (1991) has possible calculations for the random matrices. It is proved that the correlation functions $R(\omega)$ in the limit $N \to \infty$ are the same for the Gaussian and circular ensembles. For the orthogonal, unitary, and symplectic ensembles this function is described by the famous formulae

$$R_{\text{orth}}(\omega) = 1 - \frac{\sin^2 x}{x^2} - \frac{d}{dx}\left(\frac{\sin x}{x}\right) \int_1^\infty \frac{\sin xt}{t} dt \tag{6.13}$$

$$R_{\text{unit}}(\omega) = 1 - \frac{\sin^2 x}{x^2} \tag{6.14}$$

$$R_{\text{sympl}}(\omega) = 1 - \frac{\sin^2 x}{x^2} + \frac{d}{dx}\left(\frac{\sin x}{x}\right) \int_0^1 \frac{\sin xt}{t} dt \tag{6.15}$$

where $x = \pi\omega/\Delta$, and Δ is the mean energy level spacing.

Properties of Eqs. (6.13–6.15) will be discussed later. In the present section they are presented only as an illustration of the basic results of the RMT. Let us note only that the correlation functions $R(\omega)$, Eqs. (6.13–6.15), turn to zero as x^β as $x \to 0$. That means that the probability of finding a level at the distance ω from another level decays at small ω. This effect is known as "level repulsion." The agreement of the results given by the Gaussian and circular ensembles indicates that the correlation properties of the level fluctuations can be rather universal; this has been confirmed by early numerical studies of the random matrices (Porter 1965).

6.1.2 Applications of RMT

The RMT is purely phenomenological and nothing is assumed about the origin of the randomness. This property motivated Gorkov and Eliashberg (1965) to conjecture that the RMT could describe small disordered metallic particles as well. The randomness in a system of such particles can be due to both impurities inside the particles and irregularities of the shape. At the same time the volumes of the particles can be very close to each other. By now such systems have numerous experimental realizations (Halperin (1986)) and are under intensive study. The size of the particle can vary from 10 Å to 100–1000 Å, values that correspond to the mean level spacing Δ up to 10K. This makes observation of quantum size effects possible.

The thermodynamic properties of small metallic particles have been considered by Frölich (1937) who assumed that the energy levels were equally spaced, thereby naturally leading to exponential behavior of thermodynamic quantities at low temperature. Such an assumption can hold, for instance, for ideal spherical particles, but, as argued by Kubo (1962), even very small irregularities (with size of the order of atomic distances) of the shape must lead to lifting of all degeneracies of eigenstates. As a result, the position of the energy levels becomes random and later Kubo (1969) suggested that the spacing distribution had to follow the Poisson law. In the Kubo theory the mean level spacing Δ is inversely proportional to the volume V of the particle $\Delta = (\nu V)^{-1}$, where ν is the density of states at the Fermi surface of the metal. The main difference between the Poisson law and the RMT is that the level repulsion is absent in the former theory. Taking into account that the levels are separated by gaps in the Frölich theory one can conclude that all possible conjectures about the level distribution have been made.

I want to emphasize that these three approaches to the level statistics in the system of small particles are hypotheses and it is the supersymmetry technique that made it possible (Efetov (1982b)) to derive correlation functions for this system. The results are in excellent agreement with the corresponding formulae of the RMT and their derivation is presented in the subsequent sections of this chapter.

It turns out also that the zero-dimensional version of the σ-model can be derived directly from the RMT by using the models of the Gaussian ensembles (Verbaarschot, Weidenmüller, and Zirnbauer (1985)); that relationship establishes analytically the equivalence of the both theories. In systems of small metallic particles, in contrast to nuclear spectra, there is a possibility of realizing all three ensembles in practice. Spectra of particles with potential impurities only should be described by the orthogonal ensemble. By adding a magnetic field or magnetic impurities one obtains the unitary ensemble. A system with potential and spin-orbit impurities belongs to the class of the systems described by the symplectic ensemble. By changing the magnetic field one can have complete crossover between orthogonal and unitary ensembles or between symplectic and unitary ensembles.

Thinking about the application of RMT to the level statistics in small metal particles one can get an impression that this theory can always be applied to such systems. However, this is far from so. Of course, it cannot be applied, for example, to ideal spherical particles because of the absence of any randomness but it also cannot be applied in the limit of a strong disorder when the wave functions are localized near centers of localization. In this case, nondiagonal matrix elements corresponding to transitions between the states localized at different centers are very small, contrary to the main assumption of the RMT. The phenomenon of localization is the reason why the RMT cannot also be applied to the study of level statistics in one-dimensional wires of finite length however weak the disorder is. Fortunately, these systems can be successfully studied with the σ-model with proper dimensions not necessarily equal to zero. The supersymmetry method is more general than the RMT and gives the capacity to obtain results starting from the first principles.

One more popular field of research where the RMT has been applied is so-called quantum chaos. I do not want to try to present here a precise definition of this notion and imply by the words *quantum chaos* behavior of quantum systems which are in the classical limit in the regime of chaotic dynamics. It is quite clear now that many classical systems show chaotic behavior that means that trajectories are extremely unstable under both a change of initial conditions and small perturbations of the Hamiltonian. The distances between all pairs of trajectories originating in neighboring points of the phase space increase exponentially with time. This happens if the system is not integrable and a complete set of integrals of the motion does not exist.

Much is known both analytically and numerically about such classical systems (Collet and Eckmann (1980), Cornfeld, Fomin, and Sinai (1982), Lichtenberg and Lieberman (1983)), but little is known analytically about their quantum counterparts. At the same time numerical study and first analytical works revealed very interesting universal features of quantum

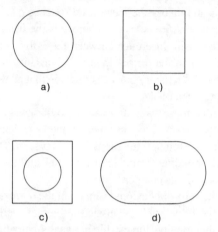

Fig. 6.1. a) and b) Billiards are integrable; c) and d) motion in the billiards is chaotic.

behavior (for reviews of chaos see Bohigas and Gianonni (1984), Casati (1985), Seligman and Nishioka (1986), Gutzwiller (1991), Bohigas (1991), Haake (1992)).

It turns out that each time a classical system is fully chaotic its quantum analogue obeys the Wigner–Dyson statistics. This was confirmed by comparison of different correlation functions calculated numerically for quantum systems with the corresponding functions calculated from RMT. In classically integrable systems characterized by a complete set of integrals of motion the level statistics are described by the Poisson law. Asymptotics of correlation functions in the semiclassical limit have also been obtained analytically, starting from the Gutzwiller trace formula (Berry (1985), Smilansky (1991), Argaman, Imry, and Smilansky (1993), Serota (1992, 1994)). The randomness in the system originates from the chaos but the system itself can be very simple.

Let us give several examples of systems that behave differently. The simplest systems for which the problem of chaotic versus regular motion can be discussed are billiards. Particles can move inside such billiards, being scattered elastically by the walls. In Fig. 6.1 four different billiards are represented.

The quantum analogues of these billiards (quantum billiards) are defined in terms of the Schrödinger equation for a free particle with the condition that the wave function is zero at the boundaries. The energy levels can be calculated easily for the circle and square billiards (Figs. 6.1a and 6.1b), and one can average over different parts of the spectra, which results in the Poisson law for the distribution of level spacings. The billiard in Fig. 6.1c is called a Sinai billiard, and it has been proved rigorously (Cornfeld, Fomin, and Sinai (1982)) that the classical motion in this system is chaotic. An intensive numerical study of the corresponding quantum system (Bohigas and Gianonni (1984)) manifested very good agreement with the predictions based on the GOE. The motion in the stadium billiard in Fig.6.1d is also chaotic and the level statistics is described by the GOE. The regular motion in the square billiard can become chaotic if one applies a magnetic field. At the same time application of the magnetic field to the circular billiard cannot change the regular character of the motion.

From the point of view of application to experiments quantum billiards are very interesting because of the modern development of nanotechnology, which made it possible to fabricate small devices (so-called quantum dots) that can be modeled by the billiards. (For reviews of mesoscopic physics see Beenakker and van Houten (1991), Altshuler, Lee, and Webb (1991),

Washburn and Webb (1986).) Another type of experiment that can be described by the quantum billiards model is microwave scattering in cavities (Stöckmann and Stein (1990), Stein and Stöckmann (1992), Doron, Smilansky, and Frenkel (1990), Sridhar (1991)). The difference between the Schrödinger and Maxwell equations is not very important although the boundary conditions can be different. Chaos in experiments on ultrasound was studied by Weaver (1989).

One more application of RMT is in the study of energy levels of the hydrogen atom in a strong magnetic field; the system is also known to be chaotic in the classical limit. Choosing the z-axis along the magnetic field $\mathbf{H}$ and writing the vector potential $\mathbf{A}$ as $\mathbf{A} = -(1/2)[\mathbf{r} \times \mathbf{H}]$ (symmetric gauge) the Hamiltonian $\mathcal{H}$ in atomic units has the form

$$\mathcal{H} = \frac{\hat{\mathbf{p}}^2}{2} - \frac{1}{r} + \frac{\gamma}{2} L_z + \frac{\gamma^2}{8} \left(x^2 + y^2 \right) \tag{6.16}$$

where γ is the reduced magnetic field H/H_c, $H_c = 2.35 \times 10^5$ T, and L_z is the angular momentum operator.

In the two limiting cases of zero and infinite field strength the system described by the Hamiltonian, Eq. (6.16), is integrable and does not show chaotic behavior. It is the region of intermediate fields where the motion is chaotic. An intensive numerical study of the system by Delande and Gay and by Friedrich (Taylor (1987)) revealed very impressive agreement with the GOE prediction.

It can take a lot of space to list all applications of RMT and it is not the task of the present book. However, it is difficult not to mention results concerning properties of the Riemann ζ-function defined for Re $z > 1$ as

$$\zeta(z) = \sum_{n=1}^{\infty} n^{-z} = \prod_{p} \left(1 - p^{-z} \right)^{-1} \tag{6.17}$$

and for other values of z by its analytical continuation. The product in Eq. (6.17) is taken over all primes. The function $\zeta(z)$ turns to zero for $z = -2n$, $n = 1, 2, \ldots$, and these are the "trivial" zeros. All the other zeros of $\zeta(z)$ lie in the strip $0 < $ Re $z < 1$ and are symmetrically located with respect to the line Re $z = 1/2$. According to the Riemann hypothesis (Riemann (1876)) these zeros actually lie on this line. Assuming that this hypothesis is correct and representing the "nontrivial" zeros in the form $z = 1/2 + i\gamma_n$ one can ask how the γ_n are distributed on the real axis. This question turns out to be very important for various mathematical applications.

Trying to answer this question, Montgomery (1973, 1974) conjectured that the two-point correlation function $R_2(r)$ of the zeros of the ζ-function on the line Re $z = 1/2$ is described by Eq. (6.14) for the unitary ensemble with $x = \pi r$. This can mean that the zeros of the ζ-function are related to the eigenvalues of a Hermitian operator. The exciting relation between the zeros of the ζ-function and RMT has been checked in a number of studies. Not long ago 79 million zeros around $N = 10^{20}$ were obtained (Odlyzko (1987, 1989)) and the agreement of the two-point correlation function with Eq. (6.14) is excellent. For more detail one can also read the book by Mehta (1991).

As concerns the problems of electron motion in random potential they can be considered a particular case of problems of quantum chaos because there can be no doubt that very complex configurations of the impurity potential must lead chaotic behavior.

I hope that the discussion presented in this section shows that many very interesting problems of nuclear, atomic, and condensed matter physics and even mathematics can be studied by using RMT. In spite of the great success of this theory there are two main shortcomings in it. First, this theory is purely phenomenological and its relevance to different phenomena is usually confirmed by computer calculations. Second, manipulations with random matrices having arbitrarily large size are not simple and require a lot of mathematical skill. It will be demonstrated in the following sections that the zero-dimensional σ-model makes it possible to consider disorder problems starting from first principles and analytically confirm the applicability of RMT at least for such systems. It will be shown also that the zero-dimensional σ-model can be derived from the Gaussian ensembles in the limit of large matrix size N. Calculations with the supersymmetry technique are much simpler than those presented, for example, by Mehta (1991) and in many cases it is convenient to use this scheme also for study of problems initially formulated in terms of random matrices.

6.2 Zero-dimensional supermatrix σ-model

6.2.1 Diffusion in a limited volume

Let us consider a metallic particle of size L. The value L is supposed to be much larger than atomic lengths and can even be arbitrarily large provided the zero temperature limit is considered. At finite temperatures size L is supposed to be less than the diffusion length $L_0 \sim (D_0 \tau_\varphi)^{1/2}$, where τ_φ is the inelastic scattering time, and D_0 is the diffusion coefficient. Inside the sample there are impurities with random positions that lead to electron scattering with a mean free path l. In principle, the metal can be so clean or, alternatively, the size can be so small that the electrons in the particle are scattered mainly by boundaries; this indicates that the mean free path l is of the same order as a. In this model the level–level correlation function $R(\omega)$ is an interesting quantity. This function has been defined already, in Eqs. (3.36, 3.37). In all cases of interest considered later the average of the product of the Green functions in Eq. (3.37) does not depend on ε and the definition Eq. (3.36) is the same as that given by Eq. (6.3). In the coordinate representation the correlation function $R(\omega)$ takes the form

$$R(\omega) = \frac{1}{16\pi^2 \nu^2 V^2} \left\langle \sum_{\sigma,\sigma'} \int \left(G^A_{\varepsilon-\omega}(y, y) - G^R_{\varepsilon-\omega}(y, y) \right) d\mathbf{r} \right.$$

$$\left. \times \int \left(G^R_\varepsilon(y', y') - G^A_\varepsilon(y', y') \right) d\mathbf{r}' \right\rangle \qquad (6.18)$$

where $y = (\mathbf{r}, \sigma)$, σ is the spin variable.

When deriving the nonlinear σ-model in the preceding chapters the inequality $\varepsilon_0 \tau \gg 1$, where ε_0 is the Fermi energy (or just energy if one speaks about properties of the spectrum of the Schrödinger equation near an energy ε), was used. To study the level statistics in a finite volume one more inequality needs to be imposed:

$$\tau^{-1} \gg \Delta \qquad (6.19)$$

where τ is the elastic mean free time. If the scattering by the boundaries is dominant, the parameter τ is estimated as $\tau \sim L/v_0$, where v_0 is the Fermi velocity.

The mean level spacing Δ in the absence of spin interactions (Models I and IIa according to the classification in Section 5.2) is equal to

$$\Delta = (\nu V)^{-1} \qquad (6.20)$$

Eq. (6.20) is valid for any shape of the metallic particle except ideally spherical, cylindrical, or cubic. In the latter cases the spacing between degenerate levels is rather large. However, any irregularities due to impurities or to a nonperfect shape, even of an atomic size, will split these degenerate levels and lead to Eq. (6.20). Recalling the discussion of the previous section, one can also say that each time the classical motion in the system is chaotic one obtains Eq. (6.20) for the mean level spacing. In Models I and IIa the level degeneracy is partially conserved because as a result of the absence of spin interactions the direction of the spin is arbitrary. In other words each level is double degenerate and Eq. (6.20) describes the distance between these degenerate levels. Spin interactions remove this spin degeneracy. As a result, in the models with spin interactions the mean level spacing $\bar{\Delta}$ is one half of the previous one

$$\bar{\Delta} = (2\nu V)^{-1} \tag{6.21}$$

Of course, Eq. (6.21) is valid provided the spin interactions are strong enough that they completely mix the states with "up" and "down" spin.

Inequality (6.19) corresponds to sufficiently strong level mixing. It is always fulfilled in two-dimensional and three-dimensional samples because $\tau^{-1} \geq v_0/L$ but $\Delta \sim L^{-d}$, where d is the dimensionality; that means that the latter quantity decreases faster with the sample size than the former quantity does. It is also valid for wires with finite length. The word *wire* will mean everywhere a sample with one-dimensional geometry but with diameter of the cross section much exceeding atomic units. For one-dimensional chains the inequality (6.19) is equivalent to the requirement that the mean free path l must be much less than the length L of the chain.

As discussed in Chapter 3 calculation of the averages $\langle G^R G^R \rangle$ and $\langle G^A G^A \rangle$ in the limit $\varepsilon_0 \tau \gg 1$ is reduced to calculation of the products of the averages $\langle G^R \rangle \langle G^R \rangle$ and $\langle G^A \rangle \langle G^A \rangle$. Calculating, for example, the quantity $\langle G_\varepsilon^{R,A} \rangle$ with standard diagrammatic technique (Abrikosov, Gorkov, and Dzyaloshinskii (1963)) one obtains

$$\langle G_\varepsilon^{R,A} (\mathbf{r}, \mathbf{r}) \rangle = \sum_m \frac{1}{\varepsilon - \varepsilon_m \pm i/(2\tau)} \tag{6.22}$$

$$= \int \frac{1}{\varepsilon - \varepsilon(\mathbf{p}) \pm i/(2\tau)} \frac{d^d \mathbf{p}}{(2\pi)^d}$$

The substitution of the sum over the energy levels by the integral over the momenta, as if an infinitely large sample was considered, is possible because of the inequality (6.19). Therefore, all physical quantities determined by the one-particle Green functions are not sensitive in the limit (6.19) to the discreteness of the levels and are well described by the corresponding classical formulae written for a macroscopic sample. The averaged Green functions $G^{R,A}$, Eq. (6.22), can be written as in Section 3.3 as

$$\langle G_\varepsilon^{R,A} (\mathbf{r}, \mathbf{r}) \rangle = \bar{A} \mp \nu \int \frac{d\xi}{\xi \mp i/(2\tau)} = \bar{A} \mp i\pi\nu \tag{6.23}$$

where $\bar{A}$ is a constant renormalizing the electron energy.

Then a natural question arises: How is the level discreteness felt if this information is lost in the averaged Green functions? The averaged Green functions give the classical formulae and then, where does quantum mechanics show? Of course, averaging over irregularities has been done but for each configuration of irregularities the spectrum remains discrete and, for example, the level–level correlation function $R(\omega)$ must "remember" the discreteness even after averaging has been done. It turns out that the diffusion modes considered in Chapter 3 are the signature of quantum mechanics on a mesoscopic scale. The electron waves interfere because they are multiply scattered chaotically and the way they scatter is sensitive to the finiteness of the volume.

In Chapter 3 we have seen that, formally, the diffusion modes manifest themselves after summation of a certain class of diagrams for averages of the type $\langle G_{\mathcal{E}}^R G_{\mathcal{E}-\omega}^A \rangle$. These are cooperons and diffusons, written for example, in Eqs. (3.52, 3.60). A typical first-order contribution of these modes can be represented in the form

$$\frac{1}{2\pi \nu V} \sum_{\mathcal{E}} \frac{1}{\mathcal{E} - i\omega} \tag{6.24}$$

In the absence of a magnetic field and magnetic or spin-orbit impurities in the limit of infinite volume V one can replace the sum over all energies $\mathcal{E}$ by the integral over momenta and in the denominator in Eq. (6.24) write

$$\mathcal{E} = D_0 \mathbf{k}^2 \tag{6.25}$$

However, in a finite volume the momentum $\mathbf{k}$ is quantized and this quantization is important as soon as

$$\omega \ll \frac{D_0}{L^2} \tag{6.26}$$

Only in the opposite limit can one replace the sum in Eq. (6.24) by the integral; then the contribution of the diffusion modes is, at least in three-dimensional space, negligible. The lowest energy $\mathcal{E}_0 = 0$ corresponds to the zero space harmonics. Higher harmonics are of the order of D_0/L^2. In the limit determined by Eq. (6.26) the main contribution is from the zero harmonics because the contribution of any higher harmonics can be estimated as

$$\frac{L^2}{\nu V D_0} \sim \left(p_0^2 L l \right)^{-1} \ll 1 \tag{6.27}$$

The sum of the higher harmonics can make a large contribution provided the effects of weak localization are important but this situation is not considered in this chapter.

Everywhere in this chapter the inequality (6.26) is assumed to be fulfilled. At this limit one can get universal statistics. At the same time to study level correlations one has to be able to consider arbitrary values of ω/Δ; that is possible when

$$\mathcal{E}_c = \frac{\pi^2 D_0}{L^2} \gg \Delta \tag{6.28}$$

The energy $\mathcal{E}_c$ is often called the Thouless energy. It was introduced by Thouless (1975, 1977) to characterize the sensitivity of levels to changing of boundary conditions. The inequality (6.28) for three-dimensional samples is the same as (6.27) and can be fulfilled easily unless the sample is very dirty and small. It is also valid in 2D in the limit $\tau \varepsilon_0 \gg 1$. The inequality (6.28) can also be valid for wires with finite length if

$$L \ll L_c \tag{6.29}$$

where $L_c \sim \nu D_0 S$, and S is the cross section of the wire. As will be seen in Chapter 11 length L_c is the localization length in the wire. Considering the first weak localization correction one can see that in two dimensions keeping only zero harmonics is justified if the sample size is much less than the two-dimensional localization length. In other words, we can say that the calculations presented in this chapter are valid when the electron motion is diffusive. Any type of electron localization makes use of the zero-dimensional σ-model unjustified.

6.2.2 Correlation functions through definite integrals over supermatrices

In Chapter 4 it was demonstrated that the diffusion modes correspond to fluctuations of the supermatrix Q, the fluctuations being described by the nonlinear σ-model, Eq. (4.55). Keeping only the zero space harmonics means that the supermatrix does not vary in space and the first term in Eq. (4.55) can be neglected. This is the zero-dimensional σ-model, which is the simplest version because when calculating physical quantities one has to calculate definite integrals over Q and not functional ones.

Now, we have to express the products $\langle G^R G^A \rangle$ entering the function $R(\omega)$, Eq. (6.18), in terms of an integral over the supermatrices Q. It can be done as in Chapter 4 by expressing the Green functions in terms of the integrals over the supervectors ψ, averaging over the random potential, and decoupling the effective interaction $(\bar{\psi}\psi)^2$ by integration over the supermatrices Q.

Then the integrals of the averages of the products of the retarded and advanced Green functions entering Eq. (6.18) can be written as

$$\int \langle G^A_{\varepsilon-\omega}(\mathbf{r}, \mathbf{r}) G^R_\varepsilon(\mathbf{r}', \mathbf{r}') \rangle d\mathbf{r} d\mathbf{r}' \tag{6.30}$$

$$= -4 \int \psi^1_\alpha(\mathbf{r}) \bar{\psi}^1_\alpha(\mathbf{r}) \psi^2_\beta(\mathbf{r}') \bar{\psi}^2_\beta(\mathbf{r}') \exp\left(-\mathcal{L}_{\mathrm{eff}}\left[\psi, \bar{\psi}, Q\right]\right)$$

$$\times \exp(-\frac{\pi\nu}{8\tau} \operatorname{str} Q^2) D\psi D Q d\mathbf{r} d\mathbf{r}'$$

where the functional $\mathcal{L}_{\mathrm{eff}}$ is described by Eq. (4.19), and α and β are arbitrary. Eq. (6.30) is very similar to Eq. (4.40) and written here to make reading easier.

The effective Lagrangian $\mathcal{L}_{\mathrm{eff}}$ is quadratic in ψ and integration over ψ in Eq. (6.30) can be performed by using Wick's theorem and the function $g_0(\mathbf{p})$, Eq. (4.22), which is proportional to the average $\langle \psi \bar{\psi} \rangle_{\mathrm{eff}}$. When calculating the Gaussian integrals in Eq. (6.30) one obtains three possible pairings

$$\langle \psi^1_\alpha(\mathbf{r}) \bar{\psi}^1_\alpha(\mathbf{r}) \rangle_{\mathrm{eff}} \langle \psi^2_\beta(\mathbf{r}') \bar{\psi}^2_\beta(\mathbf{r}') \rangle_{\mathrm{eff}}, \tag{6.31}$$

$$-k_{\beta\beta} \langle \psi^1_\alpha(\mathbf{r}) \bar{\psi}^2_\beta(\mathbf{r}') \rangle_{\mathrm{eff}} \langle \psi^2_\beta(\mathbf{r}') \bar{\psi}^1_\alpha(\mathbf{r}) \rangle_{\mathrm{eff}},$$

$$\langle \psi^1_\alpha(\mathbf{r}) \psi^2_\beta(\mathbf{r}') \rangle_{\mathrm{eff}} \langle \bar{\psi}^2_\beta(\mathbf{r}') \bar{\psi}^1_\alpha(\mathbf{r}) \rangle_{\mathrm{eff}}$$

with the matrix k given by Eq. (4.44). This leads to integrals of the form

$$\int g^{11}_{0\alpha\alpha}(\mathbf{r}, \mathbf{r}) g^{22}_{0\beta\beta}(\mathbf{r}', \mathbf{r}') d\mathbf{r} d\mathbf{r}' \tag{6.32}$$

$$\int g^{12}_{0\alpha\beta}(\mathbf{r}, \mathbf{r}') g^{21}_{0\beta\alpha}(\mathbf{r}', \mathbf{r}) d\mathbf{r} d\mathbf{r}' \tag{6.33}$$

These are the integrals corresponding to the first two lines in Eqs. (6.31). There is no necessity to write the third integral because it is analogous to the second.

The contribution of the integrals in Eqs. (6.32) and (6.33) is completely different. The function $g_0(\mathbf{r}, \mathbf{r})$ does not depend on the coordinate $\mathbf{r}$ and therefore the integral, Eq. (6.32), is proportional to the square of the volume V^2. At the same time, $g_0(\mathbf{r}, \mathbf{r}')$ decays at $|\mathbf{r} - \mathbf{r}'| \sim l$ and the integral, Eq. (6.33), is proportional to the first power of the volume V. Making a more accurate estimate we can see that the ratio of the integral in Eq. (6.33) to the integral in Eq. (6.32) is proportional to $\tau\Delta$, which is assumed by Eq. (6.19) to be small. The same estimate can be done for the third line in Eqs. (6.31) and we come to the conclusion that only the first type of pairing in Eqs. (6.31) should be kept when calculating the level–level correlation function $R(\omega)$, Eq. (6.18). Assuming that the supermatrix Q is taken at the saddle point of the free energy functional $F[Q]$, Eq. (4.45), and calculating $g_0(\mathbf{r}, \mathbf{r})$ at coinciding points we have

$$ig_0(\mathbf{r}, \mathbf{r}) = \bar{A} + i\pi\nu Q \tag{6.34}$$

where $\bar{A}$ is the same constant, as in Eq. (6.23), describing the renormalization of the energy. Eq. (6.34) differs from Eq. (4.46) only by the constant $\bar{A}$ that was omitted in the latter equation because an already renormalized value of the energy was used. Of course, it could be omitted now, too, using the same arguments. The only reason it is kept explicitly is to demonstrate that different nonuniversal contributions arise in the intermediate steps when calculating quantities like the average Green function but they cancel in the final results for the level–level correlation function $R(\omega)$.

After performing all the preceding manipulations, the integral of the average product of the retarded and advanced Green functions can be written as

$$-\left\langle \int G^A_{\varepsilon-\omega}(\mathbf{r}, \mathbf{r}) G^R_\varepsilon(\mathbf{r}', \mathbf{r}') \, d\mathbf{r} d\mathbf{r}' \right\rangle \tag{6.35}$$

$$= \frac{(\pi\nu V)^2}{64} \int \mathrm{str}\, k (1 + \Lambda)(Q - i\bar{A}/\pi\nu)\, \mathrm{str}\, k (1 - \Lambda)(Q - i\bar{A}/\pi\nu)$$

$$\times \exp(-F_0[Q]) \, dQ$$

where

$$F_0[Q] = \frac{\pi i(\omega + i\delta)}{4\Delta} \mathrm{str}(\Lambda Q) \tag{6.36}$$

Eqs. (6.35, 6.36) are written for Model I (orthogonal ensemble) when the magnetic field and the magnetic and spin-orbit interactions are absent. In addition, the subscripts α and β in Eq. (6.30) are arbitrary, and parameter Δ is determined by Eq. (6.20).

The magnetic and spin-orbit interactions can be introduced in the same way as in Section 4.3, which results in the appearance of different effective fields lowering the symmetry of the free energy functional F. In principle, the correlation function $R(\omega)$ can be calculated for any values of these interactions, but, first, let us restrict ourselves to the cases of very weak and very strong interactions, which correspond to Models I, IIa, IIb, and III. In these limits the free energy $F_0[Q]$, Eq. (6.36), retains its form (as well as the free energy $F[Q]$, Eq. (4.55)) but one has to use the proper symmetry relations for the elements of the supermatrices Q specified by Eqs. (4.33, 4.31, 4.73, 4.77). One can use 8×8 supermatrices Q with $a_i^{mn}, b_i^{mn}, \sigma_i^{mn}$, and ρ_i^{mn} just variables and not 2×2 unit matrices in Models IIb and III, too, provided the parameter Δ Eq. (6.20) in Eq. (6.36) is replaced by $\bar{\Delta}$ Eq. (6.21). As discussed, Δ is the mean level spacing when each level is doubly degenerate as a result of invariance with respect to spin rotation, whereas $\bar{\Delta}$ is the mean level spacing in the situations when the spin degeneracy is removed. Therefore, such a substitution is natural.

The characteristic fields of crossovers from Model I to Models II and III can be estimated by directly using Eqs. (4.65, 4.71, 4.75). At these crossover fields the corresponding terms $F_s[Q]$, $F_{so}[Q]$ and the term in Eq. (4.65) describing the interaction with the magnetic field become of the order of unity if one replaces Q by a number of the order of unity. Proceeding in this way one can see that the crossover from Model I to Model IIb occurs at $\tau_s^{-1} \sim \Delta$ and hence Model IIb is determined by the inequality

$$\Delta \ll \tau_s^{-1} \qquad (6.37)$$

Model III (the symplectic ensemble) is applicable provided the following inequality is fulfilled:

$$\tau_s^{-1} \ll \Delta \ll \tau_{so}^{-1} \qquad (6.38)$$

It is clear that Model I is applicable in the limit

$$\tau_s^{-1}, \tau_{so}^{-1} \ll \Delta \qquad (6.39)$$

In the same way one can estimate the characteristic magnetic field H_c of the crossover from Model I to Model IIa assuming that $A \sim H_c L$ in Eq. (4.65). It is more convenient to express the result in terms of magnetic flux ϕ through the sample; then the critical flux ϕ_c is

$$\phi_c \sim \phi_0 \left(\frac{\Delta}{\mathcal{E}_c}\right)^{1/2} \qquad (6.40)$$

where $\phi_0 = c/he$ is the flux quantum and Thouless energy $\mathcal{E}_c$ is determined by Eq. (6.28). We see that in the limit under consideration, Eq. (6.28), the critical flux for the crossover from the orthogonal to unitary ensemble is much less than the flux quantum ϕ_0.

It is not difficult to estimate parameters of all other crossovers. Let us emphasize that the boundaries, Eqs. (6.38–6.40), separating the models are correct for the region where ω is not very small. For example, even if the inequality (6.39) is fulfilled, the description using the orthogonal ensemble is correct only for $\omega \gg \tau_s^{-1}, \tau_{so}^{-1}$.

All these estimates show that in many real situations the description in terms of the zero-dimensional σ-model with three types of the symmetry is applicable. It enables us to write the final expression for the level–level correlation function $R(\omega)$ in a general form valid for all three ensembles. Substituting Eqs. (6.23, 6.35) into Eq. (6.18) one can reduce the function $R(\omega)$ to a definite integral over the supermatrix Q

$$R(\omega) = \frac{1}{2} - \frac{1}{128} \operatorname{Re} \int \operatorname{str}(k(1+\Lambda)Q) \operatorname{str}(k(1-\Lambda)Q) \exp(-F_0[Q]) \, dQ$$
$$(6.41)$$

Eq. (6.41) does not depend on energy ε and as a consequence parameter $\bar{A}$ also does not enter this expression. (When summing up all the terms the equalities $\langle Q^{11} \rangle = -\langle Q^{22} \rangle = 1$ were used. They reflect the fact that the average density of states $\langle \rho(\varepsilon) \rangle$ is a constant, as discussed at the end of Section 4.2.)

In Eq. (6.41) the symmetry of the 8×8 supermatrices Q is specified by Eqs. (4.33, 4.31, 4.73, 4.77) for the orthogonal, unitary, and symplectic ensembles. One should remember that free energy $F_0[Q]$ determined by Eq. (6.36) as the mean level spacing Δ contains the expression Eq. (6.20) for the models without spin interactions (Models I and IIa). When considering the models with magnetic or spin-orbit impurities (Models IIb and III) one should use the corresponding mean level spacing $\bar{\Delta}$ given by Eq. (6.20). The mean level spacing is the only parameter entering Eqs. (6.41, 6.36) that demonstrates the universality of these equations.

The integral, Eqs. (6.41, 6.36), contains many variables of integration and, at first glance, its calculation is not simple. However, there are variables that make the calculation of the integral feasible. In the next section a parametrization of the supermatrix Q is suggested that makes it possible to reduce the integral, Eqs. (6.41, 6.36), to a twofold (for Models II) or threefold (for Models I and III) definite integral over real numbers.

6.3 Reduction to integrals over eigenvalues

The final result of the previous section, Eq. (6.41), shows that by calculating the level–level correlation function with the supersymmetry method one comes to a definite integral over a matrix as in RMT. Then a natural question arises: Some effort has been made to develop the supersymmetry scheme but have we won anything? Of course, the nonlinear supermatrix σ-model was derived from a microscopic Hamiltonian, whereas the RMT is phenomenological and based on several hypotheses. But if calculations with the σ-model proved more complicated, this advantage would be minimal. However, one can guess from the beginning that calculation of the integral, Eq. (6.41), over the supermatrix Q can be much simpler than the corresponding calculations within RMT because matrix Q has size 8×8 whereas in RMT one has to operate with matrices of an arbitrarily large size N. The fact that half of the elements are the Grassmann variables can only simplify the calculations because integration over these variables is equivalent to differentiation.

Actually, the calculations are really not difficult provided one diagonalizes the blocks Q^{11}, Q^{22}, Q^{12}, and Q^{21} (see Eq. (4.33)) and then integrates over the eigenvalues of these blocks and the remaining variables. As a result of symmetries free energy F_0 contains only eigenvalues and integration over all other variables can be done easily. This idea is exactly the same as that used in RMT (Mehta (1991)) when integrals over the arbitrary size matrices are reduced to integrals over the eigenvalues of the matrices. When changing the variables one has to calculate the corresponding Jacobians and it has already been demonstrated in Section 6.1 how these Jacobians could be calculated, Eqs. (6.10, 6.11). The group of the symmetry of supermatrices Q is somewhat more complicated than orthogonal or unitary and contains anticommuting variables. Therefore, the procedure for changing variables is presented in some detail. The parametrization used later is central to the supersymmetry technique, and those who want to be able to calculate with this method should read the present section carefully.

As mentioned in Section 2.3, an arbitrary non-Hermitian supermatrix can be diagonalized by using two unitary supermatrices, Eq. (2.35). Using this possibility let us write the superelement Q^{12} with the symmetry specified by Eqs. (4.33, 4.31, 4.73, 4.77) in the form

$$Q^{12} = iu \sin \hat{\theta} \bar{v} \qquad (6.42)$$

where

$$\hat{\theta} = \begin{pmatrix} \hat{\theta}_{11} & 0 \\ 0 & \hat{\theta}_{22} \end{pmatrix}$$

The 2×2 elements $\hat{\theta}_{11}$ and $\hat{\theta}_{22}$ can be chosen for all three ensembles as

$$\hat{\theta}_{11} = \begin{pmatrix} \theta & 0 \\ 0 & \theta \end{pmatrix}, \qquad \hat{\theta}_{22} = i \begin{pmatrix} \theta_1 & \theta_2 \\ \theta_2 & \theta_1 \end{pmatrix}, \tag{6.43}$$

$$0 < \theta < \pi, \qquad \theta_1 > 0, \qquad \theta_2 > 0, \qquad \text{Model I}$$

$$\hat{\theta}_{11} = \begin{pmatrix} \theta & 0 \\ 0 & \theta \end{pmatrix}, \qquad \hat{\theta}_{22} = i \begin{pmatrix} \theta_1 & 0 \\ 0 & \theta_1 \end{pmatrix}, \tag{6.44}$$

$$0 < \theta < \pi, \qquad \theta_1 > 0, \qquad \text{Model II}$$

$$\hat{\theta}_{11} = \begin{pmatrix} \theta_1 & \theta_2 \\ \theta_2 & \theta_1 \end{pmatrix}, \qquad \hat{\theta}_{22} = i \begin{pmatrix} \theta & 0 \\ 0 & \theta \end{pmatrix}, \tag{6.45}$$

$$\theta > 0, \qquad 0 < \theta_1 < \pi, \qquad 0 < \theta_2 < \pi/2, \qquad \text{Model III}$$

The 4×4 supermatrices u and v in Eq. (6.42) satisfy the conditions

$$\bar{u}u = 1, \qquad \bar{v}v = 1, \qquad \bar{u} = u^+, \qquad \bar{v} = kv^+k \tag{6.46}$$

where the 4×4 superelement k is determined as before by Eq. (4.44). It is relevant to remember that the operation of the charge conjugation "$-$" in Model III differs from that in Models I and II by the replacement $c_1 \leftrightarrow c_2$. One should also remember that all elements in Model II are diagonal 2×2 matrices, so not only the elements $\hat{\theta}_{11}$ and $\hat{\theta}_{22}$, Eq. (6.44), but the corresponding elements of supermatrices u and v commute with matrix τ_3, Eq. (4.59). In the parametrization Eqs. (6.42–6.46) all Grassmann variables are situated in supermatrices u and v.

Using Eqs. (4.38, 4.47, 4.48) and diagonalizing superelement Q^{12} in Eq. (6.42) one can represent supermatrix Q in the form

$$Q = UQ_0\bar{U}, \qquad Q_0 = \begin{pmatrix} \cos\hat{\theta} & i\sin\hat{\theta} \\ -i\sin\hat{\theta} & -\cos\hat{\theta} \end{pmatrix}, \qquad U = \begin{pmatrix} u & 0 \\ 0 & v \end{pmatrix} \tag{6.47}$$

Eq. (6.47) immediately gives $Q^2 = Q_0^2 = 1$. Matrix U in Eq. (6.47) commutes with the matrix Λ, Eq. (4.9), and satisfies the conditions

$$U\bar{U} = 1, \qquad \bar{U} = KU^+K \tag{6.48}$$

where the matrix K is defined in Eq. (4.48). It can be seen directly from Eqs. (6.42–6.45) that matrix Q_0 has symmetry

$$\bar{Q}_0 = Q_0 = KQ_0^+K \tag{6.49}$$

This means that matrix Q_0 has the same symmetry properties as supermatrix Q but does not contain the anticommuting variables located in U and $\bar{U}$. In some sense matrix Q_0 plays the role of "skeleton" for supermatrix Q.

Let us notice that matrix Q_0 has a rather unusual structure: The block $\hat{\theta}_{11}$ is real but the block $\hat{\theta}_{22}$ is imaginary as a result of the presence of i in front. The fact that one should have i in $\hat{\theta}_{22}$ originates from equality $\bar{Q} = KQ^+K$. The block $\hat{\theta}_{11}$ originates from decoupling by supermatrix Q in Section 4.2 of the products of four fermion variables and hence can be called the "fermion–fermion" block of matrix Q_0. Analogously, block $\hat{\theta}_{22}$ can be referred to as the "boson–boson" block. Block $\hat{\theta}_{11}$ leads to a standard circular symmetry as in spin models or superconductivity. This is because it originated as an "order parameter" for a fermion gas with "attraction" that is very well defined. The boson–boson block originates as an "order parameter" of a Bose gas again with an "attraction." Usually it is nonsense to speak of such a system, but in the model under consideration it is possible to find the "ground state" (the saddle point) shifting the contour of integration to the complex plane. However, it results in unusual symmetry.

Substituting Eqs. (6.43–6.46) into Eq. (6.47) we see that the symmetry of the boson–boson part of matrix Q_0 is hyperbolic. In the first versions of the nonlinear σ-models based on the replica method two approaches were suggested. Schäfer and Wegner (1980) started from a boson model and got the noncompact σ-model with hyperbolic symmetry, whereas Efetov, Larkin, and Khmelnitskii (1980), starting from a fermionic model, came to the compact one with circular symmetry. Unfortunately, the replica σ-models can help only in reproducing results of perturbation theory and manipulating with the renormalization group where the difference between the compact and noncompact symmetry is not felt. Eqs. (6.42–6.47) show that, as often happens, the truth lies in between and matrix Q contains a mixture of circular and hyperbolic symmetries. Athough the supermatrix σ-model is very similar to spin models for ferromagnets the noncompact structure makes this model very peculiar. The fact that the boson–boson sector has hyperbolic symmetry is extremely important to all nonperturbative results and this will be discussed several times in the rest of the book.

Eqs. (6.43–6.46) contain inequalities for variables $\theta_1, \theta_2, \theta$ imposed to prevent double counting. However, what does it mean? Originally blocks Q^{12} contained the conventional numbers a_i, b_i and the Grassmann variables σ_i and ρ_i, Eqs. (4.31, 4.33, 4.73, 4.77). Changing to variables $\hat{\theta}$, u, v one expresses variables $\hat{\theta}$ as combinations of elements a_i and b_i and products of even numbers of Grassmann variables σ_i and ρ_i (even elements of Grassmann algebra), and for such objects the notions "more" and "less" do not exist. (See also the diagonalization of a supermatrix in Eqs. (2.28–2.33).)

To answer this question I want to repeat that, now, we are not just diagonalizing a supermatrix but changing the variables of integration. This procedure was discussed in Section 2.5, and the conclusion was that the new commuting variables can be considered in all respects as conventional numbers and one can write proper limits of integration for them provided the integrand is equal to zero at the boundaries. If it is not so, a surface contribution, Eq. (2.83), can arise but it can be taken into account in a regular way.

For explicit calculations one should choose a parametrization for matrices U of the form Eq. (6.47) satisfying conditions Eq. (6.48). Let us represent matrix U in the form

$$U = U_1 U_2, \qquad U_1 = \begin{pmatrix} u_1 & 0 \\ 0 & v_1 \end{pmatrix}, \qquad U_2 = \begin{pmatrix} u_2 & 0 \\ 0 & v_2 \end{pmatrix} \qquad (6.50)$$

where all supermatrices u_1, u_2, v_1, v_2 are unitary

$$\bar{u}_1 u_1 = \bar{u}_2 u_2 = \bar{v}_1 v_1 = \bar{v}_2 v_2 = 1$$

and put all the Grassmann variables into supermatrices u_1 and v_1. Matrices u_2 and v_2 by this definition contain the remaining commuting variables only and hence commute with matrix k. They can be written as

$$u_2 = \begin{pmatrix} \mathcal{F}_1 & 0 \\ 0 & \mathcal{F}_2 \end{pmatrix}, \qquad v_2 = \begin{pmatrix} \Phi_1 & 0 \\ 0 & \Phi_2 \end{pmatrix} \tag{6.51}$$

Eqs. (6.50, 6.51) are written for all three ensembles. Matrices $u_1, v_1, \mathcal{F}_{1,2}, \Phi_{1,2}$ entering these formulae depend on the ensemble under consideration. The unitary matrices $\mathcal{F}_{1,2}$, $\Phi_{1,2}$ are written in the form

$$\begin{cases} \mathcal{F}_1 = (1 - iM)(1 + iM)^{-1}, & \mathcal{F}_2 = \exp(i\phi\tau_3) \\ \Phi_1 = 1, & \Phi_2 = \exp(i\chi\tau_3) \end{cases} \qquad \text{Model I} \tag{6.52}$$

$$\begin{cases} \mathcal{F}_1 = \exp(i\phi\tau_3), & \mathcal{F}_2 = \exp(i\chi\tau_3) \\ \Phi_1 = 1, & \Phi_2 = 1 \end{cases} \qquad \text{Model II} \tag{6.53}$$

$$\begin{cases} \mathcal{F}_1 = \exp(i\phi\tau_3), & \mathcal{F}_2 = (1 - iM)(1 + iM)^{-1} \\ \Phi_1 = \exp(i\chi\tau_3), & \Phi_2 = 1 \end{cases} \qquad \text{Model III} \tag{6.54}$$

In Eqs. (6.52–6.54), the phases ϕ and χ vary in the intervals $0 < \phi < 2\pi$, $0 < \chi < 2\pi$. Matrices τ_3 and M are

$$\tau_3 = \begin{pmatrix} 1 & 0 \\ 0 & -1 \end{pmatrix}, \qquad M = \begin{pmatrix} m & m_1^* \\ m_1 & -m \end{pmatrix} \tag{6.55}$$

with m an arbitrary real and m_1 a complex number.

Forgetting about the anticommuting variables for a while one can check that Eq. (6.47) together with Eqs. (6.51–6.55) correctly reproduces the structure of the commuting part of matrix Q, Eqs. (4.33, 4.31, 4.73, 4.77). The symmetry specified by Eqs. (4.38, 4.48) follows immediately from the symmetry of matrix Q_0 and structure of the matrices u_2, v_2. One can also check that all the variables in matrices Q_0 and u_2, v_2 vary in such intervals so that each element in Q is counted once. Composing the matrix

$$\tilde{Q}_0 = U_2 Q_0 \bar{U}_2 \tag{6.56}$$

one can get an idea of how the commuting variables in supermatrix Q vary. The most simple is the structure of $\tilde{Q}_0$ in Model II. It is not difficult to see that the fermion–fermion compact part is equivalent to a vector on the unit sphere; the variables θ and ϕ play the role of the angles in the spherical coordinates of the vector. The noncompact part is equivalent to a vector on the hyperboloid with χ the rotational coordinate and θ_1 the corresponding hyperbolic angle.

The last thing to do, in order to have a complete specification of supermatrix Q, is to write an explicit form for supermatrices u_1 and v_1. They can be written in a general form valid for all the ensembles

$$u_1 = \begin{pmatrix} 1 - 2\eta\bar{\eta} + 6(\eta\bar{\eta})^2 & 2\eta(1 - 2\bar{\eta}\eta) \\ -2(1 - 2\bar{\eta}\eta)\bar{\eta} & 1 - 2\bar{\eta}\eta + 6(\bar{\eta}\eta)^2 \end{pmatrix} \tag{6.57}$$

$$v_1 = \begin{pmatrix} 1 + 2\kappa\bar{\kappa} + 6(\kappa\bar{\kappa})^2 & 2i\kappa(1 + 2\bar{\kappa}\kappa) \\ -2i(1 + 2\bar{\kappa}\kappa)\bar{\kappa} & 1 + 2\bar{\kappa}\kappa + 6(\bar{\kappa}\kappa)^2 \end{pmatrix} \tag{6.58}$$

Matrices η and κ depend on the ensemble and are equal to the following:

$$\eta = \begin{pmatrix} \eta_1 & \eta_2 \\ -\eta_2^* & -\eta_1^* \end{pmatrix}, \qquad \kappa = \begin{pmatrix} \kappa_1 & \kappa_2 \\ -\kappa_2^* & -\kappa_1^* \end{pmatrix}, \qquad \text{Model I} \qquad (6.59)$$

$$\eta = \begin{pmatrix} \eta_1 & 0 \\ 0 & -\eta_1^* \end{pmatrix}, \qquad \kappa = \begin{pmatrix} \kappa_1 & 0 \\ 0 & -\kappa_1^* \end{pmatrix}, \qquad \text{Model II} \qquad (6.60)$$

$$\eta = \begin{pmatrix} \eta_1 & \eta_2 \\ -\eta_2^* & \eta_1^* \end{pmatrix}, \qquad \kappa = \begin{pmatrix} \kappa_1 & \kappa_2 \\ -\kappa_2^* & \kappa_1^* \end{pmatrix}, \qquad \text{Model III} \qquad (6.61)$$

Matrices η and κ have symmetry $\bar{\eta} = \eta^+$ and $\bar{\kappa} = \kappa^+$. (One should remember that $(\eta_i^*)^* = -\eta$.) One can check by direct calculation that matrices u_1 and v_1, Eqs. (6.57, 6.58), in all cases satisfy the conditions $\bar{u}_1 u_1 = \bar{v}_1 v_1 = 1$. In the unitary ensemble the terms $(\eta\bar{\eta})^2$, $(\bar{\eta}\eta)^2$, $\eta\bar{\eta}\eta$, $\bar{\eta}\eta\bar{\eta}$, and corresponding terms for κ are equal to zero; this simplifies the form of the supermatrices u_1 and v_1.

In principle, the choice of coefficients in Eqs. (6.57, 6.58) providing the unitarity conditions, Eq. (6.50), is not unique and one has to impose one more restriction to come to the form Eqs. (6.57, 6.58). The coefficients in the parametrization for u_1, v_1, Eqs. (6.57, 6.58), can be obtained by requiring that the Jacobian arising when changing to the variables $\hat{\theta}$, u, v should be simple enough. Later, this Jacobian will be derived and it will be seen that matrices $\delta u = \bar{u} du$ and $\delta v = \bar{v} dv$ play an important role in the derivation (see also the corresponding derivation in Section 6.1 for conventional matrices). The coefficients in Eqs. (6.57, 6.58) are chosen in such a way that δu and δv can be considered as independent differentials like those defined in Eqs. (2.52, 2.53). This goal is achieved by imposing the following conditions:

$$\int \prod_{i,k} (\delta u_{12})_{ik} = \int \prod_{i,k} (\delta v_{12})_{ik} = 0 \qquad (6.62)$$

In Eq. (6.62), the products of all independent elements of matrices $(\delta u)_{12}$ and $(\delta v)_{12}$ are written down. The discussion of coefficients is important for Models I and III. In Model II one could use the most general representation

$$u_1 = \begin{pmatrix} 1 - 2|f|^2 \eta\bar{\eta} & 2f\eta \\ -2f^*\bar{\eta} & 1 - 2|f|^2 \bar{\eta}\eta \end{pmatrix} \qquad (6.63)$$

with arbitrary complex f. With this parametrization

$$(\delta u_{12})_{11} = 2f d\eta, \qquad (\delta u_{12})_{22} = -2f^* d\eta^*$$

and Eqs. (6.62) are trivially fulfilled. Of course, in Models I and III one still has the same freedom of the substitution $\eta \to f\eta$, $\kappa \to h\kappa$ with arbitrary complex numbers f and h, which does not violate Eq. (6.62).

The representation of supermatrix Q in the form Eq. (6.47) is very convenient because free energy $F_0[Q]$, Eq. (6.36), depends on the elements of matrix $\hat{\theta}$ only. In the models under consideration it has the form

$$
F_0[Q] = \begin{cases} -i(x+i\delta)(\cosh\theta_1\cosh\theta_2 - \cos\theta), & \text{Model I} \\[1em] -i(x+i\delta)(\cosh\theta_1 - \cos\theta), & \text{Model II} \\[1em] -i(x+i\delta)(\cosh\theta - \cos\theta_1\cos\theta_2), & \text{Model III} \end{cases} \tag{6.64}
$$

In Eqs. (6.64), $x = \pi\omega/\Delta$ in Models I, IIa and $x = \pi\omega/\bar{\Delta}$ in Models IIb and III.

In most cases of interest integration over the other variables can be performed very easily, and evaluation of integrals over Q is thus reduced to integration over variables θ, θ_1, θ_2. The only price one should pay for this simplification is a nontrivial Jacobian that arises as a result of the change of variables. However, this Jacobian has to be calculated only once and then can be used in calculation of many different physical quantities. The derivation of the Jacobian is presented in Appendix 1 and the result for elementary volume $[dQ]$ can be written as

$$
[dQ] = J\,dR_1\,dR_2\,d\hat{\theta}, \qquad J = J_1 J_2 \tag{6.65}
$$

where $J_1 dR_1$ is the elementary volume in the space of anticommuting variables, $J_2 dR_2$ is the elementary volume in the space of commuting, and $d\hat{\theta}$ is the volume in the space of the elements of matrix $\hat{\theta}$. The quantities entering Eq. (6.65) must be written separately for Models I, II, and III and the final result is

$$
J_1^{(\mathrm{I})} = \frac{1}{2^{20}}\left(\cosh(\theta_1+\theta_2) - \cos\theta\right)^{-2}\left(\cosh(\theta_1-\theta_2) - \cos\theta\right)^{-2}, \tag{6.66}
$$

$$
dR_1^{(\mathrm{I})} = dR_\eta^{(\mathrm{I})}dR_\kappa^{(\mathrm{I})}, \qquad dR_\eta^{(\mathrm{I})} = d\eta_1 d\eta_1^* d\eta_2 d\eta_2^*, \qquad dR_\kappa^{(\mathrm{I})} = d\kappa_1 d\kappa_1^* d\kappa_2 d\kappa_2^*,
$$

$$
J_2^{(\mathrm{I})} = \frac{2^{12}}{\pi^4}\frac{\sin^3\theta\,\sinh\theta_1\,\sinh\theta_2}{\left(1+m^2+|m_1|^2\right)^3},
$$

$$
dR_2^{(\mathrm{I})} = dm\,dm_1\,dm_1^*\,d\phi\,d\chi, \qquad d\hat{\theta}^{(\mathrm{I})} = d\theta\,d\theta_1\,d\theta_2
$$

$$
J_1^{(\mathrm{II})} = \frac{1}{2^8}\left(\cosh\theta_1 - \cos\theta\right)^2, \qquad dR_1^{(\mathrm{II})} = d\eta_1 d\eta_1^* d\kappa_1^* d\kappa_1 \tag{6.67}
$$

$$
J_2^{(\mathrm{II})} = \frac{2^2}{\pi^2}\sin\theta\,\sinh\theta_1, \qquad dR_2^{(\mathrm{II})} = d\phi\,d\chi, \qquad d\theta^{(\mathrm{II})} = d\theta\,d\theta_1
$$

$$
J_1^{(\mathrm{III})} = \frac{1}{2^{20}}\left(\cosh\theta - \cos(\theta_1+\theta_2)\right)^{-2}\left(\cosh\theta - \cos(\theta_1-\theta_2)\right)^{-2} \tag{6.68}
$$

$$
dR_1^{(\mathrm{III})} = dR_\eta^{(\mathrm{III})}dR_\kappa^{(\mathrm{III})}, \qquad dR_\eta^{(\mathrm{III})} = d\eta_1 d\eta_1^* d\eta_2 d\eta_2^*, \qquad dR_\kappa^{(\mathrm{III})} = d\kappa_1 d\kappa_1^* d\kappa_2 d\kappa_2^*,
$$

$$
J_2^{(\mathrm{III})} = \frac{2^{12}}{\pi^4}\frac{\sinh^3\theta\,\sin\theta_1\,\sin\theta_2}{\left(1+m^2+|m|^2\right)^3},
$$

$$
dR_2^{(\mathrm{III})} = dm\,dm_1\,dm_1^*\,d\phi\,d\chi, \qquad d\theta^{(\mathrm{III})} = d\theta\,d\theta_1\,d\theta_2
$$

The superscripts in Eqs. (6.66–6.68) correspond to the models considered. The Jacobians, Eqs. (6.66–6.68), do not contain anticommuting variables and this is very convenient for calculations. Notice that the Berezinians of transformations J are singular at $\theta_1 = \theta_2 = \theta = 0$. This is an essential singularity that can be removed only if the integrands turn to zero at this point. Their presence can lead to anomalous surface contributions (cf. Section 2.5), and several methods of extracting these contributions have been suggested.

This remark is important because now we have to take the next step in calculating the integral $R(\omega)$, Eq. (6.41). Using Eq. (6.47) one can rewrite the supertraces in the integrand in Eq. (6.41) in terms of the matrices $\hat{\theta}$, u, v as

$$\mathrm{str}\,(k\,(1+\Lambda)\,Q) = 2\,\mathrm{str}\left(\bar{u}ku\cos\hat{\theta}\right),$$

$$\mathrm{str}\,(k\,(1-\Lambda)\,Q) = -2\,\mathrm{str}\left(\bar{v}kv\cos\hat{\theta}\right) \tag{6.69}$$

We see that the Grassmann variables η are present only in supermatrix $\bar{u}ku$ and variables κ in $\bar{v}kv$. In order to carry out the integration over η and κ one should write the supermatrices $\bar{u}ku$ and $\bar{v}kv$ explicitly (in fact we need only the parts $(\bar{u}ku)_\parallel$ and $(\bar{v}kv)_\parallel$ commuting with k). After a simple calculation one obtains a result valid for all three ensembles

$$(\bar{u}ku)_\parallel = \bar{u}_2\left(k + 8\begin{pmatrix} -\eta\bar{\eta} + 4\,(\eta\bar{\eta})^2 & 0 \\ 0 & \bar{\eta}\eta - 4\,(\bar{\eta}\eta)^2 \end{pmatrix}\right)u_2 \tag{6.70}$$

and an analogous expression for $(\bar{v}kv)_\parallel$.

In Model II, $\eta\bar{\eta} = -\bar{\eta}\eta = \eta_1\eta_1^*$ and the terms $(\eta\bar{\eta})^2$ and $(\bar{\eta}\eta)^2$ are equal to zero. In Models I and III $(\eta\bar{\eta})^2 = -(\bar{\eta}\eta)^2 = 2\eta_1\eta_1^*\eta_2\eta_2^*$. At first glance, the term k in Eq. (6.70) cannot contribute to the integral over the anticommuting variables because it does not contain them. However, it is at just this point that the singularity of the Jacobians at $\theta_1 = \theta_2 = \theta = 0$ becomes important because one obtains a contribution of the type $0 \times \infty$. In Section 2.5 several methods to overcome this difficulty were suggested; let us now use the simplest: Instead of calculating function $R(\omega)$, Eq. (6.41), one calculates the function $\partial R(\omega)/\partial\omega$. Using Eq. (6.64) we see that the singularity in the Jacobian is now compensated for by $\mathrm{str}(\Lambda Q_0)$ and therefore the terms of zero order in the Grassmann variables cannot have a nonzero contribution.

When integrating in Eq. (6.70) over variables η, $\bar{\eta}$ only the terms $\eta\bar{\eta}$ and $\bar{\eta}\eta$ contribute in Model II, whereas in Models I and III only terms $(\eta\bar{\eta})^2$ and $(\bar{\eta}\eta)^2$ give a nonzero contribution. In all cases the result is proportional to the unit matrix and one can easily write $\partial R/\partial\omega$ in the form of an integral over variables θ_1, θ_2, and θ (θ_1 and θ in Model II). Then one can reconstruct function $R(\omega)$ from its derivative, writing

$$R(\omega) = R(\infty) + \int_\infty^\omega \frac{\partial R(\omega)}{\partial\omega} d\omega \tag{6.71}$$

The value $R(\infty)$ can be found by using the perturbation theory developed in Chapter 5. In the limit $\omega \to \infty$ only the zero-order term contributes; that means that one should simply replace Q in the integrand in Eq. (6.41) by Λ. This immediately gives

$$R(\infty) = 1 \tag{6.72}$$

and one can write the function $R(\omega)$ in the form of integrals over the variables θ_1, θ_2 and θ. The formulae are somewhat more compact in the variables

$$
\begin{cases}
\lambda_1 = \cosh\theta_1, & \lambda_2 = \cosh\theta_2, & \lambda = \cos\theta, & \text{Model I} \\[2mm]
\lambda_1 = \cosh\theta_1, & \lambda = \cos\theta, & & \text{Model II} \\[2mm]
\lambda_1 = \cos\theta_1, & \lambda_2 = \cos\theta_2, & \lambda = \cosh\theta, & \text{Model III}
\end{cases}
\tag{6.73}
$$

Then, using Eq. (6.64) and the dimensionless variable x, the function $R(x)$ takes the form

$$
R^{(\mathrm{I})}(\omega) = 1 + \mathrm{Re} \int_1^\infty \int_1^\infty \int_{-1}^1 \frac{(\lambda_1\lambda_2 - \lambda)^2 \left(1 - \lambda^2\right)}{\left(\lambda_1^2 + \lambda_2^2 + \lambda^2 - 2\lambda\lambda_1\lambda_2 - 1\right)^2}
\tag{6.74}
$$

$$
\times \exp\left[i(x + i\delta)(\lambda_1\lambda_2 - \lambda)\right] d\lambda_1 d\lambda_2 d\lambda
$$

$$
R^{(\mathrm{II})}(\omega) = 1 + \frac{1}{2}\,\mathrm{Re} \int_1^\infty \int_{-1}^1 \exp\left[i(x + i\delta)(\lambda_1 - \lambda)\right] d\lambda_1 d\lambda
\tag{6.75}
$$

$$
R^{(\mathrm{III})}(\omega) = 1 + \mathrm{Re} \int_1^\infty \int_0^1 \int_{-1}^1 \frac{(\lambda - \lambda_1\lambda_2)^2 \left(\lambda^2 - 1\right)}{\left(\lambda^2 + \lambda_1^2 + \lambda_2^2 - 2\lambda_1\lambda_2\lambda - 1\right)^2}
\tag{6.76}
$$

$$
\times \exp\left[i(x + i\delta)(\lambda - \lambda_1\lambda_2)\right] d\lambda d\lambda_2 d\lambda_1
$$

Eqs. (6.74–6.76) are the final result of the present section. We see that integrals over supermatrices Q can be reduced to integrals over the "eigenvalues" λ_1, λ_2, λ. Depending on the model, one obtains twofold or threefold integrals. Calculation of such integrals is a much simpler task than calculation of integrals over a large number N of variables encountered in RMT (Mehta (1991)). Although in this section the reduction to the integrals over the eigenvalues has been carried out for the level–level correlation function only the corresponding manipulations for studying other physical quantities are the same. In fact, we already know the Jacobians Eqs. (6.66–6.68) and the free energies Eq. (6.64). The only thing that remains to be done when calculating different physical quantities is to write a proper preexponential and carry out integration over the elements of supermatrices u and v entering the preexponential only. Then one always obtains integrals over variables λ_1, λ_2, λ analogous to those in Eqs. (6.74–6.76).

6.4 The level–level correlation function

The integrals Eqs. (6.74–6.76) completely solve the problem of calculation of the two-level correlation function $R(\omega)$, Eqs. (3.36, 6.3, 6.18), in all three cases. Although the integral Eq. (6.75) for Model II is very simple, the integrals for Models I and III, Eqs. (6.74, 6.76), look very complicated. Nevertheless, they can also be calculated analytically.

Let us first obtain the result for Model II. Integration over λ_1 and λ can be done immediately and one obtains

$$
R^{(\mathrm{II})}(\omega) = 1 + \mathrm{Re}\left(\frac{\sin x}{x}\frac{\exp(-ix)}{ix}\right)
\tag{6.77}
$$

Taking the real part in Eq. (6.77) and comparing the result with Eq. (6.14) we come to the conclusion that

$$R^{(II)}(\omega) = R_{\text{unit}}(\omega) \tag{6.78}$$

Calculation of the integrals, Eqs. (6.74, 6.76), is much less trivial. These integrals are very similar and therefore it is sufficient to present only the calculation of the integral for Model I.

So we have integral $I(x)$

$$I(x) = \int_1^\infty \int_1^\infty \int_{-1}^1 \frac{(\lambda_1\lambda_2 - \lambda)^2 (1 - \lambda^2)}{\left(\lambda_1^2 + \lambda_2^2 + \lambda^2 - 2\lambda\lambda_1\lambda_2 - 1\right)^2} \tag{6.79}$$

$$\times \exp\left[i(x + i\delta)(\lambda_1\lambda_2 - \lambda)\right] d\lambda_1 d\lambda_2 d\lambda$$

Calculation of the Fourier transform proves to be simpler and one can start with making the Fourier transformation

$$\tilde{I}(t) = \int_{-\infty}^\infty I(x) \exp(-ixt)\, dx \tag{6.80}$$

Then one easily obtains

$$\tilde{I}(t) = 2\pi \int_1^\infty \int_1^\infty \int_{-1}^1 \frac{(\lambda_1\lambda_2 - \lambda)^2 (1 - \lambda^2) \delta(\lambda - \lambda_1\lambda_2 + t)}{\left(\lambda_1^2 + \lambda_2^2 + \lambda^2 - 2\lambda\lambda_1\lambda_2 - 1\right)^2} d\lambda_1 d\lambda_2 d\lambda \tag{6.81}$$

The integration over λ is trivially done because of the presence of the δ-function. Changing the variables $u = \lambda_1\lambda_2$, $z = \lambda_1^2$ the integral $\tilde{I}(t)$ is reduced to the form

$$\tilde{I}(t) = \pi t^2 \int_1^\infty \int_1^{u^2} \frac{\left(1 - (u - t)^2\right) \theta(u - t + 1) \theta(1 - u + t)\, z\, dz\, du}{\left(z^2 + z(t^2 - u^2 - 1) + u^2\right)^2} \tag{6.82}$$

where

$$\theta(\tau) = \begin{cases} 1, & \tau > 0 \\ 0 & \tau < 0 \end{cases}$$

The integrand in Eq. (6.82) is a rational function of z that enables us to integrate over this variable. The result of the integration can be written as

$$\tilde{I}(t) = \tilde{I}_1(t) + \tilde{I}_2(t), \tag{6.83}$$

$$\tilde{I}_1(t) = 2\pi \int_1^\infty \frac{\theta(u - t + 1) \theta(1 - u + t) (u^2 - 1)}{(u + t + 1)(u + t - 1)} du \tag{6.84}$$

$$\tilde{I}_2(t) = 2\pi t^2 \int_1^\infty \frac{\theta(u - t + 1) \theta(1 - u + t) (t^2 - u^2 - 1)}{(1 + u - t)^{1/2} (u + t + 1)^{3/2} (1 - u + t)^{1/2} (u + t - 1)^{3/2}} \tag{6.85}$$

$$\times \left(\arctan \frac{t^2 - u^2 - 1}{\left(4u^2 - (t^2 - u^2 - 1)^2\right)^{1/2}} - \arctan \frac{t^2 + u^2 - 1}{\left(4u^2 - (t^2 - u^2 - 1)^2\right)^{1/2}} \right) du$$

Further integration in Eq. (6.84) is rather simple. However, in spite of the very cumbersome integrand the integration in Eq. (6.85) can also be performed analytically using the identity

$$\frac{d}{du}\left[\frac{(1+u-t)^{1/2}\,(1-u+t)^{1/2}}{(1+u+t)^{1/2}\,(t+u-1)^{1/2}}\right]$$

$$=\frac{2t\,(t^2-u^2-1)}{(1+u-t)^{1/2}\,(1+u+t)^{3/2}\,(1-u+t)^{1/2}\,(u+t-1)^{3/2}}$$

This identity makes it possible to integrate by parts in the integral $\tilde{I}_2\,(t)$, Eq. (6.85). Carrying out these transformations we find

$$\tilde{I}_2\,(t)=\pi t\int_1^\infty\frac{\theta\,(u-t+1)\,\theta\,(1-u+t)\,(3u^2-t^2+1)}{(1+u+t)\,(u+t-1)\,u}\,du \qquad (6.86)$$

Integrating the rational functions in the integrals $\tilde{I}_1\,(t)$, Eq. (6.84), and $\tilde{I}_2\,(t)$, Eq. (6.86), one can obtain a rather simple result for the Fourier transform $\tilde{I}\,(t)$ of the integral $I\,(x)$, Eq. (6.79)

$$\tilde{I}\,(t)=t\begin{cases}\frac{4}{t}-\ln\frac{t+1}{t-1}, & t>2 \\[2mm] 2-\ln\,(t+1), & 0<t<2 \\[2mm] 0, & t<0\end{cases} \qquad (6.87)$$

Finally, performing the inverse Fourier transformation, we obtain

$$R^{(\mathrm{I})}\,(\omega)=R_{\mathrm{orth}}\,(\omega) \qquad (6.88)$$

where $R_{\mathrm{orth}}\,(\omega)$ is the two-level correlation function for the orthogonal ensemble Eq. (6.13).

The same calculation can be carried out for the function $R^{(\mathrm{III})}\,(\omega)$, Eq. (6.76), and one comes to the next relation

$$R^{(\mathrm{III})}\,(\omega)=R_{\mathrm{sympl}}\,(\omega) \qquad (6.89)$$

with the level–level correlation function $R_{\mathrm{sympl}}\,(\omega)$, Eq. (6.15), for the symplectic ensemble.

Thus, the two-level correlation functions for Models I, II, and III are the same as those for the orthogonal, unitary, and symplectic ensembles known in RMT. This agreement is seen in the final result only. Nothing common arises in the intermediate steps of the calculations. In the Wigner–Dyson–Mehta theory (Mehta (1991)), Eqs. (6.13–6.15) are obtained after calculation of complicated determinants. In the theory developed previously these formulae are obtained as a result of calculation of the integrals Eqs. (6.74–6.76).

The present theory is based on a very well defined microscopic model of a particle moving in a random potential, and the results Eq. (6.78, 6.88, 6.89) first obtained by Efetov (1982b) can be considered the first analytical confirmation of RMT. Later, the zero-dimensional version of the σ-model was derived by starting directly from the statistical hypotheses of the Gaussian ensembles of RMT (Verbaarschot, Weidenmüller, Zirnbauer (1985)); that derivation helped to clarify the equivalence between RMT and the zero-dimensional supermatrix σ-model. The corresponding derivation will be presented in the next section, but now let us discuss the main properties of the level–level correlation functions Eqs. (6.13–6.15). These functions are represented in Figs. 6.2–6.4.

The behavior of the level–level correlation functions $R\,(\omega)$ is very well known and one can find a lot of information in Mehta (1991). Therefore, I want to discuss here only the most

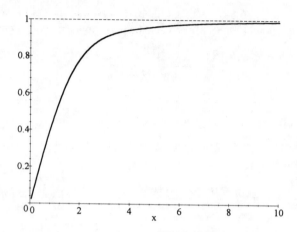

Fig. 6.2. The two-level correlation function R_{orth} for the orthogonal ensemble.

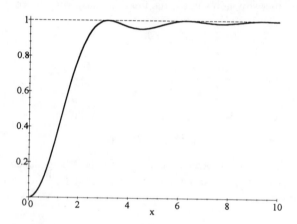

Fig. 6.3. The two-level correlation function R_{unit} for the unitary ensemble.

important properties. In the limits $\omega \to 0$ and $\omega \to \infty$ the functions $R(\omega)$, Eqs. (6.13–6.15), have quite different asymptotes. The asymptotic behavior at $\omega \to 0$ is

$$R_{\text{orth}}(\omega) \simeq \frac{\pi x}{6}$$

$$R_{\text{unit}}(\omega) \simeq \frac{x^2}{3} \tag{6.90}$$

$$R_{\text{sympl}}(\omega) \simeq \frac{x^4}{135}$$

We see that the level–level correlation function vanishes in this limit for all three ensembles. This effect is known as "level repulsion." Eqs. (6.90) show that the probability of finding a level at a distance x from the fixed one is proportional to x^β, where $\beta = 1$ for the orthogonal ensemble, $\beta = 2$ for the unitary ensemble, and $\beta = 4$ for the symplectic ensemble. The parameter β is just the same as the one entering Eqs. (6.10, 6.11) and is determined completely by the symmetry of the system. In the limit $x \to 0$ only two levels are relevant and one can consider 2×2 random matrices. Using Eqs. (6.3, 6.10, 6.11) for

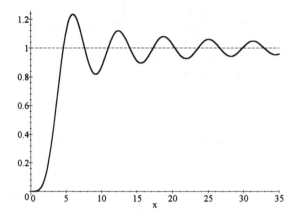

Fig. 6.4. The two-level correlation function R_{sympl} for the symplectic ensemble.

2×2 matrices one can easily reproduce the asymptotic behavior Eq. (6.90). The effect of level repulsion is strongest in the symplectic ensemble and weakest in the orthogonal ensemble.

In the opposite limit $\omega \to \infty$ the correlation functions $R(\omega)$ have the following asymptotics:

$$R_{\text{orth}}(\omega) \simeq 1 - x^{-2} + x^{-4}\left(\frac{3}{2} + \frac{1}{2}\cos 2x\right)$$

$$R_{\text{unit}}(\omega) \simeq 1 - x^{-2}\left(\frac{1 - \cos 2x}{2}\right) \tag{6.91}$$

$$R_{\text{sympl}}(\omega) \simeq 1 + (2x)^{-1}\pi \cos x$$

In all the cases the correlation functions $R(\omega)$ approach 1 as $\omega \to \infty$. This means that considering large energy intervals one loses the information about the discreteness of the energy levels and the average level density is substituted for them. However, the oscillating part decays rather slowly; the oscillations are most pronounced in the symplectic ensemble. Using the asymptotics, Eqs. (6.91), and recalling the definition of the variable x, Eqs. (6.71, 6.20, 6.21), we see that the period of the oscillations in the symplectic ensemble is the same as that in the orthogonal and unitary ensembles without spin interactions (Models I and IIa). This is because the spin-orbit interactions do not completely resolve the spin degeneracy of the system. For the same reason the period of the oscillations in the unitary ensemble with magnetic spin interactions (Model IIa) that do resolve the spin degeneracy is two times smaller than that for the other ensembles.

Now, a natural question arises: Can one obtain the asymptotic behavior Eq. (6.91) by using the perturbation theory developed in Chapter 5 or (which is equivalent) the conventional diagrammatic expansions that have to work in the limit of high frequencies? The answer is, definitely no. The perturbation theory yields a series in x^{-1} but it cannot reproduce the oscillating terms in Eq. (6.91) in any order of perturbation theory even in principle. This is the usual defect of any Taylor expansion: It cannot take into account exponentially small terms (in the case considered the exponents are imaginary). The perturbation theory can reproduce only nonoscillating terms in Eq. (6.91) that can be described in terms of the diffusion modes (diffusons and cooperons). In other words, perturbation theory can describe the quantum interference in the system but becomes useless when level discreteness is important.

The study carried out previously shows that a class of systems exists that can definitely be described by the Wigner–Dyson statistical theory. Eqs. (6.13–6.15) do not contain any parameters characterizing the disorder and look completely universal. Nevertheless, this theory works under certain conditions only. The zero-dimensional σ-model was derived in the limit of weak disorder, when electrons can move around the whole sample and feel the boundaries. Besides, the inequalities (6.19, 6.26–6.29) were essentially used. What happens if some of these inequalities are not fulfilled? The situation when the energy interval ω much exceeds the Thouless energy D_0/L^2 was considered in the work of Altshuler and Shklovskii (1986), where behavior completely different from that given by the Wigner–Dyson statistics was found. In systems where the electron localization is important the Wigner–Dyson statistics cannot work either. The physical reason is quite simple. In such systems electrons are localized at places called *localization centers*. At a sufficiently large distance from one localization center to another the electron wave functions do not overlap and levels with practically the same energy can be found. This leads to the vanishing of the level repulsion. As a result the level statistics in the systems with localization obey the Poisson law. I want to emphasize that the localization rather than a strong disorder is responsible for the breakdown of the Wigner–Dyson theory. In one-dimensional chains and wires electrons are localized in the presence of any weak disorder and the level statistics is not described by RMT. Some results concerning the level statistics in disordered chains were obtained by Molchanov (1981), Gorkov, Dorokhov, and Prigara (1983), and many others. One-dimensional wires with length L exceeding localization length L_c will be considered in Chapter 11.

Level repulsion also does not exist if electron scattering is mainly due to scattering by boundaries and the shape is such that the system is integrable. In this case one again expects that the level statistics obeys the Poisson law. It is curious that to some extent systems with localization are analogous to integrable billiards. However, in the latter models small distortions of the shape already can lead to chaotic motion and to the Wigner–Dyson statistics.

6.5 Random matrix theory and the zero-dimensional σ-model

In the previous section it was demonstrated that starting from a model with a random potential one arrives at very well known results of RMT that support the latter. However, this agreement is seen in the final formulae only and the scheme of the calculation is completely different from that developed by Wigner, Dyson, and Mehta. It is rather difficult to imagine that this agreement is just a coincidence and a desire to reveal a deeper connection between these two theories is natural. In fact, the zero-dimensional σ-model can be derived from the Gaussian ensembles in the limit of a large size N of the matrices (Verbaarschot, Weidenmüller, and Zirnbauer (1985)). The derivation is basically the same as that developed in the previous chapters for the models with random potential and let us reproduce it briefly.

One can start the calculations from Eqs. (6.1–6.3). Introducing the retarded G^R and advanced G^A Green functions as

$$G_\varepsilon^{R,A} = (\varepsilon - \mathcal{H} \pm i\delta)^{-1} \tag{6.92}$$

one can rewrite the functions $\Delta^{-1}(\varepsilon)$ and $R(\omega)$ in the following form

$$\Delta^{-1}(\varepsilon) = \frac{1}{2\pi i} \left\langle \text{tr}\big(G_\varepsilon^A - G_\varepsilon^R\big) \right\rangle_{\mathcal{D}} \tag{6.93}$$

$$R(\omega) = \frac{\Delta^2(\varepsilon)}{4\pi^2} \left\langle \text{tr}\big(G_{\varepsilon-\omega}^A - G_{\varepsilon-\omega}^R\big) \text{tr}\big(G_\varepsilon^R - G_\varepsilon^A\big) \right\rangle_{\mathcal{D}} \tag{6.94}$$

In Eqs. (6.92–6.94) the functions $G_\varepsilon^{R,A}$ are $N \times N$ matrices and the trace is calculated over the matrix elements. Let us carry out the calculations for the orthogonal ensemble. The corresponding derivations for the other two ensembles are practically the same and one can repeat them without any difficulties. Introducing eight-component supervectors ψ_m and $\bar{\psi}_m$ with the same structure as that of the supervectors ψ and $\bar{\psi}$ used in Chapter 4 we can rewrite the products $G^R G^A$ in Eq. (6.94) in the form of the following integral over the supervectors:

$$\left\langle \text{tr}\, G_{\varepsilon-\omega}^R\, \text{tr}\, G_\varepsilon^A \right\rangle_{\mathcal{D}} = \frac{1}{16} \left\langle \sum_{p,s=1}^{N} \int \text{str}\big(k(1+\Lambda)\psi_p\bar{\psi}_p\big) \right.$$

$$\times \left. \text{str}\big(k(1-\Lambda)\psi_s\bar{\psi}_s\big) \exp(-\mathcal{L})\, d\psi\, d\bar{\psi} \right\rangle_{\mathcal{D}} \tag{6.95}$$

where the Lagrangian $\mathcal{L}$ has the form

$$\mathcal{L} = i\left[\sum_{m=1}^{N} \bar{\psi}_m \left(\tilde{\varepsilon} - \frac{\omega+i\delta}{2}\Lambda\right)\psi_m - \sum_{m,n=1}^{N} \bar{\psi}_m \mathcal{H}_{mn}\psi_n \right] \tag{6.96}$$

and $\tilde{\varepsilon} = \varepsilon - \omega/2$.

The mean level density $\Delta^{-1}(\varepsilon)$ can also be written in the form of the integral with the Lagrangian $\mathcal{L}$

$$\Delta^{-1}(\varepsilon) = \frac{1}{4\pi} \left\langle \sum_{p=1}^{N} \int \text{str}\big(k\Lambda\psi_p\bar{\psi}_p\big) \exp(-\mathcal{L})\, d\psi\, d\bar{\psi} \right\rangle_{\mathcal{D}} \tag{6.97}$$

Strictly speaking, one should take in Eq. (6.97) the limit $\omega \to 0$. However, we are interested in the range of energies ω much smaller than characteristic energies at which $\Delta^{-1}(\varepsilon)$ changes and so ω can be neglected anyway when calculating this quantity. In this section the same notations are used. I repeat that for calculation of a quantity expressed in terms of one Green function one can use four-component vectors and the eight-component vectors in Eq. (6.97) are used only for representing all formulae in a unified way.

Using Eq. (6.1, 6.96) and calculating the Gaussian integrals over the symmetric matrices $\mathcal{H}_{mn}$ in Eqs. (6.95–6.97) we obtain as usual a new Lagrangian $\mathcal{L}$ with an "interaction"

$$\mathcal{L} = i \sum_{m=1}^{N} \bar{\psi}_m \left(\tilde{\varepsilon} - \frac{\omega+i\delta}{2}\Lambda\right)\psi_m$$

$$+ \frac{a^2}{4} \sum_{m,n=1}^{N} \left[(\bar{\psi}_m\psi_n)(\bar{\psi}_m\psi_n) + (\bar{\psi}_m\psi_n)(\bar{\psi}_n\psi_m) \right] \tag{6.98}$$

The interaction terms in Eq. (6.98) are very similar to the second and third terms in Eq. (4.26). Because of the structure of the supervectors ψ, Eq. (4.6), the two terms in the second line in Eq. (6.98) are equal to each other.

The next step of the calculation is to decouple the interaction term in Eq. (6.98) by Gaussian integration over supermatrices Q in the same way as in Section 4.2. As a result, we come to the effective Lagrangian $\mathcal{L}_{\text{eff}}$

$$\mathcal{L}_{\text{eff}} = i \sum_{m=1}^{N} \bar{\psi}_m \left(\bar{\varepsilon} - \frac{\omega + i\delta}{2} \Lambda - \frac{ia^2 \sqrt{N} Q}{2} \right) \psi_m \qquad (6.99)$$

To go further one should substitute the Lagrangian $\mathcal{L}_{\text{eff}}$ for $\mathcal{L}$ in Eqs. (6.95–6.97) and calculate Gaussian integrals over ψ. The result is a function of supermatrices Q and, as a final step, one has to integrate over the supermatrices with the weight

$$\exp \left(-\frac{Na^2}{8} \operatorname{str} Q^2 \right) \qquad (6.100)$$

Let us emphasize that because Q is an 8×8 supermatrix with no dependence on m, the calculation is simple. Carrying out the procedure one obtains

$$R(\omega) = \frac{N^2 \Delta^2(\varepsilon)}{128\pi^2} \operatorname{Re} \int \operatorname{str} (k(1+\Lambda) g(Q)) \left[\operatorname{str} (k(1+\Lambda) g(Q)) \right.$$

$$\left. - \operatorname{str} (k(1-\Lambda) g(Q)) \right] \exp \left(-\tilde{F}[Q] \right) dQ \qquad (6.101)$$

where

$$\tilde{F}[Q] = N \operatorname{str} \left[\frac{a^2}{8} Q^2 - \frac{1}{2} \ln \left(iy - \frac{i(\omega + i\delta)}{2\sqrt{N}} \Lambda + \frac{a^2}{2} Q \right) \right] \qquad (6.102)$$

and

$$g(Q) = N^{-1/2} \left(iy - \frac{i(\omega + i\delta)}{2\sqrt{N}} \Lambda + \frac{a^2 Q}{2} \right)^{-1}, \qquad y = \bar{\varepsilon} N^{-1/2} \qquad (6.103)$$

The corresponding formula for $\Delta^{-1}(\varepsilon)$ takes the form

$$\Delta^{-1}(\varepsilon) = \frac{N}{8\pi} \int \operatorname{str} (k\Lambda g(Q)) \exp \left(-\tilde{F}[Q] \right) dQ \qquad (6.104)$$

Up to now all the manipulations have been exact. Let us note a remarkable simplification due to the transformations made. Eqs. (6.101, 6.104) show that instead of calculating integrals over an arbitrarily large number N of variables we now have to calculate integrals over the elements of the 8×8 supermatrix Q, which is definitely a simpler task. Performing integration becomes especially simple in the most interesting limit of large N. We see that the exponent $\tilde{F}[Q]$ in the integrals Eqs. (6.101–6.104) is proportional to N and therefore one may use the saddle-point approximation. The extremum of the function $\tilde{F}[Q]$ at $\omega = 0$ is achieved at $\tilde{Q}$ satisfying the equation

$$\tilde{Q} = \left(iy + \frac{a^2 \tilde{Q}}{2} \right)^{-1} \qquad (6.105)$$

The formal solution $\tilde{q}$ of Eq. (6.105) reads

$$\tilde{q} = -\frac{iy}{a^2} \pm q, \qquad q = \frac{1}{a^2} \sqrt{2a^2 - y^2} \qquad (6.106)$$

When performing integration over the elements of Q one should shift the contours of integration in such a way that they intersect the saddle points; Section 4.2 explained how to do it. The proper saddle-point solution $\tilde{Q}$ is degenerate and can be written in the form

$$\tilde{Q} = -\frac{iy}{a^2} + qQ \tag{6.107}$$

with

$$Q = V\Lambda\bar{V}, \qquad V\bar{V} = 1 \tag{6.108}$$

Again, all notations are the same as in the previous chapters.

Using Eqs. (6.104–6.108) for the density of states $\Delta^{-1}(\varepsilon)$ we obtain

$$\Delta^{-1}(\varepsilon) = \frac{1}{\pi a^2} \left\{ \begin{array}{ll} \sqrt{2Na^2 - \varepsilon^2}, & |\varepsilon| < \sqrt{2N}a \\[2ex] 0, & |\varepsilon| > \sqrt{2N}a \end{array} \right. , \tag{6.109}$$

This is the famous Wigner semicircle law. The way it was obtained here is completely different from the derivations in Mehta (1991). The saddle-point approximation used previously is justified at large N provided the energy ε is not very close to the end points $\pm\sqrt{2N}a$. The degeneracy of the saddle-point solution described by Eq. (6.108) is not important for calculating $\Delta^{-1}(\varepsilon)$ but becomes essential when calculating $R(\omega)$. As usual, one should expand the exponent $\tilde{F}[Q]$, Eq. (6.102), in ω. Using Eqs. (6.105–6.108) one immediately obtains

$$\tilde{F}[Q] = -\frac{i\pi(\omega + i\delta)}{4\Delta(\varepsilon)} \operatorname{str}(\Lambda Q) \tag{6.110}$$

with $\Delta(\varepsilon)$ given by Eq. (6.109). Eq. (6.110) coincides with Eq. (6.36). Continuing the procedure one can reduce Eq. (6.101) to Eq. (6.41); that enables us to repeat the calculations presented in the previous section and to arrive at Eqs. (6.13–6.15). We see that although the mean level spacing $\Delta(\varepsilon)$ now depends on energy, the two-level correlation functions $R(\omega)$ preserve their form. Of course, ω must be much smaller than $\sqrt{N}a$ and ε must be not very close to the ends of the spectrum.

The derivation of this section is based on the assumption that the random matrices are distributed according to a Gaussian law. In fact, this restriction is not crucial. Recently, it has been demonstrated by Hackenbroich and Weidenmüller (1995) that in the limit $N \to \infty$ one comes to the zero-dimensional σ-model starting from an arbitrary distribution of the random matrices. Deriving the two-level correlation function from a model of a disordered metal one can systematically study corrections to the universal expressions, Eq. (6.13–6.15), which are a type of weak localization correction. These corrections are parametrically small in $\Delta/\mathcal{E}_c$. The first term of the expansion has been calculated by Kravtsov and Mirlin (1994). The oscillations in the two-level correlation function $R(\omega)$ survive even in the limit $\omega \gg \mathcal{E}_c$ although in this limit their amplitude is exponentially small. This fact was established by Altland and Fuchs (1995) and Andreev and Altshuler (1995).

The results of this section show that the supersymmetry technique can be used not only in the theory of disordered metals but also in all fields where RMT theory can be adequately applied. This means that the zero-dimensional σ-model can help in solving problems of nuclear physics and quantum chaos. The first application of supersymmetry in nuclear physics is due to Verbaarschot, Weidenmüller, and Zirnbauer (1985), who with the zero-dimensional σ-model calculated average compound-nucleus cross sections. Although the problem had been defined by various authors in the 1950s and 1960s in terms of an ensemble of random matrices coupled to a number of channels, only partial answers had been obtained in spite of numerous efforts. The general solution has been obtained only by Verbaarschot, Weidenmüller, and Zirnbauer (1985) and is also a demonstration of the effectiveness of the supersymmetry method.

7

Quantum size effects in small metal particles

7.1 Small metal particles

7.1.1 General properties

Small clusters of atoms of metallic elements have very unusual physical properties. These objects are not as small as molecules and some of their properties are reminiscent of those of bulk metals. At the same time the electron spectrum is discrete and this has many intriguing consequences. Systems of small metal particles are under very intensive study, and some of their characteristics have various technical applications. However, the physical properties of the metal clusters are no less interesting than possible applications because to describe them one needs modern quantum statistical theories analogous to those developed in nuclear physics. Moreover, some approaches invented studying the metal particles have their applications in nuclear physics and problems of quantum chaos. A complete account of experimental and theoretical results can be found in the review by Halperin (1986). An impression of the state of the art of both the experimental and the theoretical work can be gained from the proceedings of the Fourth International Meeting on Small Particles and Inorganic Clusters (Chapon, Gillet, and Henry (1989)). Some aspects of small particle physics are considered in recent reviews by Staveren, Brom, and de Jongh (1991) and Nagaev (1992).

Following the main idea of this book in this section I want to discuss several important steps in the theoretical understanding of the systems of small metal particles; then in the following sections of this chapter I present results obtained by the supersymmetry technique.

Of main interest here are electronic quantum size effects. Usually one can speak of these effects when the size of the particles varies from 10 Å to several hundred angstroms and this corresponds to clusters of several hundred to many millions of atoms. Electrons in such clusters are modeled as being confined in a box and, hence, because of the basic principles of quantum mechanics the energy levels are discrete. Of course, any finite isolated metal particle has a discrete energy spectrum. However, if the size is very large the distance between the levels becomes so small that the physical properties of the clusters are practically the same as those of the corresponding bulk metals.

Given the position of the electron energies one can calculate the partition function and, thus, all the thermodynamic quantities. For other physical phenomena, such as scattering of electromagnetic waves and nuclear magnetic resonance, information about wave functions is also necessary. The first attempt to calculate thermodynamic quantities for small metal particles was that of Fröhlich (1937), who assumed that all energy levels are equally spaced. Naturally, the result was that at temperatures smaller than the level spacing all dependencies of the thermodynamic quantities on temperature are exponential, with level spacing playing the role of activation energy.

The first question one can ask is how general this result is. It is obvious that one cannot expect equal level spacing, but is the temperature dependence exponential? For one particle or for a system of identical particles the answer is definitely, yes. At first glance, this is a general answer and it took quite a while to recognize (Kubo (1962)) that this answer is correct in exceptional cases only. This is because the experimentally available particles are never identical. For example, they can contain impurities in the bulk, and these impurities are different in different clusters. However, even particles fabricated from very clean metals have slightly different shapes. According to Kubo, the variation of the shape of the order of atomic distances is sufficient to change the spectrum completely although the mean level spacing Δ does not change. If all orbital degeneracies in the system are lifted, this quantity is equal to

$$\Delta = (\nu V)^{-1} \tag{7.1}$$

where ν is the density of states at the Fermi surface of the corresponding bulk metal, and V is the volume.

The mean level spacing Δ is of the order of ε_0 / N, where ε_0 is the Fermi energy and N is the number of the electrons in the cluster. For an aluminum particle with radius $a = 4.2 \times 10^{-7}$ cm one has $\Delta \approx 1$K. In general, for $N \sim 10^4$–10^5 the value of Δ is $\Delta \sim 0.1$–1K; that is, it is still quite large when the cluster is much larger than one molecule.

In order to describe the system, one should use a statistical theory making an assumption about the statistical properties of the energy spectrum or of the potential of the impurities. The former approach, although phenomenological, is technically simpler, and Kubo assumed that levels do not interact and any level "chooses" its position without "paying attention" to its neighbors. Then, the level spacing distribution $P(\tilde{\Delta})$ obeys the Poisson law.

$$P(\tilde{\Delta}) = \Delta^{-1} \exp\left(-\frac{\tilde{\Delta}}{\Delta}\right) \tag{7.2}$$

The assumptions of Fröhlich and Kubo are somewhat extreme because in the Kubo theory the energy levels do not interact, whereas in the Fröhlich picture they have infinite repulsion. Later Gorkov and Eliashberg (1965), using the analogy between the metal particles and complex nuclei, found it more natural to apply the random matrix statistical hypothesis and demonstrated how to calculate thermodynamic and kinetic quantities by this approach.

Let us present briefly their scheme of calculation of the response to an external electromagnetic field. The starting point is the Kubo linear response theory, which can be used provided the electric field E is weak enough that the inequality

$$eEa \ll \Delta \tag{7.3}$$

where a is the radius is fulfilled.

If the field changes little over the dimensions of the particle, the interaction $\hat{U}$ of the particle with the electric field E is written in the form

$$\hat{U} = -\mathbf{E}\hat{\mathbf{d}} \tag{7.4}$$

where $\mathbf{d}$ is the operator of the dipole moment of the cluster.

Using the linear response theory the dipole moment of the particle is written in the standard form (see Eq. (3.20))

$$d^\alpha (t) = i \int_{-\infty}^{\infty} \theta (t' - t) \left\langle \left[\hat{d}^\alpha (t), \hat{d}^\beta (t') \right] \right\rangle_T E^\beta (t') \, dt' \tag{7.5}$$

In Eq. (7.5), α, $\beta = 1, 2, 3$ numerate components of the dipole moment and electric field, and summation over the superscript β is implied. Other notations are the same as in Eq. (3.20). The operator of the dipole moment $\hat{\mathbf{d}}$ is related to the density operator $\hat{j}^0 (\mathbf{r}, t)$, Eq. (3.21), as

$$\hat{\mathbf{d}} (t) = \int \mathbf{r} \hat{j}^0 (\mathbf{r}, t) \, d\mathbf{r} \tag{7.6}$$

Using the frequency representation, Eq. (7.5) can be rewritten in terms of matrix elements of the coordinates

$$d_\omega^\alpha = -e^2 E_\omega^\beta \sum_{k,m} \frac{n (\varepsilon_k) - n (\varepsilon_m)}{\varepsilon_k - \varepsilon_m + \omega + i\delta} r_{km}^\alpha r_{mk}^\beta \tag{7.7}$$

The static part of Eq. (7.7) is large and Gorkov and Eliashberg (1965) separated it from the frequency-dependent contribution, writing for the susceptibility $\kappa (\omega)$

$$\kappa (\omega) = \kappa_0 - \frac{e^2 \omega}{3V} \sum_{k,m} \frac{n (\varepsilon_k) - n (\varepsilon_m)}{\varepsilon_k - \varepsilon_m} \frac{\mathbf{r}_{km} \mathbf{r}_{mk}}{(\varepsilon_k - \varepsilon_m)^2 - (\omega + i\delta)^2}, \tag{7.8}$$

$$\kappa_0 = -\frac{e^2}{3V} \sum_{k,m} \frac{n (\varepsilon_k) - n (\varepsilon_m)}{\varepsilon_k - \varepsilon_m} \mathbf{r}_{km} \mathbf{r}_{mk} \tag{7.9}$$

In Eqs. (7.8, 7.9) the susceptibility $\kappa (\omega)$ is defined as

$$d_\omega^\alpha = \kappa (\omega) V E_\omega^\alpha \tag{7.10}$$

and it is assumed that the shape of the particle is close to spherical.

Calculation of the sum in Eq. (7.9) is done by using the fact that the main contribution comes from states with energies close to the Fermi surface. Writing

$$n (\varepsilon_k) \approx n (\varepsilon_m) + \delta\varepsilon_k \frac{\partial n (\varepsilon_m)}{\partial \varepsilon_m}, \qquad \delta\varepsilon_k = \varepsilon_k - \varepsilon_m$$

one can immediately perform summation over m because $\partial n (\varepsilon_m) / \partial \varepsilon_m$ for $T \ll \mu$ is very close to $-\nu\delta (\varepsilon_m - \mu)$. Then one obtains

$$\kappa_0 = \frac{2}{3} e^2 \nu \langle r^2 \rangle_0 = \frac{e^2 a^2 m p_0}{5\pi^2} \tag{7.11}$$

In Eq. (7.11), the symbol $\langle \ldots \rangle_0$ stands for averaging over the volume of the particle. In fact, to obtain the static part of the susceptibility κ_0, Eq. (7.11), one does not need the discreteness of the energy spectrum and this result persists in the continuous model. Disorder is also not important for the calculation of this quantity.

The susceptibility κ_0 is very large, but that does not mean that one obtains a large electric permeability. The reason is that this is the response to a local electric field $\mathbf{E}_{loc}$ and one should use this local field in Eqs. (7.3–7.5, 7.7, 7.10). Experimentally, one measures the polarizability α_ω, which is the response to the *external* electric field. This point was discussed by Strässler, Rice, and Wyder (1972) and a more complete theory is given in Rice, Schneider, and Strässler (1973). In a rough approximation Devaty and Sievers (1980) and Halperin (1986) relate to each other the susceptibility κ_ω and polarizability α_ω of a particle as

$$\alpha(\omega) = \frac{V\kappa(\omega)}{1 + 4\pi M\kappa(\omega)} \tag{7.12}$$

where M is the depolarization factor (for spherical particles $M = 1/3$).

For large values of the susceptibility $\kappa(\omega) \gg 1$ one arrives at the classical electrostatic result $\alpha(\omega) = a^3$, which indicates that the electric field does not penetrate the metal particle. However, Eq. (7.12) is also not correct because $\mathbf{E}_{loc}$ was assumed to be homogeneous although different from the external field $\mathbf{E}$. In reality, the local field is essentially different from zero only near the surface, decaying at Fermi–Thomas screening radius $r_s = \left(8\pi e^2 \nu\right)^{-1/2}$. A proper study of the polarizability is presented in a recent publication (Efetov (1996)).

7.1.2 Gorkov and Eliashberg theory

In order to calculate the dynamic part, Eq. (7.8), one definitely needs more information about the energy spectrum as well as about the matrix elements of the coordinates. Of course, one could just start with a Hamiltonian with disorder, but such calculations became possible only with the invention of the supersymmetry technique and they are presented in the next section. Gorkov and Eliashberg suggested using the result of the RMT for the two-level correlation function $R(\omega)$, Eq. (6.13–6.15), instead of the direct scheme. With this assumption one can replace Eq. (7.8) by the expression

$$\tilde{\kappa}(\omega) = -\frac{e^2 \nu \omega^2}{3\Delta} \iint\limits_{-\infty}^{\infty} \frac{n(\varepsilon_1) - n(\varepsilon_2)}{\varepsilon_1 - \varepsilon_2} \frac{(\mathbf{r}_{12}\mathbf{r}_{21})\, R(|\varepsilon_1 - \varepsilon_2|)}{(\varepsilon_1 - \varepsilon_2)^2 - (\omega + i\delta)^2} d\varepsilon_1 d\varepsilon_2 \tag{7.13}$$

The integral over the energy difference converges at $|\varepsilon_1 - \varepsilon_2| \sim \omega$. If $\omega \ll \nu/a$, as assumed, then the dependence of $\mathbf{r}_{12}\mathbf{r}_{21}$ on energy can be neglected and this quantity can be taken at the Fermi energy. Replacing again $(n(\varepsilon_1) - n(\varepsilon_2))/(\varepsilon_1 - \varepsilon_2)$ by $-\nu\delta(\varepsilon - \mu)$, Eq. (7.13) can be reduced to the form

$$\kappa(\omega) = \kappa_0 + \frac{1}{3}e^2\,(\mathbf{r}_{12}\mathbf{r}_{21})_0\, \nu\tilde{A}(\omega), \tag{7.14}$$

$$\tilde{A}(\omega) = \frac{\omega^2}{\Delta} \int\limits_{-\infty}^{\infty} \frac{R(|\varepsilon|)\, d\varepsilon}{\varepsilon^2 - (\omega + i\delta)^2} \tag{7.15}$$

The integral over ε in Eq. (7.15) can be evaluated easily, and, at first glance, Eqs. (7.14, 7.15) completely solve the problem of the electromagnetic response. However, it is not quite so because one also has to calculate the average $\langle \mathbf{r}_{12}\mathbf{r}_{21}\rangle$. The very possibility of averaging this quantity independently is one more assumption but, in addition, RMT does not say anything about calculation of the matrix elements. Therefore, Gorkov and Eliashberg (1965) calculated the average $\langle \mathbf{r}_{12}\mathbf{r}_{21}\rangle$ using another elementary scheme. I do not write their final result here because in the next section susceptibility $\kappa(\omega)$ will be derived rigorously from the microscopic model with disorder and it is more relevant to present it there. It will be demonstrated that the frequency dependence of $\kappa(\omega)$ obtained by Gorkov and Eliashberg is correct. However, the microscopic computation shows that the static part of the susceptibility

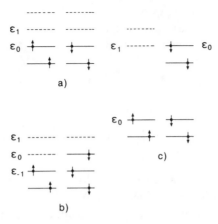

Fig. 7.1. Scheme of levels contributing to the partition function at low temperature: a) even number of electrons, b) odd number of electrons, c) levels in the magnetic field.

depends on the ensemble considered and can change when, for example, a magnetic field is applied.

Leaving discussion of these complicated effects for the next section let us consider some simple but curious results that can be obtained (Gorkov and Eliashberg (1965)) on the basis of the two-level correlation correlation function $R(x)$, Eq. (6.13–6.15). This function can be used for isolated metal particles with a discrete energy spectrum and with a fixed number of electrons in each particle. This is a quite realistic situation because in electroneutral clusters removing one electron requires the energy $W \sim e^2/a$, which is larger than other relevant energies. Although the opposite limit, when electrons can tunnel from cluster to cluster or to a supporting matrix, can apparently be experimentally achieved, too, the influence of the level statistics on the physical properties is more extreme in the former case.

In the limit $T \ll \Delta$, only the lowest electron excitations between neighboring levels are important and one can use the two-level correlation function for calculations. To calculate such a quantity as the specific heat one starts with calculation of the partition function Z. Because of the assumption of a fixed number of electrons in the cluster one should distinguish between cases with odd and even numbers of the electrons. Neglecting spin interactions and considering only several relevant levels near the Fermi energy we arrive at the pictures drawn in Fig. 7.1.

The solid lines correspond to the occupied levels; the dashed lines denote the empty ones. Each picture consists of two sequences of levels corresponding to the different spin directions. Figure 7.1a represents even filling. Taking spin degeneracy into account the partition function Z_{even} corresponding to the lowest excitations is equal to

$$Z_{even} = 1 + 4 \exp\left[-\frac{\varepsilon_1 - \varepsilon_0}{T}\right] + \exp\left[-\frac{2(\varepsilon_1 - \varepsilon_0)}{T}\right] \qquad (7.16)$$

where ε_0 is the energy of the last occupied level and ε_1 of the first empty level.

The state in Fig. 7.1b is doubly degenerate, and the lowest excitations give the following contribution to the partition function Z_{odd}:

$$Z_{odd} = 2 + 2 \exp\left[-\frac{\varepsilon_0 - \varepsilon_{-1}}{T}\right] + 2 \exp\left[-\frac{\varepsilon_{+1} - \varepsilon_0}{T}\right] \qquad (7.17)$$

In Eq. (7.17) ε_{-1} denotes the energy of the last completely filled level, ε_0 is the energy of the level with one electron, and ε_{+1} is the first empty level.

Assuming that the particles with even and odd numbers of electrons are equally probable one should average free energy $F = -T \ln Z$ over both possibilities and then integrate over the probability $P(\tilde{\Delta})$ of a given spacing $\tilde{\Delta}$ between neighboring levels. As a result one obtains

$$F = -\frac{T}{2} \ln 2 - T \int_0^\infty \left[3 \exp\left(-\frac{\tilde{\Delta}}{T} \right) + \frac{1}{2} \exp\left(-\frac{2\tilde{\Delta}}{T} \right) \right] P(\tilde{\Delta}) d\tilde{\Delta} \qquad (7.18)$$

The expression in the square brackets in Eq. (7.18) is written in the limit $T \ll \tilde{\Delta}$. When integrating over $\tilde{\Delta}$ the main contribution is from $\tilde{\Delta} \sim T$; that means that, strictly speaking, Eq. (7.18) is not accurate. However, it gives the correct temperature dependence of the free energy at low temperature $T \ll \Delta$, a dependence determined completely by the asymptotics of $P(\tilde{\Delta})$ at $\tilde{\Delta} \to 0$. The Poisson distribution suggested by Kubo, Eq. (7.2), can hardly correspond to small metal particles available experimentally, but it can describe integrable quantum billiards (see the discussion in Section 6.1). In this case $P(0) = \text{const}$ and the corresponding specific heat C_0 is proportional to temperature,

$$C_0 \propto T \qquad (7.19)$$

The specific heat C_β derived from the function $R(x)$, Eqs. (6.13–6.15) with $\beta = 1, 2, 4$ for the orthogonal, unitary, and symplectic ensembles is determined by the asymptotics, Eq. (6.90), and obeys the law

$$C_\beta \propto T^{\beta+1} \qquad (7.20)$$

A more accurate derivation of the low temperature, $T \ll \Delta$, asymptotics of specific heat C with proper coefficients as well as an approximate treatment of the region $T \sim \Delta$ was obtained by Denton, Mühlschlegel, and Scalapino (1973). In the opposite limit, $T \gg \Delta$, the discreteness of the energy spectrum is not important to the calculation of the specific heat and its average value coincides with that of the bulk metal.

Another quantity that becomes very interesting in isolated metal particles is magnetic susceptibility χ. In the presence of strong spin-orbit scattering or at high temperatures this susceptibility always coincides with the Pauli susceptibility

$$\chi_P = 2 (g\mu_B)^2 \nu \qquad (7.21)$$

where $g\mu_B$ is the coefficient determining the Zeeman splitting of electron states, and μ_B is the Bohr magneton. Unusual results can be obtained, however, for the orthogonal ensemble where spin remains a good quantum number. In this case one can speak separately about electrons with "up" and "down" spin. Let the field H have an arbitrary value ($\mu_B H$ comparable with Δ) and consider the case with an even number of electrons. The level scheme of a single particle is represented in Fig. 7.1c, where $\varepsilon_0^\uparrow$ and $\varepsilon_0^\downarrow$ stand for the energies of the last filled levels with the spins parallel (spin "up") and antiparallel (spin "down") to the field, respectively. Increasing the field from H to $H + dH$ it may turn out that the energy level $\varepsilon_0^\downarrow$ crosses the first empty spin up state with the energy $\varepsilon_1^\uparrow$, which leads to jumping of the electron from the level $\varepsilon_0^\downarrow$ to the level $\varepsilon_1^\uparrow$ and, thus, to spin flip. As a result the magnetic moment M increases by $2\mu_B gH$. The level sequences $\varepsilon_i^\uparrow$ and $\varepsilon_i^\downarrow$ are shifted one from the other by $2\mu_B gH$. This means that the probability of the crossing is equal to the probability of finding two levels in each sequence separated by the energy $2\mu_B gH$ multiplied by $\nu\mu_B gdH$ and for magnetic susceptibility χ we obtain

$$\chi = \frac{dM}{dH} = \chi_p R_{\text{orth}}(2g\mu_B H) \tag{7.22}$$

with function $R_{\text{orth}}(\omega)$ given by Eq. (6.13).

In the limit of weak magnetic fields, $g\mu_B H \ll \Delta$, susceptibility χ is small and can be obtained from Eq. (6.90)

$$\chi = \frac{\chi_p \pi^2 g\mu_B H}{3\Delta} \tag{7.23}$$

In the opposite limit of high magnetic fields, it approaches the Pauli susceptibility. Let us notice that the characteristic field $g\mu_B H_0 \sim \Delta$ of the crossover from the low field susceptibility, Eq. (7.23), to the Pauli susceptibility χ_p is different from the characteristic field H_c of the crossover from the orthogonal to the unitary ensemble given by Eq. (6.40). This fact was not noted in the phenomenological consideration of Gorkov and Eliashberg (1965), who identified H_0 with the field of the crossover between the ensembles. However, the field H_0 corresponds to interaction of the magnetic field with the electron spin, whereas the crossover between the ensembles is due to orbital motion. The ratio of the fields can be estimated for a particle with characteristic sizes L as $H_0/H_c \sim g^{-1}(l/L)^{1/2}$, which, for spherical particles, is smaller than unity. In samples with other geometries, such as rings, this ratio can become large. In the latter case, for the magnetic fields $H \geq H_c$, one should use in Eq. (7.22) $R_{\text{unit}}(\omega)$ instead of $R_{\text{orth}}(\omega)$.

In the case of an odd number of electrons an important paramagnetic contribution comes from the spin of the single electron at the last level. At low but finite temperatures $T \ll \Delta$, the susceptibility χ_{ave} averaged over both the disorder and the parity of the number of electrons was calculated by Denton, Mühlschlegel, and Scalapino (1973) and can be written as

$$\chi_{\text{ave}} = \chi_p \left[\Delta \left(4T \cosh^2\left(\frac{g\mu_B H}{2T}\right) \right)^{-1} \right. \tag{7.24}$$

$$\left. + R(2g\mu_B H) + \left[\frac{d^2 R(\varepsilon)}{d\varepsilon^2}\right]_{\varepsilon=2g\mu_B H} \frac{(\pi T)^2}{6} \right]$$

It is clear from Eq. (7.24) that the first term due to the free spin in the "odd" particles at $T \ll \Delta$ can become larger than the rest, resulting in Curi-law behavior, thus making experimental observation of the most interesting effects of the level correlation difficult. In such experiments as those using nuclear magnetic resonance (NMR), where one can measure the local magnetic susceptibility, the different contributions of the "odd" and "even" particles must lead to two different peaks on the NMR line. However, to resolve the peaks all particles must be very homogeneous in size. A detailed presentation of the experimental situation can be found in the review by Halperin (1986).

The discussion in this section demonstrates how some very interesting phenomena can be described by phenomenological theories, and it is clear that RMT is the best for this purpose. However, difficulties arise each time one is to calculate quantities containing not only energy levels but also some matrix elements. Although I do not exclude the possibility that in some cases these more complicated physical quantities can eventually be obtained within the

RMT scheme, it is clear that it would require inventing additional tricks and more work. At the same time, the presence of matrix elements does not lead to additional difficulties in the supersymmetry scheme and calculation of, for example, the electric susceptibility or some distribution functions is no more difficult than calculation of the two-level correlation function.

7.2 Electric susceptibility

7.2.1 Reduction to the σ-model

It was emphasized previously that the local electric field in the metal cluster is essentially different from zero near the surface only. Fortunately, the screening effect can rather easily be taken into account using the random phase approximation (RPA) (Efetov (1996)). To avoid complicated formulae details of this calculation are not presented here. What is important, the many-particle problem for the system of interacting electrons reduces to the one-particle problem of electron motion in an effective electric field that is to be found from the Poisson equation. The field varies slowly and one can calculate the local susceptibility neglecting space variations of the effective field. Its nontrivial dependence on coordinates can be taken into account at the end of calculation when relating the polarizability of the cluster to the local susceptibilities. So let us calculate the susceptibility as if the electric field were homogeneous. At the end it will be shown how one should change the result to take into account the screening effects.

Calculating the average local dipole moment one can start as in the preceding section with the exact formula for the linear response, Eq. (7.7), that was derived from Eq. (7.5). For a small metal particle, Eq. (7.7) taken in the limit $\omega \to 0$ describes not only the dynamic response but also the thermodynamic dipole moment. This is because the energy spectrum is discrete and one can come to the static limit of Eq. (7.7) by using the second order of the perturbation theory. By applying electric field $\mathbf{E}$ one shifts energy levels. Choosing coordinates such that the average dipole moment is zero in the absence of the external electric field and considering the interaction with the electric field, Eq. (7.4), as a perturbation we do not obtain any contribution to the shift of the eigenenergies in the first order, but the second order takes the form

$$\varepsilon_k^{(2)} = e^2 \sum_{m \neq k} \frac{|\mathbf{E}\mathbf{r}_{km}|^2}{\varepsilon_k - \varepsilon_m} \tag{7.25}$$

In Eq. (7.25) ε_k are eigenenergies of the system without the electric field, $\mathbf{r}_{km}$ are matrix elements calculated with eigenfunctions $\varphi_k(\mathbf{r})$. Writing the thermodynamic potential Ω as

$$\Omega = -T \sum_k \ln \left(1 + \exp \left(\frac{\mu - \bar{\varepsilon}_k}{T}\right)\right) \tag{7.26}$$

where $\bar{\varepsilon}_k$ are the eigenenergies in the presence of the electric field, and differentiating over the electric field we obtain for the thermodynamic dipole moment $\mathbf{d}$

$$d^\alpha = \sum_k n(\bar{\varepsilon}_k) \left(\frac{\partial \bar{\varepsilon}_k}{\partial E^\alpha} - \frac{\partial \mu}{\partial E^\alpha}\right) \tag{7.27}$$

In the case of isolated particles considered here, applying the electric field does not change the total number of electrons in the volume but can lead, in principle, to a shift of the chemical potential. This is taken into account by the second term in parentheses in Eq. (7.27). For some problems, working with a canonical ensemble rather than with the grand canonical one is important, and this question will be considered in Chapter 8 in the context of the problem of persistent currents in mesoscopic rings. However, the difference between the canonical and grand canonical ensembles is not important to the calculation of electric susceptibility. Some explanation of this fact will be given in Chapter 8, but now let us just neglect the derivative of the chemical potential in Eq. (7.27), thus performing calculations within the grand canonical ensemble.

Using Eqs. (7.25, 7.27) we can then write the thermodynamic dipole moment in the form

$$d^\alpha = -e^2 E^\beta \sum_{k \neq m} \frac{n(\varepsilon_k) - n(\varepsilon_m)}{\varepsilon_k - \varepsilon_m} r_{km}^\alpha r_{mk}^\beta \qquad (7.28)$$

It is not difficult to see that the dynamic dipole moment given by Eq. (7.7) exactly coincides in the limit $\omega \to 0$ with the thermodynamic dipole moment (the term with $k = m$ in the sum in Eq. (7.7) does not contribute to the sum for any small but finite ω). Therefore, by calculating the dynamic response one can also describe thermodynamic quantities because, again, the states are not degenerate.

To apply the supersymmetry technique one has to express, first, the quantity to be calculated in terms of Green functions. The first equation, Eq. (7.5), can easily be rewritten through the density–density correlation function $X^{00}(\mathbf{r}, \mathbf{r}', t)$, Eqs. (3.20, 3.22), multiplying the latter by $r^\alpha r^\beta$, integrating over the coordinates, and using the frequency representation. We already know how the correlation function $X^{00}(\mathbf{r}, \mathbf{r}', t)$ is expressed in terms of the retarded G_ε^R and advanced G_ε^A Green functions (cf. Eqs. (3.23, 3.24)) and can, as usual, write the average electric susceptibility $\kappa(\omega)$ as the sum of two terms

$$\kappa(\omega) = \kappa_1(\omega) + \kappa_2(\omega), \qquad (7.29)$$

$$\kappa_1(\omega) = -\frac{e^2}{2\pi i V} \sum_{\sigma, \sigma'} \int (n(\varepsilon - \omega) - n(\varepsilon))$$

$$\times \langle G_\varepsilon^R(y, y') G_{\varepsilon-\omega}^A(y', y) \rangle r^\alpha r'^\alpha d\mathbf{r} d\mathbf{r}' d\varepsilon \qquad (7.30)$$

$$\kappa_2(\omega) = \frac{e^2}{2\pi i V} \sum_{\sigma, \sigma'} \int n(\varepsilon) \langle \left[G_{\varepsilon+\omega}^R(y, y') G_\varepsilon^R(y', y) \right. $$

$$\left. - G_\varepsilon^A(y, y') G_{\varepsilon-\omega}^A(y', y) \right] \rangle r^\alpha r'^\alpha d\mathbf{r} d\mathbf{r}' d\varepsilon \qquad (7.31)$$

where the angular brackets stand for averaging over impurities, $y = (\mathbf{r}, \sigma)$. It is assumed that the dipole moment induced by the electric field is parallel to the field.

Again, as everywhere in the preceding chapters, calculation of $\kappa_2(\omega)$ is simple because the average of the products of the Green functions entering this quantity is reduced to the product of the averages determined by Eq. (3.45). Using the identity

$$2r^\alpha r'^\alpha = \left[(r^\alpha)^2 + (r'^\alpha)^2 \right] - \left[r^\alpha - r'^\alpha \right]^2 \qquad (7.32)$$

and replacing $2r^\alpha r'^\alpha$ in Eq. (7.31) by the right-hand side of Eq. (7.32) we can represent the integral in Eq. (7.31) as a sum of two terms. The integral with the second term in the right-hand side of Eq. (7.32) can be written in the momentum representation and is very small because the contribution is only from momenta p far from the Fermi surface. Only the first integral is important and $\kappa_2(\omega)$ can be written as

$$\kappa_{2\omega} = \frac{e^2}{\pi i} \left\langle \left(r^\alpha\right)^2 \right\rangle_0 \int n(\varepsilon) \left[\langle G^R_{\varepsilon+\omega}(\mathbf{p}) \rangle \langle G^R_\varepsilon(\mathbf{p}) \rangle \right.$$

$$\left. - \langle G^A_\varepsilon(\mathbf{p}) \rangle \langle G^A_{\varepsilon-\omega}(\mathbf{p}) \rangle \right] \frac{d\varepsilon d\mathbf{p}}{(2\pi)^3} \tag{7.33}$$

where $\langle \ldots \rangle_0$ denotes, as in the previous section, averaging over the metal cluster, which is, however, now assumed to have an arbitrary shape. The average Green functions entering Eq. (7.33) are determined by Eq. (3.45). The dependence of $\kappa_2(\omega)$ on ω is not important as soon as $\omega \ll \mu$ and can be neglected. Using the identity

$$\frac{\partial G^{R,A}_\varepsilon(\mathbf{p})}{\partial \varepsilon} = -\left(G^{R,A}_\varepsilon(\mathbf{p})\right)^2$$

and integrating over ε by parts we obtain with the help of Eq. (3.16)

$$\kappa_2(\omega) = \kappa_0 = 2e^2 \nu \left\langle \left(r^\alpha\right)^2 \right\rangle_0 \tag{7.34}$$

Comparing Eq. (7.34) with Eq. (7.11), which was written for spherical particles, we see that the values for κ_0 from these equations agree with each other. It is clear from the derivation that the discreteness of the energy levels is not important for $\kappa_2(\omega)$ and it could be obtained in the continuous approximation. Of course, for nonspherical particles the value of the susceptibility depends on the direction of the electric field, as is clear from Eq. (7.34).

The contribution $\kappa_1(\omega)$ is much more interesting and is the quantity that is sensitive to discreteness. The product of the retarded G^R and advanced G^A Green functions can be written as the integral of the product of four supervectors ψ over the supervectors, Eq. (4.40), with the effective Lagrangian $\mathcal{L}_{\text{eff}}[\psi, \psi, Q]$, Eq. (4.19), which is quadratic in ψ. Integrating over ψ one can reduce the integral to an integral over the supermatrices Q. Calculating Gaussian integrals over ψ with the use of the Wick theorem one can proceed in the same way as in the preceding chapters. Averaging the product of four supervectors one has three possible pairings, which lead to three terms. Representing $\kappa_1(\omega)$ in the form

$$\kappa_{1\omega} = \frac{i\omega}{V} \int R^{00}(\mathbf{r}, \mathbf{r}', \omega) r^\alpha r'^\alpha d\mathbf{r} d\mathbf{r}' \tag{7.35}$$

where

$$R^{00}(\mathbf{r}, \mathbf{r}', \omega) = \frac{e^2}{2\pi} \sum_{\sigma,\sigma'} \left\langle G^A_{\varepsilon-\omega}(y', y) G^R_\varepsilon(y, y') \right\rangle \tag{7.36}$$

(see also Eq. (3.31)), we have to write three terms for the function $R^{00}(\mathbf{r}, \mathbf{r}', \omega)$

$$R^{00}(\mathbf{r}, \mathbf{r}', \omega) = R^{00}_1(\mathbf{r}, \mathbf{r}', \omega) + R^{00}_2(\mathbf{r}, \mathbf{r}', \omega) + R^{00}_3(\mathbf{r}, \mathbf{r}', \omega) \tag{7.37}$$

When performing analogous averaging in Chapters 4 and 6, Eqs. (4.40, 6.30), it was sufficient to keep only the terms where supervectors ψ taken at the same point were paired because the other terms were small. However, now pairing of the product of the averages of ψ taken at the same point gives zero in Eq. (7.35) and the other terms become important.

Let us examine it in more detail. Using the commutation relation for the components of the vectors ψ and Eq. (4.41) one can obtain

$$R_1^{00}\left(\mathbf{r}, \mathbf{r}', \omega\right) = -\frac{1}{2\pi}\left\langle \sum_{\sigma,\sigma'}\left[\hat{\pi}_{\mathbf{r}'}^0 g_{33}^{11}\left(y', y\right)\right]\left[\hat{\pi}_{\mathbf{r}}^0 g_{33}^{22}\left(y, y'\right)\right]\right\rangle_Q \qquad (7.38)$$

$$R_2^{00}\left(\mathbf{r}, \mathbf{r}', \omega\right) = -\frac{1}{2\pi}\left\langle \sum_{\sigma,\sigma'}\lim_{2\to 1}\hat{\pi}_{\mathbf{r}_1}^0\hat{\pi}_{\mathbf{r}_1}^0 g_{33}^{12}\left(y_1', y_2'\right) g_{33}^{21}\left(y_1, y_2\right)\right\rangle_Q \qquad (7.39)$$

$$R_3^{00}\left(\mathbf{r}, \mathbf{r}', \omega\right) = -\frac{1}{2\pi}\left\langle \sum_{\sigma,\sigma'}\lim_{2\to 1}\hat{\pi}_{\mathbf{r}_1}^0\hat{\pi}_{\mathbf{r}_1}^0 g_{34}^{12}\left(y_1', y_1\right) g_{43}^{21}\left(y_2, y_2'\right)\right\rangle_Q \qquad (7.40)$$

In Eqs. (7.38–7.40) $\hat{\pi}_{\mathbf{r}}^0 = e$ is the charge operator. The notation $\lim_{2\to 1}$ means that one should take at the end $y_1 = y_2$, $y_1' = y_2'$, and $\langle\ldots\rangle_Q$ represents averaging with the nonlinear σ-model. In Chapter 8, corresponding equations derived for currents will have the same form as Eq. (7.38, 7.40) provided the charge operators $\hat{\pi}_{\mathbf{r}}^0$ are replaced by the current operators $\hat{\pi}_{\mathbf{r}}^\alpha$, Eq. (3.30). Writing operator $\hat{\pi}_{\mathbf{r}}^0$ and not just e helps to emphasize the correspondence to each other of the equations for the dipole moments and currents. The matrix function $g(y, y')$ satisfies Eq. (4.42).

Considering Models I and IIa when spin interactions are not important one does not need to write spin blocks in the supermatrices Q and in this case they have the size 8×8. Then one can omit all the spin variables σ, σ' and the sums over these variables in Eqs. (7.38–7.40), and $g_{\alpha\beta}^{12}$, $g_{\alpha\beta}^{21}$ are the matrix elements of the 4×4 supermatrices g^{12} and g^{21} (the superscripts 3 and 4 relate to the boson–boson block, the elements g_{33}, g_{44} to the diffusons, and g_{34}, g_{43} to cooperons). The presence of spin is taken into account in Models I and IIa by multiplying the final result by 2. Of course, one could use 16×16 supermatrices Q and retain these equations as they stand, but this would lead to more complicated calculations. In Models IIb and III, describing systems with magnetic and spin-orbit impurities, respectively, the spin blocks of supermatrices Q are unit matrices and again one effectively has 8×8 supermatrices Q.

For the problem of small metal particles one should use the zero-dimensional σ-model. Using Eqs. (7.35–7.40) and Eq. (4.42) calculation of $\kappa_1(\omega)$ can be reduced to a definite integral over Q. In fact, not all parts R_1, R_2, and R_3 contribute. If Q does not vary in space and the coordinates are chosen in such a way that Eq. (7.43) is fulfilled one can see immediately that R_2, Eq. (7.39), does not contribute to the term $\kappa_1(\omega)$ because the integral over the coordinates vanishes. Substituting R_1^{00}, Eq. (7.38), and R_3^{00}, Eq. (7.40), into Eq. (7.35) one can rather easily calculate the integral over the coordinates. Again, one can use the identity Eq. (7.32) and decouple the integrals into two parts. The main contribution is from the integral with the first term in the right-hand side in Eq. (7.32). Using the 8×8 supermatrices with symmetries corresponding to the orthogonal, unitary, and symplectic ensembles one can write the result in a form applicable to all the ensembles.

$$\kappa_1\left(\omega\right) = ie^2 v\left\langle\left(r^\alpha\right)^2\right\rangle_0 \omega\tau\left[1 - \left\langle Q_{33}^{11}Q_{33}^{22}\right\rangle_Q \mp \left\langle Q_{34}^{12}Q_{43}^{21}\right\rangle_Q\right] \qquad (7.41)$$

$$= 2ie^2 v\langle(r^\alpha)^2\rangle_0\omega\tau\left\{1 - \frac{1}{32}\langle\mathrm{str}\left[k(1-\tau_3)(Q-\Lambda)^{11}\right]\right. \qquad (7.42)$$

$$\times \left.\mathrm{str}\left[k\left(1-\tau_3\right)\left(Q-\Lambda\right)^{22}\right]\rangle_Q \pm \frac{1}{32}\langle\mathrm{str}\left[k\left(1-\tau_3\right) Q^{12}k\left(1+\tau_3\right) Q^{21}\right]\rangle_Q\right\}$$

Eqs. (7.41, 7.42) are two equivalent forms of the same integral. The equivalence is based on the identity $\langle Q \rangle_Q = \Lambda$ and on the symmetry of supermatrices Q. The same symmetry was used in writing the equivalent forms for the density–density correlation function, Eqs. (4.57, 4.58). The last terms in Eqs. (7.41, 7.42) are equal to zero in the unitary ensemble because they contain the "cooperon matrix elements" but have a finite contribution in the orthogonal and symplectic ensembles. The upper and lower signs should be taken in the orthogonal and symplectic case, respectively. In fact, Eqs. (7.41, 7.42) are applicable to an arbitrary magnetic field provided the proper form of free energy $F[Q]$ is taken. These equations can be used to describe the entire crossover from the orthogonal or symplectic to the unitary ensemble. We concentrate on such calculations in Chapter 8.

Eqs. (7.34, 7.41, 7.42) are written for a homogeneous electric field $\mathbf{E}$, which corresponds to electrostatic potential $\Phi_0(\mathbf{r}) = -\mathbf{E}\mathbf{r}$. In an electric field depending on coordinates in these equations one should make the replacement

$$r^\alpha \rightarrow \frac{\Phi(\mathbf{r})}{E} \tag{7.43}$$

where $\Phi(\mathbf{r})$ is the electrostatic potential corresponding to this electric field. It is assumed that at $r \rightarrow \infty$ the potential $\Phi(\mathbf{r})$ approaches the function $\Phi_0(\mathbf{r})$.

7.2.2 Calculation of integrals over supermatrix Q

According to the discussion presented in the previous section one can expect that the frequency-dependent electric susceptibility is expressed in terms of the two-level correlation function $R(\omega)$, and the imaginary part is proportional to this function. Comparing Eqs. (7.42) and (6.41) we see that the real part of the first two terms in the parentheses in the former equation coincides with the function $R(\omega)$ in the latter. But what about the third term in Eq. (7.42)? Does it add something to the contribution of the level–level correlation function $R(\omega)$?

Of course, this term is zero in the unitary ensemble, but the question is not as simple in the general case. It turns out that this term is a nontrivial function of the frequency and the magnetic field. However, in the limiting cases of the orthogonal and symplectic ensembles it is proportional to i/ω and, although contributing to static susceptibility, does not contribute to the imaginary part of the electric susceptibility.

The third term can be calculated explicitly but the result can also be obtained in a simpler way. Let us calculate it for the orthogonal ensemble. First, recall the function $Y^{00}(y, y', \varepsilon, \omega)$ introduced in Eq. (3.24). The average of this function, $Y^{00}(\mathbf{r}, \varepsilon, \omega)$, Eq. (3.49), has been represented as an integral over supermatrices Q, Eq. (4.58). The latter integral has almost the same form as the third term in Eq. (7.42). The only difference is that in Eq. (4.58) "diffuson matrix elements" are written, whereas in Eq. (7.42) only "cooperon" elements are present. However, in the orthogonal ensemble there must be symmetry between the cooperons and diffusons and one can expect that the integrals in Eqs. (4.58) and (7.42) are the same. This can be seen easily by introducing the function $\tilde{Y}^{00}(y, y', \varepsilon, \omega)$

$$\tilde{Y}^{00}(y, y', \varepsilon, \omega) = \sum_{\sigma, \sigma'} G^A_{\varepsilon-\omega}(y, y') \left(G^R_\varepsilon(y, y') - G^A_\varepsilon(y, y') \right) \tag{7.44}$$

and repeating the steps leading to Eq. (4.58). As a result, one obtains exactly the same integral as in Eq. (7.42) with the cooperon matrix elements instead of the diffuson ones.

The functions Y^{00} and $\tilde{Y}^{00}$ differ if the eigenfunctions $\varphi_k(y)$ are complex, but in the orthogonal case all eigenfunctions can be chosen real and therefore functions Y^{00} and $\tilde{Y}^{00}$ must coincide, thus proving the equivalence of the third term in Eq. (7.42) and the integral, Eq. (4.58). Using the identity Eq. (3.35), the coincidence of the functions Y^{00} and $\tilde{Y}^{00}$ for the orthogonal ensemble, and the fact that the zero-dimensional version of the σ-model is considered, we obtain

$$\left\langle \mathrm{str}\left[k\,(1-\tau_3)\,Q^{12}k\,(1+\tau_3)\,Q^{21} \right]\right\rangle_Q = \frac{32\Delta}{-i\pi\omega} \tag{7.45}$$

Of course, the equality of the integral over Q in Eq. (7.45) and the integral I_d

$$I_d = \left\langle \mathrm{str}\left[k\,(1-\tau_3)\,Q^{12}k\,(1-\tau_3)\,Q^{21} \right]\right\rangle_Q \tag{7.46}$$

that determines the density–density correlation function can be established directly by using the parametrization Eqs. (6.42–6.52) for supermatrix Q and integrating over the 4×4 supermatrices u and v. Simple integration of integrals that arise over the anticommuting variables gives

$$\int \bar{u}_1 k\,(1 \pm \tau_3)\,u_1 dR_\eta^{(\mathrm{I})} = 16\,(4 \pm (1+k)\,\tau_3) \tag{7.47}$$

and the same expression for integration over κ. We see that the difference between the two integrals under consideration can be due to different signs of τ_3. But the next step is to calculate the integral over the elements of matrix $\mathcal{F}_1$. After elementary manipulation one can obtain

$$\int \bar{\mathcal{F}}_1 \tau_3 \mathcal{F}_1 dR_2^{(\mathrm{I})} = 0$$

which shows that the difference between the integral I_d, Eq. (7.46), and the third term in Eq. (7.42) vanishes. Let us emphasize that this equality is valid only when time reversal symmetry is preserved. In Appendix 3 it is shown that in an arbitrary magnetic field one obtains

$$I_d = \frac{32\Delta}{-i\pi\omega} \tag{7.48}$$

which is a demonstration of particle conservation, Eq. (3.35). At the same time Eq. (7.45) does not hold in any nonzero magnetic field. Eqs. (7.45, 7.48) are also applicable to the symplectic ensemble, provided one replaces Δ by $\bar{\Delta} = \Delta/2$. This result can be obtained in the same way as for the orthogonal ensemble.

A contribution of the type i/ω also comes from the second term in Eqs. (7.41, 7.42). The imaginary part of this term is proportional to $\delta(\omega)$ and describes the self-correlation of a single level that is not interesting. However, the real part of the terms of type i/ω leads to an additional contribution to static susceptibility. Calculation of the integral over Q in the second terms in Eqs. (7.41, 7.42) is no more difficult than calculation of its real part presented in Chapter 6, and the final result for electric susceptibility can be written as

$$\kappa(\omega) = \kappa_0 \left(1 + \frac{1}{2\pi}\tau \Delta A(x) \right) \tag{7.49}$$

where κ_0 is given by Eq. (7.34), $x = \pi\omega/\Delta$ for Models I and IIa, and $x = 2\pi\omega/\Delta = \pi\omega/\bar{\Delta}$ for Models IIb and III. Function $A(x)$ contains both real and imaginary parts such that

$$A(x) = A_1(x) + i A_2(x) \tag{7.50}$$

The explicit formulae for the functions $A_1(x)$ and $A_2(x)$ take the form

$$A_1^{(I)}(x) = 2\left(-1 - \frac{\sin 2x}{2x} + x\,\mathrm{Ci}\,(x)\frac{d}{dx}\left(\frac{\sin x}{x}\right)\right) \tag{7.51}$$

$$A_2^{(I)}(x) = 2\left(x - \frac{\sin^2 x}{x} - x\left(\frac{\pi}{2} - \mathrm{Si}\,(x)\right)\frac{d}{dx}\left(\frac{\sin x}{x}\right)\right) \tag{7.52}$$

$$A_1^{(IIa)}(x) = -\frac{\sin 2x}{x}, \qquad A_1^{(IIb)}(x) = -\frac{\sin 2x}{2x} \tag{7.53}$$

$$A_2^{(IIa)}(x) = 2\left(x - \frac{\sin^2 x}{x}\right), \qquad A_2^{(IIb)}(x) = x - \frac{\sin^2 x}{x} \tag{7.54}$$

$$A_1^{(III)}(x) = 1 - \frac{\sin 2x}{2x} + x\,\mathrm{Si}\,(x)\frac{d}{dx}\left(\frac{\cos x}{x}\right) \tag{7.55}$$

$$A_2^{(III)}(x) = x - \frac{\sin^2 x}{x} + x\,\mathrm{Si}\,(x)\frac{d}{dx}\left(\frac{\sin x}{x}\right) \tag{7.56}$$

where $\mathrm{Ci}\,(x)$ and $\mathrm{Si}\,(x)$ stand for the integral cosine and sine

$$\mathrm{Ci}\,(x) = -\int_1^\infty \frac{\cos tx}{t}\,dt, \qquad \mathrm{Si}\,(x) = \int_0^1 \frac{\sin xt}{t}\,dt$$

Eqs. (7.49–7.56) completely describe the frequency dependence of the electric susceptibility $\kappa(\omega)$ for all ensembles. The imaginary part of this function is proportional to the two-level correlation function $R(\omega)$ multiplied by ω. This is a natural result if fluctuations of the dipole matrix elements are statistically independent of energy level fluctuations. Indeed, the imaginary part of the susceptibility determines the electromagnetic field absorption. To absorb a photon with a frequency ω one has to find two levels at distance ω from each other, with one of the states occupied, the other empty. The probability of this event is just the function $\omega R(\omega)$, which is seen from the definition of $R(\omega)$, Eq. (3.36). Forgetting about fluctuations of the dipole matrix elements, one obtains from $R(\omega)$ the frequency dependence of the imaginary part of $\kappa(\omega)$.

Thus, the agreement of the imaginary part of the susceptibility, Eqs. (7.52, 7.54, 7.56), with the two-level correlation function $R(\omega)$ signals the statistical independence of the dipole matrix elements. Nevertheless, the matrix elements and energy levels seem to be statistically independent only in the limiting cases of the orthogonal, unitary, and symplectic ensembles. In the region of the crossover between the ensembles, which can be achieved by applying a magnetic field to a system with time reversal invariance, the third term in Eqs. (7.42) becomes nontrivial and contributes to the imaginary part of the susceptibility. At the same time, the first two terms in this equation give the two-level correlation function at arbitrary fields. Any nontrivial contribution of the third term can only be attributed to a correlation between the dipole matrix elements and the energy levels. The electric susceptibility at an arbitrary magnetic field can be obtained by using the results of Chapter 8. Its dependence on the magnetic field is represented by the difference of the integrals in Eqs. (8.94, 8.96).

Eqs. (7.49–7.56) look very similar to the corresponding equations derived by Gorkov and Eliashberg (1965) from RMT (several misprints were corrected by Devaty and Sievers (1980)). However, there are some differences.

First, the coefficient before $A(x)$, Eq. (7.49), is different from that obtained by Gorkov and Eliashberg by independent averaging of the dipole matrix elements, being, for $l \ll a$, $(l/a)^2$ times smaller. The coefficient in Eq. (7.49) can be understood from the following arguments: In the limit $\omega \gg \Delta$ the discreteness of the energy spectrum is not important and one has to be able to reproduce classical formulae. Let us consider the frequency-dependent conductivity $\sigma(\omega)$ defined as

$$\sigma(\omega) = \omega \operatorname{Im} \kappa(\omega) \tag{7.57}$$

which describes currents induced by the electric field. The quantity $\sigma(\omega) E^2/2$ determines the loss of energy per unit volume. Using Eqs. (7.49–7.56) conductivity $\sigma(\omega)$ can be written as

$$\sigma(\omega) = 2e^2 \nu \left(\left\langle (r^\alpha)^2 \right\rangle_0 \omega^2 \tau \right) R(\omega) \tag{7.58}$$

where $R(\omega)$ is the two-level correlation function, Eqs. (6.13–6.15). When frequency ω much exceeds mean level spacing Δ, $x \gg 1$ and function $R(x)$ is close to unity. Then $\sigma(\omega)$ looks very similar to the classical conductivity σ_0 of a metal, Eq. (3.48), provided one replaces the classical diffusion coefficient $D_0 = v_0^2 \tau/3$ by the expression in the parentheses in Eq. (7.58). This implies, as in bulk metals, a diffusion process but now the α-component of the effective velocity of the electrons at a distance r from the center is equal to $r^\alpha \omega$ instead of v^α at the Fermi surface. In the limit $x \gg 1$ Eq. (7.58) is purely classic and can be obtained from the Boltzmann equation in the τ-approximation. Correspondingly, it can also be obtained from Eqs. (7.29, 7.30), replacing the average of the product of the retarded and advanced Green functions by the product $\langle G^R \rangle \langle G^A \rangle$ with the averaged Green functions from Eq. (3.45). The nonoscillating terms of $R(\omega) - 1$ in this limit give the weak localization quantum corrections discussed in Chapter 3.

Although the calculation within the zero-dimensional version of the σ-model does not confirm the coefficient before $A(x)$ in Eq. (7.49), agreement can be achieved by integrating out higher harmonics instead of neglecting them. By extending a scheme of Kravtsov and Mirlin (1994), Blanter and Mirlin (1995) performed such integration in the main order in $\Delta/\mathcal{E}_c$ and demonstrated equivalence of the results. However, in a realistic situation when the electric field is screened the nonzero harmonics can nevertheless be neglected (Efetov (1996)).

Another aspect completely missed in the RMT treatment is the dependence of the static susceptibility on the ensemble. One can see that function $A(x)$, Eq. (7.51–7.56), differs from function $\tilde{A}(x)$, Eq. (7.15), by its value at $x = 0$. The latter turns to zero at $x = 0$, whereas the former has different finite values A_0 depending on the ensemble, and the additional static susceptibility $\delta\kappa$ can be written as (Efetov (1996))

$$\delta\kappa = \kappa(0) - \kappa_0 = \frac{\kappa_0 \tau \Delta A_0}{2\pi} \tag{7.59}$$

$$A_0 = \begin{cases} -4, & \text{Model I} \\ -2, & \text{Model IIa} \\ -1, & \text{Models IIb, III} \end{cases}$$

We see that as a result of the discreteness of the energy spectrum the electric susceptibility is somewhat smaller than its classical value. The quantum effect is strongest in the orthogonal ensemble but decreases when applying a magnetic field. Adding magnetic or spin-orbit impurities leads to a further increase of the susceptibility. The value of the coefficients A_0 in Eq. (7.59) is rather funny but it is clear that an attempt to write something like $A_0 = 4/\beta$, with $\beta = 1, 2, 4$ for the orthogonal, unitary, and symplectic ensemble, is not correct because the value of A_0 is the same for Models IIb and III. In Chapter 11 it will be shown that the inverse localization lengths of thick wires scale in the same way when changing from Model I to IIa and then to Models IIb and III. Although these relations are curious, I do not see any deep reason for the factor of 2 that governs the crossover from Model I to Model IIa and, then, to the Models IIb and III. The crossover between Models I and IIa is especially interesting for metallic rings because the dependence on the flux ϕ of the magnetic field through the ring becomes periodic with the period $\phi_0/2$, where ϕ_0 is the flux quantum. Such behavior is analogous to periodic dependence of persistent currents in mesoscopic rings and will be discussed in more detail in Chapter 8.

The increase of the response to the electric field when applying a magnetic field and a further increase when adding spin-orbit impurities are reminiscent of the effects of weak localization discussed in Chapter 3. One can again argue that the eigenstates are formed as a result of a very complicated multiple coherent scattering by impurities or boundaries and the interference of waves is sensitive to magnetic and spin-orbit interactions. However, now we have in the limit $\omega \to 0$ a thermodynamic effect. The starting formula Eq. (7.28) was derived both from the Kubo linear response theory and by differentiation of the thermodynamic potential.

The thermodynamic origin is very important to discussion of effects of finite temperature when relaxation processes due to inelastic scattering must be taken into account (the preceding derivation shows that the smearing of the Fermi distribution alone does not lead to any effect so long as $T \ll \varepsilon_0$). This excludes a scenario in which Eq. (7.59) is valid only for $\omega \gg \tau_{inel}^{-1}$, where τ_{inel} is the mean free time of the inelastic scattering, and $\delta\kappa$ vanishes as $\omega \to 0$ for any finite τ_{inel}. Such a scenario would correspond to a system out of equilibrium. At the same time, the fact that Eq. (7.59) can be derived from a thermodynamic formula means that the quantum correction to the susceptibility corresponds to the minimum of the free energy and cannot be destroyed by, at least, a weak inelastic scattering.

When discussing effects of electron–electron interaction one should distinguish between the short-range part, which contributes additionally to electron scattering, and the long-range part, which is due to the long-range Coulomb interaction. The long-range part of the Coulomb interaction is extremely important because it leads to the screening of the electric field mentioned previously. The final expressions, Eq. (7.49–7.56), are a response to local electric field E_{loc}. Experimentally one measures a response to an external field E and, therefore, the final formulae should be written in terms of this response.

Fortunately, proper corrections can be made rather easily. As mentioned, the effect of screening can be considered properly, making the replacement Eqs. (7.43) in Eqs. (7.41, 7.42). The electrostatic potential has to be found from the classical Poisson equation. In fact, the solution has been written for a sphere and film by Rice, Schneider, and Strässler (1973). Fast decay of the potential $\Phi(\mathbf{r})$ when moving away from the surface simplifies evaluation of integrals over coordinates. As a result, the polarizability α that is the response to the external electric field E can be written as (Efetov (1996))

$$\delta\alpha \equiv \frac{\alpha}{\alpha_0 - 1} = ZA(x)\tau\Delta, \qquad Z = \frac{S}{4\pi k_s V} \tag{7.60}$$

where S is the surface of the cluster, $k_s = \left(8\pi e^2 \nu\right)^{1/2}$ is the Fermi–Thomas screening wave vector, and $A(x)$ is determined by Eqs. (7.50–7.56). The quantity α_0 is the classical polarizability; for the sphere and disk it is equal to

$$\alpha_0 = \begin{cases} a^3\left(1 - 3/ak_s\right), & \text{sphere} \\ Sd\left(4\pi\right)^{-1}\left(1 - 2/dk_s\right) & \text{disk} \end{cases} \tag{7.61}$$

where a is the radius of the sphere, and d is the thickness of the disk (the electric field is assumed to be perpendicular to the disk).

We see that as a result of screening the relative quantum correction to the classical polarizability acquires an additional factor of Z that is the ratio of the volume where the electric field is essentially nonzero to the total volume of the cluster. At the same time, the frequency dependence does not change.

7.2.3 Possibility of experimental observation

Let us write the final result in a form suitable for comparison with an experiment. To observe the change of the polarizability one can embed a system of isolated metal clusters into a nonmetal matrix with a dielectric constant ϵ_m such that the particles occupy a fraction $f \ll 1$ of the total volume. The dielectric constant $\bar{\epsilon}$ of the effective medium can be expressed through the constants α, ϵ_m, and f as

$$\epsilon = \epsilon_m \frac{1 + 4\pi\alpha f}{V} \tag{7.62}$$

Using Eq. (7.60) the variation of the effective dielectric constant $\bar{\epsilon}$ can be written as

$$\delta\epsilon \equiv \frac{\epsilon}{\epsilon_0 - 1} = Z\tau\Delta A(x)bf(1+bf)^{-1} \tag{7.63}$$

where $b = 3$ for sphere, and $b = 2$ for disk. The classical dielectric constant ϵ_0 is obtained from Eq. (7.62) replacing α by α_0. It does not vary with frequency, magnetic field, or magnetic or spin-orbit impurities. All the quantum size effects are contained in the function $A(x)$, Eqs. (7.50–7.56), and the static limit is described by Eq. (7.59). In fact, Eq. (7.63) is the first term of the expansion in the quantum effects.

The quantity $\delta\epsilon$ is rather small unless the particles are very small. Let us estimate the order of magnitude of $\delta\epsilon$ for some realistic systems, for example, for copper particles with the radius $a = 50$ Å. In the review of Halperin (1986) one can find values of Δ for different materials and particle sizes that can be directly used for estimating $\delta\epsilon$. The values of Δ for copper given in the review correspond to the Fermi momentum $p_0 \approx 1.7 \cdot 10^8$ cm^{-1}. For such a small size of the metal cluster one may assume for a rough estimate that $l = a$. Then for spherical particles one obtains

$$\delta\epsilon = \frac{f}{1+3f}\left(p_0 a\right)^{-3}\sqrt{p_0 r_B} \simeq 10^{-6} \cdot A(x) \tag{7.64}$$

where $r_B = \left(me^2\right)^{-1}$ is the Bohr radius. For the numerical estimate in Eq. (7.64) it was assumed that f is of the order of unity and the value $r_B = 0.53$ Å was used.

In principle, one can either try to measure the frequency dependence of infrared absorption, hoping to observe frequency oscillations, or study the sensitivity of the polarizability to magnetic field. The frequency oscillations have not been observed as yet (Halperin (1986)). At the same time, nobody has tried to study the dielectric constant experimentally as a function of magnetic field. It follows from Eqs. (7.64, 7.59) that applying a magnetic field one can increase the relative static dielectric constant $\delta\epsilon$ of a system consisting of copper particles with radius $a = 50$ Å by the amount $\delta\tilde{\epsilon}$

$$\delta\tilde{\epsilon} = \delta\epsilon_{\text{unit}} - \delta\epsilon_{\text{orth}} \simeq 10^{-6} \tag{7.65}$$

Although 10^{-6} is a small number, it can be detected in high-sensitivity experiments (Foote and Anderson (1987)). At first glance, one can win a lot by making the particles smaller and this is true for optical measurements for the frequency-dependent $A(x)$. However, in regard to studying the magnetic field dependence of the static dielectric constant, the magnetic field H_c of the crossover from the orthogonal to the unitary ensemble grows with decrease in the size of the particles and can reach excessively large values, leading, thus, to other experimental difficulties.

Again, to give some numbers let us evaluate the field H_c for the same system of copper particles. This field can be obtained by using Eq. (6.40), but to be more precise, one can take the representation of supermatrix Q in the form of Eq. (5.1) and estimate the contribution of the "cooperon degrees" of freedom $B_\perp$ averaging $(B_\perp)_\beta(\bar{B}_\perp)_\beta$, where β stands for a component of the block $B_\perp$, with the quadratic part of Eq. (5.3). The field H_c can be defined from the equation

$$\langle (B_\perp)_\beta(\bar{B}_\perp)_\beta\rangle_2 = 1 \tag{7.66}$$

where $\langle\cdots\rangle_2$ is the Gaussian averaging with the quadratic part F_2 of the free energy functional from Eq. (5.3). This gives the following equation:

$$\frac{1}{\pi\nu D_0}\left(\frac{c}{2e}\right)^2\left(\int A^2 d\mathbf{r}\right)^{-1} = 1 \tag{7.67}$$

In Eq. (7.67), one should use the vector potential $\mathbf{A}$ in the London gauge

$$\text{div}\,\mathbf{A} = 0, \qquad A_n = 0 \tag{7.68}$$

where A_n is the component of the vector potential perpendicular to the surface of the particle and taken at the surface. This condition is exactly the same as that used in superconductors (De Gennes (1966)). For a spherical particle one can write

$$\mathbf{A} = \frac{1}{2}[\mathbf{H}\times\mathbf{r}] \tag{7.69}$$

Substituting Eq. (7.69) into Eq. (7.67) one obtains

$$H_c = \frac{\phi_0}{a^2\pi}\sqrt{\frac{15a\Delta}{8\pi\nu_0}\frac{a}{l}} = \frac{\phi_0}{2\pi a^2}\frac{3.6}{p_0 a}\left(\frac{a}{l}\right)^{1/2} \tag{7.70}$$

Using the same parameters as before we obtain $H_c \approx 1T$. Both $\delta\epsilon$ and H_c grow fast with decrease in the size of the particles and, therefore, making the clusters smaller does not seem to be a good way to make experiments easier. However, the dependence of H_c on p_0 is weaker than that of $\delta\epsilon$ and one can try to choose a metal with a smaller p_0. Possibly, the best results can be achieved for semimetals such as Bi where p_0 is very small.

The effect of the increase of the dielectric constant with the magnetic field is not sensitive to variation of the particle size, and therefore there is no need to take special care to fabricate equal size particles. At the same time, observation of the oscillations in the frequency dependence in optical experiments is possible only when the distribution of the particle sizes is very narrow. The failure to observe the latter effects was attributed by Devaty and Sievers (1980) to the smearing of the oscillations by variation of the particle size.

Thus, we have considered the effects of the long-range part of the Coulomb interaction. As a result of this type of interaction the contribution of the quantum size effects to the dielectric constant is additionally reduced and this is reflected by the presence of small Z in Eq. (7.63). Another important contribution can come from the short-range part of the interaction. Usually, in considering interacting fermions, one applies the Landau theory of Fermi liquid. Then, one can replace the strongly interacting electrons by weakly interacting quasi particles. These quasi particles are not exact eigenstates and can decay with time. Therefore, the quasi particle levels acquire a finite width Γ

$$\Gamma \sim \frac{T^2}{\mu} \qquad (7.71)$$

In bulk metals the characteristic energy of the quasi particles is of the order of T and Γ is much less than this energy so long as $T \ll \mu$, which justifies the Fermi liquid picture. In small mesoscopic systems, as we have seen, two other energies, the mean level spacing Δ and the Thouless energy $\mathcal{E}_c$, are important, too, and one may ask at what temperatures the quantum size effects considered in this chapter can be observed. In fact, different effects depend on temperature in different ways. The nontrivial effects in the thermodynamic quantities of completely isolated particles, considered in the preceding section, are washed out at $T \geq \Delta \sim 1$ K (estimates are given for the particle radius ~ 50 Å, such that $\Delta \sim 1$ K).

The quantum size effects considered in the present section are not sensitive to the smearing of the Fermi distribution so long as $T \ll \mu$ and they can disappear only as a result of interactions leading to a finite Γ. It is quite clear that in order to observe an oscillating frequency dependence in optical experiments the level width Γ should not exceed the mean level spacing Δ and this leads to the inequality

$$T \ll (\mu \Delta)^{1/2} \sim 100 \text{ K} \qquad (7.72)$$

If the inequality (7.72) is fulfilled, the quasi-particle levels are well resolved and one can use Eqs. (7.51–7.56) for frequencies $\omega \gg \Gamma$.

The question about the range of temperatures where the static quantum size effects, Eq. (7.59), can be observed is more difficult. Of course, these effects do not disappear at $\Gamma \ll \Delta$. Therefore, they can definitely be observed at temperatures satisfying the inequality Eq. (7.72). However, the origin of the static quantum size effects is different from that of dynamic ones. The dynamic effects are due to interlevel transitions whereas the change of the static electric susceptibility with, for example, a magnetic field is a one-level effect. The susceptibility changes because the eigenfunction changes when applying the magnetic field. But the wave functions are formed by multiple scattering by impurities or boundaries. All the coherence effects are smeared only if the quasi-particle decay time is shorter than the diffusion time $\tau_D \sim a^2/D_0$. If this guess is correct, one should use the following inequality for the existence of the static quantum size effects:

$$T \ll (\mu \mathcal{E}_c)^{1/2} \sim 1000 \text{ K} \qquad (7.73)$$

That means that one does not need to worry about temperature at all. However, even if this guess is wrong and one should use the inequality (7.72), the restriction on temperature imposed by this inequality is not important and, possibly, the corresponding experiments will have been done soon.

7.3 Local density of states distribution function and NMR

7.3.1 NMR intensity line and the σ-model

In this section it will be demonstrated that chaotic electron motion in small mesoscopic metal particles also results in very interesting properties of nuclear magnetic resonance (NMR). It is well known (Slichter (1980)) that the position of the NMR line in metals is shifted as a result of interaction of spins of conduction electrons with spins of nuclei (Knight shift), where the shift is proportional to the spin susceptibility χ of the conduction electrons. By studying the position of the resonance line one can get information about susceptibility.

Unusual dependence of spin susceptibility χ on the magnetic field and temperature in isolated particles has been discussed in Section 7.1. An essential assumption of that consideration was that the number of electrons was fixed. Depending on whether the number of electrons is even or odd, one can obtain quite different values of susceptibility, which would lead to two different NMR peaks provided the system contains particles with both even and odd numbers of electrons. Of course, to observe these effects experimentally one has to have a system of very well isolated particles with a very narrow size distribution; otherwise the two-peak structure cannot be resolved.

However, the NMR in a system of small metal particles is even more interesting than one would expect from such a consideration because the very notion of susceptibility χ needs some clarification. In direct magnetic experiments one measures a susceptibility averaged over disorder, particle size, and so on. At the same time, the Knight shift probes a local spin susceptibility $\chi(\mathbf{r})$ that can vary from point to point. As a result, one has in NMR experiments a sum of different contributions from different points and the observable line is not necessarily sharp. Nevertheless, when studying NMR in a bulk metal, it is assumed that the local paramagnetic susceptibility $\chi(\mathbf{r})$ does not fluctuate considerably, and therefore one can replace this quantity by the Pauli susceptibility χ_p. In fact, such a substitution is correct provided the mean free path l is large, $lp_0 \gg 1$ (the one-dimensional case is an exception due to localization (Altshuler and Prigodin (1989))).

The situation changes in a system of small metal particles because the small particle size leads to strong fluctuations of the electron wave function. As a result of the chaotic character of motion these fluctuations can be very strong even in clean particles where electrons can move without collisions from boundary to boundary. The local paramagnetic susceptibility $\chi(\mathbf{r})$ strongly depends on the wave function $\varphi_\kappa(\mathbf{r})$ at the point $\mathbf{r}$ and also fluctuates. Therefore, the NMR line shape that coincides with the distribution of the local susceptibilities can become highly nontrivial. The effect is possible in a system of nonisolated particles and one can obtain a complicated NMR line shape working within the grand canonical ensemble assuming that electrons can tunnel from one particle to another or exit into the environment. This possibility was suggested by Efetov and Prigodin (1993) and the presentation of this section follows these works.

Let us consider the simplest model that demonstrates interesting behavior. A homogeneous line broadening that may be due to inelastic processes is neglected. In the limit of low temperatures this can be a good approximation and one can start from a conventional formula for the shift $\Delta\omega\,(\mathbf{r}_a)$ of the resonance frequency of a nuclear spin located at a point $\mathbf{r}_a$

$$\Delta\omega\,(\mathbf{r}_a) = J\,(g\mu_B)^{-1}\,\chi\,(\mathbf{r}_a)\,H \tag{7.74}$$

In Eq. (7.74), it is assumed that the shift $\Delta\omega\,(\mathbf{r}_a)$ is due to the Fermi contact interaction of the nuclear spin with spins of conduction electrons, and J represents the hyperfine coupling constants. The function $\chi\,(\mathbf{r}_a)$ in Eq. (7.74) is the local spin susceptibility of the conduction electrons. To simplify the model, electron–electron interactions and interactions with magnetic or spin-orbit impurities are neglected. Then the susceptibility $\chi\,(\mathbf{r}_a)$ can be expressed through the one-particle local density of states $\rho\,(\varepsilon, \mathbf{r}_a)$ in a standard way

$$\chi\,(\mathbf{r}_a) = 2\,(g\mu_B)^2 \int \rho\,(\varepsilon, \mathbf{r}_a)\,n'\,(\varepsilon)\,d\varepsilon \tag{7.75}$$

where $n'\,(\varepsilon) = (4T)^{-1}\cosh^{-2}\,((\varepsilon - \mu)\,/2T)$ is the derivative of the Fermi function. Let us start the consideration from the case $T = 0$. Then the function $n'(\varepsilon)$ in Eq. (7.75) is just the δ-function and the local spin susceptibility $\chi\,(\mathbf{r}_a)$ is proportional to the local density of states $\rho\,(\varepsilon, \mathbf{r}_a)$.

In a traditional bulk metal the shift $\Delta\omega$ is the same for all nuclear spins and the NMR line is narrow. The local density of states $\rho\,(\varepsilon, \mathbf{r}_a) = 2\nu$ (the spin degeneracy is taken into account) does not depend on coordinates. The local susceptibility $\chi\,(\mathbf{r}_a)$ coincides with the Pauli susceptibility χ_p. In disordered metals in the limit $lp_0 \gg 1$, where l is the mean free path, the local density of states and the susceptibility are again given by the values ν and χ_p. Therefore, in this approximation, the NMR line is infinitely narrow. What I want to discuss now is what will happen to the NMR line shape if one replaces the bulk metal by a system of small particles of the same metal and of equal size?

Let us start the consideration from an exact expression for the resonance line shape, $I\,(\omega)$, corresponding to a system of nuclear spins

$$I\,(\omega) = A \sum_a \delta\,(\omega - \omega_0 - \Delta\omega\,(\mathbf{r}_a)) \tag{7.76}$$

where the constant A stands for a weight and ω_0 is the nonshifted position of the resonance. In Eq. (7.76) one should sum over positions $\mathbf{r}_a$ of all nuclear spins in the system and let us assume that all the nuclear spins are located in a system of macroscopically equal metal particles. At the same time, the particles may have nonperfect shapes that may differ from particle to particle on atomic scale or have impurities inside. Therefore, one should average $I\,(\omega)$, Eq. (7.76), over this disorder to obtain the physical line shape. Earlier discussions of NMR in small metal particles (Halperin (1986)) were based on substitution of $\Delta\omega\,(\mathbf{r}_a)$ by an average $\langle\Delta\omega\,(\mathbf{r}_a)\rangle$ over the disorder. However, it will be shown later by direct calculation of $I\,(\omega)$ that this substitution is not justified and can lead to wrong results.

Because of the proportionality at $T = 0$ of the susceptibility $\chi\,(\mathbf{r})$ to the local density of states the line shape is, after some rescaling, nothing more than the local density of states distribution function $P\,(s)$ defined as

$$P\,(s) = \left\langle \delta\left(s - \frac{\rho\,(\varepsilon, \mathbf{r})}{\nu}\right)\right\rangle \tag{7.77}$$

A system of almost isolated metal particles is considered and although the particles are assumed to be of the same size they may have slightly different shapes. As argued earlier, the averaging over impurities should be equivalent to averaging over the random shapes of the particles. Then, the regime of chaotic dynamics can be achieved in very clean particles with the bulk mean free path exceeding the diameter.

The local density of states $\rho(\varepsilon, \mathbf{r})$ entering Eq. (7.77) can be written in terms of the retarded G^R and advanced Green functions G^A in the usual way, Eq. (3.16). Because of the assumed possibility for the electrons to leave particles to a nonmetallic matrix or to other particles one should use, in principle, Green functions of the entire system of the particles, and these Green functions are different from those for the closed particles. Of course, it is not easy to find a solution for such a system, and therefore it is more reasonable to use a phenomenological approach assuming that the presence of the surrounding leads to smearing of the levels in the particle. Neglecting electron–electron interaction the Green functions are written in the form

$$G_\varepsilon^{R,A}(\mathbf{r}, \mathbf{r}') = \sum_k \frac{\varphi_k(\mathbf{r})\,\varphi_k^*(\mathbf{r}')}{\varepsilon - \varepsilon_k \pm i\gamma/2} \tag{7.78}$$

where φ_k and ε_k are eigenfunctions and eigenenergies of an isolated metal particle. The level smearing is described by the parameter γ entering Eq. (7.78), which is assumed to be independent of k.

Let us explain briefly why the Green functions can be written in the form of Eq. (7.78) and why a dependence of γ on k can be neglected. An effective Hamiltonian H of a particle in a matrix leading to the Green functions $G_\varepsilon^{R,A}$, Eq. (7.78), can be written as

$$\mathcal{H} = \mathcal{H}_c \mp \frac{i\alpha}{2\pi\nu S} a_S(\mathbf{r}) \tag{7.79}$$

where $\mathcal{H}_c$ is the Hamiltonian of the corresponding closed particle. The second term in Eq. (7.79) describes the possibility of tunneling from the particle into its surroundings; α is a dimensionless parameter and S stands for the surface area. One should choose the $+$ sign for the retarded Green function and $-$ for the advanced one. The function $a_S(\mathbf{r})$ is zero everywhere except at the surface, where it is equal to infinity. It can be considered as a surface δ-function. For an arbitrary function $b(\mathbf{r})$ one has by definition

$$\int b(\mathbf{r}) a_S(\mathbf{r})\,d\mathbf{r} = \int b(\mathbf{r})\,dS \tag{7.80}$$

where $\int \ldots dS$ stands for the integral over the surface.

The second term in Eq. (7.79) can be derived by starting from a Hamiltonian with tunneling from the particle to the nonmetallic matrix written in the form

$$t_{pm} \int \left[\psi_p^+(\mathbf{r})\,\psi_m(\mathbf{r}) + \text{c. c.}\right] a_S(\mathbf{r})\,d\mathbf{r}$$

where ψ_p and ψ_m are the field operators of the particle and the matrix, respectively, t_{pm} is the tunneling amplitude, and c.c. is the complex conjugate. Taking the trace over the operators ψ_m, ψ_m^+ one obtains the imaginary term in Eq. (7.79). A more general derivation of the imaginary part due to tunneling from an otherwise closed metallic system is discussed in Chapter 9. Provided the parameter α is not very large, one can calculate the level width γ due to the second term in Eq. (7.79) in the first order of perturbation theory

$$\gamma = \frac{\alpha}{\pi\nu S} \int |\varphi_k(\mathbf{r})|^2 dS \tag{7.81}$$

Formally, the integral in Eq. (7.81) depends on k. However, if the eigenfunctions $\varphi_k(\mathbf{r})$ are random enough, the integration over the surface in Eq. (7.81) is equivalent to averaging over states. Of course, it is important that the electrons can leave the particle through any point of the surface, which corresponds to the situation under consideration. Under this assumption the level width γ is the same for all levels and equals

$$\gamma = \frac{\alpha \Delta}{\pi} \tag{7.82}$$

The possibility of expressing the distribution function $P(s)$ in terms of the Green functions, Eqs. (7.77, 3.16, 7.78), enables us to use the supersymmetry technique, although a slight modification has to be made. Until now, the supersymmetry method has been applied for calculation of correlation functions that can be expressed in terms of a product of no more than two Green functions. In order to calculate the distribution function $P(s)$ one has to be able to average a function of the Green functions or, using Taylor expansions, a product of an arbitrary number of Green functions. At first glance, when calculating a product of $2m$ Green functions, one has to use the same scheme but with $8m \times 8m$ supermatrices Q. Certainly, one can proceed in this way, but we have seen in the previous chapters that often the integrals obtained are not simple even for $m = 1$. As concerns larger m, it is not at all clear how to carry out nonperturbative calculations. Fortunately, for some physical quantities including $\rho^n(\varepsilon, \mathbf{r})$, explicit calculations can be performed for arbitrary n without increasing the size of the standard 8×8 supermatrices Q.

Using Eq. (3.16) an arbitrary power of $\rho(\varepsilon, \mathbf{r})$ can be written in the form

$$M_n = \left\langle \left(\frac{\rho(\varepsilon, \mathbf{r})}{\nu} \right)^n \right\rangle \tag{7.83}$$

$$= \frac{1}{(2\pi i)^n} \sum_{k=0}^{n} \frac{(-1)^k n!}{(n-k)! k!} \left\langle \left(G_\varepsilon^R(\mathbf{r}, \mathbf{r}) \right)^k \left(G_\varepsilon^A(\mathbf{r}, \mathbf{r}) \right)^{n-k} \right\rangle$$

The next step is to express the product of the Green functions in Eq. (7.83) in terms of functional integrals over the supervectors ψ, Eqs. (4.6). We know already very well how to write the product $G^R G^A$

$$G_\varepsilon^R(\mathbf{r}, \mathbf{r}) \, G_\varepsilon^A(\mathbf{r}, \mathbf{r}) = \int \left| S^1(\mathbf{r}) \right|^2 \left| S^2(\mathbf{r}) \right|^2 \exp(-\mathcal{L}) \, D\psi \tag{7.84}$$

where

$$\mathcal{L} = i \int \bar{\psi}(\mathbf{r}) \left(\varepsilon - \mathcal{H}_c - \frac{i\gamma\Lambda}{2} \right) \psi(\mathbf{r}) d\mathbf{r}$$

However, Eq. (7.84) can immediately be generalized for arbitrary powers of the Green functions. It is very important that we want to calculate $\rho^n(\varepsilon, \mathbf{r})$ such that all the Green functions in the product are taken at the same coordinate $\mathbf{r}$ and energy ε. Using this fact and carrying out Gaussian integration one can easily show that

$$\left(G_\varepsilon^R(\mathbf{r}, \mathbf{r}) \right)^k \left(G_\varepsilon^A(\mathbf{r}, \mathbf{r}) \right)^{n-k} \tag{7.85}$$

$$= \frac{(-1)^k i^n}{k!(n-k)!} \int \left| S^1(\mathbf{r}) \right|^{2(n-k)} \left| S^2(\mathbf{r}) \right|^{2k} \exp(-\mathcal{L}) \, D\psi$$

We see from Eq. (7.85) that the product of arbitrary powers of the Green functions taken at the same coordinates and energies can be expressed in terms of a functional integral over eight-component supervector ψ with the same Lagrangian $\mathcal{L}$ as before. Therefore, the calculational scheme developed previously can be used now. All calculations become more simple if time reversal symmetry is broken. In NMR experiments this symmetry can be destroyed by external magnetic field. In this section it is assumed that the magnetic field is strong enough that we are in the limiting case of the unitary ensemble.

The effective Lagrangian $\mathcal{L}_{\mathrm{eff}}\left[\psi, \bar{\psi}, Q\right]$ that is obtained after averaging over the impurities and decoupling the "interaction" term by the 8×8 supermatrix Q is, as in Eq. (4.19), quadratic in ψ and so, the integrals over ψ in Eq. (7.85) are reduced to products of Gaussian averages. Proceeding in this way we obtain for the nth moment M_n of the density of states

$$M_n = \frac{1}{(2\pi v)^n} \sum_{k=0}^{n} \sum_{0 \leq l \leq \min(n-k,k)} \frac{n!\,(-1)^k}{(n-k-l)!\,(k-l)!\,(l!)^2}$$

$$\times \left\langle \left(g_{33}^{11}\right)^{n-k-l} \left(g_{33}^{22}\right)^{k-l} \left(g_{33}^{12} g_{33}^{21}\right)^{l} \right\rangle_Q \tag{7.86}$$

In Eq. (7.86), the brackets $\langle \ldots \rangle_Q$ stand for averaging with the free energy functional $F[Q]$

$$F[Q] = \frac{\pi v}{8} \int \mathrm{str}\left[D_0\,(\nabla Q)^2 - 2\gamma \Lambda Q\right] d\mathbf{r} \tag{7.87}$$

The matrix Green function g in Eq. (7.86) is the solution of the equation

$$\left(-i\mathcal{H}_c + \frac{Q}{2\tau}\right) g\left(\mathbf{r}, \mathbf{r}'\right) = \delta\left(\mathbf{r} - \mathbf{r}'\right) \tag{7.88}$$

In principle, one should write in Eq. (7.88) the Hamiltonian $\mathcal{H}$, Eq. (7.79), but in the limit of interest, $\gamma \ll \tau^{-1}$, it can be approximated by the Hamiltonian of the closed particle $\mathcal{H}_c$.

With the constraint $Q^2 = 1$, one obtains Eq. (6.34) and Eq. (7.86) can be rewritten as

$$M_n = 2^{-n} \sum_{k=0}^{n} \sum_{0 \leq l, l' \leq \min(n-k,k)} \frac{n!\,(-1)^k}{(n-k-l)!\,(k-l')!\,l!\,l'!}$$

$$\times \left\langle \left(Q_{33}^{11}\right)^{n-k-l} \left(Q_{33}^{22}\right)^{k-l'} \left(Q_{33}^{12}\right)^{l} \left(Q_{33}^{21}\right)^{l'} \right\rangle_Q \delta_{ll'} \tag{7.89}$$

Representing the δ-function $\delta_{ll'}$ in the form

$$\delta_{ll'} = (2\pi)^{-1} \int_0^{2\pi} \exp\left(i\,(l - l')\,t\right) dt$$

one can reduce Eq. (7.89) to the following integral:

$$M_n = \frac{2^{-n}}{2\pi} \int_0^{2\pi} \left\langle \left(Q_{33}^{11} - Q_{33}^{22} + Q_{33}^{12} e^{-it} - Q_{33}^{21} e^{it}\right)^n \right\rangle_Q dt \tag{7.90}$$

Having expressed all the moments M_n by Eq. (7.90), we can write the entire distribution function $P(s)$, Eq. (7.77), in terms of the following integral over the supermatrices Q:

$$P(s) = \int_0^{2\pi} dt \left\langle \delta\left(s - \frac{1}{2}\left(Q_{33}^{11} - Q_{33}^{22} + Q_{33}^{12} e^{-it} - Q_{33}^{21} e^{it}\right)\right)\right\rangle_Q \tag{7.91}$$

Although the small metal particles are described by the zero-dimensional σ-model, Eq. (7.91) is more general and can be used in an arbitrary dimension.

7.3.2 Density of states distribution function

Let us notice that Eq. (7.87) differs from the corresponding Eq. (4.55) by the presence of the level width γ instead of the frequency ω. Therefore one can immediately write the condition for zero dimensionality (compare with Eqs. (6.26, 6.28))

$$\mathcal{E}_c = \frac{\pi^2 D_0}{L^2} \gg \Delta, \qquad \gamma \ll \mathcal{E}_c \qquad (7.92)$$

In the zero-dimensional limit one should replace the free energy functional $F[Q]$, Eq. (7.87), by $F_0[Q]$

$$F_0(Q) = -\frac{\pi \gamma}{4\Delta} \, \text{str} \, (\Lambda Q) \qquad (7.93)$$

Using the parametrization for the supermatrix Q, Eqs. (6.42–6.54), one can calculate the distribution function $P(s)$ explicitly, first integrating over the supermatrices u and v, which do not enter Eq. (7.93), and then over the variables λ and λ_1. The energy $F_0(Q)$ is written in these variables as

$$F_0 = \frac{\pi \gamma}{\Delta} (\lambda_1 - \lambda) \qquad (7.94)$$

After integration over supermatrices u and v and variable t we can represent the distribution function $P(s)$, Eq. (7.77), in the form

$$P(s) - P_0(s) = \frac{1}{2\pi} \frac{d^2}{ds^2} \int \frac{s \exp\left[-\pi\gamma(\lambda_1 - \lambda)/\Delta\right] d\lambda d\lambda_1}{\left(2s\lambda_1 - s^2 - 1\right)^{1/2}(\lambda_1 - \lambda)} \qquad (7.95)$$

where $P_0(s) = \delta(s-1)$ is the distribution function of the bulk. In Eq. (7.95) one should integrate over λ in the interval $-1 < \lambda < 1$ and over $\lambda_1 > (s+1/s)/2$. The contribution $P_0(s)$ originates from the term of the zeroth order in the Grassmann variables η, κ and can be obtained by replacing supermatrix Q in Eq. (7.91) by Λ (see explanations in Sections 2.5 and 6.3). The integral in the right-hand side of Eq. (7.95) is obtained in the regular way from the term with the product $\eta_1 \eta_1^* \kappa_1 \kappa_1^*$.

Further calculation can be carried out, integrating first over $\lambda_1 + \lambda$ and then over $\lambda_1 - \lambda$. After somewhat cumbersome manipulations we come to the final result

$$P(s) = (\alpha/8\pi)^{1/2} s^{-3/2} \exp\left[-\alpha(s+1/s)/2\right]$$
$$\times [2\cosh\alpha + (s+1/s - 1/\alpha)\sinh\alpha] \qquad (7.96)$$

where α is given by Eq. (7.79).

Eq. (7.96) is the complete solution of the problem of calculation of the local density of states distribution function. Let us describe the basic properties of the function $P(s)$. It is drawn in Fig. 7.2 for several values of α. It depends on the parameter α only and does not depend on the mean free path l, but, of course, the condition $\tau\Delta \ll 1$ was essential when deriving Eq. (7.96). The limit $\alpha \to \infty$ corresponds to large samples where the discreteness of the levels is not felt. In this limit Eq. (7.96) gives

$$P(s) = P_0(s) = \delta(s-1) \qquad (7.97)$$

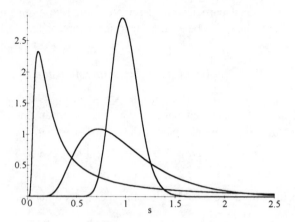

Fig. 7.2. The distribution function of the local density of states (NMR line shape) for $\alpha = 0.5, 5, 50$. When decreasing the parameter α the position of the maximum of the distribution moves to zero.

The NMR line corresponding to the function $P_0(s)$ is in this case very narrow and its position gives the Knight shift of the bulk metal.

In the limit $\alpha \ll 1$, the discreteness of the levels becomes important. The function $P(s)$ is very broad and its maximum moves to zero when $\alpha \to 0$. In the region of large density of states it has a long tail. The discrete character of the spectrum can help in explaining the main features of the function $P(s)$.

In the limit $s \ll \alpha$ the function $P(s)$ can be written as

$$P(s) \approx \left(\frac{\alpha^3}{8\pi}\right)^{1/2} s^{-5/2} \exp\left(-\frac{\alpha}{2s}\right) \tag{7.98}$$

This asymptotics can be interpreted by writing the local density of states $\rho(\varepsilon)$ in terms of the exact eigenfunctions $\varphi_k(\mathbf{r})$ and eigenvalues ε_k of the closed particle

$$\rho(\mathbf{r},\varepsilon) = \frac{\gamma}{2\pi} \sum_k \frac{|\varphi_k(\mathbf{r})|^2}{(\varepsilon - \varepsilon_k)^2 + (\gamma/2)^2} \tag{7.99}$$

A low local density of states can be due to large fluctuations of the wave functions $\varphi_k(\mathbf{r})$. Let us assume that the spectrum is very rigid such that the levels are approximately equidistant

$$\varepsilon - \varepsilon_n = \Delta\left(n + \frac{1}{2}\right) \tag{7.100}$$

where n are all possible positive and negative integers (the energy is counted from the Fermi energy). Then, we have to make an assumption about fluctuations of the wave functions $v_n = V|\varphi_n|^2$. It turns out that the fluctuations of the wave functions obey the Gaussian law and this will be demonstrated later explicitly. So, we take v_n independent for different n and distributed as

$$W(v_n) = \exp(-v_n) \tag{7.101}$$

Using Eqs. (7.99–7.101) we write the density of states distribution function $\tilde{P}(s)$ in the limit $\alpha \ll 1$ in the following form:

$$\tilde{P}(s) = \int \delta\left(s - \frac{\alpha}{2\pi^2} \sum_{n=-\infty}^{\infty} \frac{v_n}{(n+1/2)^2}\right) \exp\left(-\sum_{n=-\infty}^{\infty} v_n\right) \prod_n dv_n \qquad (7.102)$$

Representing the δ-function in Eq. (7.102) as

$$\delta(s) = \frac{1}{2\pi} \int_{-\infty}^{\infty} \exp(iqs)\, dq \qquad (7.103)$$

and integrating over all v_n we obtain

$$\tilde{P}(s) = \frac{1}{2\pi} \int_{-\infty}^{\infty} \exp(ius) \prod_{n=1}^{\infty} \left[1 + \frac{i\alpha u}{2\pi^2 (n+1/2)^2}\right]^{-2} du \qquad (7.104)$$

Using the formula

$$\prod_{n=1}^{\infty} \left(1 + \frac{t^2}{\pi^2 (n+1/2)^2}\right) = \cosh t \qquad (7.105)$$

and changing the variables of integration $u = z\alpha/s^2$, Eq. (7.104) can be reduced to the form

$$\tilde{P}(s) = \frac{\alpha}{2\pi s^2} \int_{-\infty}^{\infty} \exp\left[\frac{iz\alpha}{s} - 2\ln\left[\cosh\left(\left(\frac{z}{2}\right)^{1/2} \frac{\alpha}{s}\right)\right]\right] dz \qquad (7.106)$$

In the limit $s \ll \alpha$ the integral, Eq. (7.106), can be calculated by the saddle-point method and we obtain

$$\tilde{P}(s) = \left(\frac{2}{\pi}\right)^{3/2} \alpha^{1/2} s^{-3/2} \exp\left(-\frac{\alpha}{2s}\right) \qquad (7.107)$$

One can see from Eq. (7.107) that the exponent in Eq. (7.98) is correctly reproduced. Although this rough estimate does not give the correct power of s in the preexponential, the physical explanation for the asymptotics, Eq. (7.98), looks very reasonable.

In the region $\alpha \ll s \ll 1/\alpha$, Eq. (7.96) gives

$$P(s) \approx \left(\frac{\alpha}{8\pi}\right)^{1/2} s^{-3/2} \qquad (7.108)$$

and let us reproduce this behavior. In order to explain the behavior in the limit $s \ll \alpha$ it was assumed that the energy ε was located somewhere between levels. Larger values of the density of states can be obtained if there is a level at a distance $\delta\varepsilon \ll \Delta$ from the energy ε that gives the main contribution. Assuming that $\delta\varepsilon \gg \gamma$ and that the characteristic values $|\varphi_\beta(\mathbf{r})|^2$ are of the order of $1/V$ the contribution of this level in Eq. (7.99) is estimated as

$$\rho \sim \gamma (\delta\varepsilon)^{-2} V^{-1} \qquad (7.109)$$

If the probability of finding the level at the distance $\delta\varepsilon$ is proportional to Δ^{-1}, the distribution function is proportional to $\Delta^{-1} d(\delta\varepsilon)/d\rho$. Then using Eq. (7.109) we recover Eq. (7.108).

In the region of large density of states $s \gg 1/\alpha$, the function $P(s)$ takes the form

$$P(s) \approx \left(\frac{\alpha^3}{8\pi}\right)^{1/2} s^{-1/2} \exp\left(-\frac{\alpha s}{2}\right) \qquad (7.110)$$

We see from Eq. (7.110) that distribution function $P(s)$ has a very long tail in the region $s \gg 1/\alpha$. A large value of s can be obtained if an eigenvalue ε_k coincides with energy ε. Then we have from Eq. (7.99)

$$s = 2(\pi \nu \gamma)^{-1} |\varphi_k(\mathbf{r})|^2 \tag{7.111}$$

The asymptotic behavior, Eq. (7.110), together with Eq. (7.111) shows that there is a probability, proportional to $\exp(-V/\Omega)$, that the eigenfunction $\varphi_k(\mathbf{r})$ is concentrated in a volume $\Omega < V$. This is a very interesting result because it shows that, even in the limit $\tau \varepsilon_0 \gg 1$ when the wave functions of the bulk are almost plane waves, strong fluctuations of the wave functions are highly probable. The concentration of the wave function in the volume $\Omega < V$ corresponds to a finite probability that the electron will be almost localized in some region of the sample.

Thus, it has been shown that the local density of states distribution function $P(s)$ gives important information about statistical properties of wave functions and coincides with the NMR line shape at zero temperature. Recently Eq. (7.96) was obtained by Beenakker (1994), using an approach based on RMT. The local density of states distribution function can also be computed for $\gamma \gg \mathcal{E}_c$. In this limit the calculation can be done diagrammatically, although one has to sum a rather complicated series (Lerner (1988)).

The unusual form of the NMR line shape, Eq. (7.96), can manifest itself experimentally, but before discussing experiments let us consider two more theoretical quantities, namely, moments M_n of the distribution function and coefficients b_n of the inverse participation ratio. These quantities can give important information about wave functions and can also be compared with numerical simulations.

As concerns calculation of the moments M_n, one can use Eq. (7.90) or perform the averaging of s^n with the distribution function $P(s)$, Eq. (7.96). The calculation is straightforward and one obtains

$$M_n(\alpha) = 1 + n(n-1) \int_\alpha^\infty \sqrt{\frac{2}{\pi \alpha'}} K_{n-1/2}(\alpha') \, d\alpha' \tag{7.112}$$

In the limit of a large sample, when $\alpha \gg 1$, all the moments approach 1. In the opposite limit $\alpha \ll 1$, the asymptotics of $M_n(\alpha)$ take the form

$$M_n(\alpha) = \frac{n}{(2\alpha)^{n-1}} \frac{(2n-2)!}{(n-1)!} \tag{7.113}$$

The coefficients b_n of the inverse participation characterize fluctuations of the wave functions at energy ε and are defined as

$$b_n = \nu^{-n} \left\langle \sum_k |\varphi_k(\mathbf{r})|^{2n} \delta(\varepsilon - \varepsilon_k) \right\rangle \tag{7.114}$$

In situations when the energy spectrum is discrete, Wegner (1980) established a direct relation between the coefficients b_n and the moments of M_n of the density of states. This is very important because, as we have seen, the moments M_n can be expressed in terms of the Green functions and, hence, can in some cases be computed analytically.

The relation between the coefficients b_n and M_n is rather simple and can be derived by considering the function $D(\epsilon_1, \epsilon_2, \ldots, \epsilon_n)$

$$D(\epsilon_1, \epsilon_2, \ldots, \epsilon_n) = \left\langle \prod_{i=1}^{n} y(\epsilon_i) \right\rangle \tag{7.115}$$

where

$$y(\epsilon) = v^{-1} \sum_k |\varphi_k(\mathbf{r})|^2 \delta(\epsilon - \varepsilon_k) \tag{7.116}$$

is the reduced local density of states of a system with a discrete spectrum and n is an arbitrary integer. As a result of the discreteness of the spectrum the function D can be written in the form

$$
\begin{aligned}
D(\epsilon_1, \epsilon_2, \ldots, \epsilon_n) &= c_1 + c_2 \sum_{i \neq j} \delta(\epsilon_i - \epsilon_j) \\
&+ c_3 \sum_{i \neq j \neq k} \delta(\epsilon_i - \epsilon_j) \delta(\epsilon_i - \epsilon_k) + \cdots + c_n \prod_{i=2}^{n} \delta(\epsilon_i - \epsilon_1)
\end{aligned} \tag{7.117}
$$

The identity Eq. (7.117) can be obtained by substituting Eq. (7.116) into Eq. (7.115) and rewriting the product of the sums for each $y(\epsilon_i)$ as a sum of all possible products. The first term in Eq. (7.117) contains a product of terms corresponding to all different energy levels ε_k; in the second one level is taken in two multipliers, and so on. The last term originates from the product of all the terms coming from one level.

It is not difficult to realize that the coefficient c_n is equal to the coefficient b_n of the inverse participation ratio, Eq. (7.114). In order to express it through the momenta M_n for a system with smeared levels, let us multiply both sides of Eq. (7.117) by the function $K(\epsilon_1, \epsilon_2, \ldots, \epsilon_n)$

$$K(\epsilon_1, \epsilon_2, \ldots \epsilon_n) = \prod_{i=1}^{n} \frac{\gamma}{2\pi} \frac{1}{(\epsilon_i - \epsilon)^2 + (\gamma/2)^2} \tag{7.118}$$

with γ the level width and integrate over all ϵ_i. In the limit $\gamma \to 0$, only the last term in the right-hand side of Eq. (7.117) contributes, and we obtain, using Eq. (7.82)

$$b_n = \lim_{\gamma \to 0} (2\alpha \Delta)^{n-1} ((n-1)!)^2 ((2n-2)!)^{-1} M_n \tag{7.119}$$

Eq. (7.119) demonstrates that the moments M_n completely determine the coefficients of the participation ratio. In the situation considered the discreteness of the spectrum is due to the finite volume of the particle. At the same time, the result, Eq. (7.119), holds for an infinite system in the regime of localization as well and this characteristic was used by Wegner (1980) and Altshuler and Prigodin (1989) to study the localization.

Using Eqs. (7.113, 7.119) one immediately obtains

$$b_n = \Delta^{n-1} n! \tag{7.120}$$

The coefficients b_n in Eq. (7.120) grow with n faster than exponentially and this reflects the existence of the strong fluctuations mentioned before. Moreover, the proportionality of the coefficients b_n to $n!$ proves the Gaussian distribution of the wave functions, Eq. (7.101), which was assumed to explain the low and large density of states asymptotics of the local density of states distribution function, Eqs. (7.107, 7.110). So we see that by calculating the moments M_n of the local density of states and reconstructing the coefficients b_n of the inverse participation ratio one can obtain the distribution function of the wave functions.

Although this procedure gives the final result, it is not straightforward. In Chapter 9 a more direct scheme that makes it possible to express the distribution function of the wave functions in the form of an integral over the supermatrices is suggested. As a result, the distribution function of the wave functions will be calculated for an arbitrary magnetic field, thus describing the entire crossover from the orthogonal to the unitary ensemble.

7.3.3 NMR line shape-comparison with experiments

Now, let us discuss the possibility of observing the discovered effects experimentally. The NMR line shape $I(\omega)$ at $T = 0$ can easily be obtained from Eqs. (7.74–7.77)

$$I(\omega) = BP\left(\frac{\omega - \omega_0}{\delta\omega_0}\right) \tag{7.121}$$

where $P(s)$ is the local density of states distribution function, B is a constant, ω_0 is the unshifted position of the resonance, and $\delta\omega_0$ is the Knight shift of the corresponding bulk metal. The distribution function drawn in Fig. 7.2 is just the NMR line shape. The asymmetric broadening seems to be more readily observable than, for instance, the oscillating response to an electromagnetic field considered in the preceding sections because it does not require a very narrow distribution of particle size. Experimental observation of the NMR effects is simplified by its apparent sensitivity to temperature at very low temperatures $T \sim \Delta$, which is due to the discreteness of the levels. There are no other temperature-dependent effects at such low temperatures and it is very important to generalize the theory developed to $T \neq 0$.

At first glance, an extension of the theory to nonzero temperatures cannot be difficult because one can just use Eq. (7.75) relating the zero temperature quantity $\rho(\varepsilon, \mathbf{r})$ to the susceptibility at arbitrary temperature. However, the presence of Green functions with different energies would demand construction of a σ-model with $8n \times 8n$ supermatrices for the nth moment. Although the second moment can be calculated easily in this way, explicit calculations with the σ-model with an arbitrary n do not seem to be possible. Therefore, I cannot say anything certain about the possibility of calculating exactly the dependence of the NMR line shape on temperature. The only thing one can do without difficulties is to invent an interpolation that would enable us to make a rough comparison with experiments. The simplest approximate way is to replace the function $f(\varepsilon)$ in Eq. (7.75) by a normalized Lorentzian $f_L(\varepsilon)$

$$f_L(\varepsilon) = \pi^{-1}\left(\frac{\tilde{\gamma}}{2}\right)\left[\varepsilon^2 + \left(\frac{\tilde{\gamma}}{2}\right)^2\right]^{-1} \tag{7.122}$$

As written in Eq. (7.122), the function $f_L(\varepsilon)$ contains an arbitrary parameter $\tilde{\gamma}$ that must be fixed by the requirement of the best fit to the function $f(\varepsilon)$ in Eq. (7.75). This cannot be done in a unique way, but for simplicity, let us fix $\tilde{\gamma}$ by the following equality:

$$\int f^n(\varepsilon)\,d\varepsilon = \int f_L^n(\varepsilon)\,d\varepsilon \tag{7.123}$$

for $n \neq 1$.

Of course, the choice depends on n. For example, for $n = 2$ one gets $\tilde{\gamma}_2 = 1.91T$ and for $n = \infty$, $\tilde{\gamma}_\infty = 2.55T$. Nevertheless, the difference is not very large and it is reasonable to assume that the best interpolation is achieved by $\tilde{\gamma} = CT$ with

$$1.91 < C < 2.55 \qquad (7.124)$$

Because of the Lorentzian form of the function f_L one can repeat at finite temperatures all the manipulations done at $T = 0$. This is because $\rho\,(\varepsilon, \mathbf{r})$ in Eq. (7.75) is also Lorentzian and integration with $f_L\,(\varepsilon)$ results only in the substitution of $\gamma\,(T)$ for γ in Eqs. (7.78, 7.99)

$$\gamma\,(T) = \gamma\,(0) + CT \qquad (7.125)$$

Then, in order to use all the results derived, one should replace γ by $\gamma\,(T)$ from Eq. (7.125).

In the limit $\gamma\,(0) \gg \Delta$, the NMR line is sharp and is not sensitive to temperature. This case corresponds to large α in Fig. 7.2. In the opposite limit $\gamma\,(T) \ll \Delta$ (small α), the line is very asymmetric and has a long tail. The form of the line is very sensitive to temperature and changes considerably at $T \sim \Delta$. Although, in principle, the line can be broad as a result of other effects (such as surface interactions in platinum), the variation of the shape at $T \sim \Delta$ can only be due to the mesoscopic effect considered previously.

Experimentally, if the line shape is complicated, the Knight shift is defined as the position of the maximum. The maximum of the function $P\,(s)$, Eq. (7.96), is achieved in the limit $\alpha \ll 1$ at

$$s_m = 0.22\alpha = \frac{0.22\pi\gamma\,(T)}{\Delta} \qquad (7.126)$$

Substituting Eq. (7.125) into Eq. (7.126) one obtains the explicit dependence of the maximum $s_m\,(T)$ on temperature

$$s_m\,(T) = s_m\,(0) + \frac{0.68CT}{\Delta} \qquad (7.127)$$

Eq. (7.127) shows that the maximum of the resonance line changes linearly with temperature with the slope f

$$f = \frac{0.68C}{\Delta} \qquad (7.128)$$

A few NMR experiments on small metal particles were carried out long ago (Kobayashi, Takahashi, and Sasaki (1971, 1972), Yee and Knight (1975); for a review, see also Halperin (1986)). The measured line shape in Kobayashi, Takahashi, and Sasaki (1971, 1972) looks very similar to that presented in Fig. 7.2 for $\alpha = 0.5$. At higher temperature the NMR line is smoother and typically looks like the curve in Fig. 7.3. This corresponds to larger values of α in Fig. 7.2. The line broadens when the temperature and the particle size decrease, is very asymmetric, and has a long tail.

The temperature dependence of the maximum of the resonance line K_{obs} related to K_N, where K_N is the Knight shift of the bulk metal, observed by Yee and Knight (1975) on copper particles obeyed an approximately linear law. The experimental data are displayed in Fig. 7.4. The linear behavior agrees well with the prediction given by Eq. (7.127). Good results were obtained by Yee and Knight (1975) for particles with diameter 40 Å. The mean level spacing Δ for the particles of this size is 29 K and the slope f in the linear temperature dependence of the maximum of the resonance line can be estimated as 0.053. Substituting these values into Eq. (7.128), one obtains $C_{\mathrm{exp}} = 2.26$, which is in very good agreement

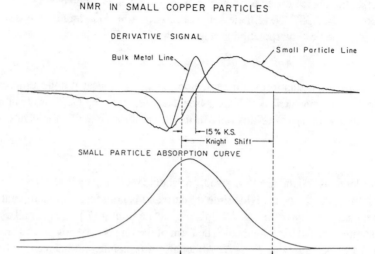

Fig. 7.3. The NMR line deduced from the NMR derivative spectrum by Yee and Knight (1975) at a temperature of 0.4 K. The copper particles have average diameter 100 Å.

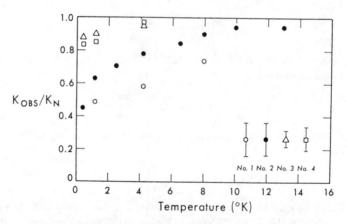

Fig. 7.4. The Knight shifts of copper particles observed as a function of temperature by Yee and Knight (1975) for four samples with different sample sizes. The data correspond to average sizes of 25 Å (○); 40 Å (●); 100 Å (△); and 110 Å (□).

with the estimate Eq. (7.124). The agreement with the data for the particles of other sizes is not as good and it would be interesting to perform new experiments.

It is important to the applicability of the theory developed in this section that the tunneling into the surrounding volume should not be small; otherwise Coulomb blockade effects can become important. In the latter case one would have to consider separately particles with odd and even numbers of electrons; this would give experimentally, as discussed in Section 7.1, a two-peak resonance. Variations of the particle size can lead experimentally to smearing of the two-peak structure into a broad line, which seems to be a similar prediction. However, let us emphasize that the effects considered in the present section are of a completely different nature. They are due to multiple scattering of electron waves on impurities or boundaries of the particles. The NMR line predicted here must be broad even for a system of equal particles; therefore, NMR experiments can be a very interesting tool for studying mesoscopic fluctuations in small metal particles.

Recently, mesoscopic fluctuations were studied by the NMR technique in monodisperse Pt-clusters (Brom et al. (1994)). The compounds investigated in this work consist of tightly packed platinum particles of very small size. Apparently, electrons can quite easily tunnel from particle to particle and thus, the system corresponds to the situation considered in the present section. Very good agreement between the experimental NMR line shape and the curve in Fig. 7.2 was observed. However, in platinum the density of states is determined by the contribution of not only s- but also d-electrons. The density of states of the d-band depends on the distance from the surface, and this leads to an additional broadening of the NMR line even at zero temperature. This additional broadening can be distinguished from mesoscopic broadening by its independence of temperature at low temperatures, and, possibly, a careful study of the temperature effects is still necessary to separate the mesoscopic effect unambiguously from the contribution of d-electrons.

Anyway, the main purpose of this section is to draw attention to study of mesoscopic effects by NMR technique rather than to discuss experimental details. The very existence of the effects considered in this section was discovered by the supersymmetry technique, and this shows that this technique can be useful not only in explicit calculations for well-understood physical phenomena but in prediction of new effects. Another effect predicted for the first time by supersymmetry, namely, the dependence of the electric susceptibility of small metal particles on magnetic field and magnetic or spin-orbit impurities, has been considered in the preceding section. In fact, quite different phenomena can be considered by supersymmetry, and the next chapter is devoted to discussion of a currently very popular problem, persistent currents in mesoscopic objects.

8

Persistent currents in mesoscopic rings

8.1 Basic properties

The recent observations of persistent currents in small metallic rings by Lévy et al. (1990), Chandrasekhar et al. (1991), and Mailly, Chapelier, and Benoit (1993) have opened a new field of research. Although small mesoscopic systems have been under intensive study for quite a long time, they have usually been studied by making contacts with metallic leads. As a result, one could obtain a finite conductance of the system that corresponded to a finite current in the presence of a finite voltage only. The experiments by Lévy et al. (1990) and by Chandrasekhar et al. (1991) were carried out in such a way that the metallic rings remained isolated. A slowly varying magnetic field was applied and a magnetic response was measured. In this situation it became possible to observe persistent currents in quite dirty samples in which the elastic mean free path l was much shorter than the circumference L. The existence of the persistent current is possible in isolated rings only, and Mailly, Chapelier, and Benoit (1993) demonstrated how the value of the persistent current decreases when increasing a weak coupling to leads.

The intriguing question of persistent currents in metal rings enclosing a magnetic flux was discussed in the 1960s by Byers and Yang (1961), F. Bloch (1965, 1968, 1970), Schick (1968), Gunter and Imry (1969). Although in the these works the properties of superconducting rings were mainly discussed, the possibility of circulating currents in sufficiently small and clean metal rings was also mentioned. However, the idea of persistent currents in normal metal rings containing elastic scatterers is more recent and is based on the fact that such scatterers do not prevent the electron wave function from extending coherently over the whole circumference of the ring. For one-dimensional loops, the existence of such currents was proposed by Büttiker, Imry, and Landauer (1983) and further discussed by Büttiker (1985).

The main underlying idea of such discussions is that the electron motion in a ring is equivalent to the motion in an infinite system with a periodic potential. Independently of how complicated the impurity potential in the ring is, going around the circumference one obtains the same potential, although the wave function can change its phase, depending on the magnetic flux through the ring.

The presence of the magnetic field through the ring represented schematically in Fig. 8.1 can be reduced simply to flux-modified azimuthal boundary conditions

$$\psi(L) = \exp\left(\frac{2\pi i\phi}{\phi_0}\right)\psi(0), \qquad \left.\frac{d\psi}{dx}\right|_L = \exp\left(\frac{2\pi i\phi}{\phi_0}\right)\left.\frac{d\psi}{dx}\right|_0 \qquad (8.1)$$

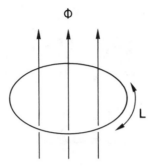

Fig. 8.1. One-dimensional ring threaded by a magnetic flux ϕ.

where, as before, $\phi_0 = hc/|e|$ is the flux quantum, x is the coordinate along the circumference, and L is the length of the circumference. Eqs. (8.1) imply that the eigenstates and eigenenergies and, therefore, all equilibrium properties of the ring are periodic with the period ϕ_0. This is true also in the presence of disorder. Eqs. (8.1) are equivalent to the corresponding equations determining the Bloch waves of electrons moving in a periodic potential, where the parameter $2\pi\phi/L\phi_0$ plays the role of the momentum. The energy levels of the ring form microbands as a function of ϕ with period ϕ_0 analogous to the Bloch electron bands in the extended k-zone picture.

A nonzero persistent current in the ring is equivalent to a nonzero total electron flow in the periodic potential. Although the potential can be quite complicated as a result of the impurities, it is its periodicity that leads to the possibility of such finite flow. Let us emphasize that, here, only thermodynamic properties are considered and, hence, any nonzero current corresponding to the minimum of the free energy does not decay in time at finite temperatures.

The total current, $I(\phi)$, is the sum over the contributions i_n of all states weighted with the appropriate occupation probability. The current i_n carried by a level ε_n is

$$i_n = \frac{2ev_n}{L}, \qquad v_n = \frac{\partial \varepsilon_n}{\partial k} \tag{8.2}$$

(the factor of 2 is due to the spin). Using the analogy with the electron motion in a periodic potential one can rewrite this equation as

$$i_n = -2c\frac{\partial \varepsilon_n}{\partial \phi} \tag{8.3}$$

At finite temperatures, instead of summing the currents i_n over all levels with weight $n(\varepsilon_n)$ one can calculate the total current I from free energy F of the system

$$I(\phi) = -c\frac{\partial F}{\partial \phi} \tag{8.4}$$

In Fig. 8.2, the energies ε_n of the eigenstates are shown schematically as a function of the magnetic flux for a one-dimensional chain. In the absence of disorder in the ring, the curves form intersecting parabolas. Adding disorder leads, as discussed in Chapter 6, to level repulsion. As a result, gaps open at the points of intersection in the same way as bands form in the band structure problem. We see from Fig. 8.2 that the picture is symmetric in flux; this means that the energy levels do not change with changing the sign of the magnetic

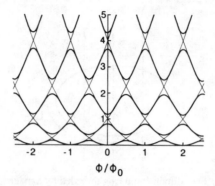

$$\phi/\phi_0$$

Fig. 8.2. Electron energy levels (arbitrary units) as function of the magnetic flux ϕ/ϕ_0 in a one-dimensional ring. Thin and thick lines correspond to a clean and a disordered ring, respectively.

field. It is clear from Eq. (8.3) that the current carried by an eigenstate is proportional to the slope of the curves describing the dependence of the energy on flux. At an integer or half-integer flux quantum, the energy reaches a maximum or minimum, and, hence, at these values of ϕ the current is zero.

The basic properties of the total persistent currents in the rings can be understood by considering the simplest model of one-channel ideal rings without disorder (Cheung et al. (1988a,b)) threaded by the magnetic flux ϕ. In such a model energy ε_n and current i_n of the nth eigenstate are

$$\varepsilon_n = \frac{1}{2m}\left[\frac{2\pi}{L}\left(n + \frac{\phi}{\phi_0}\right)\right]^2, \qquad i_n = \frac{4\pi e}{mL^2}\left(n + \frac{\phi}{\phi_0}\right) \qquad (8.5)$$

with $n = 0, \pm 1, \pm 2, \ldots$.

The total current at $T = 0$ is obtained by adding all contributions from levels with energies less than the chemical potential μ. In the one-dimensional case with a fixed number of electrons, the chemical potential μ is determined by the relation $\mu = (N_e\pi)^2/2mL^2$, where $2N_e$ is the total number of electrons in a ring, and the total persistent current is

$$I(\phi) = \begin{cases} -2I_0\,(2\phi/\phi_0) & \text{for } N_e \text{ odd,} & -1/2 \le \phi/\phi_0 < 1/2 \\ -2I_0\,(2\phi/\phi_0 - 1) & \text{for } N_e \text{ even,} & 0 \le \phi/\phi_0 < 1 \end{cases} \qquad (8.6)$$

where

$$I_0 = \frac{|e|v_0}{L} \qquad (8.7)$$

and v_0 is the Fermi velocity.

The current $I(\phi)$ is periodic with the period ϕ_0 for both even and odd numbers of electrons and is represented in Figs. 8.3a and 8.3b, respectively. However, an important general property of a system of rings can be obtained from this simple model. Suppose that the number of rings with an even number of electrons is equal to the number of rings with an odd number of electrons. Then, to calculate the total current, one has to take the average $\langle I(\phi)\rangle$ of the two currents in Eq. (8.6). The result is represented in Fig. 8.3c. One can see that the average current $\langle I(\phi)\rangle$ has a period $\phi_0/2$ instead of ϕ_0. In other words, the averaging leads to halving the period, a result that holds quite generally. It will be shown in the next sections that averaging over impurities leads to the same halving of the period.

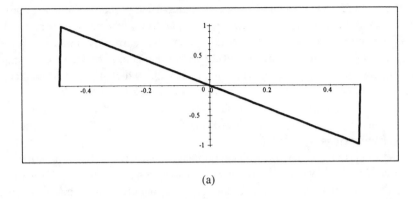

(a)

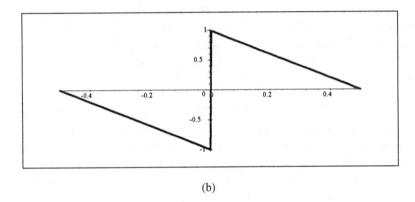

(b)

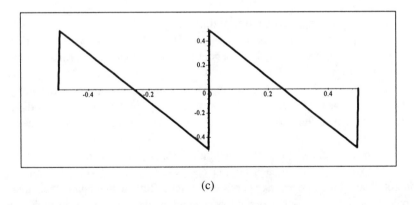

(c)

Fig. 8.3. Persistent current $I\phi/2I_0$ over one period of the magnetic flux ϕ/ϕ_0. The chemical potential corresponds to a) even number of filled levels N_e, b) odd number N_e, and c) current averaged over even and odd N_e.

From Eqs. (8.5, 8.6), one can learn one more important property. The maximum of the modulus of the average total current $\langle I(\phi)\rangle$ is equal to I_0, whereas the current of a level at the Fermi surface equals $4I_0$. This means that the total current is obtained as a result of a rather tricky cancellation of the one-level currents and adding one level can completely change the final result. Therefore, the assumptions under which one performs calculations are very important. For example, one may not replace sums over the levels by integrals because the total current can just vanish. This demonstrates that the persistent currents can exist only in small rings and vanish in the limit $L \to \infty$. That is not the case for superconducting rings.

In order to make a comparison with experiments one has to be able to describe disorder that is always present. Of course, it is not possible to obtain results for an arbitrary impurity potential, and, as usual, one has to introduce an averaging procedure. However, depending on the type of experiment one should average different quantities. In the experiment of Lévy et al. (1990) the magnetization of 10^7 copper rings was measured and, for this system, one has to calculate the average current. In the experiments of Chandrasekhar et al. (1991) and of Mailly et al. (1993) the magnetic response of a single ring was studied. In this case not only average current but higher moments have to be known. The amplitude of the current in a single ring can be much larger than the average over the disorder and, to estimate its value, the typical current I_{typ} is used

$$I_{\text{typ}} = \langle I^2 \rangle^{1/2} \tag{8.8}$$

Calculation of the quantity I_{typ} is usually simpler because one has to average a positive quantity that is less sensitive to the approximations made. As concerns $\langle I \rangle$, the averaging in this quantity is highly nontrivial and requires special care. Using Eq. (8.6) we can see how different averaging procedures work. Neglecting the small difference between the Fermi velocities of systems with $2N_e$ and $2(N_e + 1)$ electrons we can rewrite Eq. (8.6) as

$$I(\phi) = \sum_{n=1}^{\infty} \frac{4I_0}{n\pi} \cos(n p_0 L) \sin\left(\frac{2\pi n\phi}{\phi_0}\right) \tag{8.9}$$

where $p_0 = \pi N_e / L$ is the Fermi momentum.

Averaging over integer N_e in an interval ΔN, $1 \ll \Delta N \ll N_e$, we see that only even harmonics contribute because the Fourier coefficients alternate in sign for odd n but do not change sign for even n. This implies that if we average over an ensemble of isolated rings, where the number of electrons in each ring is fixed but varies randomly from one ring to the other, only terms with even n survive. As a result one obtains a $\phi_0/2$ periodicity of the persistent current. A different result would be obtained if the averaging were done over only even or odd N_e.

Now, suppose that the chemical potential μ (or p_0) fluctuates in an interval $\Delta \mu$, $p_0/mL \ll \Delta \mu \ll \mu$. Then each Fourier coefficient in Eq. (8.9) averages to zero and we do not obtain any finite average persistent current. But we understand that such a result is wrong. This means that one may not forget about the discreteness of the number of electrons in each ring. In performing different types of averaging one has to keep an integer number of electrons in each ring. Alternatively, one may carry out an averaging procedure with a fixed number of electrons (canonical averaging) and average over this number in the end; the last averaging is usually not important. The importance of variation of the chemical potential was first realized by Cheung et al. (1988a,b) for one-dimensional rings and then by Bouchiat and Montambaux (1989) for multichannel systems.

At the same time calculation of the typical current I_{typ} is less sensitive to the averaging procedure and the same order of magnitude

$$I_{\text{typ}} \sim I_0 \qquad (8.10)$$

can be obtained independently of whether the averaging is carried out with or without the restriction of an integer number of electrons. This result can be obtained by taking the square of the sum in Eq. (8.9) and averaging $\cos^2 (n p_0 L)$. Averaging without the condition of an integer number of electrons (grand canonical averaging) one should replace this function by $1/2$; this leads immediately to Eq. (8.10). The canonical averaging gives a similar result, although the numerical coefficient can be different. Let us mention that in strongly disordered rings when the wave functions are localized on a length $L_c \ll L$ the sensitivity of the phase to the boundary condition, Eq. (8.1), is exponentially small and the persistent current in this case, although remaining finite, is small and proportional to $\exp\left(-L/L_c\right)$.

The preceding discussion in terms of the model of clean one-channel rings demonstrates all basic properties of the persistent currents surprisingly well. The only thing that remains to be done is to extend the results to the case of multichannel disordered mesoscopic rings. In the experiment by Lévy et al. (1990) the circumference L of the rings was $L \sim 22,000$ Å, the cross section $S \sim 300$ Å $\times$ 600 Å, and the elastic mean free path $l \sim 200$ Å. This means that at distances smaller than geometrical sizes electrons move diffusively. The persistent currents in such a system are again due to the changing of the phase of the electron function when going around the rings, but now the electrons diffuse around the rings rather than fly ballistically. We will see that although the effects considered in the present section hold for the diffusive regime qualitatively, the values of the average and typical currents are much smaller than I_0 and one should be very careful calculating these quantities (especially the average current). Concerning mesoscopic objects that are available experimentally the diffusion regime seems to be most interesting and in the rest of the chapter all calculations will be performed in this limit.

To conclude the section let us emphasize that the very existence of persistent currents is possible only if the rings are very weakly connected to a reservoir or are completely isolated. The dependence of the persistent current on the connection to the reservoir was studied experimentally by Mailly, Chapelier, and Benoit (1993).

8.2 Diffusion modes and persistent currents

8.2.1 Typical current

Working in the diffusion regime as usual one has to express physical quantities in terms of the retarded G^R and advanced G^A Green functions and then to try to apply well-developed methods. The total current along the circumference of a ring, I, is written in the standard form

$$I = \frac{1}{i\pi L} \int \lim_{\mathbf{r}' \to \mathbf{r}} \hat{\pi}_{\mathbf{r}} \left[G_\varepsilon^A \left(\mathbf{r}, \mathbf{r}' \right) - G_\varepsilon^R \left(\mathbf{r}, \mathbf{r}' \right) \right] n \left(\varepsilon \right) d\varepsilon d\mathbf{r} \qquad (8.11)$$

where $\hat{\pi}_{\mathbf{r}} = e\partial\varepsilon \left(-i\nabla - e/cA \right) / \partial \left(-i\nabla \right)$ is the current operator. Let us calculate the current, averaging over the impurities with a fixed chemical potential. Then, we come

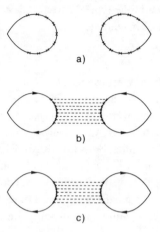

Fig. 8.4. Diagrams for the square of the current, I^2: a) nonaveraged I^2, b) and c) the square of the typical current $\langle I^2 \rangle$ with one cooperon and one diffuson, respectively.

immediately to the average Green functions, Eq. (3.45), which can be written in the presence of the magnetic field as

$$\left\langle G_\varepsilon^{R,A}(\mathbf{p}) \right\rangle = \left[\varepsilon - \varepsilon \left(\mathbf{p} - \frac{e}{c}\mathbf{A} \right) \pm \frac{i}{2\tau} \right]^{-1} \tag{8.12}$$

Eq. (8.12) is written in the momentum representation and this simple expression is valid provided $\Delta \ll \tau^{-1}$, with Δ the mean level spacing. Substituting Eq. (8.12) into Eq. (8.11), one obtains

$$\langle I \rangle = 2eS \int n(\varepsilon) \, \frac{\partial \varepsilon (\mathbf{p} - e\mathbf{A}/c)}{\partial \mathbf{p}} \tag{8.13}$$

$$\times \left\{ \left[\varepsilon - \varepsilon \left(\mathbf{p} - \frac{e}{c}\mathbf{A} \right) + \frac{i}{2\tau} \right]^{-1} - \text{c.c.} \right\} \frac{d\mathbf{p}}{(2\pi)^3} \frac{d\varepsilon}{2\pi i}$$

Integration over continuous $\mathbf{p}$ in Eq. (8.13) gives zero. A nonzero value of the current averaged in such a way can be obtained by taking into account discreteness of the levels. However, in this case one can obtain only values exponentially small in $(\tau \Delta)^{-1}$. Does that mean that the current in each ring is so small? The answer is, definitely no. In the preceding section it was demonstrated that grand canonical averaging can easily give zero in a situation when both the properly averaged $\langle I \rangle$ and typical currents I_{typ} are relatively large. Therefore, for calculations in the diffusive regime, one should try to use, as in the preceding section, more sophisticated approaches. The typical current I_{typ} is simpler because it is less sensitive to averaging procedure, and, so, let us start with evaluating this quantity by following the scheme of Cheung, Riedel, and Gefen (1989).

In the diffusion regime it is convenient to use the cross technique presented in Chapter 3. Using the expression for the nonaveraged current in terms of the Green functions, Eq. (8.11), one can represent I^2 as in Fig. 8.4a. Having Eq. (8.13), we understand that, when calculating $\langle I^2 \rangle$, averaging each loop independently does not give any noticeable contribution and one should consider diagrams connected by impurity lines. We understand that the persistent currents in normal rings can exist in small systems only, and, hence, one should take into account diagrams that can give a large contribution that is due only to the small size. Such diagrams have been discussed many times throughout the book and they are the diffusion modes. The main contribution in I_{typ} is given by first-order diagrams in the diffusion modes, and these diagrams (with one cooperon and one diffuson) are represented in Figs. 8.4b and 8.4c. To obtain a nonzero result, one of the Green functions must be retarded and the other advanced. Using Eqs. (3.53, 3.54, 3.60) one can express the contribution of Figs. 8.4b and 8.4c in the form

$$\langle I^2 \rangle = \frac{2e^2}{(2\pi\nu\tau)^2\,\pi^2 L^2} \sum_k \int [D(\mathbf{k}, \varepsilon_1 - \varepsilon_2) - C(\mathbf{k}, \varepsilon_1 - \varepsilon_2)] \tag{8.14}$$

$$\times \frac{\partial\varepsilon(\mathbf{p})}{\partial\mathbf{p}} \frac{\partial\varepsilon(\mathbf{p}-\mathbf{k})}{\partial\mathbf{p}} \left\langle G^R_{\varepsilon_1}(\mathbf{p}) \right\rangle^2 \left\langle G^A_{\varepsilon_2}(\mathbf{p}-\mathbf{k}) \right\rangle^2 n(\varepsilon_1)\,n(\varepsilon_2) \frac{d\mathbf{p}\,d\varepsilon_1\,d\varepsilon_2}{(2\pi)^3}$$

As in Chapter 3, it is convenient to integrate first over large momenta p and reduce calculation of the integral in Eq. (8.14) to a summation over $\mathbf{k}$ in the resulting expression containing the cooperon and diffuson. To simplify the integration over ε_1 and ε_2 one can use the identity

$$\int_{-\infty}^{\infty} C(\mathbf{k}, \varepsilon_1 - \varepsilon_2)\,G^A_{\varepsilon_2}(\mathbf{p} - \mathbf{k}, \varepsilon_2)\,d\varepsilon_2 = 0 \tag{8.15}$$

which is a simple consequence of the fact that both functions in the integrand have poles in the upper half-plane of the complex variables. Then, in Eq. (8.14) one may replace $n(\varepsilon_2)$ by $n(\varepsilon_2) - 1$ and now, the integral over the energies converges explicitly on a scale much smaller than the Fermi energy. Changing to the variables $\varepsilon = (\varepsilon_1 + \varepsilon_2)/2$, $\omega = \varepsilon_1 - \varepsilon_2$ one can perform integration over ε and Eq. (8.14) is thus reduced to an expression containing integration over ω and summation over $\mathbf{k}$, the latter being rather special as a result of the ring geometry.

Assuming that the ring is thin enough, one can neglect all but the zero transversal $k_\perp$ and perform summation over longitudinal k only. This is correct provided the characteristic ω are much less than D_0/S, where S is the cross section (both sizes of the cross section are assumed to be of the same order). At the same time, the thickness may exceed atomic distances such that the Fermi surface is three-dimensional. The longitudinal momenta k are quantized and can take the values $2\pi n/L$, $n = 0, \pm 1, \pm 2, \ldots$. The magnetic field does not enter the diffuson. The cooperon can be obtained from the diffuson by the substitution $n \to n + 2\phi/\phi_0$ for all n and we have

$$\langle I^2 \rangle = \frac{4e^2 v_0 l}{\pi^2 L^2} \int_0^{\infty} \omega \coth \frac{\omega}{2T} \left[B_\omega \left(\frac{2\phi}{\phi_0} \right) - B_\omega(0) \right] d\omega \tag{8.16}$$

where

$$B_\omega(y) = \sum_{n=-\infty}^{\infty} \frac{4\mathcal{E}_c\,(n+y)^2}{16\mathcal{E}_c^2\,(n+y)^4 + \omega^2} \tag{8.17}$$

and $\mathcal{E}_c = \pi^2 D_0/L^2$. Eq. (8.16) shows that the quantity $\langle I^2 \rangle$ is periodic in flux with the period $\phi_0/2$ and this reflects the periodicity of the nonaveraged current with the period ϕ_0. Formally, the integral in Eq. (8.16) logarithmically diverges at the upper limit, but that means only that one should make the cutoff at $\omega \sim 1/\tau$. To analyze the dependence of $\langle I^2 \rangle$ on magnetic flux it is convenient to expand this quantity in a Fourier series and calculate coefficients in this expansion. Experimentally, one studies the harmonics content of the nonaveraged current I (Chandrasekhar et al. (1991), Mailly et al. (1993)). Of course, making the Fourier transform of $\langle I^2 \rangle$, unambiguously reconstructing the typical value of each harmonics is a difficult task. The simplest approximation that can be made is to neglect correlations between different harmonics of the nonaveraged current. Then, the first term in the expansion for I, which is proportional to $\sin(2\pi\phi/\phi_0)$, corresponds to the term with $\cos(4\pi\phi/\phi_0)$ in the expansion for $\langle I^2 \rangle$.

Introducing dimensionless units $p = \omega/(4\mathcal{E}_c)$, $t = 2\phi/\phi_0$, $\Theta = T/(4\mathcal{E}_c)$, Eq. (8.16) can be written as

$$\langle I^2 \rangle = \frac{16e^2 v_0 l \mathcal{E}_c}{\pi^2 L^2} \left(g(0) - g(t) \right), \tag{8.18}$$

$$g(t) = -\int_0^\infty \coth\left(\frac{p}{2\Theta}\right) \sum_{n=-\infty}^\infty \frac{p(n+t)^2}{p^2 + (n+t)^4} dp$$

Representing the function $g(t)$ as

$$g(t) = g_0 + 2 \sum_{m=1}^\infty g_m \cos(2\pi m t) \tag{8.19}$$

and calculating corresponding integrals we reduce the coefficients g_m to the form

$$g_m = \frac{1}{2\pi^2 m^3} \int_0^\infty z^2 \coth\left(\frac{z^2}{\Theta_m}\right) \exp(-z)(\sin z - \cos z)\, dz \tag{8.20}$$

where $\Theta_m = T/T_m$, and $T_m = \mathcal{E}_c/(\pi m)^2$. At $T = 0$ the coefficients g_m are equal to $\left(2\pi^2 m^3\right)^{-1}$. Substituting Eqs. (8.19–8.20) into Eq. (8.18) one obtains

$$\langle I^2 \rangle = \frac{64}{3} \left(I_0 \frac{l}{L} \right)^2 \sum_{m=1}^\infty g_m \sin^2(\pi m t) \tag{8.21}$$

with I_0 defined in Eq. (8.7). At $T = 0$ the first harmonics of the typical current reads (Cheung, Riedel, and Gefen (1989))

$$I_{\text{typ}} = \langle I^2 \rangle^{1/2} = \pm \frac{2\sqrt{2}}{\pi\sqrt{3}} \left(\frac{l}{L} \right) I_0 \sin\left(\frac{2\pi\phi}{\phi_0} \right) + \text{higher harmonics} \tag{8.22}$$

We see that the typical current of a disordered ring in the diffusion regime is much smaller than the current I_0 obtained for the one-channel clean rings. So, although the disorder does not completely destroy the persistent currents, it reduces their value. The amplitude of the current in Eq. (8.22) can also be written as being proportional to $|e|/\tau_D$, where $\tau_D \sim L^2/D_0$ is the diffusion time around the ring. There is no dependence of this current on the cross section (in other words, on the number of channels due to the transversal quantization) and no significant (e.g., oscillatory) dependence on chemical potential.

The smearing of the Fermi distribution at a finite temperature leads to reduction of the size of the typical current. This is because at finite temperatures, many different energies contribute. The nonaveraged current oscillates as a function of energy and, hence, the total current decreases. The crossover from the region of low temperature to the region of high temperature can be characterized by the thermal diffusion length

$$l_T = \left(\frac{2\pi D_0}{T}\right)^{1/2} \tag{8.23}$$

The low-temperature regime is characterized by the inequality $l_T > L$; the high-temperature regime is determined by the opposite inequality. Eq. (8.22) is written for $l_T \gg L$. In the opposite limit the typical current takes the form

$$I_{\text{typ}} = \pm 8\sqrt{\pi/3}\frac{l}{L}\left(\frac{L}{l_T}\right)^{3/2} \exp\left(-\frac{\pi L}{l_T}\right) I_0 \sin\left(\frac{2\pi\phi}{\phi_0}\right) + \text{h.h.} \tag{8.24}$$

where h.h. represents higher harmonics. Eq. (8.24) is written assuming that not only the inequality $L \gg l_T$ but also $L \gg S/l_T$ is fulfilled. In the limit $L \ll S/l_T$, but still $L \gg l_T$, the prefactor is slightly modified. An exponential decay is usual in situations when one has to average oscillating quantities; therefore, Eq. (8.24) indicates oscillations of the nonaveraged current as a function of energy. The exponential in Eq. (8.24) can be rewritten as $\exp\left[-\text{const}\,(T/\mathcal{E}_c)^{1/2}\right]$; that implies that groups of single-level currents with alternating sign are separated at the scale of $\mathcal{E}_c$. In the ballistic regime, $\mathcal{E}_c \sim v_0/L$. For a one-dimensional ring this energy is equal to the mean level spacing, and so the results of this section match those of the preceding one.

The basic difference between the one-channel and many-channel thick rings is that in the former one has either ballistic transport ($l \gg L$) or localization ($l \ll L$). In the thick rings, one has the additional possibility $l \ll L \ll L_c$, where L_c is the localization length proportional to $l\left(p_0^2 S\right)$. In terms of experiments on metallic rings this regime is most realistic. In this limit electrons diffuse around the ring. It is the inequality $L \ll L_c$ that allows the possibility of performing calculations taking into account only one cooperon or diffuson. The evaluation of the typical current is so easy because one obtains a good approximation by averaging over impurities with a fixed chemical potential.

It is interesting to note that practically the same equations as Eqs. (8.16–8.25) are obtained when calculating the average current in the presence of an electron–electron interaction (Ambegaokar and Eckern (1990)). The final result for the average current in the work of Ambegaokar and Eckern (1990) is proportional to the typical current I_{typ}, Eqs. (8.22, 8.24), multiplied by the interaction amplitude.

8.2.2 Canonical versus grand canonical

Now, let us try to calculate the average current, neglecting the electron–electron interaction. In the simplest case considered in the preceding section one could simply carry out summation over all electrons under the Fermi surface. In a disordered system this is not possible, and one should invent a new approach to treat the problem. Considering the one-channel clean ring it was demonstrated that the main reason for the failure of standard methods is that in an isolated system by changing an impurity configuration, one changes the chemical potential because the number of particles must be an integer. Of course, the chemical potential is fixed if the system is not isolated, but then the persistent current is really vanishing.

It is quite natural to consider a situation when the system is connected to the surrounding environment such that the energy levels are smeared. However, if the level width is not larger than all other relevant energies one can hope that, although small, there is some correlation between the disorder potential and the chemical potential, and this could lead to finite (not exponentially small) persistent currents. Using the assumption that fluctuations of the chemical potential are small one can try to make an expansion in these fluctuations. This idea has been used in several works (Altshuler, Gefen, and Imry (1991), Schmid (1991), von Oppen and Riedel (1991), Akkermans (1991), Oh, Zyuzin, and Serota (1991)). Following the work of Altshuler, Gefen, and Imry (1991) one should start by writing the free energy $F(N, \phi)$ in the canonical ensemble (N is the number of electrons) and the thermodynamic potential $\Omega(\mu, \phi)$ in the grand canonical one. Using simple thermodynamic relations one has

$$\left(\frac{\partial F}{\partial \phi}\right)_N = \left(\frac{\partial \Omega}{\partial \phi}\right)_{\mu=\partial F/\partial N|_\phi} = -\frac{1}{c} I(\phi) \tag{8.25}$$

Assuming the number of electrons in the ring to be fixed, one arrives at a fluctuating chemical potential μ, which is written in the form

$$\mu = \langle \mu \rangle + \delta\mu(\phi) \tag{8.26}$$

where $\langle \cdots \rangle$ stands for the disorder average. If $\delta\mu$ is small, one can make expansions in this quantity and the first terms of the expansion for the free energy F from Eq. (8.25) can be written as

$$\left(\frac{\partial F}{\partial \phi}\right)_N = \left(\frac{\partial \Omega}{\partial \phi}\right)_{\mu=\langle\mu\rangle} + \delta\mu \left(\frac{\partial}{\partial \mu}\left(\frac{\partial \Omega}{\partial \phi}\right)_\mu\right)_{\mu=\langle\mu\rangle} \tag{8.27}$$

Averaging both sides of Eq. (8.27) over disorder one has to study only the contribution of the second term in the right-hand side because the first term gives the current averaged with a fixed chemical potential, which is exponentially small. Changing the order of the derivatives in the second term we obtain

$$\left\langle \left(\frac{\partial F}{\partial \phi}\right)_N \right\rangle = \left\langle \delta\mu \frac{\partial}{\partial \phi}\left(\frac{\partial \Omega}{\partial \mu}\right)_\phi \right\rangle_{\mu=\langle\mu\rangle} = -\left\langle \delta\mu \left(\frac{\partial N}{\partial \phi}\right)_\mu \right\rangle_{\mu=\langle\mu\rangle} \tag{8.28}$$

where the identity $N = -\partial\Omega/\partial\mu$ was used. In order to obtain the fluctuating part $\delta\mu$ of the chemical potential one should write the equation for the total number of particles

$$N = V \int \rho(\varepsilon) n_\mu(\varepsilon) d\varepsilon \tag{8.29}$$

where $\rho(\varepsilon)$ is the density of states, taking into account the spin, and $n_\mu(\varepsilon)$ is the Fermi function with the chemical potential μ. Expressing the density of states as

$$\rho(\varepsilon) = \langle \rho(\varepsilon) \rangle + \delta\rho(\varepsilon), \qquad \langle \rho(\varepsilon) \rangle = 2\nu \tag{8.30}$$

and using the assumption that N does not fluctuate we obtain in the linear approximation

$$\delta\mu = -(2\nu)^{-1} \int \delta\rho(\varepsilon) n(\varepsilon) d\varepsilon \tag{8.31}$$

where $n(\varepsilon)$ without a subscript stands for the Fermi function with the chemical potential $\langle \mu \rangle$. Substituting Eq. (8.31) into Eqs. (8.28) and using the fact that $\langle \rho(\varepsilon) \rangle$ does not depend on flux we come, with the help of Eq. (8.25), to the following expression for the current:

$$I = -\frac{Vc}{4\nu} \frac{\partial}{\partial \phi} \int \langle \delta\rho(\varepsilon_1) \delta\rho(\varepsilon_2) \rangle n(\varepsilon_1) n(\varepsilon_2) d\varepsilon_1 d\varepsilon_2 \tag{8.32}$$

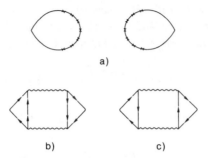

a)

b)   c)

Fig. 8.5. Diagrams for the square of fluctuating number of electrons N: a) nonaveraged N^2, b) and c) contribution of two cooperons and two diffusons, respectively.

where averaging on the right-hand side has to be performed at fixed $\mu = \langle\mu\rangle$.

8.2.3 Average current

The last thing that remains to be done is to calculate the expression on the right-hand side of Eq. (8.32) using known schemes. It turns out that, again, the diffusion modes make a crucial contribution. The density of states $\rho(\varepsilon)$ can be written using Eq. (3.16) in terms of the Green functions. In order to take into account level smearing one may use the Green functions in the form given by Eq. (7.78) assuming that the level width $\gamma = (2\tau_\varphi)^{-1}$, where τ_φ is a phase coherence time due to inelastic processes. The nonaveraged quantity N^2 is represented in Fig. 8.5a, which differs from Fig. 8.4a by replacement of the current operators in the vertices by 1. Averaging over the impurities one should connect the loops by impurity lines. Two disconnected averaged loops give the average number of electrons N_0^2 and this quantity does not contribute to the correlation function in Eq. (8.32). As usual, one of the Green functions must be retarded and the other advanced. The graphs in Figs. 8.5b, c have the most important contribution but only the two-cooperon graph in Fig. 8.5b depends on the magnetic field and contributes to the persistent current.

Although the nonaveraged squares of the current and density are represented by similar diagrams, the main contribution for the averaged quantities is from different diagrams (the diagrams in Figs. 8.4b, c contain one diffusion mode, whereas those in Figs. 8.5b, c have two). This is because the triangles in the diagrams with two diffusion lines can be very far from each other and such configurations give an important contribution. At the same time, when calculating the square of the current, the triangles contain the velocity operators in the vertices and, therefore, the otherwise most important contribution of low momenta is small and one is left with the diagrams with one diffusion mode. Calculating the square of the density one may neglect the latter diagram provided $\Delta \ll \mathcal{E}_c$. Neglecting the temperature smearing of the Fermi distribution one obtains

$$I = -\frac{c}{\Delta}\frac{\partial}{\partial\phi}\int n(\varepsilon_1) n(\varepsilon_2) R(\varepsilon_1 - \varepsilon_2) d\varepsilon_1 d\varepsilon_2 \qquad (8.33)$$

where the correlation function of the density of states $R(\omega)$ was introduced in Eq. (6.18). This correlation function has been calculated exactly in Chapter 6, in the limiting cases of the orthogonal, unitary, and symplectic ensembles. However, now we have to describe the crossover between the orthogonal and unitary or between the symplectic and unitary ensembles. In the next section it will be described precisely, but in this section let us write, following Altshuler, Gefen, and Imry (1991) and Schmid (1991), the two-cooperon contribution corresponding to Fig. 8.5b. In this approximation the function $R(\omega)$ equals

$$R(\omega) = 1 - \frac{\Delta^2}{2\pi^2} \operatorname{Re} \sum_{n_a} \left(\omega + \frac{i}{\tau_\varphi} + i D_0 \mathbf{k}^2 \right)^{-2} \tag{8.34}$$

Here a can be x, y, or z, $\omega = \varepsilon_1 - \varepsilon_2$; and, for a sample with dimensions $L \times L_y \times L_z$ (L is the ring perimeter),

$$\mathbf{k}^2 = 4\pi^2 \sum_{a=x,y,z} \left(\frac{n_y^2}{L_y^2} + \frac{n_z^2}{L_z^2} + \frac{(n_x + 2\phi/\phi_0)^2}{L^2} \right) \tag{8.35}$$

with $n_{x,y,z} = 0, \pm 1, \pm 2, \ldots$.

When calculating the derivative over the flux in Eq. (8.33) with the use of Eqs. (8.34, 8.35) the main contribution in the integral over the energies is from ω close to the Fermi level. It is clear from Eqs. (8.34, 8.35) that I is periodic in ϕ with the period $\phi_0/2$ and hence this quantity can be represented as

$$I = -i \sum_{m=-\infty}^{\infty} I_m \exp \left(4\pi i \frac{m\phi}{\phi_0} \right) \tag{8.36}$$

Assuming for simplicity that $L_y, L_z \ll L$ we neglect, as when calculating the typical current, all $n_y, n_z \neq 0$ in the sum in Eq. (8.35). For the coefficients of the Fourier expansion I_m, neglecting the temperature smearing of the Fermi distribution, one obtains

$$I_m = \frac{c\Delta i}{\pi^2 \phi_0} \int_{-\infty}^{\mu} \int_{-\infty}^{\mu} d\varepsilon_1 d\varepsilon_2 \int_{-\infty}^{\infty} d\phi \tag{8.37}$$

$$\left[\frac{\partial}{\partial \phi} \operatorname{Re} \left(\varepsilon_1 - \varepsilon_2 + \frac{i}{\tau_\varphi} + 16 i \mathcal{E}_c \frac{\phi^2}{\phi_0^2} \right)^{-2} \right] \exp \left(-4\pi i m \frac{\phi}{\phi_0} \right)$$

The most important contribution is that of the energies ε_1, ε_2 near the Fermi energy. Integration over the energies ε_1 and ε_2 can be performed using the variables $\varepsilon = (\varepsilon_1 + \varepsilon_2)/2$ and $\omega = \varepsilon_1 - \varepsilon_2$. Integrating in Eq. (8.37) over the variable ω, then over the flux ϕ, and finally over ε one gets as a result

$$I_m = \frac{|e|\Delta}{\pi^2} \exp \left(-\frac{|m|\pi}{(\mathcal{E}_c \tau_\varphi)^{1/2}} \right) \operatorname{sgn} m \tag{8.38}$$

Eq. (8.38) shows that for $m^2 < \mathcal{E}_c \tau_\varphi$, I_m is independent of $\mathcal{E}_c$ (i.e., of disorder) and m, and is of the order of Δ. Taking into account the temperature smearing of the Fermi distribution one should replace τ_φ^{-1} by $\max \{ \tau_\varphi^{-1}, T \}$.

The average persistent current calculated within the canonical ensemble in the diffusion regime is periodic in the magnetic flux with the period $\phi_0/2$. This is in agreement with the results of Section 8.1 obtained for one-channel clean rings. However, the amplitude of the current I_m is of the order of $I_0 (p_0^2 S)^{-1}$, which is much smaller than the amplitude of the persistent currents I_0 for one-channel rings. The value $p_0^2 S$ is proportional to the number of transversal channels. For the rings used in the experiments of Lévy et al. (1990)

and Chandrasekhar et al. (1991) this value can reach 10^4. The value of I_m is also much smaller than the typical current I_{typ}, Eq. (8.22), where the ratio I_{typ}/I_m is of the order of $\mathcal{E}_c/\Delta \sim 10^2$.

In the region of temperatures τ_φ^{-1}, $T \sim \Delta$, the first $(\mathcal{E}_c/\Delta)^{1/2}$ harmonics contribute to the average persistent current. With increasing the temperature the number of contributing harmonics decreases and, at high temperatures, only the first harmonic is important. The maximal amplitude of the average persistent current decreases from $|e|(\mathcal{E}_c\Delta)^{1/2}$ to $|e|\Delta$ when T and τ_φ^{-1} increase from 0 to $\mathcal{E}_c$.

The size of the persistent currents found here is about two orders of magnitude smaller than observed experimentally. Another theory including electron–electron interaction was proposed by Ambegaokar and Eckern (1990) even before the work with canonical averaging. However, this theory also gives a smaller value of the persistent current than the experimental one. From a theoretical point of view it is important to note that, again, the diffusion modes are responsible for the finite persistent currents.

Calculation of the average persistent current is a rather difficult task because one should be careful about the averaging procedure. Therefore, one could try to attribute the disagreement with the experiment to some delicate features not taken into account by the approximate treatments. As concerns the typical current, it does not seem to be sensitive to whether the averaging is performed within the canonical or grand canonical ensemble. An electron–electron interaction also does not play a crucial role. In this situation the typical current I_{typ} should be given by Eq. (8.22). It comes as a great surprise that the experiment by Chandrasekhar et al. (1991) gives values of the typical current that are of order I_0 rather than I_{typ}, Eq. (8.22). At the same time, the experiment by Mailly, Chapelier, and Benoit (1993) made with a sample with parameters corresponding to the ballistic regime gives a current of order I_0 in agreement with the theoretical prediction. The result of this experiment shows that electron–electron interaction does not play an important role at least in the ballistic regime.

The reason for the disagreement between theory and experiment in the diffusion regime is not clear yet and I do not want to discuss this question more. One may ask a question that can be more important to the contents of this book: Is there any need for using the supermatrix σ-model for the problem of persistent currents? Concerning a description of the three existing experiments the use of supersymmetry does not seem to be necessary. Only the experiments by Lévy et al. (1990) and by Chandrasekhar et al. (1991) were performed on rings with parameters corresponding to the diffusion regime where the σ-model helps to produce nonperturbative results. However, these experiments were performed at temperatures $T \gg \Delta$ where the perturbation theory in diffusion modes works well. The σ-model can give new results at temperatures $T \leq \Delta$ and one can hope to reach this regime only in future experiments. At the same time, a theoretical study at such low temperatures can nevertheless be interesting because it helps to establish the upper limit for the persistent currents. Besides, the calculational scheme developed for the persistent currents can help in studying other physical phenomena.

The fact that the results of this subsection cannot be used at low temperatures is already clear from Eq. (8.38) because at $T = 0$ all harmonics have the same amplitudes, which lead to divergencies. This signals that the two-cooperon contribution used for function $R(\omega)$ is not sufficient and one has to calculate function $R(\omega)$ exactly. A more difficult question concerns the applicability of the starting formulae, Eq. (8.32). No doubt, this formula should work well in the temperature regime T, $\tau_\varphi^{-1} \geq \Delta$. However, evaluating the next terms of the expansion over the fluctuating chemical potential $\delta\mu$ leads to very complicated correlation functions. Moreover, we know already that expansions in the diffusion modes in the region $T, \tau_{\varphi}^{-1} \leq \Delta$ can fail because the discreteness of the energy spectrum is not perceived in such a treatment. The only possibility for gaining reliable knowledge of the value of persistent currents is to try to develop an alternative independent scheme and compare results.

Two different treatments of persistent currents will be presented in the rest of the chapter. The first approach is based on Eq. (8.33) with $R(\omega)$ the exact level–level correlation function and corresponding calculations are presented in Section 8.3. Another, completely different dynamic scheme is the subject of Section 8.4.

To conclude the present Section let us return briefly to a question raised in Section 7.2. In that section, considering the electric susceptibility of small metal particles, it was assumed that one could neglect a difference between canonical and grand canonical ensembles, whereas in the present section this difference is shown to be important. What is the reason for such a discrepancy?

In fact, there is no discrepancy. The difference between the canonical and grand canonical ensembles in the problem of persistent currents is well described by Eq. (8.32). The derivative over the magnetic flux is finite because cooperons contain this quantity. When calculating the electric susceptibility the difference between the canonical and grand canonical ensembles can be described by a formula similar to Eq. (8.32) but with the derivative over the electric field instead of over the magnetic flux. Neither diffusons nor cooperons contain the electric field and so the difference between the canonical and grand canonical averages vanishes in this approximation. One can learn from this example that in discussing the difference between canonical and grand canonical ensembles one should be specific about what physical quantity is under study.

8.3 Nonperturbative calculations at arbitrary magnetic field

8.3.1 Integral over supermatrices

Now, let us calculate the average persistent current given by Eq. (8.33) at $T \ll \Delta$. In this limit one may neglect the temperature smearing of the Fermi distribution as well as inelastic processes such that $\tau_{\varphi}^{-1} = 0$. The correlation function of the density of states, or, in other words, the level–level correlation function $R(\omega)$ entering this equation has been studied in Chapter 6. Introduced in a general way in Eqs. (6.3, 6.18) this function has been calculated in the limiting cases of the orthogonal, unitary, and symplectic ensembles. To describe the persistent currents one needs to know this function at arbitrary magnetic field and hence one has to be able to calculate the integrals, Eq. (6.41), over the supermatrices Q with the zero-dimensional version of the σ-model, Eq. (4.65), containing an arbitrary magnetic field.

Some words have to be said about the the zero-dimensional version of the σ-model at a nonzero magnetic field. Eq. (4.65) is gauge-invariant and therefore the vector-potential $\mathbf{A}$ can be chosen in an arbitrary gauge. However, keeping only the zero space harmonics is not a gauge-invariant approximation and one should find a gauge with which this approximation is valid. This problem was mentioned in the previous chapter. One encounters the same problem in the theory of superconductivity, where the most convenient method is to work in the London gauge determined by the following relations

$$\operatorname{div}\mathbf{A} = 0, \qquad \mathbf{A}_n|_S = 0 \tag{8.39}$$

where $A_n|_S$ is the component of the vector-potential perpendicular to the surface S and taken at this surface. Using the similarity between the functional $F[Q]$, Eq. (4.65), and the Ginzburg–Landau free energy functional in the theory of superconductivity (cf. De Gennes (1966), Abrikosov (1988)) we can conclude that with the London gauge one can keep the zero harmonics of Q only. This is correct if the magnetic field is weak enough that the energy $\mathcal{E}_\perp$, Eq. (5.18), is much less than the Thouless energy $\mathcal{E}_c$. In other words, the magnetic length $l_H = (2|e|H/c)^{-1/2}$ must be much larger than the sizes of the sample.

The notion of "zero harmonics" needs some clarification for the ring geometry. In the absence of the magnetic field it means, of course, that Q does not depend on coordinates. However, Eqs. (4.65, 6.41) are invariant with respect to the following substitution:

$$\tilde{\phi} = \phi - \frac{n\phi_0}{2}, \qquad \tilde{Q}(x) = \exp\left(\frac{i\pi n\tau_3 x}{L}\right) Q(x) \exp\left(-\frac{i\pi n\tau_3 x}{L}\right) \tag{8.40}$$

where x is the coordinate along the circumference, n is an integer, and $\phi = AL$ is the magnetic flux through the ring (vector potential A with a constant absolute value A directed along the perimeter is a solution of the Eq. (8.39)). As a result of this invariance the total response must be periodic in flux ϕ with the period $\phi_0/2$. Using the periodicity one may choose n such that $|\phi - n\phi_0/2|$ is minimal and replace Q by $\tilde{Q}$ according to Eq. (8.40). Then the main contribution is from integration over $\tilde{Q}$, which does not depend on coordinates, and this is the zero harmonics for the ring. Using the periodicity, one may consider only the interval $0 < \phi < \phi_0/2$ and hence the value $n = 0$.

Taking into account the zero harmonics one can rewrite Eqs. (4.65, 6.41) as follows:

$$R(\omega) = \frac{1}{2} - \frac{1}{128} \operatorname{Re} \int \operatorname{str}(k(1+\Lambda)Q) \operatorname{str}(k(1-\Lambda)Q) \exp(-F[Q]) \, dQ \tag{8.41}$$

$$F[Q] = F_1[Q] + F_0[Q], \qquad F_0[Q] = \frac{i(x+i\delta)}{4} \operatorname{str}(\Lambda Q) \tag{8.42}$$

$$F_1[Q] = -\frac{X^2}{16} \operatorname{str}[Q, \tau_3]^2$$

where

$$X^2 = 2\pi\nu D_0 \left(\frac{e}{c}\right)^2 \int A^2 d\mathbf{r}, \qquad x = \frac{\pi\omega}{\Delta} \tag{8.43}$$

For the ring geometry the parameter X entering Eq. (8.43) can be written in a simpler form

$$X = \frac{2\pi\phi}{\phi_0} \left(\frac{2\mathcal{E}_c}{\pi\Delta}\right)^{1/2} \tag{8.44}$$

So, the problem of evaluation of the level–level correlation function $R(\omega)$ has been reduced as in Chapter 6 to calculation of definite integrals over supermatrices Q.

In the preceding chapters integrals over supermatrices Q were evaluated using the parametrization specified by Eqs. (6.42–6.61). This parametrization is very convenient for calculations of integrals in the unitary ensemble. Calculations for the orthogonal and symplectic ensembles are more complicated and manipulating, for example, with the integrals, Eqs. (6.74, 6.76) demands high mathematical skill. As concerns the regime of the crossover between the orthogonal and unitary or symplectic and unitary ensembles, analytical calculation of the corresponding integrals using the same parametrization does not seem to be feasible at all. We know already very well that violating the time reversal invariance leads to breaking of the symmetry between the diffusons and cooperons. The parametrization of Section 6.3 "mixes" both types of degrees of freedom, and therefore one comes to very complicated integrands. Fortunately, another parametrization was suggested recently by Altland, Iida, and Efetov (1993). In this parametrization, the cooperon and diffuson degrees of freedom form two separate classes of the integration variables, and symmetries within each class of the variables are preserved. As a result, integrals become much simpler and can be evaluated in most cases analytically. Of course, for the unitary ensemble, when the cooperon degrees of freedom are absent, both parametrizations coincide. To distinguish between the parametrization of Section 6.3 and that suggested in Altland, Efetov, and Iida (1993) let us call the latter one "magnetic field parametrization" because it is most convenient for problems with a magnetic field. As concerns the former parametrization, it can be called "standard parametrization."

The supermatrix Q in the parametrization of Altland, Iida, and Efetov for a system without spin-orbit interactions is written in the form

$$Q = V_d Q_c \bar{V}_d \tag{8.45}$$

$$Q_c = \begin{pmatrix} u_c & 0 \\ 0 & v_c \end{pmatrix} \begin{pmatrix} \cos \hat{\theta}_c & iC_0 \sin \hat{\theta}_c \\ -iC_0^T \sin \hat{\theta}_c & -\cos \hat{\theta}_c \end{pmatrix} \begin{pmatrix} \bar{u}_c & 0 \\ 0 & \bar{v}_c \end{pmatrix} \tag{8.46}$$

$$V_d = \begin{pmatrix} u_d & 0 \\ 0 & v_d \end{pmatrix} \begin{pmatrix} \cos \left(\hat{\theta}_d/2\right) & -i \sin \left(\hat{\theta}_d/2\right) \\ -i \sin \left(\hat{\theta}_d/2\right) & \cos \left(\hat{\theta}_d/2\right) \end{pmatrix} \begin{pmatrix} \bar{u}_{2d} & 0 \\ 0 & 1 \end{pmatrix} \tag{8.47}$$

where the matrix C_0 is defined in Eqs. (4.8, 4.10) and the "charge conjugate" $\bar{V}$ of supermatrix V is given by the same formulae as in Eq. (4.35). Supermatrix Q_c can also be written as

$$Q_c = V_c \Lambda \bar{V}_c \tag{8.48}$$

$$V_c = \begin{pmatrix} u_c & 0 \\ 0 & v_c \end{pmatrix} \begin{pmatrix} \cos \left(\hat{\theta}_c/2\right) & -iC_0 \sin \left(\theta_c/2\right) \\ -iC_0^T \sin \left(\hat{\theta}_c/2\right) & \cos \left(\hat{\theta}_c/2\right) \end{pmatrix} \begin{pmatrix} \bar{u}_{2c} & 0 \\ 0 & 1 \end{pmatrix}$$

An equivalent form for supermatrix Q is

$$Q = V \Lambda \bar{V}, \qquad V = V_d V_c \tag{8.49}$$

In Eqs. (8.45–8.49), all variables with the subscripts "c" and "d" relate to the cooperon and diffuson degrees of freedom, respectively. Both types of the variables enter Eqs. (8.45–8.49) in a symmetric way. All block matrices in these equations except C_0 commute with τ_3 and C_0 anticommutes with this matrix. The matrices $\hat{\theta}_c$ and $\hat{\theta}_d$ have the same structure

$$\hat{\theta}_{c,d} = \begin{pmatrix} \theta_{c,d} & 0 \\ 0 & i\theta_{1c,d} \end{pmatrix}, \qquad \begin{matrix} 0 < \theta_c < \pi/2, \qquad 0 < \theta_d < \pi \\ 0 < \theta_{1c,d} < \infty \end{matrix} \tag{8.50}$$

as well as the matrices u_c, u_d and v_c, v_d:

$$u_{c,d} = u_{1c,d} u_{2c,d}$$

$$u_{1c,d} = \begin{pmatrix} 1 - 2\eta_{c,d}\bar{\eta}_{c,d} & 2\eta_{c,d} \\ -2\bar{\eta}_{c,d} & 1 + 2\eta_{c,d}\bar{\eta}_{c,d} \end{pmatrix}$$

$$u_{2c,d} = \exp\left(i\hat{\phi}_{c,d}\tau_3\right), \qquad \hat{\phi} = \begin{pmatrix} \phi_{c,d} & 0 \\ 0 & \chi_{c,d} \end{pmatrix}, \qquad 0 < \phi_{c,d}, \chi_{c,d} < \infty$$

$$v_{c,d} = \begin{pmatrix} 1 + 2\kappa_{c,d}\bar{\kappa}_{c,d} & 2i\kappa_{c,d} \\ -2i\bar{\kappa}_{c,d} & 1 - 2\kappa_{c,d}\bar{\kappa}_{c,d} \end{pmatrix} \qquad (8.51)$$

$$\eta_{c,d} = \begin{pmatrix} \eta_{c,d} & 0 \\ 0 & -\eta_{c,d}^* \end{pmatrix}, \qquad \kappa_{c,d} = \begin{pmatrix} \kappa_{c,d} & 0 \\ 0 & -\kappa_{c,d}^* \end{pmatrix}$$

$$\bar{u}_{c,d} = u_{c,d}^{-1}, \qquad \bar{v}_{c,d} = v_{c,d}^{-1}, \qquad \bar{\eta}_{c,d} = \eta_{c,d}^+, \qquad \bar{\kappa}_{c,d} = \kappa_{c,d}^+$$

In order to be able to carry out integration with parametrization one has to write a corresponding Jacobian (Berezinian). The derivation of the Jacobian for the parametrization given by Eqs. (8.45–8.51) is rather involved and is presented, as well as some explanation of Eqs. (8.45–8.51), in Appendix 2. The final result for the elementary volume $[dQ]$ can be written as follows:

$$[dQ] = 4\left(\frac{\lambda_c}{\lambda_{1c} + \lambda_c}\right)^2 J_c J_d d R_c d R_d d\lambda_c d\lambda_{1c} d\lambda_d d\lambda_{1d} \qquad (8.52)$$

where

$$J_{c,d} = \frac{1}{2^6 \pi^2} \frac{1}{\left(\lambda_{1c,d} - \lambda_{c,d}\right)^2}, \qquad \lambda_{c,d} = \cos\theta_{c,d}, \qquad \lambda_{1c,d} = \cosh\theta_{1c,d}$$

$$dR_{c,d} = d\phi_{c,d} d\chi_{c,d} d\eta_{c,d} d\eta_{c,d}^* d\kappa_{c,d}^* d\kappa_{c,d}$$

Using Eqs. (8.45, 8.52) one can calculate the integrals over Q in Eqs. (8.41, 8.42). It is convenient to integrate first over u_d, v_d, then over u_c, v_c. After these two integrations one obtains conventional definite integrals over the variables λ_c, λ_d, λ_{1c}, and λ_{1d}. As a result of the symmetries of the function $F[Q]$, Eq. (8.42), there are some simplifications. For example, $F_1[Q]$ does not depend on V_d and u_c, v_c; $F_0[Q]$ does not depend on u_d, v_d. Although the integration is rather straightforward it is lengthy and therefore it is useful to write down some intermediate expressions.

The functions F_1 and F_0 take the form

$$F_1[Q] = X^2\left(\lambda_{1c}^2 - \lambda_c^2\right) \qquad (8.53)$$

$$F_0[Q] = -i(x + i\delta)\left[\lambda_{1d}\lambda_{1c} - \lambda_d\lambda_c \right.$$
$$\left. + 2(\lambda_{1c} - \lambda_c)(\lambda_{1d} - \lambda_d)\left(\eta_c\eta_c^* - \kappa_c\kappa_c^*\right)\right]$$

When calculating the integrals over Q with this parametrization one encounters a difficulty related to the necessity to integrate singular terms proportional, for example, to $(\lambda_{1c} - \lambda_c)^{-2}$ or $(\lambda_{1d} - \lambda_d)^{-2}$ that originate from $J_{c,d}$ in the Jacobian, Eq. (8.52). Although such terms do not contain the Grassmann variables η_c, κ_c, the integral over $\lambda_{1c,d}$, $\lambda_{c,d}$ diverges at $\lambda_{1c,d} = \lambda_{c,d} = 1$ and special care must be taken to avoid this singularity. Analogous singularities have been discussed in Sections 2.5 and 6.3 in calculations with standard parametrization.

This difficulty can be overcome by using an already familiar trick. An integral $I(X)$ of the form

$$I(X) = \int \Phi(Q) \exp(-F_1[Q]) \, dQ \tag{8.54}$$

can be rewritten as

$$I(X) = I(X_0) + \int_{X_0}^{X} \frac{\partial I(X')}{\partial X'} \, dX' \tag{8.55}$$

In Eqs. (8.54, 8.55), Φ is an arbitrary function of Q, $F_1[Q]$ is specified by Eq. (8.53), and X_0 is an arbitrary parameter.

The second term in Eq. (8.55) is not singular in the limit λ_{1c}, $\lambda_c \to 1$ because $\partial F_1/\partial X$ compensates for the singularity. As concerns the first term, one can take the limit $X_0 \to \infty$. In this limit there is no need to use the parametrization, Eqs. (8.45–8.51), because this is the limit of the unitary ensemble (Model II). All cooperon degrees of freedom are frozen out and integrals can be calculated in a standard way. Using this fact we can rewrite the level–level correlation function $R(\omega)$ as

$$R(\omega) = R_{\text{unit}}(\omega) - \frac{1}{32} \lim_{X_0 \to \infty} \text{Re} \int \text{str}\left(kQ^{11}\right) \text{str}\left(kQ^{22}\right) \exp(-F_0[Q])$$

$$\times \left(\exp(-F_1[Q, X]) - \exp(-F_1[Q, X_0])\right) dQ \tag{8.56}$$

where $R_{\text{unit}}(\omega)$ is the two-level correlation function for the unitary ensemble.

8.3.2 Level–level correlation function and persistent current

In Eq. (8.56) the function $F_1[Q, X]$ is the same as $F_1[Q]$ in Eqs. (8.43, 8.53), but now the dependence on X is explicitly marked. The integrand in Eq. (8.56) does not contain any singularity at $\lambda_{1c} = \lambda_c = 1$ but still contains the singularity at $\lambda_{1d} = \lambda_d = 1$, and so a similar manipulation should be done for the diffuson variables. The corresponding procedure can be performed as follows: Instead of calculating integrals with $F_1[Q]$ let us calculate integrals with a more general function $\tilde{F}[Q, X, Y]$

$$\tilde{F}[Q, X, Y] = F_1[Q, X] - Y \, \text{str}\left(\Lambda V_d \Lambda \bar{V}_d\right) \tag{8.57}$$

Then the exponentials in Eq. (8.56) can be written as

$$\exp(-F_1[Q, X]) - \exp(-F_1[Q, X_0]) = Z_1(Q) + Z_2(Q) \tag{8.58}$$

$$Z_1(Q) = \exp\left(-\tilde{F}_1[Q, X, 0]\right) - \exp\left(-\tilde{F}_1[Q, X_0, 0]\right)$$

$$- \exp\left(-\tilde{F}_1[Q, X, Y_0]\right) + \exp\left(-\tilde{F}[Q, X_0, Y_0]\right)$$

$$Z_2(Q) = \exp\left(-\tilde{F}[Q, X, Y_0]\right) - \exp\left(-\tilde{F}[Q, X_0, Y_0]\right)$$

Integrals with $Z_1(Q)$ do not contain any singularities and one can proceed by integrating over both commuting and anticommuting variables in a regular way. Taking the limit $Y_0 \to \infty$ simplifies calculation of integrals with $Z_2(Q)$ because now the diffuson degrees of freedom are frozen. To calculate the integral in Eq. (8.56) with the second line replaced by $Z_2(Q)$ one should simply replace all Q in the integrand by Q_c and integrate over the cooperon variables using the cooperon part of Eq. (8.53). Then, one can integrate over supermatrices u_c and v_c because they enter the preexponential only. Using Eq. (6.70) one easily obtains

$$\int \bar{u}_c k u_c d\eta_c d\eta_c^* = \int \bar{v}_c k v_c d\kappa_c^* d\kappa_c = 8 \tag{8.59}$$

and the integral with $Z_2(Q)$ is reduced in the limit $Y_0 \to \infty$ to the form

$$2\,\mathrm{Re} \int\limits_1^\infty \int\limits_0^1 \exp\left[X^2\left(\lambda_c^2 - \lambda_{1c}^2\right) + i\,(x+i\delta)(\lambda_{1c}-\lambda_c) \right] \frac{\lambda_c^2 d\lambda_{1c} d\lambda_c}{(\lambda_{1c}+\lambda_c)^2} \tag{8.60}$$

In the integral with $Z_1(Q)$, one has to integrate first over u_d, v_d. Eq. (8.59) holds for the diffuson variables as well and the integral is thus reduced to

$$\frac{1}{2} \int \left(\mathrm{str}\left(\tilde{Q}_{d0} Q_c \right) \right)^2 Z_1(Q) \exp\left(-F_0[Q] \right) dQ \tag{8.61}$$

where $\tilde{Q}_{d0} = \bar{V}_{d0} \Lambda V_{d0}$, and V_{d0} is the central part in Eq. (8.47). Let us notice that the term F_0 in Eq. (8.42) can also be written as

$$F_0[Q] = \frac{1}{4} i\,(x+i\delta)\,\mathrm{str}\left(\tilde{Q}_{d0} Q_c \right) \tag{8.62}$$

and therefore the preexponential in Eq. (8.61) is proportional to the second derivative of the $\exp(-F_0[Q])$. This fact somewhat simplifies the calculation because now the anticommuting variables enter $F_0[Q]$ only. By expanding $\exp(-F_0[Q])$ in the variables η_c, η_c^*, κ_c, κ_c^* and integrating over them one obtains a term proportional to $(\lambda_{1c}-\lambda_c)^2(\lambda_{1d}-\lambda_d)^2$ that compensates for the corresponding terms in the Jacobian. As a result, integration over λ_{1d} and λ_d becomes trivial and one obtains for the integral with $Z_1(Q)$

$$-\frac{1}{2}\,\mathrm{Re} \int\limits_1^\infty \int\limits_0^1 \frac{(\lambda_{1c}+\lambda_c)^2 \exp(ix(\lambda_{1c}+\lambda_c)) - (\lambda_{1c}-\lambda_c)^2 \exp(ix(\lambda_{1c}-\lambda_c))}{\lambda_{1c}\lambda_c}$$

$$\times \exp\left(X^2\left(\lambda_c^2 - \lambda_{1c}^2\right) \right) \left(\frac{\lambda_c}{\lambda_{1c}+\lambda_c} \right)^2 d\lambda_{1c} d\lambda_c \tag{8.63}$$

Adding Eqs. (8.60) and (8.61) one obtains the final expression for the level–level correlation function $R(\omega)$

$$R(\omega) = 1 - \frac{\sin^2 x}{x^2} \tag{8.64}$$

$$+ \int\limits_1^\infty \frac{\sin(\lambda_1 x)}{\lambda_1} \exp\left(-X^2 \lambda_1^2 \right) d\lambda_1 \int\limits_0^1 \lambda \sin(\lambda x) \exp\left(X^2 \lambda^2 \right) d\lambda$$

where, as everywhere before, $x = \pi\omega/\Delta$, and subscripts c of the variables λ and λ_1 are omitted. Eq. (8.64) describes the two-level correlation function at an arbitrary magnetic field. In the limit of large $X \gg 1$ the second term in Eq. (8.64) is small and one comes to Eq. (6.14) for the unitary ensemble. Without a magnetic field one has $X = 0$, integration over λ can be carried out easily, and one recovers Eq. (6.13) for the orthogonal ensemble. At any finite magnetic field, the low-frequency asymptotics $x \ll X^2$ obeys the law

$$R(\omega) = C(X)x^2 \tag{8.65}$$

with the function $C(X)$ that can be directly obtained from Eq. (8.64)

$$C(X) = \frac{1}{3} + \int_1^\infty \exp\left(-X^2\lambda_1^2\right) d\lambda_1 \int_0^1 \lambda^2 \exp\left(X^2\lambda^2\right) d\lambda$$

At high magnetic fields this function approaches $1/3$ (cf. Eq. (6.90)), whereas in the low field limit it is proportional to X^{-1}, thus diverging in the limit $X \to 0$. The vanishing of the correlation function $R(\omega)$ in the limit $\omega \to 0$ corresponds to the level repulsion at arbitrary magnetic fields. In other words, the energy levels as functions of the magnetic field do not cross each other.

It is interesting that even in the limiting case of the orthogonal ensemble calculation with the magnetic field parametrization is more straightforward and simple than with the standard one (cf. Section 6.4). However, in some cases the standard parametrization can be more convenient because of a simpler expression for $F_0[Q]$, and so these two parametrizations are somehow complementary.

The correlation function, Eq. (8.64), was first obtained by Pandey and Mehta (1983) and by Mehta and Pandey (1983) from random matrix theory (RMT) using a completely different mathematical scheme. Of course, a magnetic field cannot be included in the scheme of RMT microscopically and therefore a phenomenological way of gradually breaking time reversal symmetry was used. The crossover between orthogonal and unitary ensembles was described considering random matrices $\mathcal{H}$ of the form

$$\mathcal{H} = \mathcal{H}_s + i\alpha\mathcal{H}_a, \qquad 0 \le \alpha \le 1 \tag{8.66}$$

where $\mathcal{H}_s$ and $\mathcal{H}_a$ are N-dimensional real symmetric and antisymmetric statistically independent matrices with the same Gaussian distribution, Eq. (6.1). The crossover is driven by the parameter α, $\alpha = 0, 1$ corresponding to the case of pure orthogonal or pure unitary ensembles respectively.

Although the zero-dimensional σ-model in the presence of a magnetic field was derived from a model with impurities, it can be derived from the random matrix model described by Eq. (8.66) as well. The derivation is presented in Altland, Iida, and Efetov (1993) and is practically the same as that given in Section 6.5 for the orthogonal ensemble. The parameter α in Eq. (8.66) is related to the parameter X in the σ-model, Eq. (8.42), as

$$X^2 = 2N\alpha^2 \tag{8.67}$$

This establishes the correspondence between RMT and the zero-dimensional σ-model for an arbitrary magnetic field or another parameter breaking time reversal symmetry and also proves the applicability of RMT to a broad class of microscopic models.

Now, let us return to calculation of the persistent current using Eq. (8.33). Because of the presence of the derivative over the flux, only the term in the second line in Eq. (8.64) contributes. It is convenient to integrate first over the energies ε_1 and ε_2. In the preceding section two methods of integration over these variables were presented. When calculating

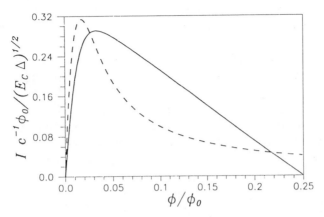

Fig. 8.6. The current, $Ic^{-1}\phi_0/(E_c\Delta)^{1/2}$, for $E_c/\Delta = 100$ as a function of the flux ϕ/ϕ_0. The dashed line is calculated from Eq. (8.68). The solid line corresponds to results of Section 8.4 obtained with the dynamic approach. (From Efetov and Iida (1993).)

the typical current I_{typ}, the Fermi function $n\,(\varepsilon_2)$ was replaced by $n\,(\varepsilon_2) - 1$ in the integral over ε_2. Then one could integrate first over $\varepsilon = (\varepsilon_1 + \varepsilon_2)/2$ and then over $\omega = \varepsilon_1 - \varepsilon_2$. Calculating the average persistent current with the approximate correlation function $R\,(\omega)$ integration over ω was carried out before integration over ε. We can follow the latter procedure also for calculation with the exact two-level correlation function $R\,(\omega)$. Integrating the second term in Eq. (8.64) over ω in the interval $-2\varepsilon < \omega < 2\varepsilon$ and then over $\varepsilon > 0$ one can reduce the current $I\,(\phi)$ to the form

$$I\,(\phi) = \frac{4|e|}{\pi^2}\left(\frac{2E_c\Delta}{\pi}\right)^{1/2} X \int\limits_1^\infty \int\limits_0^1 \frac{\lambda^2}{\lambda_1^2 - \lambda^2}\exp\left(X^2\left(\lambda^2 - \lambda_1^2\right)\right)d\lambda d\lambda_1 \qquad (8.68)$$

The characteristic value of the current given by Eq. (8.68) is of the order of $|e|\,(E_c\Delta)^{1/2}$ and this agrees with what one could expect from the perturbative results of Section 8.2. Eq. (8.68) is applicable in the region $\phi \ll \phi_0$. Provided $\mathcal{E}_c \gg \Delta$, this gives a good description of the current in the most interesting region of fluxes $\phi \sim \phi_0\,(\Delta/\mathcal{E}_c)^{1/2} \ll \phi_0$. At the same time, Eq. (8.68) is somewhat disadvantageous with respect to Eqs. (8.36, 8.38) because it is not explicitly periodic and therefore is not valid near the points $\phi_0/4 + n\phi_0/2$, $n = 0, 1, \pm2, \ldots$ At small fluxes $\phi \ll \phi_0\,(\Delta/\mathcal{E}_c)^{1/2}$, the current is linear in flux and the sign corresponds to a paramagnetic response. In this region the current can be written as

$$I\,(\phi) = \frac{16\mathcal{E}_c|e|\phi}{3\pi^2\phi_0} \qquad (8.69)$$

The dependence of the current on flux, Eq. (8.68), is represented by the dashed line in Fig. 8.6.

The origin of the orbital paramagnetism, Eq. (8.69), can be understood from simple arguments. The nonzero susceptibility is due to the difference between canonical and grand canonical averaging, the latter giving zero. This difference arises because energy levels can cross a given chemical potential μ from above or from below and their contribution to the canonical energy should, respectively, be subtracted from or added to the grand potential. Since the curvatures of the levels crossing μ from above or below are opposite, they give contributions of the same sign to the susceptibility, which is easily seen to be paramagnetic. Increasing the temperature or the sample size leads to reducing the persistent currents and at high temperatures one is left with Landau diamagnetism only. Therefore, orbital susceptibility as a function of temperature changes sign at some temperature.

The nonperturbative calculation of the persistent current starting from Eq. (8.33) and using the zero-dimensional σ-model was done first by Altland et al. (1992, 1993). However, these authors did the calculations using the standard parametrization. As a result they could obtain the result only numerically. The same problem was also addressed by Serota and Zyuzin (1993), using an interpolation scheme to treat effects of the magnetic field. It is the magnetic field parametrization that enables us to represent the dependence of the current on magnetic flux in a simpler form, Eq. (8.68). Let us notice that Altland et al. (1992, 1993) came to Eq. (8.32) by using completely different arguments. However, it was argued later by Efetov and Iida (1993) that approximations used in that work could be equivalent to considering temperatures $T \geq \Delta$. Then, it is no surprise that their result agreed with Eq. (8.32), which was derived by using a perturbative scheme that seems to be correct in the high temperature regime.

8.4 Dynamic approach

8.4.1 Thermodynamics through a dynamic response

In this section a different possibility for calculating persistent currents at low temperatures is considered. This method was developed by Efetov (1991) and Efetov and Iida (1993) and is based on the disorder averaging of the derivative of the current over the magnetic flux. We have seen in the preceding section that the difference between canonical and grand canonical behavior is due to levels crossing a fixed chemical potential. Physical averaging implies that one has a system of rings with different impurity configurations. However, in each single ring, the chemical potential must always be located somewhere between levels. This means that, in a theoretical consideration, one must carry out averaging over impurities keeping the chemical potential between the levels that corresponds to an integer number of electrons. We have seen in Section 8.1 a simple example of how the total current may have the same order of magnitude as the single-level current (see also Bouchiat and Montambaux (1989)). Averaging current with a fixed chemical potential is not a good procedure because one "illegal" intersection of a level with the chemical potential can drastically change the result. The elementary calculation, Eqs. (8.11–8.13), shows that such averaging leads to an exponentially small current. Let us explain in more detail why it is so. Assuming that N electrons occupy N lowest discrete levels one can write the equilibrium current density J as

$$J = \sum_{m=1}^{N} j_m = 2 \sum_{m} n_\mu (\varepsilon_m) j_m \qquad (8.70)$$

where $n_\mu (\varepsilon)$ is the Fermi function (the factor 2 is due to spin), μ is the chemical potential that depends on the number of particles N and the disorder potential, and j_m is the current density of a one-particle state m. The current density j_m of a level m, as well as its derivative over A, can be found by substituting in the initial Hamiltonian $\mathcal{H}$

$$\mathcal{H} = \sum_{i=1}^{N} \left[\varepsilon \left(\hat{\mathbf{p}}_i - \frac{e}{c} \mathbf{A} \right) + V (\mathbf{r}_i) \right]$$

A by $\mathbf{A} + \delta \mathbf{A}$ and computing the corresponding correction within the standard quantum mechanical perturbation theory. Assuming that the energy levels are not degenerate one obtains in the first and the second order

$$j_m = -\frac{c}{V} \frac{\partial \varepsilon_m (A)}{\partial A} \tag{8.71}$$

$$\frac{\partial j_m (A)}{\partial A} = -\frac{c}{V} \frac{\partial^2 \varepsilon_m}{\partial A^2} = \frac{1}{cV} \left(\sum_{n \neq m} \frac{|\hat{\pi}_{mn}|^2}{\varepsilon_n - \varepsilon_m} - \frac{e^2}{mc} \right)$$

where $\hat{\pi}$ is the current operator, Eq. (3.30).

Let us differentiate the total density J over the vector potential A keeping the number of electrons fixed (for simplicity, a ring with the vector potential A directed along the circumference is considered). From Eqs. (8.70, 8.71) one has

$$\left(\frac{\partial J}{\partial A} \right)_N = 2 \sum_m n_\mu (\varepsilon_m) \frac{\partial j_m}{\partial A} \tag{8.72}$$

$$= \frac{2}{cV} \left[\sum_{m \neq n} |\hat{\pi}_{mn}|^2 \frac{n_\mu (\varepsilon_n) - n_\mu (\varepsilon_m)}{\varepsilon_m - \varepsilon_n} - \frac{e^2 N}{m} \right]$$

At the same time, the corresponding derivative with fixed chemical potential can be written by using Eqs. (8.70) as

$$\left(\frac{\partial J}{\partial A} \right)_\mu = \left(\frac{\partial J}{\partial A} \right)_N + 2 \frac{V}{c} \sum_m (j_m)^2 \delta (\varepsilon_m - \mu) \tag{8.73}$$

Of course, both expressions, Eqs. (8.72, 8.73), coincide for any discrete set of eigenenergies $\{\varepsilon_m\}$ because μ is always between levels and the second term in Eq. (8.73) is equal to zero. However, analytical calculations for an arbitrary impurity potential are not possible and averaging over impurities can usually be carried out only by keeping the chemical potential fixed. In this case, we immediately see a difference between $\langle (\partial J / \partial A)_N \rangle$ and $\langle (\partial J / \partial A)_\mu \rangle$ because by changing impurity configurations one allows levels to move and cross the chemical potential μ if it is fixed. Then, the second term in Eq. (8.73) can give a considerable contribution because the coefficient in front of the δ-function is always positive and one-level current has the same order of magnitude as the total averaged current.

Are all physical quantities as sensitive to the averaging procedure? Would not it be more reasonable to find a physical quantity less sensitive to the averaging procedure? We already understand that such a quantity should not change much if a level crosses the chemical potential. The first term in the second line of Eq. (8.72) apparently has this property. However, calculation of the second term by averaging in the grand canonical ensemble leads to large errors because, as shown by Altshuler and Shklovskii (1986), the average fluctuation of the number of particles $(\delta N)^2$ in an energy band $\mathcal{E} \gg \mathcal{E}_c$ is large

$$\left\langle (\delta N)^2 \right\rangle \sim \left(\frac{\mathcal{E}}{\mathcal{E}_c} \right)^{3/2} \gg 1 \tag{8.74}$$

The result, Eq. (8.74), was obtained by writing $(\delta N)^2$ as an integral of the function $R(\omega) - 1$ over energies in the interval $(-\mathcal{E}/2, \mathcal{E}/2)$, where the correlation function $R(\omega)$ is taken in the form given by Eq. (8.34). Taking $\mathcal{E} \sim \mu$ we see that fluctuations of the number of electrons at fixed chemical potential are really large. Nevertheless, the difficulty due to the large fluctuations of the number of electrons N can be avoided. Let us recall that a sample with a ring geometry is under consideration and the magnetic field is weak such that it affects only the phases of wave functions. Using periodicity of the wave functions in the coordinate along the circumference one can see that all quantities, for example, eigenenergies and matrix elements, are periodic functions of magnetic flux ϕ through the ring with period ϕ_0 (in fact, averaged quantities are periodic with period $\phi_0/2$, but this is not important to the subsequent discussion). The periodicity in flux means particularly that the current is also periodic and we have

$$\overline{\left(\frac{\partial J}{\partial A}\right)_N} = 0 \tag{8.75}$$

where the overbar denotes averaging over the period of the flux performed with a fixed number of particles. Eq. (8.75) enables us to rewrite Eq. (8.72) in the following equivalent form:

$$\left(\frac{\partial J}{\partial A}\right)_N = \frac{2}{cV}\left[\sum_{m\neq n}|\hat{\pi}_{mn}|^2\frac{n_\mu(\varepsilon_n)-n_\mu(\varepsilon_m)}{\varepsilon_m-\varepsilon_n}\right.$$
$$\left. -\overline{\sum_{m\neq n}|\hat{\pi}_{mn}|^2\frac{n_\mu(\varepsilon_n)-n_\mu(\varepsilon_m)}{\varepsilon_m-\varepsilon_n}}\right] \tag{8.76}$$

The only assumption used to derive Eq. (8.76) is that levels do not cross (level crossings can occur if there are spin-orbit impurities in the system and therefore some modifications are necessary for this case; this question is considered in Section 8.5). Although Eq. (8.76) is exact, it is complicated. It can be considerably simplified by using its relation to a linear dynamic response. An analogous scheme has been used in Chapter 7 for calculation of the electric susceptibility. The dynamics is introduced by assuming that the magnetic field $H(t)$ depends on time as

$$H(t) = H + H_\omega\cos(\omega t) \tag{8.77}$$

where H is a static part.

Then one calculates the linear response to the oscillating part of the magnetic field. For an arbitrary electron system one has for current density J

$$J(\omega) = \frac{1}{c}K(\omega)A_\omega, \qquad K(\omega) = K_p(\omega) + K_d \tag{8.78}$$

where A_ω is the oscillating part of the vector potential, and K_d is the diamagnetic part of the response

$$K_d = -\frac{e^2 N}{mV} \tag{8.79}$$

The paramagnetic part $K_p(\omega)$ is equal to

$$K_p(\omega) = \frac{2}{V}\sum_{m,n}|\hat{\pi}_{mn}|^2\frac{n_\mu(\varepsilon_m)-n_\mu(\varepsilon_n)}{\varepsilon_n-\varepsilon_m+\omega+i\delta} \tag{8.80}$$

Comparing Eqs. (8.78–8.80) with Eq. (8.72) we see that in the limit $\omega \to 0$ the response $K(\omega)/c$ coincides with the derivative of the thermodynamic current. The presence of a small but finite ω in Eq. (8.80) is very important because in this case the term with $m = n$ does not contribute. Of course, the spectrum is assumed to be discrete without degeneracies.

The diamagnetic part of the response K_d can be expressed again in terms of an average over the period of the flux of the paramagnetic part. Repeating the arguments that led to Eq. (8.76) for the averaged derivative of the current over magnetic flux we obtain

$$\frac{\partial I}{\partial \phi} = \frac{S}{L}\left\langle\frac{\partial J}{\partial A}\right\rangle_N = \frac{S}{cL}\lim_{\omega\to 0}\left\langle\left[K_p(\omega) - \overline{K_p(\omega)}\right]\right\rangle_N \tag{8.81}$$

where S is the cross section. As will be shown, the average current I in the dynamic approach is periodic with the period $\phi_0/2$ and therefore can be written as a Fourier series

$$I = \sum_{m=1}^{\infty} I_m \sin\left(4\pi m\frac{\phi}{\phi_0}\right) \tag{8.82}$$

$$I_m = \frac{1}{\pi m}\frac{S}{cL}\lim_{\omega\to 0}\left\langle\int_0^{\phi_0/2} K_p(\omega, \phi)\cos\left(4\pi m\frac{\phi}{\phi_0}\right)d\phi\right\rangle_N$$

In Eq. (8.82), the dependence of $K_p(\omega)$ on ϕ is stressed by writing the second argument ϕ. One should remember that not only matrix elements and energies but chemical potential depend on flux. Changing the order of integration over the magnetic flux, averaging over the impurities with the fixed number of particles, and using Eq. (8.80), one can reduce the Fourier coefficients I_m to the form

$$I_m = \frac{1}{\pi m}\frac{S}{cL}\lim_{\omega\to 0}\int_0^{\phi_0/2}\int\frac{n_0(\varepsilon - \Omega) - n_0(\varepsilon)}{\Omega - \omega - i\delta}$$

$$\times \mathcal{L}_N(\varepsilon, \Omega, \phi)\cos\left(\frac{4\pi m\phi}{\phi_0}\right)d\varepsilon d\phi d\Omega \tag{8.83}$$

$$\mathcal{L}_N(\varepsilon, \Omega, \phi) = \left\langle\frac{2}{V}\sum_{m,n}|\hat{\pi}_{mn}|^2\delta(\varepsilon + \mu_N(\phi) - \varepsilon_m)\delta(\varepsilon_m - \varepsilon_n - \Omega)\right\rangle_N$$

where $n_0(\varepsilon) = (\exp(\varepsilon/T) + 1)^{-1}$, and dependence of the chemical potential $\mu_N(\phi)$ on the magnetic flux and number of electrons N is expressed explicitly to remind us that all operations are done with the fixed number of electrons N.

Up to now no approximations have been made and Eq. (8.83) is still exact. However, to proceed further one has to simplify the averaging procedure because averaging directly in Eq. (8.83) is difficult. This can be done by using a hypothesis that seems intuitively reasonable. It can be formulated as follows: Averaging over impurities in Eq. (8.83) one may replace the canonical average by the grand canonical average or, in other words, carry out the averaging procedure keeping the chemical potential $\mu(\phi)$ fixed. Although the chemical potential may still depend on the magnetic flux, this dependence will not enter the final formulae.

At first glance, this hypothesis looks strange because we understand that changing the disorder potential at a fixed number of particles has to lead to very strong fluctuations of the chemical potential because all levels below the chemical potential contribute to these fluctuations. Using Eq. (8.74) we see that the characteristic fluctuation $\delta\mu$ much exceeds the mean level spacing Δ. At the same time, characteristic fluctuations of the distance between levels is of the order of Δ. For this reason we can assume that the distances between levels and the current matrix elements entering Eq. (8.83) fluctuate independently of the fluctuations of the chemical potential.

If this is correct, one can average in Eq. (8.83) first for a fixed chemical potential and only then average over the chemical potential. However, it will be seen that the result does not depend on any given chemical potential (strictly speaking, the result depends on the chemical potential through the average density of states, Fermi velocity, etc., but this dependence is smooth and does not lead to any interesting effects). Hence, averaging the result over the fluctuations of the chemical potential does not yield anything new, and the final result is the same as if all calculations were carried out with a fixed chemical potential.

In fact, the correlation function $\mathcal{L}(\Omega)$, Eq. (8.83), is very similar to the two-level correlation function $R(\omega)$. It is well known that the function $R(\omega)$ is also universal with respect to the averaging procedure (cf. discussion in Chapter 6). For example, instead of averaging over the disorder potential one can average over different parts of the energy spectrum and the final result is the same. There is no reason to think that function $\mathcal{L}(\Omega)$ is different in this respect. Strong fluctuations of chemical potential correspond somehow to averaging over different parts of the spectrum.

One may argue also that the result of averaging over the disorder potential with a fixed number of electrons should not depend on N if $N \gg 1$. Therefore, one can also average Eq. (8.83) over the number of electrons in an interval $(N - N_1/2, N + N_1/2)$, with $1 \ll N_1 \ll N$, without changing the result. Inverting the averaging procedure one can average first over the number of electrons keeping the disorder potential fixed. This seems to give the final results and the averaging over impurities is not important. At the same time, instead of averaging over the number of electrons in a certain interval one may average over the energies in a certain energy interval. If both procedures give the same result, one can then invert the order of averaging over the energies and impurities. Assuming again that after averaging over the disorder potential the result does not depend on energy we come once more to the conclusion that one may perform averaging in Eq. (8.83) keeping the chemical potential fixed. Of course, the preceding arguments are not rigorous and can be checked analytically for simple models only. In the diffusion regime their validity can apparently be checked only numerically.

Replacing in Eq. (8.83) the canonical averaging over the disorder potential by grand canonical averaging makes it possible to proceed further. Eq. (8.81) can be written as

$$I(\phi) = \frac{S}{cL} \int_0^\phi \lim_{\omega \to 0} \left\langle \left[K_p(\omega) - \overline{K_p(\omega)} \right] \right\rangle d\phi \qquad (8.84)$$

with the angle brackets representing grand canonical averaging. The function $I(\phi)$ given by Eq. (8.84) is explicitly periodic in flux. I would like to stress again that the absence of level crossing is a necessary condition for the dynamic approach. For example, the model of the ideal one-dimensional ring in Section 8.1 does not possess this property and therefore Eq. (8.84) cannot be used directly.

8.4.2 Calculation of current

Eq. (8.80) is very similar to Eq. (7.7) and so one can repeat all manipulations of Chapter 7 with small modifications. First, one subdivides the averaged paramagnetic response $\langle K_p \rangle$

into two parts

$$\langle K_p(\omega) \rangle = K_1(\omega) + K_2(\omega) \tag{8.85}$$

which is analogous to Eq. (7.29). The function $K_2(\omega)$ contains products of two retarded or two advanced Green functions. After averaging over impurities this function does not depend on the magnetic flux and therefore does not contribute in Eq. (8.84). The function $K_1(\omega)$ contains a product of one retarded and one advanced Green function and is, as usual, of chief interest. Writing this function as

$$K_1(\omega) = i\omega R^{11}(\omega), \qquad R^{11}(\omega) = \int R^{11}(\mathbf{r}, \mathbf{r}', \omega) \, d\mathbf{r}' \tag{8.86}$$

and introducing as usual supervectors ψ and supermatrices Q we may write the function $R^{11}(\mathbf{r}, \mathbf{r}',\omega)$ as a sum of three terms

$$R^{11}(\mathbf{r}, \mathbf{r}', \omega) = R_1^{11}(\mathbf{r}, \mathbf{r}', \omega) + R_2^{11}(\mathbf{r}, \mathbf{r}', \omega) + R_3^{11}(\mathbf{r}, \mathbf{r}', \omega) \tag{8.87}$$

The functions R_1^{11}, R_2^{11}, and R_3^{11} are described by Eqs. (7.38–7.39) provided one replaces the charge operators $\hat{\pi}^0$ by the current operators $\hat{\pi}$. Averaging over Q should be done with the free energy $F[Q]$, Eq. (4.65). The next step is to integrate over the coordinates in the matrix functions g. This integration is not difficult and for the model without spin interactions one obtains for the functions

$$R_i^{11}(\omega) = \int R_i^{11}(\mathbf{r}, \mathbf{r}', \omega) \, d\mathbf{r}', \qquad i = 1, 2, 3$$

the following relations:

$$R_1^{11}(\omega) = 2e^2 D_0 \nu \tag{8.88}$$

$$\times \left\{ 1 - \frac{1}{32} \left\langle \mathrm{str}\left[k(1-\tau_3)(Q-\Lambda)^{11}\right] \mathrm{str}\left[k(1-\tau_3)(Q-\Lambda)^{22}\right] \right\rangle_Q \right\}$$

$$R_2^{11}(\omega) = -2e^2 D_0^2 \nu^2 \frac{\pi}{32} \int \left\langle \mathrm{str}\left[\mathbf{j}^{12}(\mathbf{r})\mathbf{j}^{21}(\mathbf{r}')\right] \right\rangle_Q d\mathbf{r}' \tag{8.89}$$

where

$$\mathbf{j}(\mathbf{r}) = k(1-\tau_3) Q(\mathbf{r}) \{-i\nabla Q(\mathbf{r}) - (e\mathbf{A}/c)[Q(\mathbf{r}), \tau_3]\}$$

and

$$R_3^{11}(\omega) = -2e^2 D_0 \nu \frac{1}{32} \left\langle \mathrm{str}\left[k(1-\tau_3) Q^{12}(1+\tau_3) k Q^{21}\right] \right\rangle_Q \tag{8.90}$$

The terms $R_1^{11}(\omega)$ and $R_3^{11}(\omega)$ are proportional to the first and second terms in Eq. (7.42), respectively. However, notice that $R_3^{11}(\omega)$ has the opposite sign with respect to the second term in Eq. (7.42). This is a consequence of the substitution of the current operator $\hat{\pi}$ for π^0. A term corresponding to $R_2^{11}(\omega)$ did not appear in Eq. (7.42). The supermatrix $\mathbf{j}$ looks like a current operator. To derive the expression for the current $\mathbf{j}$ from $\lim_{\mathbf{r}_1 \to \mathbf{r}} \nabla_r g(\mathbf{r}, \mathbf{r}_1)$ one should differentiate Eq. (4.42) over the coordinate $\mathbf{r}$ and then solve the equation obtained for $\nabla_{\mathbf{r}} g(\mathbf{r}, \mathbf{r}_1)$, making an expansion in ∇Q and $\mathbf{A}$.

When deriving Eqs. (8.87–8.89), the fact that persistent currents are under study was not used. One arrives at the same formulae in the problem of localization. Using these equations and Eqs. (8.78–8.80, 8.85, 8.86) one can get a frequency-dependent response. The term $K_2(\omega)$ cancels the diamagnetic term K_d; Eq. (8.79) and $K_1(\omega)$ determine the total response. It is seen easily from Eq. (8.86) that the real part of $R^{11}(\omega)$ gives alternating current conductivity. The first term in Eq. (8.88) is the classical conductivity. At high frequencies deviations of Q from Λ are small. Expanding Eq. (4.65, 8.88–8.90) in these deviations, one can get all the effects of weak localization considered in Chapters 3 and 5.

It can be instructive to demonstrate how one obtains the first quantum correction to the classical conductivity form Eqs. (8.88–8.90) because in Chapter 5 this correction was obtained by calculating the density–density correlation function and not directly calculating conductivity. Using the parametrization, Eq. (5.1), and expanding Q in P in the quadratic approximation (cf. Eq. (5.3)) we get for $F[Q]$

$$F[P] = \frac{\pi \nu}{2} \int \mathrm{str} \left[D_0 \left(\nabla P - \frac{ie}{c} \mathbf{A}[P, \tau_3] \right)^2 - i\omega P^2 \right] d\mathbf{r} \qquad (8.91)$$

The same expansion must be made in Eqs. (8.88–8.90). As a result of the structure of the supermatrix P the first-order quantum correction comes from the function $R_3(\omega)$ only because only this term contains P^2. Substituting Eq. (5.1) into Eq. (8.90) and using Eq. (8.86) we have in the lowest order for conductivity σ

$$\sigma = \sigma_0 \left\{ 1 - \frac{1}{16} \left\langle \mathrm{str}\left[k(1 - \tau_3) P k (1 + \tau_3) P \right] \right\rangle_P \right\} \qquad (8.92)$$

where $\langle \dots \rangle_P$ stands for averaging with the free energy, Eq. (8.91), and $\sigma_0 = 2e^2 D_0 \nu$ is the classical conductivity. Representing supermatrix P as the sum of matrices P_d and P_c, where P_d commutes with τ_3 and P_c anticommutes with τ_3, we see that only P_c contributes in Eq. (8.92) and this is the one-cooperon contribution. As a result we have (cf. Eq. (5.23))

$$\sigma = \sigma_0 \left(1 - \frac{1}{\pi \nu} \sum_{\mathcal{E}_\perp} \frac{1}{\mathcal{E}_\perp - i\omega} \right) \qquad (8.93)$$

where $\mathcal{E}_\perp = (4D_0|e|H/c)(n + 1/2) + D_0(p_H)^2$, with the momentum p_H parallel to the magnetic field H. Eq. (8.93) is written for an infinite three-dimensional sample and can be generalized easily to any dimensionality. We already know very well that in small samples one should quantize the momenta and the most important contribution is from the zero space harmonics. Nonzero space harmonics can be neglected provided $L \ll L_c$, where $L_c = \pi \nu S D_0$ is the localization length in thick wires. When deriving the σ-model the inequality $l \ll L$ was also used. These two inequalities define the diffusion regime.

As concerns the problem of persistent currents, we expect that a finite limit $K_1(0)$ for the response exists, implying that $R(\omega) \sim 1/i\omega$ for small $\omega \ll \Delta$. In this limit expansion in cooperons and diffusons does not work and one must calculate integrals in Eqs. (8.88–8.90) exactly. Using the zero mode approximation (with some peculiarities due to the ring geometry described in Section 8.3), all calculations can be done in the same way as in Section 8.3. One should again use magnetic field parametrization for supermatrices Q, Eqs. (8.45–8.51); the Jacobian, Eq. (8.52); and the free energy F, Eq. (8.53). To calculate persistent currents one should take the limit $\omega \to 0$. In this limit all calculations simplify considerably because the main contribution is from large $\lambda_{1d} \sim \Delta/\omega \gg 1$; then integration over λ_{1d} and λ_d can trivially be carried out. As a result one obtains expressions that contain integrals over λ_{1c} and λ_c only

$$i\omega R_1^{11}(\omega) = -\frac{2e^2 D_0}{\pi V} \tag{8.94}$$

$$i\omega R_2^{11}(\omega) = -\frac{8e^2 D_0 X^2}{\pi V} \int_1^\infty \int_0^1 \frac{\lambda_c^2}{\lambda_{1c}} \exp\left[X^2\left(\lambda_c^2 - \lambda_{1c}^2\right)\right] d\lambda_{1c} d\lambda_c \tag{8.95}$$

$$i\omega R_3^{11}(\omega) = \frac{4e^2 D_0}{\pi V} \int_1^\infty \int_0^1 \frac{\left(\lambda_{1c}^2 + 1\right)\lambda_c^2}{\lambda_{1c}^3\left(\lambda_{1c}^2 - \lambda_c^2\right)} \exp\left[X^2\left(\lambda_c^2 - \lambda_{1c}^2\right)\right] d\lambda_{1c} d\lambda_c \tag{8.96}$$

Eqs. (8.94–8.96) give the complete solution of the problem. We see that the computation of the persistent current is very similar to the corresponding procedure for the static electric susceptibility performed in Chapter 7. The terms $R_1^{11}(\omega)$ and $R_3^{11}(\omega)$ are proportional to the first and second terms in Eq. (7.42). Using Eq. (8.90) and taking the limit $X \to 0$ in Eq. (8.96) one recovers Eq. (7.45). However, because of the opposite sign of the function $R_3^{11}(\omega)$ and of the second term in Eq. (7.42) the terms $R_1^{11}(\omega)$ and $R_3^{11}(\omega)$ cancel each other in the limit $\omega \to 0$, $X \to 0$, whereas the corresponding terms in Eq. (7.42) are added to each other. With the help of Eq. (8.96) one can describe the crossover between the results for the electric susceptibility κ for the orthogonal and unitary ensembles in Eq. (7.59).

Using Eqs. (8.84–8.86) we obtain for the persistent current I the following expression:

$$I = \frac{8|e|\mathcal{E}_c}{\pi^2 \phi_0} \int_0^\phi \left[K_0(X) - \overline{K_0}\right] d\phi \tag{8.97}$$

where

$$K_0(X) = \int_1^\infty \int_0^1 \left\{\frac{\left(\lambda_1^2 + 1\right)\lambda^2}{\lambda_1^3\left(\lambda_1^2 - \lambda^2\right)} - \frac{2X^2\lambda^2}{\lambda_1}\right\} \exp\left[X^2\left(\lambda^2 - \lambda_1^2\right)\right] d\lambda_1 d\lambda \tag{8.98}$$

The variables X and ϕ are related to each other by Eq. (8.44) and $\overline{K_0}$ is the average over the flux. Further calculations in Eq. (8.98) can be done only numerically. However, in the limit under consideration, $\mathcal{E}_c \gg \Delta$, the shape of the dependence of I on ϕ can be understood quite easily. In this limit the function K_0 vanishes at fluxes $\phi \geq \phi_0(\Delta/\mathcal{E}_c)^{1/2}$ and the main contribution to the right-hand side of the Eq. (8.97) comes from $\overline{K_0}$. Therefore, the current in this region is linear in flux. The shape of the flux dependence looks like a sawtooth. Calculation of $\overline{K_0}$ is not difficult. When integrating $K_0(X)$ over the flux, one may extend the integration from $-\infty$ to ∞ and the result reads

$$\overline{K_0} = \frac{\pi}{8}\left(\frac{\Delta}{2\mathcal{E}_c}\right)^{1/2} \tag{8.99}$$

Eq. (8.99) together with Eq. (8.97) determines the slope of the linear dependence of the current not very close to $n\phi_0/2$, where n is an integer.

Close to $n\phi_0/2$ the dependence of the current on the flux is also linear but the slope is positive. In this regime, the current is equal to

$$I = \frac{4|e|\mathcal{E}_c\left(\phi - n\phi_0/2\right)}{\pi^2 \phi_0} \tag{8.100}$$

Eq. (8.100) shows that close to $n\phi_0/2$ the system is paramagnetic. The maximum of the current is of the order of

$$I_{\max} \sim \frac{c \, (\mathcal{E}_c \Delta)^{1/2}}{\phi_0} \tag{8.101}$$

Comparing Eqs. (8.97–8.101) with Eqs. (8.68, 8.69) we see that the dynamic approach gives the results that are compatible with the results obtained by the canonical averaging. The maximum currents and the slopes at the points $n\phi_0/2$ have the same order of magnitude although they are different numerically. The shape of the current given by Eqs. (8.97, 8.98) is represented by the solid line in Fig. 8.6 such that the results of both approaches can easily be compared.

The shape of the current described by Eqs. (8.97–8.98) is very similar to that of ideal rings (cf. Fig. 8.3c), where the average current for the one-channel ring is represented. Both currents represented in Fig. 8.6 are still too small to explain the results of the experiment by Lévy et al. (1990). They become even smaller if the inelastic mean free time τ_φ is short enough. The reason for the large experimental values of the current is not clear and, therefore, it is not possible yet to check the ideas of the last two sections by a comparison with an experiment.

A comparison of results of the dynamic approach with results obtained by considering the thermodynamics with a fixed number of particles has been made by Kopietz and Efetov (1992) for low concentrations of impurities, such that $L \ll l \ll LM$, where $M \sim p_0^2 S$ is the number of transversal channels in the ring. This limit can be thought of as quasi-ballistic because electrons can traverse the ring without being scattered. At the same time, the last inequality is equivalent to $\tau^{-1} \gg \Delta$, indicating that the disorder is important. For example, grand canonical averaging of the current would give a result exponentially small in $(\tau \Delta)^{-1}$. In the limit $L \ll l \ll LM$ both canonical thermodynamic averaging and grand canonical averaging of dynamic quantities can be carried out exactly and give the same result. The result obtained by canonical averaging also agrees with the result of Altland et al. (1992), and so this limit does not help to clarify the difference between the approaches.

8.4.3 Comparison with results of numerical simulations

For the diffusion regime only numerical simulations are complementary to the preceding analytical calculations. Montambaux et al. (1990) and Bouchiat, Montambaux, and Sigeti (1991) calculated a quantity $\langle I \rangle$ defined as

$$\langle I \rangle = \frac{2\pi}{\phi_0} \int\limits_0^{\phi_0/4} \tilde{I}(\phi) \, d\phi \tag{8.102}$$

where $\tilde{I}(\phi)$ is the current per one degree of freedom. It is two times smaller than the current calculated previously.

As has been discussed, the current has the form of a smeared sawtooth. In order to compute the integral in Eq. (8.102) in the limit $\mathcal{E}_c \gg \Delta$ one can replace this form by an ideal sawtooth with the slope given by Eqs. (8.97, 8.99). This gives immediately for $\langle I \rangle$

$$\langle I \rangle = \frac{|e|}{32} \left(\frac{\mathcal{E}_c \Delta}{2} \right)^{1/2} = \frac{\pi}{32} I_0(\mu) \left(\frac{\tau(\mu)}{6\nu(\mu) V} \right)^{1/2} \tag{8.103}$$

where $I_0(\mu) = |e| v_0 / L$.

Montambaux et al. (1990) and Bouchiat, Montambaux, and Sigeti (1991) considered a model with the Hamiltonian

$$\mathcal{H} = \sum_{r,r'} t a_r^+ a_{r'} + \sum_r u_r a_r^+ a_r \tag{8.104}$$

where u_r is a random energy distributed homogeneously in an interval between $-W/2$ and $W/2$, and the pair (r, r') denotes nearest neighbors. In the Born approximation used in these numerical works one has

$$\tau^{-1}(\mu) = 2\pi v(\mu) \langle u^2 \rangle V_0 = \frac{\pi v(\mu) W^2}{6} V_0 \tag{8.105}$$

where $V_0 = b^3$ is the elementary volume, and b is the distance between nearest neighbors.

Subsituting Eq. (8.105) into Eq. (8.103) one has

$$\langle I \rangle = \frac{\sqrt{\pi}}{32} \frac{I_0(\mu)}{\sqrt{MK} W v(\mu)} \tag{8.106}$$

where $MK = V/V_0$ and $K = L/b$ is the circumference in the units of the lattice length.

The quantity $\langle I \rangle$ depends on the Fermi energy μ. The spectrum without disorder is described by

$$\varepsilon(\mathbf{p}) = 2t \left(\cos p_x + \cos p_y + \cos p_z \right) \tag{8.107}$$

For the spectrum $\varepsilon(\mathbf{p})$, Eq. (8.107), $\langle I \rangle$ can be expressed for an arbitrary μ in terms of a complicated integral. Simpler formulae can be obtained for the quantity $\langle \langle I \rangle / I_0(\mu) \rangle_N$, where $\langle \rangle_N$ stands for averaging over all possible numbers of electrons. Using the relation

$$dN = V v(\mu) \, d\mu$$

one can reduce integration over the number of electrons to integration over μ, which can vary from $-6t$ to $6t$. Carrying out this integration, one obtains

$$Z \equiv \left\langle \frac{\langle I \rangle}{I_0(\mu)} \right\rangle_N \frac{W\sqrt{MK}}{t} = \frac{3\pi^{1/2}}{8} \tag{8.108}$$

We see from Eq. (8.108) that in the diffusive regime the function Z is a constant provided one uses the Born approximation and this makes a comparison with numerical results more reliable.

The dependence of the quantity Z on disorder and number of transversal channels was studied numerically by Montambaux et al. (1990) and Bouchiat, Montambaux, and Sigeti (1991). In the earlier work (Montambaux et al. (1990)) it was found that for samples with $M \le 100$ the function Z is almost constant in the region of W corresponding to the diffusive regime, and this gives support to the results of the dynamic approach. The value of $Z \simeq 0.5$ can be extracted from the computation and this is in reasonable agreement with the value given by Eq. (8.108). Surprisingly, for the larger sample with $M = 196$ studied by Bouchiat, Montambaux, and Sigeti the function Z grows with W in the diffusive regime and is in better agreement with the results of Section 8.3.

It is rather difficult to imagine what can happen when M varies between 100 and 196 because these channel numbers are already very large. My guess is that, possibly, considering the very large samples with $M = 196$ accuracy of the computation was to some extent sacrificed to spare computer time. But the loss of accuracy may be equivalent to the effects of level smearing. If the smearing is large enough the perturbative approach is no doubt correct, whereas the dynamic approach loses accuracy. This means that by changing the accuracy of the numerical computation one can obtain a crossover between the two curves in Fig. (8.6). At the same time, for the values $\mathcal{E}_c/\Delta \sim 20$ that are achievable numerically the curves are very close to each other. In order to distinguish finally between the results of the two approaches one apparently needs more computer power.

8.5 Effects of spin-orbit interactions

So far in this chapter, the spin degrees of freedom of the electron have been neglected except for the trivial multiplier 2. Magnetic impurities cannot be interesting because they act like a magnetic field. If their concentration is high enough, such that $\tau_s^{-1} \gg \Delta$, where τ_s is the mean free time for the magnetic scattering, the persistent currents just vanish. Effects of spin-orbit interactions turn out to be much more spectacular, and this section is devoted to considering the persistent currents in the presence of spin-orbit impurities.

The most intriguing feature of the persistent currents in the presence of spin-orbit interactions is the prediction of a jump in the dependence of the current on magnetic flux in rings with an odd number of electrons made by Entin-Wohlman et al. (1992) and by Kravtsov and Zirnbauer (1992). The jump is possible as a result of the well-known Kramers degeneracy (Kramers (1930), Landau and Lifschitz (1959)) of the ground state at $\phi = 0$ for a system with an odd number of electrons that is violated by any nonzero flux. Of course, even in the absence of spin-orbit interactions the ground state of any system with time reversal invariance and an odd number of electrons is degenerate because the spin direction is arbitrary. However, in this case electrons with "up" and "down" spin can be considered independently. Then, at $\phi = 0$ all wave functions can be chosen real, which results in zero current.

Applying a weak magnetic field one can consider its effect by using the standard quantum mechanical perturbation theory. The first-order correction to the eigenenergies gives zero because the magnetic field changes sign under time reversal whereas wave functions do not. A nonzero result for the eigenenergies can be obtained in the second order of the perturbation theory. This contribution is proportional to $-\phi^2$, which leads to a linear dependence of the current on the flux and a finite susceptibility, in agreement with the result of the preceding sections.

What happens if spin-orbit impurities are present? Looking at Eq. (4.74) describing the interaction of electrons with these impurities we see that in changing spin direction one has to change the direction of the momentum in order to conserve the form of the Hamiltonian. This means that, in the ground state, the total current of a system with an odd number of electrons can be finite although the states with opposite directions of currents have the same energy and are equally probable. In applying a weak magnetic field one has to use degenerate perturbation theory because matrix elements of the magnetic field between the states are not necessarily zero. The first order of the degenerate perturbation theory is linear in the perturbation. Hence, we come to a linear dependence of the eigenenergies on flux with a cusp at $\phi = 0$ because the energy does not change with changing of the sign of the magnetic field. Differentiating the energies over the flux one obtains a jump in the dependence of the current on the magnetic field.

How can this singularity be obtained by the supersymmetry technique? If the levels cross at $\phi = 0$ and the energies vary linearly in flux for small fluxes there must be a maximum at $\omega \sim \phi$ in the two-level correlation function and one can use the supersymmetry technique to evaluate this function. We know from Chapter 4 that the structure of supermatrix Q in the symplectic ensemble is very similar to that in the orthogonal ensemble (cf. Eqs. (6.42–6.46)). How does the singularity nevertheless appear?

To answer this question and calculate the dependence of the persistent current on the magnetic field it is convenient to use a parametrization for supermatrices Q analogous to the magnetic field parametrization introduced in Section 8.3. This can be done by writing supermatrix Q in the presence of a strong spin-orbit scattering in the form of Eqs. (8.45–8.47). The only differences lie in the forms of the matrices C_0, $\eta_{c,d}$, $\kappa_{c,d}$ and the elementary volume $[dQ]$, which have to be written as

$$
C_0 = \begin{pmatrix} c_2 & 0 \\ 0 & c_1 \end{pmatrix}, \qquad \eta_{c,d} = \begin{pmatrix} \eta_{c,d} & 0 \\ 0 & \eta_{c,d}^* \end{pmatrix}, \qquad \kappa_{c,d} = \begin{pmatrix} \kappa_{c,d} & 0 \\ 0 & \kappa_{c,d}^* \end{pmatrix} \tag{8.109}
$$

$$
[dQ] = 4 \left(\frac{\lambda_{1c}}{\lambda_{1c} + \lambda_c} \right)^2 J_c J_d d R_c d R_d d\lambda_c d\lambda_{1c} d\lambda_d d\lambda_{1d} \tag{8.110}
$$

The quantities $\lambda_{c,d}$, $\lambda_{1c,d}$, $J_{c,d}$, and $dR_{c,d}$ are the same as those in Eqs. (8.52) and the variables $\theta_{c,d}$ and $\theta_{1c,d}$ obey the inequalities written in Eqs. (8.50).

To calculate the two-level correlation function one can start again from the general expression, Eq. (6.18). Proceeding in the same way as in Section 8.3 one comes to Eqs. (8.41, 8.42, 8.44). The only difference with respect to the system without spin-orbit interactions is that the quantity Δ must be replaced by $\bar{\Delta} = \Delta/2$, which now plays the role of mean level spacing. Further manipulations are a repetition of the calculations carried out in Section 8.3. The elementary volume $[dQ]$, Eq. (8.110), is obtained from the corresponding expression, Eq. (8.52), by the substitution $\lambda_c^2 \to \lambda_{1c}^2$ in the numerator, and this is the only difference. As a result one gets an expression that is only slightly modified with respect to Eq. (8.64)

$$
R(\omega) = 1 - \frac{\sin^2 x}{x^2} \tag{8.111}
$$

$$
+ \int_1^\infty \lambda_1 \sin(\lambda_1 x) \exp\left(-X^2 \lambda_1^2\right) d\lambda_1 \int_0^1 \frac{\sin(\lambda x)}{\lambda} \exp\left(X^2 \lambda^2\right) d\lambda
$$

where $x = \pi\omega/\bar{\Delta}$. The function $R(\omega)$, Eq. (8.111), was obtained for the first time by Mehta and Pandey (1983) by a completely different approach of random matrix theory. The parameter X in their theory describes a crossover between unitary and symplectic matrices.

Although function $R(\omega)$, Eq. (8.111), is very similar to Eq. (8.64), it is not analytical at $X = 0$ (Kravtsov and Zirnbauer (1992)). In the limit $x \ll 1$, $X \ll 1$ function $R(\omega)$ takes the form

$$
R(\omega) = R_{\text{sympl}}(\omega) + \frac{\sqrt{\pi} x^2}{4X^3} \exp\left(-\frac{x^2}{4X^2}\right) \tag{8.112}
$$

where $R_{\text{sympl}}(\omega)$ is the two-level correlation function for the symplectic ensemble, Eq. (6.15). The second term in Eq. (8.112) has a maximum at $x \sim X$, and this shows that it is highly probable that there is a level at a distance from a given level proportional to the magnetic field. This agrees with the qualitative arguments based on degenerate perturbation theory. With the same arguments one can write for the energies $\varepsilon_\pm(\phi)$ of a split level

$$\varepsilon_{\pm}(\phi) = \varepsilon(0) \pm \frac{I\phi}{c} \tag{8.113}$$

where I is the current at $\phi = 0$. From Eqs. (8.112, 8.113) one can immediately derive the current distribution function $P(I)$ at $\phi = 0$

$$P(I) = \frac{4I^2}{\sqrt{\pi} \bar{I}_{typ}^3} \exp\left(-\frac{I^2}{\bar{I}_{typ}^2}\right) \tag{8.114}$$

where

$$\bar{I}_{typ} = \frac{|e|}{\pi} \left(\frac{2\mathcal{E}_c \bar{\Delta}}{\pi}\right)^{1/2} \tag{8.115}$$

is the typical value of the current.

With the current distribution function $P(I)$ one can obtain the average current in the limit $\phi \to +0$, which can be denoted by $\bar{I}$

$$\bar{I} = \frac{2|e|}{\pi^2}\sqrt{2\mathcal{E}_c \bar{\Delta}} \tag{8.116}$$

In the limit $\phi \to -0$ the average current has the opposite sign. Any finite temperature leads to smearing of the jump. The dependence of the average current on temperature can be understood by using an analogy of the current with an Ising spin that can take the values $+\bar{I}$ and $-\bar{I}$. Then the average current $\bar{I}(\phi)$ can be written without difficulty in the form

$$\bar{I}(\phi) = \bar{I} \tanh\left(\frac{\bar{I}\phi}{cT}\right) \tag{8.117}$$

Thus, the singular behavior of the persistent current is rounded off by finite temperature, but the typical slope of $I(\phi)$ diverges as $\bar{I}_{typ}^2/T$ for $T \to 0$.

The singularity exists in rings with odd numbers only. In a system consisting of rings with odd and even number of electrons in half the average jump of the current is equal to $\bar{I}/2$.

Now, let us determine the dependence of the average current at an arbitrary magnetic flux. With the two approaches discussed in the preceding sections in hand we can make the computation in the same way. Using canonical averaging we may start with Eq. (8.33), which is applicable to a system with spin interactions, too, and substitute Eq. (8.111) for the function $R(\omega)$ in it. Then for the average current $I(\phi)$ we obtain

$$I(\phi) = \frac{2|e|}{\pi^2} \left(\frac{2\mathcal{E}_c \bar{\Delta}}{\pi}\right)^{1/2} X \int_1^{\infty} \int_0^1 \frac{\lambda_1^2}{\lambda_1^2 - \lambda^2} \exp\left(X^2\left(\lambda^2 - \lambda_1^2\right)\right) d\lambda d\lambda_1 \tag{8.118}$$

In the limit $X \to 0$ large $\lambda_1 \sim 1/X$ give the main contribution, the integral simplifies considerably, and one obtains the finite current

$$I(0+) = \frac{|e|}{\pi^2} \left(2\mathcal{E}_c \bar{\Delta}\right)^{1/2} \tag{8.119}$$

which is two times smaller than the current $\bar{I}$, Eq. (8.116). This means that canonical averaging somehow also includes averaging over the parity of the number of particles. In the limit $X \gg 1$ the current given by Eq. (8.118) is suppressed by the factor $1/4$ compared

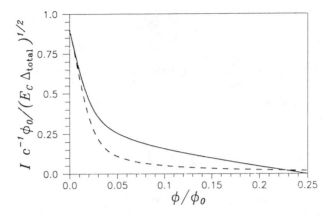

Fig. 8.7. The current, $Ic^{-1}\phi_0/(E_c\bar\Delta)^{1/2}$, in the presence of spin-orbit interactions for $E_c/\bar\Delta = 100$ as a function of the flux ϕ/ϕ_0. The dashed line is calculated from Eq. (8.118); the solid line is obtained from the dynamic calculation, Eq. (8.127). (From Efetov and Iida (1993).)

with the corresponding current in the absence of spin-orbit interactions and this agrees with the results derived in this limit from Eq. (8.32) by Mathur and Stone (1991) and by Entin-Wohlman et al. (1992). The dependence of the current on flux calculated from Eq. (8.118) is represented by the dashed line in Fig. 8.7.

In order to extend the dynamic approach to the case of a system with spin-orbit interactions one has an obvious difficulty. The relation between the thermodynamic current and the linear dynamic response, Eq. (8.84), is based on the assumption that the energy levels do not cross each other. But we already know that in the presence of spin-orbit interactions levels can cross each other and the assumption does not hold when $\phi = 0$ (and in general $\phi = n\phi_0/2$ for an arbitrary integer n). However, since nonzero magnetic fields resolve these degeneracies one can circumvent this difficulty because one may still use the dynamic approach everywhere else in the interval $(0, \phi_0/2)$ excluding the terminal points. The only peculiarity arises when one tries to subtract the constant term from the derivative of the current.

Calculating the linear response for a system with spin-orbit interactions one arrives at integrals analogous to Eqs. (8.88–8.90). The expression for $R_1^{11}(\omega)$ is the same as in Eq. (8.88) and the only modification is that one should write the proper supermatrices using Eqs. (8.109, 8.110). The term $R_3^{11}(\omega)$ retains its form but changes the sign analogously to changing the sign in the expression for electric susceptibility, Eqs. (7.41, 7.42). The term $R_2^{11}(\omega)$ is obtained from Eq. (8.88) by multiplying the latter equation by 2. In the limit $\omega \to 0$ integration over variables λ_{1d}, λ_d is again simple and one obtains the following equations corresponding to Eqs. (8.94–8.96):

$$-i\omega R_1^{11}(\omega) = \begin{cases} \frac{2e^2 D_0}{\pi V}, & \phi = 0 \\ \frac{e^2 D_0}{\pi V}, & \phi \neq 0 \end{cases} \qquad (8.120)$$

$$-i\omega R_2^{11}(\omega) = \begin{cases} 0, & \phi = 0 \\ \frac{2e^2 D_0}{\pi V}\int \exp\left[X^2\left(\lambda_c^2 - 1\right)\right]d\lambda_c, & \phi \neq 0 \end{cases} \qquad (8.121)$$

$$- i\omega R_3^{11}(\omega) = \begin{cases} \frac{e^2 D_0}{\pi V}, & \phi = 0 \\ 0, & \phi \neq 0 \end{cases} \tag{8.122}$$

At first glance, Eqs. (8.120–8.122) are very strange because they have discontinuities at $\phi = 0$. However, these discontinuities do not mean anything. Only the sum $R^{11}(\omega)$ of the terms Eqs. (8.120–8.122) determines physical quantities, but this function itself is continuous, having the form

$$- i\omega R^{11}(\omega) = \frac{e^2 D_0}{\pi V}\left(1 + 2\int_0^1 \exp\left[X^2\left(\lambda_c^2 - 1\right)\right] d\lambda_c\right) \tag{8.123}$$

Using Eq. (8.86) for the flux derivative of the current we obtain

$$\frac{dI(\phi)}{d\phi} = \frac{8e\mathcal{E}_c}{\pi^2 \phi_0} K_0(X) + f \tag{8.124}$$

$$K_0(X) = -\frac{1}{2}\int_0^1 \exp\left[X^2\left(\lambda_c^2 - 1\right)\right] d\lambda_c \tag{8.125}$$

where f does not depend on field. As has been mentioned, the preceding expression is valid in the interval $(0, \phi_0/2)$ excluding the end points. In order to obtain the constant f, we must know the integral of the response over the period (cf. Eq. (8.81)). Integration over flux gives

$$\int_0^{\phi_0/2} \frac{dI}{d\phi} d\phi = I\left(\frac{\phi_0}{2} -\right) - I(0+) = I(0-) - I(0+) = -2I(0+) \tag{8.126}$$

The last two equalities of the preceding equations come from the fact that the current is an odd periodic function of flux with $\phi_0/2$ period. This gives the relation between the constant f in Eq. (8.124) and the jump $2I(0+)$ of the average current at $\phi = 0$. The jump cannot be calculated by the dynamic method, but it has been obtained in Eq. (8.116) from quite general arguments for a ring with an odd number of electrons. Assuming that the system is equally represented by rings with odd and even number of electrons and using Eqs. (8.124–8.126) one obtains

$$I(\phi) = I(0+)\left(1 - 4\frac{\phi}{\phi_0}\right) + \frac{8|e|\mathcal{E}_c}{\pi^2 \phi_0}\left(\int_{0+}^{\phi} K_0(X') d\phi' - \overline{K_0}\phi\right) \tag{8.127}$$

where $X' = 2\pi\phi'/\left(\phi_0\sqrt{2\mathcal{E}_c/\pi\bar{\Delta}}\right)$, and $I(0+)$ is given by Eq. (8.119). The current described by Eq. (8.127) is represented by the solid line in Fig. 8.7. Again, the difference between the results of the two approaches is not large. The main difference between the curves in Fig. 8.6 and 8.7 is the finiteness of the current $I(0+)$ in the latter case. The order of magnitude of the currents is in all the cases the same. The value of the amplitudes of the current is the maximum one can obtain from a model of noninteracting electrons in the diffusive regime. The fact that these values are still considerably smaller than the experimental values observed by Lévy et al. (1990) apparently means that more sophisticated models should be considered.

9

Transport through mesoscopic devices

9.1 Universal conductance fluctuations

9.1.1 Diffusion modes and conductance fluctuations

During the last 10–15 years study of transport through very small conductors has been a very popular topic of both theoretical and experimental research (for reviews see, e.g., Altshuler, Lee, and Webb (1991); Kirk and Reed (1992); Beenakker and van Houten (1991)). These systems are much larger (typically of the size 100–10,000 Å) than atomic distances, and, naturally, they cannot be considered as microscopic objects. Then, what can be special in their properties with respect to properties of macroscopic conductors? The answer is related to the quantum interference that proved to be so crucial to the localization problem. Although the conductors may contain internal defects, such disorder does not destroy the coherence of the wave functions and quantum effects can become very important, leading to completely new physical phenomena. The last two chapters were devoted to study of some exotic effects in isolated or almost isolated samples.

The quantum effects in isolated metal particles can be destroyed by inelastic scattering only. It is only at the classical limit that both elastic and inelastic scattering play equivalent roles in transport. Considering inelastic processes one should distinguish between the cases $l_\varphi < L$ and $l_\varphi > L$, where l_φ is the inelastic mean free path and L is the size of the sample. In the former limit, one has just a macroscopic sample with the usual well-known properties. However, if sample size is smaller than the inelastic mean free path, the quantum nature of electrons is essential and such objects are classified as being mesoscopic.

Transport through a small conductor can be studied in practice by attaching leads that are inevitably macroscopic. It is well known that the nature of a measuring process in quantum mechanics is highly nontrivial because an experimental device interacting with a measured object changes the latter. By now, everybody is accustomed to hearing these words in relation to measuring a state of an atom or molecule. But what about the mesoscopic samples?

An isolated sample has a discrete set of energy levels, but attaching the leads necessary for measuring transport properties causes smearing of the levels. If the level width due to the coupling to the leads is larger than the Thouless energy $\mathcal{E}_c$ one may forget about the level discreteness. In this region of parameters, by varying the strength of the coupling to the leads, one does not change transport coefficients. So, if the leads are really large and their coupling is strong, the transport properties are not very dependent on the leads. This limit corresponds to the situation when the intrinsic conductance of the sample is much smaller than the conductance of the contacts.

In the opposite limit, the discreteness of the levels can be important and all transport properties become dependent on level width. Changing the form of the potential at the contact can result in essentially different total conductance. In this case one may say that the result of the measurement depends on the measuring process. Moreover, if the coupling to the leads is weak enough, charging effects become important. Adding or subtracting one electron leads to a change of charging energy of the order of e^2/C, where C is the capacitance. This energy can be much larger than other energies in the system and the charging effects must be taken into account.

All the effects make study of transport through small conductors that are often called "quantum dots" very nontrivial and interesting. It is very important that for transport through mesoscopic objects, it is not sufficient to know quantities averaged over irregularities. It turns out that most results are sample-specific. Of course, all possible information would be available if one were able to solve the problem for an arbitrary configuration of irregularities. However, this is almost never possible and, in fact, is not interesting. Using an analogy with conventional statistical physics one may argue that although knowledge of how a single molecule of a gas moves can be extracted from different computer simulations, this knowledge is of no interest because physical quantities are expressed in terms of averages over motion of a large number of molecules.

If one wants to characterize transport through a mesoscopic sample in a complete way, instead of trying to calculate a physical quantity for a given configuration of impurities, one should know not only the average conductance or other averaged quantities but moments and distribution functions. We have learned from Chapter 8 that, for example, the average persistent current can be considerably different from the typical current. The same is true of transport quantities, and the present chapter is devoted to description of several interesting effects.

It is clear from the preceding chapters that in situations when the energy levels are smeared one may use perturbation theory in the diffusion modes. The supersymmetry technique can be useful but is not necessary in such cases. Therefore, if a mesoscopic sample is strongly connected to the leads, all calculations can be done in terms of the cooperons and diffusons. Although several mesoscopic effects were predicted qualitatively (Büttiker, Imry, and Landauer (1983), Gefen, Imry, and Azbel (1984), Landauer (1985), Azbel (1984)), quantitative computation using perturbation theory was done first by Altshuler (1985) and Lee and Stone (1985). In the next sections more complicated cases when the level discreteness is important will be considered. To describe those effects one really has to use supersymmetry technique. In the rest of the present section, let us discuss the "universal conductance fluctuations" (UCFs) considered by Altshuler (1985), Lee and Stone (1985), Altshuler and Khmelnitskii (1985), and Lee, Stone, and Fukuyama (1987).

The main quantity that will be calculated is the correlation function $B(H^+, H^-)$ of the total conductances of the sample $G(H^{\pm})$ taken at $\omega = 0$ and at different magnetic fields $H^{\pm} = H_0 \pm H/2$

$$B\left(H^+, H^-\right) = \left(\frac{h}{2e^2}\right)^2 \left(\langle G\left(H^+\right) G\left(H^-\right)\rangle - \langle G\rangle^2\right) \tag{9.1}$$

In order to compute this quantity one can start with a standard linear response formula for the current density $\mathbf{J}(\mathbf{r}, \omega)$ that has been used several times in different forms in the preceding chapters:

$$\mathbf{J}(\mathbf{r}, \omega) = \mathbf{J}_p(\mathbf{r}, \omega) + \mathbf{J}_d(\mathbf{r}, \omega) \tag{9.2}$$

where

$$\mathbf{J}_p (\mathbf{r}, \omega) = \mathbf{J}_1 (\mathbf{r}, \omega) + \mathbf{J}_2 (\mathbf{r}, \omega), \qquad \mathbf{J}_d (\mathbf{r}, \omega) = -\frac{e^2 N}{mcV} \mathbf{A}_\omega (\mathbf{r}) \qquad (9.3)$$

In Eq. (9.3), $\mathbf{J}_d (\mathbf{r}, \omega)$ is the diamagnetic part of the current density and the functions $\mathbf{J}_1 (\mathbf{r}, \omega)$ and $\mathbf{J}_2 (\mathbf{r}, \omega)$ are equal to

$$\mathbf{J}_1 (\mathbf{r}, \omega) = \frac{i}{\pi c} \int (n (\varepsilon) - n (\varepsilon - \omega)) \qquad (9.4)$$

$$\times \left[G_\varepsilon^R (\mathbf{r}, \mathbf{r}') \overleftrightarrow{\hat{\pi}_r} \overleftrightarrow{\hat{\pi}_{r'}} G_{\varepsilon-\omega}^A (\mathbf{r}', \mathbf{r}) \right] \mathbf{A}_\omega (\mathbf{r}') \, d\mathbf{r}' d\varepsilon$$

$$\mathbf{J}_2 (\mathbf{r}, \omega) = \frac{i}{\pi c} \int \left[G_{\varepsilon+\omega}^R (\mathbf{r}, \mathbf{r}') \overleftrightarrow{\hat{\pi}_r} \overleftrightarrow{\hat{\pi}_{r'}} G_\varepsilon^R (\mathbf{r}', \mathbf{r}) \right.$$

$$\left. - G_\varepsilon^A (\mathbf{r}, \mathbf{r}') \overleftrightarrow{\hat{\pi}_r} \overleftrightarrow{\hat{\pi}_{r'}} G_{\varepsilon-\omega}^A (\mathbf{r}', \mathbf{r}) \right] n (\varepsilon) \mathbf{A}_\omega (\mathbf{r}') \, d\mathbf{r}' d\varepsilon \qquad (9.5)$$

where the operator $\overleftrightarrow{\hat{\pi}_r}$ acts on arbitrary functions f and g as

$$f (\mathbf{r}) \overleftrightarrow{\hat{\pi}_r} g (\mathbf{r}) = \frac{1}{2} \left(f (\mathbf{r}) \hat{\pi}_r g (\mathbf{r}) + \left(\hat{\pi}_r^* f (\mathbf{r}) \right) g (\mathbf{r}) \right) \qquad (9.6)$$

and $\hat{\pi}_r$ is the current operator.

In the same way as Altshuler (1985) and Lee and Stone (1985) let us consider the conductance of noninteracting electrons in the following system: a finite disordered region of volume $V = L_x L_y L_z$, extended to $\pm\infty$ in the z-direction by attachment of ideal "leads." It is assumed that the electric field is applied only to the disordered region and, being parallel to the z-axis, does not depend on the coordinates in this region. Then the conductance G is related to the current density $\mathbf{J}$ by the following equation:

$$i\omega G L_z^2 \mathbf{A}_\omega = c \int \mathbf{J} (\mathbf{r}, \omega) \, d\mathbf{r} \qquad (9.7)$$

To compute the correlation function $B (H^+, H^-)$, one should substitute Eqs. (9.2–9.7) into Eq. (9.1), taking two currents at different magnetic fields H^+ and H^-, and perform averaging over the disorder. Even at weak disorder, several sequences of diagrams have a large contribution. These diagrams are represented in Fig. 9.1. They contain the diffusion modes that have been discussed here many times. The four electron lines represent Green functions at two different magnetic fields and energies. Two of the Green functions must be retarded and two advanced and one should consider all such combinations. The arrows on the Green functions can be both parallel and antiparallel; that means that the contribution of both cooperons $C (\mathbf{k}, \omega)$ and diffusons $D (\mathbf{k}, \omega)$ is important. As in the problem of persistent currents considered in Chapter 8 the diffusion modes depend on a discrete set of momenta. Looking at Eqs. (9.2–9.7) one can get the impression that everything is the same as in the problem of persistent currents. But here we expect to obtain finite conductance, whereas in the problem of persistent currents the conductance must be infinity. Where is the difference?

To answer this question one should recall that in the present chapter we consider a sample connected to bulk leads, whereas the persistent currents were studied in an isolated system. This results in a difference in boundary conditions for the diffusion modes. The fact that the leads are made of a clean metal means that the diffusion modes do not exist at the

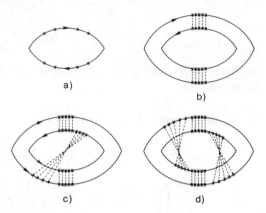

Fig. 9.1. Diagrams for the autocorrelation function of conductances: a) unaveraged conductance, b)–d) diagrams contributing to the averaged correlation function.

leads and, hence, both the functions $C(\mathbf{k}, \omega)$ and $D(\mathbf{k}, \omega)$ must turn to zero at the leads. At the same time a current in the transversal direction must be zero at the walls and this gives the boundary conditions $\partial D/\partial x = \partial D/\partial y = 0$ at the walls (the same equations apply for the cooperons). These boundary conditions were first suggested by Altshuler and Aronov (1981). Because the electron Green functions depend on different magnetic fields and energies, the diffusons contain the differences H and ω of these quantities and satisfy the following equation

$$\left[-i\omega + D_0\left(-i\nabla - \frac{e}{c}\mathbf{A}\right)^2 + \tau_\varphi^{-1}\right] D(\mathbf{r}, \mathbf{r}', \omega, H) = 2\pi\nu\delta(\mathbf{r} - \mathbf{r}') \qquad (9.8)$$

where the vector potential $\mathbf{A}$ corresponds to the the difference of the magnetic fields H, τ_φ^{-1} is an inelastic scattering rate. We know that the density–density correlation function $Y^{00}(\mathbf{k}, \omega)$, Eqs. (3.24, 3.49), has a similar form, Eq. (3.52). However, because of the particle conservation, no inelastic processes could lead to the appearance of τ_φ^{-1} in the function $Y^{00}(\mathbf{k}, \omega)$. Formally, this is because all self-energy contributions to the Green function are canceled by the corresponding interaction lines connecting the Green functions. It is not the case for the present problem of mesoscopic fluctuations. The two loops in Fig. (9.1) refer to two different measurements and therefore may not be connected by interaction lines. Hence, the usual cancellation of self-energy and vertex correction does not occur; the result is a nonzero τ_φ^{-1} in Eq. (9.8).

Eq. (9.8) can be solved by using a spectral representation for the function $D(\mathbf{r}, \mathbf{r}', \omega, H)$

$$D(\mathbf{r}, \mathbf{r}', \omega, H) = 2\pi\nu\sum_m \frac{\phi_m(\mathbf{r})\phi_m^*(\mathbf{r}')}{\lambda_m} \qquad (9.9)$$

where $\phi_m(\mathbf{r})$ and λ_m are the eigenfunctions and eigenvalues of the differential operator in the left-hand side of Eq. (9.8). With the chosen boundary conditions, the eigenfunctions for the rectangular sample at $H = 0$ have the form

$$\phi_m(\mathbf{r}) = \left(\frac{8}{L_x L_y L_z}\right)^{1/2} \sin\left(\frac{m_z \pi z}{L_z}\right) \cos\left(\frac{m_x \pi x}{L_x}\right) \cos\left(\frac{m_y \pi y}{L_y}\right) \qquad (9.10)$$

where $m_z = 1, 2, \ldots, \infty$, $m_{x,y} = 0, 1, 2, \ldots, \infty$, and the eigenvalues are $\lambda_m = D_0 (\pi/L_z)^2 \tilde{\lambda}_m$, with $\tilde{\lambda}_m$ determined by the relation

$$\tilde{\lambda}_m = m_z^2 + m_x^2 \left(\frac{L_z}{L_x}\right)^2 + m_y^2 \left(\frac{L_z}{L_y}\right)^2 + \gamma_\varphi - i\eta \tag{9.11}$$

In Eq. (9.11), $\gamma_\varphi = (L_z/\pi L_\varphi)^2$, where $L_\varphi = (D_0 \tau_\varphi)^{1/2}$ is the inelastic diffusion length and $\eta = \omega L_z^2 / (\pi^2 D_0) \equiv \omega/\mathcal{E}_{cz}$. The energy $\mathcal{E}_{cz}$ is the Thouless energy in the z-direction. We see that the zero mode with $m_x = m_y = m_z = 0$ that was so important in the problem of the persistent currents must be discarded here. An equation similar to Eq. (9.9) can be written for the cooperons. The only difference is that in the equation for the cooperons one should write $2\mathbf{A}_0$ instead of $\mathbf{A}$, where $\mathbf{A}_0$ corresponds to the field H_0.

9.1.2 Correlation function of conductances

It is convenient to write at $T = 0$ a more general correlation function $\tilde{B}(H^+, H^-, \omega)$ of conductances at different energies $\varepsilon + \omega/2$ and $\varepsilon - \omega/2$

$$\tilde{B}(H^+, H^-, \omega) = \langle \delta g (H^+, \varepsilon^+) \delta g (H^-, \varepsilon^-) \rangle \tag{9.12}$$

where $\delta g = g - \langle g \rangle$, $\varepsilon^\pm = \varepsilon \pm \omega/2$. The final result of the summation of the diagrams in Fig. 9.1 can be written at $H^\pm = 0$ as

$$\tilde{B}(0, 0, \omega) = \left(\frac{8}{\pi^4}\right) (B_b + B_c + B_d) \tag{9.13}$$

where B_b, B_c, B_d are the contributions represented in Figs. 9.1b, c, and d, respectively. These quantities are equal to

$$B_b = 2 \sum_{m_x, m_y=0}^{\infty} \sum_{m_z=1}^{\infty} \left[\mathrm{Re}\left(\tilde{\lambda}_m^{-1}\right)\right]^2 \tag{9.14}$$

$$B_c = -8\,\mathrm{Re}\left[\sum_{m_x, m_y=0}^{\infty} \sum_{m_z=1,3\ldots}^{\infty} \sum_{n_z=2,4\ldots}^{\infty} \frac{(f_{mn})^2}{\tilde{\lambda}_m \tilde{\lambda}_n} \left(\frac{1}{\tilde{\lambda}_m} + \frac{1}{\tilde{\lambda}_n}\right)\right] \tag{9.15}$$

$$B_d = 24\,\mathrm{Re}\left[\sum_{m_x, m_y=0}^{\infty} \sum_{m_z,q_z=1,3\ldots}^{\infty} \sum_{n_z,q_z=2,4\ldots}^{\infty} \frac{f_{mn} f_{np} f_{pq} f_{qm}}{\tilde{\lambda}_m \tilde{\lambda}_n \tilde{\lambda}_p \tilde{\lambda}_q}\right] \tag{9.16}$$

where $f_{mn} = 4m_z n_z/\pi (m_z^2 - n_z^2)$. The second summations in Eqs. (9.15, 9.16) should be performed over odd numbers, whereas the third ones are over even numbers. The eigenvalues $\tilde{\lambda}_m$ should be taken from Eq. (9.11).

Putting $\omega = 0$ one can see from Eqs. (9.13–9.16) that the autocorrelation function of conductances $B(0, 0) = \tilde{B}(0, 0, 0)$ does not depend on the size of the conductor. Only factors relating to its shape enter these equations. At $\gamma_\varphi = 0$, the function $B(0, 0)$ gives the variance

$$\mathrm{Var}\,(g) = \langle g^2 \rangle - \langle g \rangle^2 \tag{9.17}$$

of the dimensionless conductance $g = (h/2e^2) G$. We see that this quantity is just a number independent of the sample size, disorder, and dimensionality. Therefore, Lee and Stone

(1985) called mesoscopic fluctuations in the regime considered "universal conductance fluctuations" (UCFs). The variance Var (g), Eq. (9.17), depends on the geometry of the sample. If $L_z \gg L_x, L_y$ (geometry of a wire), Var $(g) = 0.133$. For a two-dimensional square Var $(g) = 0.186$ and for a three-dimensional cube Var $(g) = 0.296$. (Note that the dimensionless conductance g used here is the conductance per one spin degree of freedom, which differs from g used by Lee and Stone (1985) by factor of 2.)

The second result that can be extracted from Eqs. (9.13–9.16) is the value of a typical spacing between peaks and valleys in g as a function of the Fermi energy. This energy correlation range is simply the half-width of $\tilde{B}\,(0,\,0,\,\omega)$ and is approximately determined by the condition $\eta = 1$, which corresponds to energies of the order of $\mathcal{E}_{cz} = \pi^2 D_0/L_z^2$. This energy is just the inverse time for the particle to diffuse across the sample in the current direction. Since the correlation function $\tilde{B}\,(0,\,0,\,\omega)$ decays as ω goes to infinity, one may argue that the same variance of the conductance can be obtained not by ensemble averaging but by summing over many energies for a given impurity configuration. Such an ergodic hypothesis is confirmed by numerical simulations.

Considering the effects of the magnetic field one should distinguish between changing the magnetic field H_0 that enters the cooperons and the field difference H that enters the diffusons. If the field H_0 is large enough, the cooperon contributions can be neglected and we are in the unitary ensemble. Then, the variance Var (g) is two times smaller than the corresponding value for systems invariant under time reversal provided one uses the approximation represented by the graphs Fig. 9.1b–d. The symmetry between the cooperons and diffusons ensures that $\langle [g\,(H) - g\,(-H)]^2 \rangle = 0$, which corresponds to the invariance of the conductance under the reversal of the direction of the magnetic field. In the limit of the unitary ensemble, the correlation function $B\left(H^+, H^-\right)$, Eq. (9.1), depends on the field difference H only and decreases with increasing H. It is not difficult to estimate the field correlation range, H_c, which determines the typical spacing of the fluctuations of H. To compute $B\left(H^+, H^-\right)$ for arbitrary H one has to solve Eq. (9.8) (choosing a gauge with $\mathbf{A}$ parallel to the z-axis) with $A_z = (e/c)\,Hy$. However, for a simple estimate one may consider the lowest eigenvalue and use perturbation theory for H. Then, we obtain for the quasi-one-dimensional case

$$\tilde{\lambda}\left(m_z = 1, m_x = m_y = 0\right) = 1 + \frac{1}{12}\left(\frac{L_z L_y e H}{c\pi}\right)^2 \tag{9.18}$$

At the characteristic field H_c the second term in the right-hand side in Eq. (9.18) equals unity. This gives a result for H_c that may be written as

$$\frac{\phi_c}{\phi_0} = \sqrt{3} \tag{9.19}$$

where ϕ_c is the change in the magnetic flux through the area of the sample normal to the field and $\phi_0 = hc/e$ is the flux quantum. The field of the crossover between the orthogonal and unitary ensembles can be obtained in the same way and is described by a similar equation.

Using the zero temperature correlation function $\tilde{B}\left(H^+, H^-, \omega\right)$, Eq. (9.12), one can derive the correlation function $B\left(H^+, H^-\right)$ at arbitrary temperature. Temperature effects due to the smearing of the Fermi distribution function can be studied (Altshuler and Khmelnitskii (1985) and Lee, Stone, and Fukuyama (1987)) through the relation

$$g = -\int n'\,(\varepsilon)\,g\,(\varepsilon)\,d\varepsilon \tag{9.20}$$

where $n'(\varepsilon)$ stands for the derivative of the Fermi function and $g(\varepsilon)$ is the dimensionless conductance at an energy ε. The correlation function $B\left(H^{+}, H^{-}\right)$, Eq. (9.1), can be written at arbitrary temperature as

$$B\left(H^{+}, H^{-}\right) = \int_{-\infty}^{\infty} M(\omega)\,\tilde{B}\left(H^{+}, H^{-}, \omega\right)d\omega \tag{9.21}$$

where $M(\omega)$ is the convolution integral

$$M(\omega) = \int_{-\infty}^{\infty} n'\left(\varepsilon^{+}\right) n'\left(\varepsilon^{-}\right)d\varepsilon$$

The integral over ε in this expression for $M(\omega)$ can be calculated explicitly and the function $M(\omega)$ takes the form

$$M(\omega) = T^2 \frac{d}{dT}\left(2T \sinh\left(\frac{\omega}{2T}\right)\right)^{-2} \tag{9.22}$$

The function $M(\omega)$ decays at energies ω of the order of temperature T. Using Eqs. (9.12, 9.21, 9.22) one can see that temperature smearing becomes important at temperatures $T \geq \mathcal{E}_{cz}$. At lower temperatures one can use the zero temperature results discussed for the conductance fluctuations. At temperatures $T \gg \mathcal{E}_c$ the variance of the conductances decreases according to a power law. The effect of the vanishing of the fluctuations due to the smearing of the Fermi distribution is easily understood. Changing the energy is similar to changing the impurity configuration and leads to large fluctuations. At finite temperatures all energies in the interval T near the Fermi energy contribute in a random way and this is equivalent to averaging of the conductance. Let us emphasize that this smearing is important to the conductance fluctuations but does not play any role in the weak localization corrections.

Temperature effects due to inelastic scattering can be included in the scheme by writing a finite temperature-dependent γ_φ in Eqs. (9.11–9.16). A general formula for the temperature dependence of the correlation function $B(0, 0)$, Eq. (9.21), can be written, within a numerical factor, as (Altshuler and Khmelnitskii (1985))

$$B(0, 0) \sim \frac{1}{L_z^3}\begin{cases} L_x L_y L_T, & d = 3\ (L_T < L_{x,y,z}) \\ L_y L_T^2, & d = 2\ (L_x < L_T < L_{y,z}) \\ L_T^2 L_\epsilon, & d = 1\ (L_{x,y} < L_T < L_z) \end{cases} \tag{9.23}$$

where $L_T = \sqrt{D_0/T}$, $L_\epsilon = \min\left\{L_\varphi, L_z\right\}$. In the limit $L_{x,y,z} \ll L_T$ that corresponds to temperatures much lower than all the Thouless energies $\mathcal{E}_{cx}$, $\mathcal{E}_{cy}$, $\mathcal{E}_{cz}$, one obtains the zero temperature universal numbers discussed previously.

The discussion presented in this section helps one to understand the main ideas used in a study of mesoscopic fluctuations. Since the first works by Umbach et al. (1984), Webb et al. (1985), and Lucini et al. (1985) the effects have been observed in numerous experiments. Of course, experimentally it is difficult to average over disorder and, usually, one studies the fluctuations changing the magnetic field, the Fermi energy, or other physical parameters. The number of works on this topic is no less than the number of works on weak localization. For references one can read reviews by Washburn and Webb (1986), Aronov and Sharvin (1987), Beenakker and van Houten (1991), reviews in Altshuler, Lee, and Webb (1991);

and proceedings of conferences in Kramer, Bergmann, and Bruynseraede (1985) and Ando and Fukuyama (1988).

The analysis in this section is based on the assumption that a small conductor under investigation is strongly connected to bulk metallic leads. Then, only one characteristic energy, namely, the Thouless energy $\mathcal{E}_c$, enters the correlation function at different energies. Characteristic temperatures leading to smearing of the mesoscopic fluctuations are of the same order of magnitude, whereas the characteristic magnetic fluxes ϕ are of the order of ϕ_0, Eq. (9.19). All levels are well smeared by the connections with the leads, and, therefore, the mean level spacing Δ does not enter the theory.

Making the connection to the leads weaker or the sample size smaller one can reach the limit when the level width becomes much smaller than the energy $\mathcal{E}_c$. This limit corresponds to what are now called "quantum dots." The only thing that has to be changed in the theory is the boundary conditions. However, this change drastically influences the theoretical consideration. If the level width is of the order of the mean level spacing, the perturbation theory analysis in the present section is no longer valid and the necessity of using the supersymmetry technique arises. Several interesting problems relevant to quantum dots are discussed in the rest of the chapter.

9.2 Nonlinear σ-model for a system with leads

9.2.1 Landauer approach

In the preceding section, the role of the diffusion modes in quantum transport through a disordered conductor was explicitly demonstrated. We have been accustomed to the fact that as soon as the diffusion modes come into play, one may start deriving the nonlinear σ-model and, in the present chapter, I want to follow this habit. However, for the problem of mesoscopic conductance fluctuations the procedure developed previously should be modified to include the connections to the leads properly. In Section 9.1 the presence of the leads was reduced to omitting the zero mode in the diffusion modes, as is correct provided the coupling to the leads is very strong. In the general case a more sophisticated treatment considering the leads explicitly is necessary.

Of course, Eqs. (9.2–9.5) hold for an arbitrary disordered metal and one may try to find a solution for the system including the mesoscopic conductor and leads. However, such an approach would be very difficult because the electric field entering the equations may depend on coordinates. In the geometry considered in Section 9.1 one can neglect space variations of the electric field, but for devices used in most experiments the electric field in a system consisting of a conductor and leads is not homogeneous. In fact, experimentally one maintains voltages on the leads and measures currents without knowing anything about the space dependence of the electric field. As a result one obtains total conductance whereas local conductivities are of no interest. Can one reformulate the linear response theory in such a way that only voltages and currents on the leads enter equations?

Another reason to try to represent the conductance of the system in a different way follows from the computation of the conductance fluctuations carried out in the preceding section. The paramagnetic part $\mathbf{J}_p$ of the current density, Eqs. (9.2, 9.3), consists of the two parts $\mathbf{J}_1$ and $\mathbf{J}_2$, Eqs. (9.4–9.7). The current $\mathbf{J}_2$ contains the products $G^R G^R$ and $G^A G^A$. When calculating averaged responses the current $\mathbf{J}_2$ does not give any interesting contributions. The diffusion modes and the σ-model are derived when considering the term $\mathbf{J}_1$.

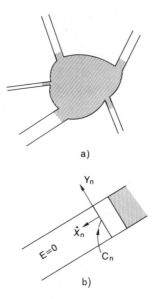

Fig. 9.2. a) An arbitrary multiprobe structure; a possibly disordered region (hatched) is connected to N_L straight, ideal leads; b) asymptotic region of lead n, where C_n is a cross section line in lead n located in the asymptotic region where the electric field is zero.

However, when calculating the variance of the conductance one can see from Fig. 9.1 that contributions from $\langle \mathbf{J}_2^2 \rangle$ should also be considered, making the computation more cumbersome. For example, in Fig. 9.1c and d three- and four-diffusion modes arise from $\langle \mathbf{J}_2^2 \rangle$, but when calculating the variance of the total conductance all such terms can be neglected (Altshuler and Shklovskii (1986), Kane, Serota, and Lee (1988)). Is that a general property valid for higher moments of conductance? If yes, then can one formulate the theory in such a way that terms with the products $G^R G^R$ and $G^A G^A$ are absent from the beginning?

Fortunately, another formulation of linear response theory helps to circumvent the difficulties described and in many interesting cases leads to a drastic simplification of calculations. It is based on an idea proposed long ago by Landauer (1957, 1970) of viewing transport in solids as a consequence of the incident carrier flux. Since 1980 this approach has become very popular and a lot of theoretical work using this method has been carried out. The method became especially useful as a result of the development of mesoscopic physics. One can find a complete list of references in a review article by Stone and Szafer (1988). What is very important, as established later by Fisher and Lee (1981) and Langreth and Abrahams (1981), the Landauer approach and Kubo linear response theory were equivalent at zero frequency. The theory is suitable for an arbitrary multiprobe structure (Büttiker (1986, 1988)) like the one represented in Fig. 9.2a.

If the leads are ideal metals with one-dimensional geometry in the regions asymptotically far from the scattering area, electron states in those regions are plane waves in the longitudinal direction while being quantized in the transversal direction. It is usual to speak of the transversal quantization in terms of channels. If at zero longitudinal momentum $2M_n$ states (including spin) are under the Fermi energy of the nth lead, then the lead is said to have $2M_n$ channels. Suppose that voltages V_n are applied to leads and in the asymptotic regions they do not depend on coordinates such that in those regions the electric field is zero. The total current response I_m through lead m is determined completely in terms of the voltages

V_n independently of the electric field in the scattering regions as

$$I_m = \sum_{n=1}^{N_L} G_{mn} V_n \tag{9.24}$$

where N_L is the number of leads. The multichannel Landauer–Büttiker formula relates the conductances g_{mn} to transmission amplitudes $t_{mn}^{\alpha\beta}$ by a simple equation

$$G_{m \neq n} = \frac{e^2}{h} \sum_{\alpha,\beta} |t_{mn}^{\alpha\beta}|^2 \equiv \frac{e^2}{h} \operatorname{tr}\left(t_{mn} t_{mn}^{+}\right) \tag{9.25}$$

The coefficients $t_{mn}^{\alpha\beta}$ denote transmission amplitudes of an electron incident in lead m in channel β to exit in probe n in channel α and t_{mn} is the corresponding matrix. Using current conservation one has

$$\sum_{m=1}^{N_L} G_{mn} = \sum_{n=1}^{N_L} G_{mn} = 0 \tag{9.26}$$

and this equation gives the possibility of finding G_{mm}. We see that each channel in Eq. (9.25) contributes with equal probability. This is a consequence of the assumption that the leads in the asymptotic regions are quasi-one-dimensional. The longitudinal electron motion for channel α is given by plane wave χ_α

$$\chi_\alpha = \frac{1}{\sqrt{v_\alpha}} \exp{(ikx)} \tag{9.27}$$

where v_α is the Fermi velocity of channel α. The flux carried by this state is equal to $v_\alpha |\chi_\alpha|^2$, which is unity. This is the reason the channels contribute with the equal probability.

It is remarkable that using the same system with the leads as in Fig. 9.2a one can arrive at Eq. (9.25) from the Kubo linear response theory. A derivation can also be done in the presence of an arbitrary magnetic field (Baranger and Stone (1989)), and the final result for the conductances G_{mn} takes the form

$$G_{mn} = -\frac{1}{2\pi} \int d\varepsilon\, n'(\varepsilon) \tag{9.28}$$

$$\times \int_{C_m} dy \int_{C_n} dy'\, G_\varepsilon^R (r, r') \left(\overleftrightarrow{\pi}_r \hat{\mathbf{x}}_m\right) \left(\overleftrightarrow{\pi}_{r'} \hat{\mathbf{x}}_n\right) G_\varepsilon^A (r', r)$$

where operators $\overleftrightarrow{\pi}_r$ are defined in Eq. (9.6), $\hat{\mathbf{x}}_{m,n}$ are unit vectors along leads m, n directed outward from the scattering area (see Fig. 9.2b), and integration over y_m and y_n should be done over the cross sections C_n and C_m in the asymptotic regions of leads m and n ($r_{m,n}$ stand for both the coordinate and spin variables, and integration over $y_{m,n}$ also includes summation over the spin). Cross sections $C_{m,n}$ can be chosen at arbitrary longitudinal coordinates as a consequence of the current conservation. The Green functions $G^{R,A}$ in Eq. (9.28) satisfy the usual equation

$$(\varepsilon - \mathcal{H})\, G_\varepsilon^{R,A} (r, r') = \delta (r - r') \tag{9.29}$$

where $\mathcal{H}$ is the Hamiltonian of the system consisting of the conductor and leads. At infinity (far from the scattering area) the wave functions are plane waves describing free motion, and this determines the boundary conditions for the Green functions.

We see from Eq. (9.28) that the total conductances are expressed in terms of a product of $G^R G^A$ only and nothing is assumed about space dependence of the electric field in the scattering region. The only condition is that the field is zero in the asymptotic regions of the leads. Nothing is assumed about the region where the leads are connected to the mesoscopic conductor. There can be, for example, a potential barrier such that tunneling probability is small. All these details are included in the Green functions G^R and G^A in Eq. (9.28) but the form of the equation is most general.

Although Eq. (9.28) is already much more convenient for studying mesoscopic problems than the conventional Kubo formulae there is still a drawback, which is due to the boundary conditions for the Green functions. It is natural to put the cross sections of integration C_m just at the connections of the leads to the conductor and try to consider the conductor only. But we may forget about the region outside the scattering area only if we are able to write boundary conditions for the Green functions at the surfaces C_m. However, we know boundary conditions for the Green functions at infinity only and this forces us to seek the solution for the whole system.

9.2.2 Elimination of the leads

This shortcoming can rather easily be circumvented by a procedure known as R-matrix theory (Lane and Thomas (1958)) in nuclear physics. Its application to the conductance problem was described by Zirnbauer (1992). Following this procedure one looks for values of the Green functions on the surfaces C_m that fulfill the boundary conditions at infinity. As soon as the values of the Green functions on these surfaces are known one may study only the interior of the conductor using the values obtained as boundary conditions. They can be found in an elegant way by using the current conservation only.

The boundary conditions on C_m can be derived for each lead independently. Let us consider a lead m. As a first step, it is convenient to introduce on the lead m auxiliary retarded and advanced Green functions $\bar{G}_\varepsilon^{R,A}(r,r')$ satisfying the same equation Eq. (9.29) everywhere in the asymptotic region of the lead limited by the surface C_m. As concerns the boundary conditions, the functions $\bar{G}_\varepsilon^{R,A}(r,r')$ are supposed to turn to zero if either r or r' is on this surface. The functions $\bar{G}_\varepsilon^{R,A}(r,r')$ differ from functions $G_\varepsilon^{R,A}(r,r')$ just in the boundary conditions. Let us consider $G_\varepsilon^{R,A}(r,r')$ with r belonging to the lead and r' outside the lead (in other words, r and r' are in different regions separated by the surface C_m; the point r' can be in the scattering region or in other leads).

Current conservation implies the validity of the relation

$$\nabla_r \sum_\sigma \left[\left(\bar{G}_\varepsilon^A(r,r'') \right)^* \hat{\mathbf{v}}_r G_\varepsilon^R(r,r') + \left(\hat{\mathbf{v}}_r \bar{G}_\varepsilon^A(r,r'') \right)^* G_\varepsilon^R(r,r') \right]$$
$$= 2i\delta(r-r'') G_\varepsilon^R(r,r') \tag{9.30}$$

where $\hat{\mathbf{v}}_r = \hat{\pi}_r/e$ is the velocity operator. In the presence of spin-orbit interactions it can be written as

$$\hat{\mathbf{v}}_r = \frac{1}{m}\left(-i\nabla_r - \frac{ie\mathbf{A}}{c} \right) + \nabla u_{so} \times \sigma \tag{9.31}$$

In Eq. (9.30), r, r', r'' stand for both coordinates and spins, $r = (\mathbf{r}, \sigma)$. The validity of Eq. (9.30) can be checked directly by using Eq. (9.29) for the Green functions. Integrating both parts of Eq. (9.30) over all r in the volume of the lead (restricted from one side by

C_m) and using the boundary conditions for the function $\left(\bar{G}_\varepsilon^A \left(r, r''\right)\right)$ on the surface C_m one obtains

$$G_\varepsilon^R \left(r'', r'\right) = -\frac{i}{2} \int_{C_m} \left((\mathbf{x}_m \hat{\mathbf{v}}_z) \bar{G}_\varepsilon^A \left(z, r''\right)\right)^* G_\varepsilon^R \left(z, r'\right) dz \qquad (9.32)$$

where $z = (\mathbf{r}, \sigma)$ and $\mathbf{r}$ is on the surface C_m. Integration in Eq. (9.32) is performed over C_m supplemented with summation over spin. Applying the velocity operator $\hat{\mathbf{v}}_{r''}$ to both sides of Eq. (9.32), taking the point $\mathbf{r}''$ to C_m, and using the relation $\left(\bar{G}_\varepsilon^A \left(z, y\right)\right)^* = \bar{G}_\varepsilon^R \left(y, z\right)$ we can rewrite Eq. (9.32) in the form

$$\left(\mathbf{x}_m \hat{\mathbf{v}}_y\right) G_\varepsilon^R \left(y, r\right) = \int_{C_m} B_m \left(y, z\right) G_\varepsilon^R \left(z, r\right) dz, \qquad (y \in C_m) \qquad (9.33)$$

where

$$B_m \left(y, z\right) = \frac{i}{2} \left(\left(\mathbf{x}_m \hat{\mathbf{v}}_y\right) (\mathbf{x}_m \hat{\mathbf{v}}_z) \bar{G}_\varepsilon^R \left(y, z\right)\right), \qquad (y, z \in C_m) \qquad (9.34)$$

Eq. (9.33) can be considered as a boundary condition satisfied by the Green function $G_\varepsilon^R \left(r, r'\right)$ and we thus arrive at an exact reformulation of the original problem: Instead of solving the Schrödinger equation for the Green functions $G_\varepsilon^{R,A} \left(r, r'\right)$ for the total system with metallic boundary conditions at infinity we may solve the same equation in the region of the mesoscopic conductor itself (restricted by the surfaces C_m) supplemented with the boundary conditions, Eq. (9.33). Of course, one has to find, first, the functions $B_m \left(y, z\right)$, but these functions are determined completely by properties of the leads that are assumed to be known.

In some special cases, the functions $B_m \left(y, z\right)$ can be found rather easily. Of its experimental applications one of the most interesting is the case when the leads have the geometry of a straight wire. Assuming that the wires are ideal in the sense that they are clean and the confining potential does not depend on coordinates one can expand the Green functions $\bar{G}_\varepsilon^R \left(\mathbf{r}, \mathbf{r}'\right)$ on lead m in a Fourier series (spin variables can be omitted)

$$\bar{G}_\varepsilon^R \left(\mathbf{r}, \mathbf{r}'\right) = \frac{2}{\pi} \sum_n \int_0^\infty \frac{\varphi_n \left(y\right) \varphi_n^* \left(y\right) \sin k \left(x - x_m\right) \sin k \left(x' - x_m\right)}{\varepsilon - k^2/2m + \mu_n + i/2\tau_L} dk \qquad (9.35)$$

In Eq. (9.35), $\mu_n = \mu - \varepsilon_n$ is the Fermi energy for band n, $\varphi_n \left(y\right)$ are normalized eigenfunctions describing transversal motion, and ε_n are corresponding eigenenergies. The chemical potential μ is assumed to be the same in all leads because we are calculating within the limit of the linear response, and the longitudinal coordinate x_m stands for the position of the surface C_m. The parameter τ_L is the electron mean free time, which is assumed to be very large. The function $\bar{G}_\varepsilon^R \left(\mathbf{r}, \mathbf{r}'\right)$ turns to zero if one of its arguments $\mathbf{r}$ or $\mathbf{r}'$ is on C_m.

As will be seen soon, the real part of the function $B_m \left(y, z\right)$ is most important. Substituting Eq. (9.35) into Eq. (9.34) and integrating over momentum k in the limit $\tau_L \to \infty$ we obtain

$$\operatorname{Re} B_m \left(z, y\right) = \sum_{n=1}^{M_n} \alpha_n v_n \varphi_n \left(z\right) \varphi_n^* \left(y\right) \qquad (9.36)$$

In Eq. (9.36), the sum is performed over all states with $\varepsilon_n < \mu$, $v_n = (2\mu_n/m)^{1/2}$ is the Fermi velocity of the band n, and all $\alpha_n = 1$. The variables z and y denote two points on the surface C_m. The function $\operatorname{Re} B_m \left(z, y\right)$ at $y = z$ gives the flux carried by all states in the lead below the chemical potential. The fact that $\alpha_n = 1$ for all n is a consequence of the assumption that the leads have the geometry of a straight wire and electrons move freely in the longitudinal direction. In other words, all channels are open.

What happens if the wire has a potential barrier between the surface C_m and infinity? In this case one has to write more complicated eigenfunctions in Eq. (9.35). However, if the potential depends only on the coordinate x, the longitudinal and transversal motion decouple and one comes again to Eq. (9.36), but now, the coefficients α_n are determined by transmission probabilities through the barrier and can be different from unity. If the shape of the barrier is smooth, these probabilities and the coefficients α_n are exponentially small in the thickness of the barrier. Although in each case one can try to find the coefficients α_n, this question will not be discussed in the rest of the book. Everywhere later these coefficients are introduced phenomenologically except in the case with all $\alpha_n = 1$, which clearly corresponds to having all channels open.

As we have understood the functions $\mathrm{Re}\, B_m\,(y,z)$ are related to the flux carried by the states in the leads. Then, looking at Eq. (9.25) one can guess that the conductance G_{mn} has to be related to this function. This relation can really be established by using the following identities for the boundary conditions:

$$\left(\mathbf{x}_m \hat{\mathbf{v}}_y\right) G_\varepsilon^R\,(y,r) = \int_{C_m} B_m\,(y,z)\,G_\varepsilon^R\,(z,r)\,dz = \left(\left(\mathbf{x}_m \hat{\mathbf{v}}_y\right) G_\varepsilon^A\,(r,y)\right)^* \tag{9.37}$$

$$\left(\mathbf{x}_m \hat{\mathbf{v}}_y\right) G_\varepsilon^A\,(y,r) = -\int_{C_m} B_m^*\,(z,y)\,G_\varepsilon^A\,(z,r)\,dz = \left(\left(\mathbf{x}_m \hat{\mathbf{v}}_y\right) G_\varepsilon^R\,(r,y)\right)^* \tag{9.38}$$

Eq. (9.38) can be derived in the same way as Eq. (9.37) substituting in Eq. (9.30) the retarded (advanced) Green functions by the advanced (retarded) functions. Using Eqs. (9.28, 9.37, 9.38) and introducing the dimensionless conductances g_{mn} per one spin as

$$g_{mn} = \frac{h}{2e^2} G_{mn} \tag{9.39}$$

one obtains

$$g_{mn} = -\int n'\,(\varepsilon)\,g_{mn}\,(\varepsilon)\,d\varepsilon, \tag{9.40}$$

$$g_{mn}\,(\varepsilon) = \frac{1}{2}\int_{C_m} dz \int_{C_n} dy\,\hat{B}_m G_\varepsilon^R\,(z,y)\,\hat{B}_n G_\varepsilon^A\,(y,z)$$

where the operator $\hat{B}_m$ acts on an arbitrary function $f\,(z)$ on the surface C_m as

$$\hat{B}_m f\,(z) = \int_{C_m} \left(\mathrm{Re}\, B_m\,(z,u)\right) f\,(u)\,du \tag{9.41}$$

In Eqs. (9.40, 9.41) integration over the variables z and u is performed over the surface C_m whereas integration over y is done over C_n. The integrations are supplemented by summation over spin. In the absence of spin interactions this gives an additional factor of 2.

Thus, we have really gotten rid of the leads. They determine the operators $\hat{B}_m$ but as soon as we know these operators we can solve the Schrödinger equation for the Green functions in the scattering region with the boundary conditions given by Eqs. (9.37, 9.38) and calculate the integrals in Eq. (9.40). For simple cases this program can be executed directly but for the disorder problems our final goal is to derive a nonlinear σ-model. For this purpose it is convenient to take one more simple step. Instead of using the Hamiltonian $\mathcal{H}$ for the scattering region and the rather complicated boundary conditions, Eqs. (9.37, 9.38), one can use an effective Hamiltonian $\mathcal{H}_{\text{eff}}$ defined as

$$\mathcal{H}_{\text{eff}} = \mathcal{H} \mp \frac{i}{2} \sum_{m=1}^{N_L} \hat{B}_m \delta_{C_m} \tag{9.42}$$

where δ_{C_m} is a δ-function with uniform support on the surface C_m such that integrating over the volume of the system one obtains an integral over the surface C_m

$$\int \delta_{C_m} f(r)\, dr = \int_{C_m} f(z)\, dz \tag{9.43}$$

for an arbitrary function $f(r)$. Then the problem of solving Eq. (9.29) with the boundary conditions Eqs. (9.37, 9.38) is replaced by seeking solutions of the equation

$$(\varepsilon - \mathcal{H}_{\text{eff}})\, G_\varepsilon^{R,A}\left(r, r'\right) = \delta\left(r - r'\right) \tag{9.44}$$

with boundary conditions

$$(\mathbf{x}_m \hat{\mathbf{v}})\, G_\varepsilon^{R,A}\left(r, r'\right)|_{C_m+0} = 0 \tag{9.45}$$

where $C_m + 0$ means that the derivatives in Eq. (9.45) should be calculated near the surface C_m from the side of the lead. The minus sign $-$ (or plus sign $+$) in $\mathcal{H}_{\text{eff}}$ should be taken when writing the equation for G_ε^R (G_ε^A). Integrating Eq. (9.44) over the longitudinal coordinate x_m along an infinitesimal piece of a curve intersecting surface C_m and using Eq. (9.45) one arrives at Eqs. (9.29, 9.37, 9.38), which prove Eqs. (9.44, 9.45).

Until now nothing has been said about conditions at boundaries between the quantum dot and the surrounding medium, which can be just the vacuum. Of course, if the confining potential is very sharp, the wave functions vanish at such a boundary. However, if the size of the dot much exceeds the electron wavelength, all physical quantities are not very sensitive to boundary conditions and it is quite reasonable to use the more general condition described by Eq. (9.45) not only at the contacts with leads but everywhere at boundaries of the quantum dot. This condition corresponds to vanishing of the component of the current normal to the boundary, and this is precisely what one would like to have from general arguments.

Only the real part of the function $B_m(y, z)$ is written in $\mathcal{H}_{\text{eff}}$, Eq. (9.42). This term violates the hermiticity of the Hamiltonian and corresponds to an "absorption" of electrons that is the inevitable consequence of the presence of the leads. The absorption term describes the possibility of the electrons' escaping to the leads. The imaginary part of $B_m(z, y)$ only renormalizes the confining potential. This does not lead to interesting effects and will not be considered.

9.2.3 σ-Model for a system with leads

With Eqs. (9.40, 9.42, 9.44) in hand a proper σ-model can be obtained in exactly the same way as in other situations in preceding chapters. The only new difficulty is the presence of the second term in Eq. (9.42), which depends on coordinates and deserves a little more care. One can derive the σ-model either by averaging over disorder inside the conductor or by using the random matrix hypothesis if electrons are scattered mainly by the walls. The latter case corresponds to quantum chaotic billiards, which are widely used for description of ballistic microstructures. The σ-model for closed systems has been derived through both types of averaging in Chapters 4 and 6. Models with impurity disorder are more general because one can obtain not only the zero-dimensional σ-model but the corresponding models in higher dimensions.

As usual, one should write, first, the Green functions in terms of integrals over supervectors ψ, substitute them into Eq. (9.40) for the conductance $g_{mn}(\varepsilon)$, and average over the disorder. Then one decouples quartic terms in the effective Lagrangian by integration over supermatrices Q, which allows one to integrate out supervectors ψ and reduce the computation to an integral over supermatrices Q. The next step is to use the saddle-point approximation, thus fixing the eigenvalues of the supermatrices. The result of these manipulations can be written for the average conductance $\langle g_{mn}(\varepsilon)\rangle$ as

$$\langle g_{mn}(\varepsilon)\rangle = -\int dQ \exp\left(-F[Q]\right) \tag{9.46}$$

$$\times \int_{C_n} dy \hat{B}_n g_{33}^{12}(y, y, Q)\int_{C_m} dz \hat{B}_m g_{33}^{21}(z, z, Q)$$

For simplicity, in Eq. (9.46), spin interactions are neglected and integration over contours C_m, C_n does not include summation over spin. The operators $\hat{B}_n$ and $\hat{B}_m$ act on the first arguments of the functions g_{33}^{12} and g_{33}^{21}, respectively, and one should integrate over the coordinates $y \in C_n$ and $z \in C_m$. The matrix function g obeys the following equation (cf. Eq. (4.42)):

$$\left(\tilde{\mathcal{H}}_0 + \frac{iQ(\mathbf{r})}{2\tau} + \frac{i}{2}\Lambda\sum_{l=1}^{N_L}\hat{B}_l\delta_{C_l}\right)g(\mathbf{r}, \mathbf{r}', Q) = i\delta(\mathbf{r} - \mathbf{r}') \tag{9.47}$$

The supermatrix Q is assumed to correspond to the saddle point and, to express it explicitly, one can use one of the parametrizations considered in preceding chapters. For such supermatrices the free energy functional $F[Q]$ can be written in the form (cf. Eq. (4.45))

$$F[Q] = -\frac{1}{2}\,\text{str}\int\left[\ln\left(-i\tilde{\mathcal{H}}_0 + \frac{Q(\mathbf{r})}{2\tau} + \frac{1}{2}\Lambda\sum_{m=1}^{N_L}\hat{B}_l\delta_{C_l}\right)\right]d\mathbf{r} \tag{9.48}$$

The presence of the new terms does not influence the saddle-point solution because the characteristic energy γ due to the absorption terms is much smaller than τ^{-1}. Eqs. (9.46–9.48) are valid for Q slowly varying in space and make it possible to obtain a σ-model in arbitrary dimension.

For further calculations let us assume that all the leads are separated by distances much larger than the mean free path. This is not a very strong restriction even for ballistic microstructures because electrons traveling from one lead to another may be reflected many times by the walls. In fact, in order to calculate the conductance we need the functions $g(\mathbf{r}, \mathbf{r}', Q)$ with both coordinates located on one of the surfaces C_m, C_n and this simplifies the computation.

Let us carry out calculations for the lead n. Eq. (9.47) can be solved by expansion in a series in the absorption term. This can be written as

$$-i g (y_1, y_2) = -i g_0 (y_1, y_2)$$

$$+ \int_{C_n} g_0 (y_1, y') \frac{1}{2} \Lambda \hat{B}_n g_0 (y', y_2) dy' \tag{9.49}$$

$$-i \int_{C_n} \int_{C_n} g_0 (y_1, y') \frac{1}{2} \Lambda \hat{B}_n g_0 (y', y'') \frac{1}{2} \Lambda \hat{B}_n g_0 (y'', y_2) dy'dy'' + \cdots$$

In Eq. (9.49) all the arguments of the functions g_0 are located on the surface C_n and the integrations are performed over this surface (the argument Q is not written explicitly). Because of the assumption that the distance between the leads much exceeds the mean free path l, only lead n contributes to the expansion. Contributions from the other leads contain the functions $g_0 (\mathbf{r}, \mathbf{r}')$ with $\mathbf{r}$ and $\mathbf{r}'$ located on different leads, but these functions decay exponentially on the mean free path l, and so the contribution from the other leads is exponentially small.

The function $g_0(y, y')$ is the solution of Eq. (9.48) written in the absence of the absorption term. This function must satisfy the same boundary conditions as those in Eq. (9.45) for the exact functions $G_\varepsilon^{R,A}$ and therefore can be written near the surface C_n as

$$g_0 (\mathbf{r}, \mathbf{r}') = \frac{2}{\pi} \sum_p \int_0^\infty \frac{\varphi_p (y) \varphi_p^* (y) \cos k (x - x_n) \cos k (x' - x_n)}{\varepsilon - k^2/2m + i Q/2\tau} dk \tag{9.50}$$

Eq. (9.50) is very similar to Eq. (9.35) and contains the same wave functions $\varphi_p (y)$ of the transversal motion. The longitudinal coordinates x_n denote the position of surface C_n and supermatrix Q should be taken near the surface (supermatrix Q is assumed to vary on distances much exceeding l). To calculate the integrals in Eq. (9.49) one should recall Eq. (9.36, 9.41) for the operators $\hat{B}_n$. Using the orthogonality of the functions $\varphi_p (y)$ entering both functions $g_0(y, y')$ and operator $\hat{B}_n$ one can easily calculate the integrals. Summing up the series obtained we write the function $g (y, y')$ for $y, y' \in C_n$ as

$$g (y, y') = 2 \sum_p v_p^{-1} \varphi_p (y) \varphi_p^* (y') Q (1 + \alpha_p \Lambda Q)^{-1} \tag{9.51}$$

where α_p are coefficients characterizing the lead n. The same equations can be written for the other leads, but, of course, the functions $\varphi_p (y)$, coefficients α_p, and Fermi velocities v_p for channels can be different in different leads.

Similar manipulations can be done for the free energy functional $F [Q]$, Eq. (9.48), which can be written in the form

$$F [Q] = F_d [Q] + F_L [Q] \tag{9.52}$$

where $F_d [Q]$ is the free energy functional for the conductor itself (without leads). One may use any corresponding expressions derived in the previous chapters. In the absence of magnetic and spin-orbit impurities one may take, for example, Eq. (4.65). The term $F_L [Q]$ is due to the leads and is equal to

$$F_L = -\frac{1}{2} \text{str} \sum_{m,p} \ln \left(1 + \alpha_p^{(m)} \Lambda Q_m \right) \tag{9.53}$$

In Eq. (9.53) one has to perform summation both over the channels p in each lead and over all leads m; Q_m denotes supermatrix Q at the contact between the conductor and lead m. If the zero-dimensional version of the σ-model can be used, all Q_m are equal to each other.

Substituting Eq. (9.51) into Eq. (9.46) and using Eq. (9.40) one reduces the average conductance g_{mn} between leads m and n to the form

$$\langle g_{mn} \rangle = - \int \sum_{p,q} \bar{f}_{33}^{12} \left(\alpha_q^{(n)}, \, Q_n \right) \bar{f}_{33}^{21} \left(\alpha_p^{(m)}, \, Q_m \right) \exp \left(- F \left[Q \right] \right) DQ \qquad (9.54)$$

where

$$\bar{f} \left(\alpha, \, Q \right) = 2 \alpha Q \left(1 + \alpha \Lambda Q \right)^{-1} \qquad (9.55)$$

Eqs. (9.52–9.55) were derived for the first time by Iida, Weidenmüller, and Zuk (1990) using a model originally developed in nuclear physics by Nishioka et al. (1986). They started from an S-matrix formulation of the problem and reduced the problem to calculation of Green functions. A randomness was introduced by using the random matrix hypothesis, which was discussed in Chapter 6. The coefficients α_p in their approach are phenomenological parameters.

The derivation given in the present section demonstrates how to derive the same formulae from the Kubo linear response theory. It has an advantage because it enables us, at least in principle, to compute all the parameters of the σ-model from the microscopic Hamiltonians for concrete devices. Besides, one can study the crossover from mesoscopic to macroscopic disordered samples within the same microscopic models. Let us note that different channels in a lead and leads themselves enter Eqs. (9.53, 9.54) on equal footing; this is quite similar to what one can see in the scattering formula Eq. (9.25). With a slight modification one can reduce computation to an integral over supermatrices Q and this can be done not only for average conductances but for variance and even for the entire distribution function. This will be done in the next sections.

For further calculations it is convenient to write Eqs. (9.53–9.55) in a somewhat different form

$$F_L = - \frac{1}{4} \, \mathrm{str} \sum_{p,m} \ln \left(2 - t_p^{(m)} + t_p^{(m)} \Lambda Q_m^{\parallel} \right), \qquad (9.56)$$

$$\langle g_{mn} \rangle = - \int \sum_{p,q} f_{33}^{12} \left(t_q^{(n)}, \, Q_n \right) f_{33}^{21} \left(t_p^{(m)}, \, Q_m \right) \exp \left(- F \left[Q \right] \right) DQ \qquad (9.57)$$

where the new matrix functions $f \left(t, \, Q \right)$ have the form

$$f \left(t, \, Q \right) = t Q^{\perp} \left(2 - t + t \Lambda Q^{\parallel} \right)^{-1} \qquad (9.58)$$

In Eqs. (9.56–9.58) supermatrices $Q^{\parallel}$ and $Q^{\perp}$ are defined as

$$Q = Q^{\parallel} + Q^{\perp}, \qquad Q^{\parallel} \Lambda - \Lambda Q^{\parallel} = 0, \qquad Q^{\perp} \Lambda + \Lambda Q^{\perp} = 0 \qquad (9.59)$$

and the coefficients $t_p^{(m)}$ are related to the coefficients $\alpha_p^{(m)}$ by a simple formula

$$t_p^{(m)} = \frac{4 \alpha_p^{(m)}}{\left(1 + \alpha_p^{(m)} \right)^2} \le 1 \qquad (9.60)$$

The coefficients $t_p^{(m)}$ were called "sticking probabilities" by Iida, Weidenmüller, and Zuk because, in the language of scattering theory, they measure that part of the flux incident in channel p on lead m, which is not reemitted instantaneously into the same channel. The relation of the coefficients $t_p^{(m)}$ to the conductance of the leads is discussed in the next section. We know already that if $\alpha_p^{(m)} = 1$, channel p is open and we see that the corresponding coefficient $t_p^{(m)} = 1$. The coefficients $t_p^{(m)}$ are always smaller than unity, consistently with their treatment as a probability.

9.3 Semiclassical theory of transport

9.3.1 Addition of resistances

Eqs. (9.56–9.60) give a new formulation of transport phenomena in terms of the nonlinear σ-model. The main advantage of this approach with respect to the standard approach based on conventional Kubo linear response theory is that now one can compute conductances for an arbitrary geometry of the scattering region without going into detail about space distribution of the electric field. Besides, one can consider different couplings to the leads and, thus, give a microscopic description of any multiprobe mesoscopic device.

Since Eqs. (9.56–9.60), being more general, are somewhat different from those used in the preceding chapters it is desirable to reproduce basic known results and find their extension to situations when the approach of the preceding chapters cannot be used. As in Chapter 5, it very useful to compute, first, physical quantities in situations when one can use a perturbation theory. In the language of the σ-model this means considering small deviations of supermatrix Q from Λ. Calculations of these type are done in the present section. We know that the perturbation theory works well in a macroscopic sample provided the system is in the regime of a good metal. As soon as localization effects become important one should apply more sophisticated schemes. We know also that in an isolated system perturbation theory does not work even in the limit of weak disorder because energy levels are quantized. What does govern the crossover from a small isolated sample to a macroscopic metal in the absence of inelastic scattering?

The answer is clear: Such a crossover can be due to coupling to leads. It turns out that description based on perturbation theory is possible provided the level width γ due to the coupling to the leads is much larger than the mean level spacing Δ. Intuitively, one can argue that as soon as $\gamma \gg \Delta$ one loses information about the discreteness of the levels and enters the limit of the good metal. This limit can be considered as semiclassical, and perturbation theory, in fact, describes deviations from classical physics.

A semiclassical approach based on a formula expressing the scattering matrix in terms of an integral over all classical trajectories has been developed for studying problems of chaotic motion in ballistic cavities (for a review, see Smilansky (1991)). The region of the applicability of the latter approach is also $\gamma \gg \Delta$, and there is evidence that results obtained with this approach are confirmed by perturbation theory starting from the σ-model. However, for the supersymmetry technique it is not important whether chaos is due to nontrivial shape of the cavity or to impurities inside the system. Therefore, from the point of view of applications of the nonlinear σ-model, it seems to be quite reasonable to use the term *semiclassical theory* in a broader sense to indicate a theory based on perturbation theory. In this sense all weak localization effects and universal conductance fluctuations are obtained within the semiclassical theory.

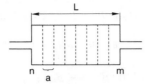

Fig. 9.3. A conductor with a one-dimensional geometry connected to two leads.

In spite of a limited region of applicability of semiclassical theory one can obtain very interesting effects because there is one more energy scale $\mathcal{E}_c$, which is usually much larger than Δ. These effects can be observed experimentally, so studying the semiclassical limit can be important in itself and yield new physics insights.

As the first step, let us obtain the average conductance in the zero order of the perturbation theory. It is convenient to use, as in Chapter 5, parametrization for supermatrices Q given by Eq. (5.1). The Jacobian for this parametrization is equal to unity, thus simplifying manipulations. Let us assume that the disordered conductor has a one-dimensional geometry like that in Fig. 9.3 and consider a two-probe experiment. Two leads in Fig. 9.3 serve as a connection to source and drain reservoirs and may contain arbitrary numbers of conducting channels. Subsituting Eq. (5.1) into Eqs. (9.52, 9.56–9.58) and making an expansion in B in the quadratic approximation we obtain

$$F_2 = \pi \nu S D_0 \int \mathrm{str} \left(\frac{dB}{dx} \frac{d\bar{B}}{dx} \right) dx + \frac{1}{2} \sum_{l=m,n} \mathrm{str} \left(\bar{\Gamma}_l \bar{B}_l B_l \right) \qquad (9.61)$$

where S is the cross section of the conductor and

$$\bar{\Gamma}_m = \sum_p t_p^{(m)}, \qquad \bar{\Gamma}_n = \sum_q t_q^{(n)} \qquad (9.62)$$

In Eq. (9.61), only the zero harmonics of the transversal quantization is taken into consideration and therefore the σ-model in Eq. (9.61) is one-dimensional. The conductance g_{mn} is written in the same approximation as

$$\langle g_{mn} \rangle = - \int \bar{\Gamma}_m \bar{\Gamma}_n \left(B_{33} \right)_n \left(\bar{B}_{33} \right)_m \exp \left(-F_2 \right) DB D\bar{B} \qquad (9.63)$$

where the subscripts m and n indicate that the elements B_{33} and $\bar{B}_{33}$ should be taken at the points where the leads are attached to the conductor.

If the σ-model were zero-dimensional, we would be able to write the result of the integration immediately. However, it is interesting to describe the crossover to the classical conductivity, so let us assume that the one-dimensional case is studied and that the distance between the leads is equal to L. The functional integral in Eq. (9.63) can be calculated as usual by subdividing the conductor into small pieces with length a (see Fig. 9.3) such that $L = aK$, where K is an integer. Then, one has to replace the gradients in the first term in Eq. (9.61) by finite differences and the integral by a sum over all sites. As a result, the first term F_2' is written in the form

$$F_2' = \sum_{i=1}^{K-1} F_2^{(i)} + F_2^{(L)} + F_2^{(R)}, \qquad (9.64)$$

$$F_2^{(i)} = M \operatorname{str} (B_i - B_{i+1}) \left(\bar{B}_i - \bar{B}_{i+1} \right),$$

$$F_2^{(L)} = M \operatorname{str} (B_n - B_1) \left(\bar{B}_n - \bar{B}_1 \right), \qquad (9.65)$$

$$F_2^{(R)} = M \operatorname{str} (B_{K-1} - B_m) \left(\bar{B}_{K-1} - \bar{B}_m \right)$$

where

$$M = \frac{\pi \nu D_0 S}{a} \qquad (9.66)$$

In Eqs. (9.65), $B_{n,m}$ represents the supermatrices at the left and right points of the conductor. These are the matrices entering Eq. (9.63). It is natural to integrate, first, over all the matrices except B_m and B_n, to obtain a function $\Pi (B_n, B_m, K)$. In terms of this function, Eq. (9.63) can be rewritten in the form

$$\langle g_{mn} \rangle = -\bar{\Gamma}_m \bar{\Gamma}_n \int (B_{33})_n \left(\bar{B}_{33} \right)_m \Pi (B_n, B_m, K) \qquad (9.67)$$

$$\times \exp \left(-\frac{1}{2} \operatorname{str} \sum_{l=n,m} \bar{\Gamma}_l \bar{B}_l B_l \right) d B_n d \bar{B}_n d B_m d \bar{B}_m$$

When integrating over all supermatrices B_i, $\bar{B}_i$, $1 \le i \le K - 1$, let us make, as the first step, the shift

$$B_i \to B_i + B_n$$

After this shift we have the same expression for $F_2^{(i)}$ in Eqs. (9.65) whereas $F_2^{(L)}$ and $F_2^{(R)}$ take the form

$$F_2^{(L)} = M \operatorname{str} \left(B_1 \bar{B}_1 \right),$$

$$F_2^{(R)} = M \operatorname{str} (B_{K-1} + B_n - B_m) \left(\bar{B}_{K-1} + \bar{B}_n - \bar{B}_m \right) \qquad (9.68)$$

Now, we integrate over supermatrices B_1 and have

$$\int \exp \left(-F_2^{(L)} - F_2^{(1)} \right) d B_1 d \bar{B}_1 = \exp \left(-\frac{M}{2} \operatorname{str} \left(\bar{B}_2 B_2 \right) \right)$$

The next step is to integrate over supermatrices B_2

$$\int \exp \left(-\frac{M}{2} \operatorname{str} \left(\bar{B}_2 B_2 \right) - F_2^{(2)} \right) d B d \bar{B} = \exp \left(-\frac{M}{3} \operatorname{str} \left(\bar{B}_3 B_3 \right) \right)$$

Continuing the procedure we can reach the site $K - 1$. It is clear that integration over all the sites located to the left of the site $K - 1$ gives

$$\exp \left(-\frac{M}{K-1} \operatorname{str} \left(\bar{B}_{K-1} B_{K-1} \right) \right) \qquad (9.69)$$

As the last step, one should integrate $\exp \left(-F_2^{(R)} \right)$ with $F_2^{(R)}$ from Eq. (9.68), multiplied by the exponential from Eq. (9.69) over B_{K-1}. The result of the last integration is the function $\Pi (B_n, B_m, K)$, which takes the form

$$\Pi (B_n, B_m, K) = \exp \left(-\frac{M}{K} \operatorname{str} \left((\bar{B}_n + \bar{B}_m) (B_n + B_m) \right) \right) \qquad (9.70)$$

Let us notice that the function $\Pi\left(B_n, B_m, K\right)$ has a finite limit as $K \rightarrow \infty$ (keeping the sample length L finite); that means that the functional integral is properly computed. Substituting Eq. (9.70) into Eq. (9.67), integrating over the remaining supermatrices B_n and B_m, and using Eq. (9.66) we reach the final result

$$\langle g_{mn} \rangle = \frac{g_0 \bar{\Gamma}_m \bar{\Gamma}_n}{\bar{\Gamma}_m \bar{\Gamma}_n + g_0 \left(\bar{\Gamma}_m + \bar{\Gamma}_n\right)} = \left(g_0^{-1} + \bar{\Gamma}_m^{-1} + \bar{\Gamma}_n^{-1}\right)^{-1} \tag{9.71}$$

where

$$g_0 = \frac{2\pi \nu D_0 S}{L} = \frac{2\mathcal{E}_c}{\pi \Delta} \tag{9.72}$$

is the classical conductance in the units $2e^2/h$.

Eq. (9.71) is nothing but the classical formula for addition of resistances connected in series; that implies that $\bar{\Gamma}_m$ and $\bar{\Gamma}_n$ are conductances of the contacts. According to Eq. (9.62) these conductances are the sums of the coefficients $t_{m,n}^{(p)}$. Therefore, these coefficients can be thought of as channel conductances. Then, one may represent the channels as resistors connected in parallel and in series to the disordered area. Although Eq. (9.71) is classical, the dependence of the conductance on the sample length is, because of the presence of the contacts, unusual. One obtains the standard proportionality of the resistance to the sample length only in the limit

$$g_0 \ll \bar{\Gamma}_m, \bar{\Gamma}_n \tag{9.73}$$

when

$$\langle g_{mn} \rangle \simeq g_0 \tag{9.74}$$

In the opposite limit

$$g_0 \gg \bar{\Gamma}_m, \bar{\Gamma}_n \tag{9.75}$$

the main contribution to the total resistance is from the resistances of the contacts; one obtains

$$\langle g_{mn} \rangle = \frac{\bar{\Gamma}_m \bar{\Gamma}_n}{\bar{\Gamma}_m + \bar{\Gamma}_n} \tag{9.76}$$

This is the analogue of what is called the *Hauser–Feshbach formula* in the statistical theory of nuclear reactions (Hauser and Feshbach (1952)). Eq. (9.76) can be obtained immediately from the zero-dimensional version of the σ-model, Eq. (9.61), whereas to obtain the conventional classical conductance, one has to use the σ-model in higher dimensions. The criterion for the applicability of the zero-dimensional σ-model can also be written in terms of parameters Γ_m, Γ_n

$$\Gamma_{m,n} = \bar{\Gamma}_{m,n} \Delta \tag{9.77}$$

characterizing the energy level width. Using Eqs. (9.72, 9.75, 9.77) one obtains the following inequality:

$$\Gamma_{m,n} \ll \mathcal{E}_c \tag{9.78}$$

That means that the level width must be much less than Thouless energy $\mathcal{E}_c$.

The semiclassical approximation used in this section implies small characteristic values of supermatrices B. It can be extracted from Eq. (9.61) that these fluctuations are small provided

$$\langle g_{mn} \rangle \gg 1 \tag{9.79}$$

The inequality (9.79) is violated if either $\bar{\Gamma}_m^{-1} + \bar{\Gamma}_n^{-1} \geq 1$ or $g_0 \leq 1$. In the former case the discreteness of the energy levels becomes important and one should work with the zero-dimensional σ-model exactly. Some problems of this type are considered in Sections 9.5 and 9.7. In the latter situation one comes to the limit of Anderson localization that has to be studied within the σ-model in higher dimensions. Localization problems are considered in Chapters 11 and 12.

9.3.2 Weak localization effects

Eq. (9.71) shows how the classical Drude formula for conductance changes with decrease in the size of the sample. This formula is not very sensitive to varying such experimentally tunable parameters as, for example, a magnetic field, and therefore it is not easy to check it experimentally. However, as we have learned from Chapter 5, the first quantum correction to the classical conductivity is very sensitive to the magnetic field that has been observed in numerous experiments. Recently, it has been discovered that the conductance of ballistic quantum dots also changes when a weak magnetic field is applied (Marcus et al. (1993), Berry et al. (1994a, b)). This shows that the weak localization effects survive even in the ballistic microstructures in the regime of chaotic motion. So, it is very important to have a formula describing the weak localization effects in the situation when the conductance due to leads is comparable with the conductance of the conductor itself.

To calculate the first quantum correction one can follow a procedure similar to the one described in Section 5.1, where the correction was calculated for a large sample. First, using the parametrization, Eq. (5.1), one should make an expansion in P in the free energy F, Eq. (9.52), and in the preexponential, Eqs. (9.57, 9.58). The higher-order terms arising when making the expansion in the bulk part F_d of the free energy are given in Section 5.1. It is important to note that in the computation in the first order in Section 5.1, the quartic terms with gradients did not contribute. The only averages that were allowed by the symmetry of supermatrices B to be nonzero had the form $\langle \nabla B M B \rangle$ or $\langle \nabla \bar{B} M \bar{B} \rangle$. These terms had to be zero in the bulk, because they changed sign under reversal of the coordinates. However, in the formulation of the present chapter, such terms can contribute because the surface has to be taken into account explicitly. Quartic terms F_{L4} arising in F_L can be written as

$$F_{L4} = -\frac{1}{2} \mathrm{str} \left[\sum_{l=m,n} \sum_p \left(t_p^{(l)} - \frac{1}{2} \left(t_p^{(l)} \right)^2 \right) \left(B_l \bar{B}_l \right)^2 \right] \tag{9.80}$$

whereas the functions f entering the preexponential in Eq. (9.57) are equal to

$$f = \frac{it \Lambda P}{1 + (1 - t) P^2} \tag{9.81}$$

Substituting the first nonlinear terms into Eq. (9.57) one can compute the correction for an arbitrary relation between $\bar{\Gamma}$ and g_0. The result, as usual, depends on the ensemble. In the unitary ensemble, the first-order correction is zero. In the orthogonal ensemble, a general formula can be found in Iida, Weidenmüller, and Zuk (1990). Because it is rather cumbersome, it is more relevant to present here its particular form for all $t_m^{(p)} = 1$ and $\bar{\Gamma}_m = \bar{\Gamma}_n = \bar{\Gamma}$. In this limit, the parametrization Eq. (5.1) for supermatrices Q is especially convenient because not only is the Jacobian of the transformation unity but the function f, Eq. (9.81), is simply proportional to P. The final result for the first weak localization correction δg_{mn} can be written as

$$\delta g_{mn} = -\frac{1}{3}\left(1 + \sum_{n=1}^{3}\frac{c_n}{\left(2 + \bar{\Gamma}/g_0\right)^n}\right) \tag{9.82}$$

where $c_1 = -3/8$, $c_2 = 25/32$, $c_3 = -33/16$. We see that the first quantum correction in the case under consideration is a function of $\bar{\Gamma}/g_0$ only. In the Ohmic regime, $\bar{\Gamma} \gg g_0$, the second term in Eq. (9.82) is small and one obtains $\delta g_{mn} = -1/3$. This value also can be obtained from Eq. (5.14) taking the limit $\omega \to 0$, and substituting the integral by a sum over momenta $k_n = n\pi/L$, $n = 1, 2, 3, \ldots$, which follows from the boundary conditions $Q = \Lambda$ at the leads. In the opposite limit, $\bar{\Gamma} \ll g_0$, the second term in Eq. (9.82) is a number that does not depend on any parameters and one obtains $\delta g_{mn} = -1/4$. Thus, the well-known effects of weak localization exist not only in disordered bulk metals but also in ballistic microstructures.

As has been mentioned, because the weak localization correction is zero in the unitary case, a magnetic field leads to an increase in the conductance. In principle, one can derive a formula for the dependence of the conductance on the magnetic field for arbitrary $\bar{\Gamma}/g_0$. However, most interesting are the limiting cases. For the limit of a bulk disordered conductor, the result has been written in Chapter 5, Eq. (5.23). To write a corresponding formula for the quantum billiards one should use as F_d in Eq. (9.52) the zero-dimensional σ-model in the presence of the magnetic field, Eq. (8.42). Again, using the parametrization, Eq. (5.1), and writing out the free energy up to quartic terms, one easily finds the dependence of the first quantum correction δg on the magnetic field

$$\delta g = -\frac{1}{4}\frac{1}{1 + 4X^2/N} \tag{9.83}$$

where X is determined by Eq. (8.43). Eq. (9.83) is written for the case when all $t_m^{(p)} = 1$ and $\bar{\Gamma}_m = \bar{\Gamma}_n = \bar{\Gamma} = N$, where N is the number of channels in each lead. The validity of the semiclassical approximation can also be written as $N \gg 1$. The coefficient of proportionality between the parameter X and the magnetic field depends only on the geometry of the scattering area. So Eq. (9.83) shows that in changing the magnetic field proportionally to changing the square root of the number of the channels, one does not change the quantum correction δg. This property is convenient for an experimental test. A Lorentzian dependence of the weak localization correction on the magnetic field for a nonintegrable ballistic cavity has been observed by Chang et al. (1994), and this is consistent with Eq. (9.83). In contrast, a nonchaotic cavity showed linear dependence.

Recently, the dependence of the weak localization correction on the magnetic field has been calculated for the quantum dots exactly (without using the expansion in the diffusion modes) (Pluhar, Weidenmüller, and Zuk (1994)). The result can be expressed in terms of an integral over the variables λ, λ_1, and λ_2 (cf. Section 6.3). Numerical evaluation of the integral shows that even for $N = 1$ the difference between the exact result and the Lorentzian, Eq. (9.83), is within 1 percent, which is much smaller than experimental errors.

Transport quantities cannot be easily expressed in terms of random Hamiltonians. Therefore, the Wigner–Dyson random matrix theory discussed in Chapter 6 cannot be applied
directly to calculating conductances. It was the supersymmetry technique that helped to
compute the transport quantities starting from this type of the random matrix hypothesis.
Another random matrix hypothesis for mesoscopic transport through ballistic cavities was
proposed in works by Blümel and Smilansky (1988, 1989, 1990). These authors conjectured that chaotic scattering in the quantum regime had to be described by a random matrix
theory for the scattering matrix S. The emphasis in those works is on the properties of the
eigenphases of S, which also are not directly connected to transport properties because they
involve both reflection and transmission.

Very recently it was shown by Baranger and Mello (1994) and by Jalabert, Pichard, and
Beenakker (1994) how one could calculate transport properties using this type of random
matrix theory. They started expressing the S-matrix for scattering involving two leads each
with N channels as

$$S = \begin{pmatrix} r & t' \\ t & r' \end{pmatrix} \tag{9.84}$$

where r, t are $N \times N$ reflection and transmission matrices for particles from the left and
r' and t' for those from the right. The conductance can be written in terms of the S-
matrix using Eq. (9.25). Because of current conservation the matrix S is unitary and one
may try to describe the system assuming that the matrix should be random and described
in terms of physical symmetry by one of the Dyson circular ensembles. This hypothesis
makes it possible to calculate average conductance as well as its moments. In particular,
those authors obtained $\delta g = -1/4$ for the difference between the conductances for the
orthogonal and unitary ensembles. This value agrees with the weak localization correction,
Eq. (9.83). Results for the conductance variance obtained in the next sections can also serve
as a demonstration that the random matrix theory for the scattering matrix is reasonable,
at least, for situations when the channels are open and the numbers of the channels in the
leads are equal.

Recently, the equivalence between the random matrix theory for the scattering matrix and
that for Hamiltonians of a system with leads was proved by using more general arguments
(Brouwer (1995)). But we know already that the supermatrix σ-model can be derived from
models with random Hamiltonians. Therefore, the agreement of the results obtained in
the next sections with the corresponding results obtained by random matrix theory for the
scattering matrix is no surprise.

9.4 Conductance fluctuations in the semiclassical limit

9.4.1 Reduction of correlation functions of conductances to integrals over supermatrices

In the preceding section, it was demonstrated how one can obtain averaged physical quantities in mesoscopic devices in the whole region of parameters in the semiclassical limit
when one may make an expansion in the diffusion modes. The final formulae were derived
and written in a unified manner valid in disordered conductors as well as in ballistic cavities
in the regime of chaotic dynamics. The same can be done for conductance fluctuations, and
one can get formulae for conductance variance and higher moments in a closed form. In

many interesting cases it is possible to calculate the conductance distribution function and even the distribution function of wave functions.

Recently, there has been great interest in the effects of quantum-mechanical interference on electronic transport through ballistic quantum dots, and therefore I want to restrict the analysis in the rest of the chapter to the case when the resistance of the dot itself is much smaller than the resistance due to the leads. In this limit one may use the zero-dimensional σ-model, which simplifies all calculations. Besides, if the contact between the leads and the dot is good (potential barriers are absent) and the number of channels in the leads is large, one can use perturbation theory in the diffusion modes. In fact, the semiclassical approximation is quite good even if each lead has only one channel and to get formulae that can be relevant for experiments like those in Marcus et al. (1992, 1993), Berry et al. (1994a, b), Keller et al. (1994) one can make the expansion in the diffusion modes.

It is important to note that these experiments are made on very clean samples and electrons are scattered mainly by the boundaries of the dot. So one is dealing here with quantum chaos. However, we understand from the discussion of Chapter 6 that this case should be described by the zero-dimensional σ-model as well. Another type of experiment on quantum chaos is based on the study of microwave scattering in cavities (Stöckmann and Stein (1990), Doron, Smilansky, and Frenkel (1990), Sridhar (1991)).

Using the zero-dimensional σ-model one can calculate many different quantities characterizing the conductance fluctuations, but most interesting are those that can be compared with experiments. A very important correlation function often studied experimentally is the correlation function of conductances at different magnetic fields $B(H^+, H^-)$ defined in Eqs. (9.1, 9.12, 9.20, 9.21). This function gives information about the value of the fluctuations as well as their correlations at different magnetic fields. Using Eqs. (9.40–9.43) one can reduce the problem of calculation of the correlation function $B(H^+, H^-)$ to an integral over supermatrices with a σ-model by using a procedure similar to that of the previous section.

However, to be able to write the correlation function at different magnetic fields in a form of integral over supervectors and then over supermatrices one has to double the number of components of the supervectors as well as the size of the supermatrices. Generally speaking it makes explicit calculations with this new σ-model more difficult but in the semiclassical limit, when one has to calculate only Gaussian integrals, these difficulties are not essential. One needs only to compute several Gaussian integrals for integration over supermatrices once and then use these values everywhere by applying the Wick theorem.

Here, let us calculate the correlation function $B(H^+, H^-)$ for the unitary ensemble, which implies that the field H_0 is large enough. Experimentally, already rather weak magnetic fields correspond to the unitary ensemble and so, as the easiest for calculations, the unitary ensemble is very suitable for describing the experiments. To obtain the correlation function $B(H^+, H^-)$ at arbitrary temperature one can use Eqs. (9.20, 9.21, 9.12) and calculate, first, the zero-temperature correlation function $\tilde{B}(H^+, H^-, \omega)$ at different magnetic fields and energies. In the preceding chapters, manipulating with the σ-model we used the effectively 4×4 supermatrices Q, whereas now we come to 8×8 supermatrices.

Later, a model with three leads is considered. Two of them (1 and 2) are the source and drain leads analogous to those considered in the preceding sections and the conductance is measured between these leads. The additional lead can serve as a voltage probe. A basic notion of mesoscopic physics is that the measurement of a voltage at some point in the sample is an invasive act that may destroy the phase coherence throughout the whole

sample. Therefore, to describe the measuring process properly the voltage probe should be
included in the model. The mechanism by which the measurement of the voltage destroys
the phase coherence is that electrons entering the voltage lead are reinjected into the system
without any phase relationship.

To simplify formulae let us assume that the channel conductances obey the relation
$t_m^{(1,2)} = t_{1,2}$, for all m; that means that all the conductances in each channel are equal. Then,
the free energy F describing the correlation function $\tilde{B}(H^+, H^-, \omega)$ can be written as

$$F = -\frac{1}{2} \sum_{l=1}^{2} N_l \operatorname{str} \ln \left(2 - t_l + t_l \Lambda Q^{\|}\right)$$

$$+ \operatorname{str}\left(-\frac{\alpha}{4} \Lambda Q + \frac{ix\tau_3}{2} Q - \frac{X^2}{16} [Q, \tau_3]^2\right) \qquad (9.85)$$

In Eq. (9.85) Q is an 8×8 supermatrix with the constraint $Q^2 = 1$, Λ, and τ_3 are diagonal
supermatrices. In fact, they are the same used in all previous cases. In supermatrix Λ the
first four elements at the diagonal are equal to 1 and the rest to -1; in the supermatrix τ_3
the elements with odd subscripts are 1 and those with even ones are -1. The matrix Λ
originates as everywhere before from the difference between G^R and G^A, and matrix τ_3
is due to the different magnetic fields and frequencies in the two different conductances.
Let us emphasize that considering the unitary case there are no cooperons, so the meaning
of the matrix τ_3 used in this section is different from that used when calculating averaged
quantities for the orthogonal ensemble. Supermatrix $Q^{\|}$ is the part of Q commuting with
Λ, but the structure of supermatrix Q is different from that used for the 8×8 supermatrices
for the orthogonal ensemble. Supermatrix Q has the following symmetry:

$$\bar{Q} \equiv K Q^+ K = Q \qquad (9.86)$$

where the superscript $+$ stands for the Hermitian conjugation and the supermatrix K is the
same as in Eq. (4.48) (let us emphasize that $\bar{Q}$ is *defined* in Eq. (9.86) and nothing like
Eq. (4.35) is implied).

The first term in Eq. (9.85) is analogous to Eq. (9.56) describing the source and drain
leads. The term with α stands for the third lead. It has a form different from that of the
first term. Such a form for a contact of a metallic sample with a surrounding medium has
been used in Section 7.3, where it was derived from a Hamiltonian with all levels having
the same width. The term with α in Eq. (9.85) can also be obtained from an expression
analogous to the first term in the limit of a large number of channels when the logarithm
may be expanded. At the same time the channel conductances t may be small and so the
value of α can be arbitrary. Thus, the form of this term corresponds to an extended contact
that can be quite relevant for describing experiments.

The coefficients x and X are related to the frequency ω and the vector-potential $\mathbf{A}$ corresponding to the difference of the magnetic fields H as

$$x = \frac{\pi\omega}{\Delta}, \qquad X^2 = \pi\nu D_0 \left(\frac{e}{c}\right)^2 \int \mathbf{A}^2 d\mathbf{r} \qquad (9.87)$$

Using Eq. (9.40) for the nonaveraged conductance and introducing functions $f(t, Q)$
in the same way as in Eq. (9.58) one can write a formula for the correlation function
$\tilde{B}(H^+, H^-, \omega)$. Now, in contrast to Eq. (9.57), one obtains products of four functions

$f(t, Q)$ and the result can be written as

$$\tilde{B}\left(H^+, H^-, \omega\right) = \tilde{B}'(X, x) - \tilde{B}_0, \tag{9.88}$$

$$\tilde{B}'(X, x) = \int [\tilde{f}_{12}^{21}(t_1) \tilde{f}_{21}^{21}(t_1) \tilde{f}_{12}^{12}(t_2) \tilde{f}_{21}^{12}(t_2) N_1 N_2 \tag{9.89}$$

$$+ \sum_{i \neq k} \tilde{f}_{11}^{21}(t_i) \tilde{f}_{22}^{21}(t_i) \tilde{f}_{12}^{12}(t_k) \tilde{f}_{21}^{12}(t_k) N_i^2 N_k$$

$$+ \tilde{f}_{11}^{21}(t_1) \tilde{f}_{22}^{21}(t_1) \tilde{f}_{11}^{12}(t_2) \tilde{f}_{22}^{12}(t_2) N_1^2 N_2^2] \exp\left(-F[Q]\right) dQ,$$

$$\tilde{B}_0 = N_1^2 N_2^2 \left(\int \tilde{f}_{11}^{21}(t_1) \tilde{f}_{11}^{12}(t_2) \exp\left(-\bar{F}[Q]\right) dQ \right)^2 \tag{9.90}$$

where $\bar{F}$ does not include the terms with x and X in Eq. (9.85) (However, one may use the whole function F instead of $\bar{F}$. The results are the same anyway.) The functions $\tilde{f}_{rs}^{ik}(t)$ are expressed through the elements of the supermatrices $f(t, Q)$ as

$$\tilde{f}_{rs}^{ik}(t) = \left(f_{22}^{ik}(t, Q)\right)_{rs} \tag{9.91}$$

In Eqs. (9.88, 9.91) the superscripts denote as everywhere before the blocks that originate from the difference between the retarded and advanced Green functions, the subscript 22 in the functions f stands for the boson–boson block, and the subscripts r and s in the functions $\tilde{f}$ originate from the fact that two different conductances are considered. The quantity $\tilde{B}_0$ is the average conductance.

9.4.2 Temperature effects

Eqs. (9.85–9.91) give a general expression of the correlation function $\tilde{B}(H^+, H^-, \omega)$ for the unitary ensemble in terms of an integral over the 8×8 supermatrices Q. In principle, exact evaluation of this integral is possible, but here only the semiclassical limit $\bar{\Gamma}_{1,2} \sim N_{1,2} t_{1,2} \gg 1$ is considered. In this limit, using the parametrization Eq. (5.1), one may make an expansion in the supermatrices P. At first glance, it is sufficient in the limit $\bar{\Gamma}_{1,2} \ll \mathcal{E}_c$ just to repeat the calculations presented in Section 9.1 considering a two-diffuson contribution. Calculations in this approximation could be done by using Eqs. (9.85–9.91) by keeping the quadratic in the P part in the free energy F. As concerns the functions f one might retain the linear term in P only. In this approximation, only the first term in Eq. (9.89) contributes and one obtains just the two-diffuson contribution. Such a procedure is perfectly correct in the limit $\bar{\Gamma}_{1,2} \gg \mathcal{E}_c$ (strictly speaking, Eqs. (9.85, 9.89)) are written for the zero-dimensional case, but a generalization to higher dimensionality is straightforward) and higher-order terms as well as the other terms in Eq. (9.89) give a small contribution in the parameter $\mathcal{E}_c/\bar{\Gamma}_{1,2}$.

The calculation in the opposite limit $\bar{\Gamma}_{1,2} \ll \mathcal{E}_c$ is more difficult because the small parameter of the expansion at $t_{1,2} \sim 1$ is $1/N_{1,2}$. At the same time, we see that the terms in the second and third lines in Eq. (9.89) contain large factors $N_1^2 N_2$, $N_1 N_2^2$ or $N_1^2 N_2^2$, and, therefore, nonquadratic terms in the expansion of the energy F in P can have an important contribution. To see explicitly how many terms should be taken into account let us write the expansion of the free energy F up to the sixth order in P. It is convenient to express supermatrix P as

$$P = P_\parallel + P_\perp, \qquad P_\parallel \tau_3 - \tau_3 P_\parallel = 0, \qquad P_\perp \tau_3 + \tau_3 P_\perp = 0 \qquad (9.92)$$

The free energy F in this approximation takes the form

$$F = F_2 + F_4 + F_6, \qquad (9.93)$$

$$F_2 = \mathrm{str}\left[\tilde{a}_\parallel P_\parallel^2 + \tilde{a}_\perp P_\perp^2\right] \qquad (9.94)$$

$$F_4 = -\mathrm{str}[\frac{\tilde{b}}{2} P_\parallel^4 + \left(\frac{\tilde{b}}{2} + 2X^2 - ix\tau_3\Lambda\right) P_\perp^4 \qquad (9.95)$$

$$+2\left(\tilde{b} + X^2 - ix\tau_3\Lambda\right) P_\parallel^2 P_\perp^2 + \left(\tilde{b} - ix\tau_3\Lambda\right)\left(P_\perp P_\parallel\right)^2]$$

$$F_6 = \mathrm{str}[\frac{\tilde{c}}{2} P_\parallel^6 + \left(\frac{\tilde{c}}{2} + 3X^2 - ix\tau_3\Lambda\right) P_\perp^6 \qquad (9.96)$$

$$+ \left(3\tilde{c} + 2X^2 - 2ix\tau_3\Lambda\right) P_\parallel^4 P_\perp^2 + \left(3\tilde{c} + 8X^2 - 4ix\tau_3\Lambda\right) P_\parallel^2 P_\perp^4$$

$$+ 3\tilde{c} P_\parallel^3 P_\perp P_\parallel P_\perp + \left(3\tilde{c} + 2X^2 - 2ix\tau_3\Lambda\right) P_\perp^3 P_\parallel P_\perp P_\parallel$$

$$+ \left(\frac{3}{2}\tilde{c} - 2ix\tau_3\Lambda\right) P_\perp^2 P_\parallel P_\perp^2 P_\parallel + \left(\frac{3}{2}\tilde{c} + X^2 - ix\tau_3\Lambda\right) P_\perp P_\parallel^2 P_\perp P_\parallel^2]$$

The parameters $\tilde{a}_\parallel$, $\tilde{a}_\perp$, $\tilde{b}$, and $\tilde{c}$ entering Eqs. (9.94–9.96) equal

$$\tilde{a}_\parallel = \frac{1}{2}\sum_{i=1}^{2} N_i t_i, \qquad \tilde{a}_\perp = \frac{1}{2}\sum_{i=1}^{2} N_i t_i + X^2 - ix\tau_3\Lambda, \qquad (9.97)$$

$$\tilde{b} = \sum_{i=1}^{2} N_i t_i\left(1 - \frac{t_i}{2}\right), \qquad \tilde{c} = \sum_{i=1}^{2} N_i t_i\left(1 - t_i + \frac{t_i^2}{3}\right)$$

Considering the quadratic form F_2, Eq. (9.96), as the bare part of the energy function, one can expand $\exp\left(-F_4 - F_6\right)$ in F_4 and F_6. The first nonvanishing contribution to the correlation function of conductances is of the order of N^0, and to take into account all terms of this order one should retain F_4^2 and F_6 in the expansion of the exponential. The corresponding expansions should also be done in the functions f. The manipulations, although very lengthy, are straightforward. One can compute Gaussian integrals by using the Wick theorem. Then one needs to know integrals of different pairs of the supermatrices P. These supermatrices anticommute with Λ and have symmetry $P = \bar{P}$ (cf. Eq. (9.86)). For arbitrary supermatrices $M = \bar{M}$ and $N = \bar{N}$ of the form

$$M = \begin{pmatrix} 0 & M^{12} \\ M^{21} & 0 \end{pmatrix}, \qquad N = \begin{pmatrix} N^{11} & 0 \\ 0 & N^{22} \end{pmatrix}$$

one can derive the following formulae:

$$\langle P_s M P_s\rangle_2 = 0, \qquad \langle \mathrm{str}\left(P_s M_1\right) \mathrm{str}\left(P_s M_2\right)\rangle_2 = \mathrm{str}\left(\frac{1}{2\tilde{a}_s} M_{1s} M_{2s}\right) \qquad (9.98)$$

$$\langle P_s N P_s\rangle_2 = \frac{1}{8\tilde{a}_s}\left(\mathrm{str}\, N \pm \tau_3\, \mathrm{str}\,\left(\tau_3 N\right) - \Lambda\, \mathrm{str}\,\left(\Lambda N\right) \mp \tau_3\Lambda\, \mathrm{str}\,\left(\tau_3\Lambda N\right)\right) \qquad (9.99)$$

where $\langle\ldots\rangle_2$ stands for the averaging with the quadratic part F_2 of the free energy, Eq. (9.94), and $s = \parallel, \perp$. In Eq. (9.99), the upper signs should be chosen for $s = \parallel$, the lower ones for $s = \perp$.

When writing out the expansion it is more convenient to do it from the beginning for the function $\tilde{B}\left(H^+, H^-, \omega\right)$ and not for the functions $\tilde{B}'(X, x)$ and B_0 separately. All terms containing only $P_\parallel$ cancel each other, corresponding to the cancellation of disconnected diagrams in the language of conventional diagrammatic expansions. It clear from Eqs. (9.98, 9.99) why one should expand the free energy up to terms of the sixth order. For example, the last term in Eq. (9.89) multiplied by F_6 contains a product of ten supermatrices P. At the same time, this product is multiplied by a factor N^5. Each pairing using Eqs. (9.98, 9.99) is of the order of N^{-1} and thus, one obtains a result of the order N^0.

The final result for the correlation function $\tilde{B}\left(H^+, H^-, \omega\right)$ can be found for arbitrary N_1, N_2, t_1, t_2 and is written out in Efetov (1995). It is very cumbersome and I will not write it out here. Experimentally, it became possible to make very good contacts between the quantum dots and the leads (Marcus et al. (1992, 1993), Berry et al. (1994a, b), Keller et al. (1994)) such that the channel conductances t_m are unities. One can also change the number of open channels in each lead and make them, for example, equal. The third lead is not necessarily present. In such a situation ($N_1 = N_2 = N$, $t_1 = t_2 = 1$, $\alpha = 0$) the final formula for the correlation function $\tilde{B}\left(H^+, H^-, \omega\right)$ drastically simplifies and can be written as

$$\tilde{B}\left(H^+, H^-, \omega\right) = \frac{1}{16} \frac{1}{\left(1 + X^2/N\right)^2 + (x/N)^2} \qquad (9.100)$$

At $X = 0$, $x \neq 0$, and at $x = 0$, $X \neq 0$, the function $\tilde{B}$, Eq. (9.100), agrees with corresponding formulae obtained from a consideration of path integrals (see Smilansky (1991) and Jalabert, Baranger, and Stone (1990)). In recent works by Baranger and Mello (1994) and by Jalabert, Pichard, and Beenakker (1994) the variance of the conductance Var (g) was calculated by using the phenomenological approach based on random scattering matrices (see also Section 9.3). For the unitary ensemble in the large N limit the result was Var $(g) = 1/16$. The variance can be obtained from the function $\tilde{B}\left(H^+, H^-, \omega\right)$ putting $X = 0$ and $x = 0$ and we see that Var (g) obtained from Eq. (9.100) agrees with the results of the theory of the random scattering matrices.

Eq. (9.100) looks as if it has been obtained only in the two-diffuson approximation. However, it is not difficult to check that two-diffuson approximation gives the factor $1/4$ instead of $1/16$ and is definitely not sufficient to obtain the correct result, Eq. (9.100). The difference is even more essential at arbitrary t_1, t_2, N_1, N_2 (see Efetov (1995)). Using a similar method the variance Var (g) has also been calculated for the orthogonal ensemble by Iida, Weidenmüller, and Zuk (1990), who got Var $(g) = 1/8$, which is two times larger than the value for the unitary ensemble.

Substituting Eq. (9.100) into Eqs. (9.21, 9.22) and integrating over ω one can represent the correlation function $B\left(H^+, H^-\right)$, Eq. (9.1), for an open system with equal leads in the form

$$B\left(H^+, H^-\right) = \frac{\bar{T}^2}{8\left(Y^2 + 1\right)} \sum_{n=1}^{\infty} \frac{n}{\left(Y^2 + 1 + n\bar{T}\right)^3} \qquad (9.101)$$

where $Y^2 = X^2/(2\bar{g})$, $\bar{T} = T\pi^2/(\bar{g}\Delta)$, and $\bar{g} = N/2$ is the average conductance of the dot. Eq. (9.101) shows that the temperature-dependent correlation function of the

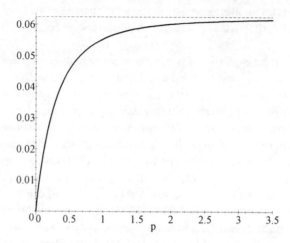

Fig. 9.4. Dependence of the conductance variance Var(g) on the inverse reduced temperature $p = 1/\bar{T}$. This quantity approaches $1/16$ as $\bar{T} \to 0$.

conductances at different magnetic fields $B\left(H^+, H^-\right)$ is, after proper rescaling, remarkably universal, decaying monotonously when increasing the magnetic field and temperature. Let us note that the characteristic temperature T_c at which the correlation function changes is of the order of the level width $\Gamma \sim N\Delta$. Such a temperature is much smaller than the energy $\mathcal{E}_c$ that determined the temperature dependence in the limit $\Gamma \gg \mathcal{E}_c$ (cf. Section 9.1). The same change of scale occurs for the magnetic field. Now, the characteristic magnetic flux ϕ through the dot is of the order of $\phi_0 \left(\Gamma/\mathcal{E}_c\right)^{1/2}$ and is much smaller than ϕ_0, which was characteristic in the regime $\Gamma \gg \mathcal{E}_c$. At the same time, in the closed systems considered in Chapter 8, all quantities changed on fluxes $\phi_0 \left(\Delta/\mathcal{E}_c\right)^{1/2}$.

Experimentally, one can check Eq. (9.101) by changing the magnetic field, temperature, and number of channels in the leads. In the limit of low temperatures $\bar{T} \ll 1 + Y^2$ one can replace the sum in Eq. (9.101) by the integral, and $B\left(H^+, H^-\right)$ has the form of the square of a Lorentzian

$$B\left(H^+, H^-\right) = \frac{1}{16}\left(Y^2 + 1\right)^{-2} \tag{9.102}$$

In the opposite limit of high temperatures $\bar{T} \gg 1 + Y^2$ the function $B\left(H^+, H^-\right)$ is also rather simple

$$B\left(H^+, H^-\right) = \frac{\pi^2}{48\bar{T}}\left(Y^2 + 1\right)^{-1} \tag{9.103}$$

Eq. (9.103) shows that the correlation function $B\left(H^+, H^-\right)$ decays with temperature rather slowly. Now, one has a Lorentzian instead of the square of the Lorentzian but surprisingly with the same width as in Eq. (9.102). The function $B\left(H^+, H^-\right)$ is less sensitive to temperature at high magnetic fields. This is because, in this limit, the main contribution is from short trajectories, which are less dependent on energy.

The conductance variance Var $\left(g\left(H\right)\right)$ is equal to the correlation function at coinciding fields, $B\left(H, H\right)$. The dependence of this function on the reduced temperature $\bar{T}$ is represented in Fig. 9.4 for the unitary ensemble. By changing the temperature T or the number of

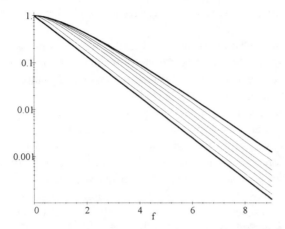

Fig. 9.5. The correlation function $S(f)/S(0)$ for the values of reduced temperature $\bar{T} =$ ∞, 50, 10, 5, 3, 2, 1, 0 (the lowest curve corresponds to the highest temperature).

channels one can study the dependence of the variance on these parameters experimentally. In a recent experiment (Clarke et al. (1995)) linear dependence of the variance on average conductance $\bar{g}$ is clearly seen at not very large $\bar{g}$, in agreement with Eq. (9.103).

Experimentally, one can also study the Fourier transform $S(f)$ (power spectrum) of the correlation function $B\left(H^+, H^-\right)$

$$S(f) = \int B\left(H^+, H^-\right) \cos\left(Yf\right) dY \qquad (9.104)$$

The Fourier transforms of Eqs. (9.102, 9.103) are exponential, with the same exponents. This means that the slope of the power spectrum is not very sensitive to temperature. This is clearly seen from Fig. 9.5, where the function $S(f)/S(0)$ is represented for several values of the reduced temperature $\bar{T}$. Therefore, an experimental study of the slope of the power spectrum only cannot give much information about the temperature dependence and one should try to fit the experimental data onto the exact curve, Eqs. (9.101, 9.103).

A finite temperature can not only result in smearing of the Fermi distribution but also lead to inelastic scattering of electrons. However, in a recent paper, Sivan, Imry, and Aronov (1994) came to the conclusion that the inverse inelastic scattering time τ_φ^{-1} in the quantum dots is proportional to T^2, and one might argue that thermal smearing had to be the most important temperature effect. Of course, this is true, for example, for the variance. The question about the dependence of the function $B\left(H^+, H^-\right)$ on the magnetic field is more complicated because the exponent in the exponential form of the power spectrum (Fourier transform of the function $B\left(H^+, H^-\right)$ in the magnetic field) measured experimentally is rather insensitive to temperature. The temperature smearing does not influence average conductance at all and, hence, the weak localization correction, Eq. (9.83), is not affected by this phenomenon.

Therefore, in many cases one has to include the effects of inelastic scattering. This can most easily be done in the model with the third lead by writing $\alpha = \alpha_0 + \pi\left(\tau_\varphi\Delta\right)^{-1}$, where α_0 describes the real lead that can result in changing the particle number in the dot. In the limit $0 < \alpha \ll N$, Eq. (9.101) is still valid provided one makes the substitution

$Y^2 \to Y^2 + \alpha / (2\bar{g})$. Nonzero values of α can reduce the conductance variance considerably below $1/16$.

The fact that the smearing of the Fermi distribution does not affect the slope of the power spectrum much allows the possibility of obtaining the inelastic mean free path τ_φ experimentally because this quantity gives an important contribution to the variation of the slope. This was done in work by Clarke et al. (1995).

An experimental check of Eq. (9.100) is presented in a recent publication by Chan et al. (1995). These authors measured the conductance of dots as a function of the magnetic field H and a gate voltage V_g that controlled distortion of the shape of the dots. Then the correlation function of the conductances taken at different magnetic fields and voltages was extracted from the results of the measurements by averaging over the magnetic field or the shape distortions. These two correlation functions are expected to correspond to the limiting cases of the function $\tilde{B}$, Eq. (9.100), at $x = 0$, $X \neq 0$, and $X = 0$, $x \neq 0$. The corresponding magnetic field S_B (f_B) and shape distortion S_V (f_V) power spectra have the form

$$S_B (f_B) = S_B (0) (1 + 2\pi a \phi_0 f_B) \exp (-2\pi a \phi_0 f_B) \tag{9.105}$$

$$S_V (f_V) = S_V (0) \exp (-2\pi \kappa f_V) \tag{9.106}$$

In Eqs. (9.105, 9.106), a is the characteristic inverse area of semiclassical trajectories. It includes inelastic scattering in the way described. The quantity κ is proportional to a. The "generalized frequencies" f_B and f_V are variables of the Fourier transformation of functions of the magnetic field and the gate voltage, respectively. The results of the measurements and the comparison with the theoretical predictions are represented in Fig. 9.6.

Another quantity that can be sensitive to inelastic processes is the weak localization correction δg. To include this effect, Eq. (9.83) can also be modified by the substitution $4X^2 \to 4X^2 + \alpha$. Nevertheless, the modeling of the inelastic scattering by an additional lead is not always correct. For example, Eq. (9.71) for the average conductance in the classical limit contains only real leads. This is more or less the density–density correlation function and its form is determined by particle conservation. The inelastic processes do not violate particle conservation and therefore cannot enter the formula for average conductance.

The results of the present section are perturbative and could in principle be obtained by the conventional diagrammatic technique making expansion in diffusons and cooperons. In particular, one can describe some relevant experiments on chaotic billiards in this way (Falko (1995)). Of course, in proceeding in this way the equivalence between chaotic ballistic billiards and disordered systems is implied. Some estimates that can be used for the experiments were obtained diagrammatically long ago (Serota et al. (1987)). Recently, Argaman (1995) reproduced Eq. (9.100) for classically chaotic systems (with the correct residue) making a summation over classical orbits.

Although the results of this section are formally applicable in the limit $N \gg 1$ only, one can hope that they are not numerically bad even at $N = 1$. The accuracy can be checked by calculating the variance Var (g) for arbitrary N, which is a simpler task than calculating the correlation function $B (H^+, H^-)$. Calculation of this quantity is interesting because it also allows a comparison with the results of other approaches. The variance at arbitrary N as well as the entire conductance distribution function (at $N = 1$) in the unitary ensemble are obtained in the next sections.

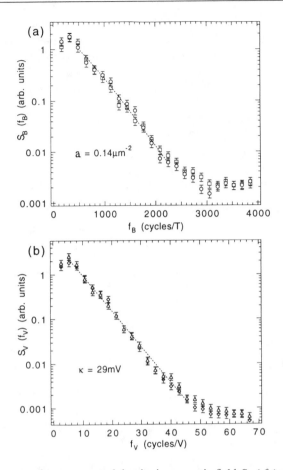

Fig. 9.6. a) Shape-averaged power spectral density in magnetic field S_B (f_B) away from $H = 0$, with arbitrary units. The dashed line is a fit of the circles to Eq. (9.105). The squares are averaged over a different region and shown for comparison. b) Field-averaged ($H > 0$) power spectral density in shape (V_g), S_V (f_V). The dashed line is a fit of the triangles to Eq. (9.106). The diamonds are averaged over a different region for comparison. (From Chan et al. (1995).)

9.5 Average conductance and variance in quantum limit

Comparing the last two sections one can certainly recognize that computation of the average conductance is simpler than calculation of its moments because one needed a smaller size of supermatrix Q. Although any moment can be expressed in terms of a definite integral over the supermatrix, explicit calculation of the integral can be very difficult because of the large number of variables of the integration. Fortunately, in some important cases one can express all moments of the conductance and even the conductance distribution function in terms of the same 8×8 (in the unitary case 4×4) supermatrices as used everywhere in the book except the last section. The possibility of calculating all moments of the local density of states in such a way was demonstrated in Chapter 7.

In the present section I want to show how the average conductance as well as the conductance variance can be calculated in a unified way for an arbitrary number N of channels in each lead in a quantum dot with two leads.

Let us calculate, first, the average conductance $\langle g \rangle$. For simplicity, assume that the leads have the same number of channels and all the channels are open, such that $t_m = 1$ for all channel conductances t_m. Using Eq. (9.57) suitable for this case one can write $\langle g \rangle$ as

$$\langle g \rangle = -N^2 \int f_{33}^{12}(Q)\, f_{33}^{21}(Q)\, \exp\left(-F[Q]\right) dQ \tag{9.107}$$

where $F[Q]$ is written in Eq. (9.56) and the function $f(Q)$ is defined in Eq. (9.58). Evaluation of the integral in Eq. (9.107) can be carried out in exactly the same way as when calculating, for example, the two-level correlation function in Chapter 6. The calculation is most simple in the unitary ensemble; let us restrict ourselves to this case. Using the parametrization of Section 6.3 for the unitary ensemble, Eqs. (6.42–6.44, 6.50, 6.51), one can rewrite the free energy $F[Q]$, Eq. (9.56), for the case considered as

$$F[Q] = 2N \ln\left(\frac{1 + \cosh\theta_1}{1 + \cos\theta}\right) \tag{9.108}$$

The blocks $f^{12}(Q)$ and $f^{21}(Q)$ of the function $f(Q)$ are written in this parametrization in the form

$$f^{12}(Q) = iu f_0 \bar{v}, \qquad f^{21}(Q) = -iv f_0 \bar{u}, \tag{9.109}$$

$$f_0 = \begin{pmatrix} q & 0 \\ 0 & iq_1 \end{pmatrix}, \qquad q = \frac{\sin\theta}{1 + \cos\theta}, \qquad q_1 = \frac{\sinh\theta_1}{1 + \cosh\theta_1} \tag{9.110}$$

Substituting Eqs. (9.108–9.110) into Eq. (9.107), using the explicit form for the matrices u and v, Eqs. (6.57–6.60), and Eq. (6.67) for the Jacobian we can immediately integrate over all variables except θ and θ_1 and reduce Eq. (9.107) to the following integral:

$$\langle g \rangle = N^2 \int_0^\infty \int_0^\pi \left(q^2 + q_1^2\right) \exp\left(-F[Q]\right) (\cosh\theta_1 - \cos\theta)^{-2}$$

$$\times \sinh\theta_1 \sin\theta\, d\theta_1 d\theta = 2N^2 \int_1^\infty \int_{-1}^1 \frac{(1+\lambda)^{2N-1}}{(\lambda_1+1)^{2N+1}} \frac{d\lambda_1 d\lambda}{\lambda_1 - \lambda} \tag{9.111}$$

The last integral in Eq. (9.111) can be evaluated exactly. As the first step, one should integrate over λ using the identity

$$(1+\lambda)^{2N-1} = \sum_{m=0}^{2N-1} (\lambda_1+1)^m (\lambda-\lambda_1)^{2N-1-m} \frac{(2N-1)!}{m!\,(2N-1-m)!} \tag{9.112}$$

As a result of the integration, one obtains a sum such that each term of the sum can be integrated explicitly over λ_1. The procedure is not complicated and one finally obtains a simple formula

$$\langle g \rangle = \frac{N}{2} \tag{9.113}$$

Surprisingly, the exact quantum formula for the average conductance, Eq. (9.113), coincides in the case of leads with an equal number of channels with the classical Hauser–Feshbach formula, Eq. (9.76). This means that in the case under consideration the resistivities of the leads just add and there is no interference. Apparently, this property is accidental because in the case $N = 1$ but t arbitrary the average conductance can be found to be

$$\langle g \rangle = \frac{t}{3} + \frac{t^2}{6} \tag{9.114}$$

which coincides with $t/2$ for $t = 1$ only.

In order to calculate $\langle g^2 \rangle$ one can start with Eq. (9.40), which contains a product of retarded G^R and advanced G^A Green functions that should be expressed in terms of integrals over supervectors similar to those in Eq. (4.4). At the same time, we do not want to double the number of components of the supervectors when writing the product of four Green functions; this goal can be achieved at $T = 0$ when all the Green functions are taken at one energy ε. Since, in Eq. (4.4), one may write in the preexponential arbitrary components of the supervectors, the product of four Green functions entering $\langle g^2 \rangle$ can be represented as

$$G_\varepsilon^R (z_1, y_1) G_\varepsilon^A (y_1, z_1) G_\varepsilon^R (z_2, y_2) G_\varepsilon^A (y_2, z_2) \tag{9.115}$$

$$= \int S^2 (z_1) S^{2*} (y_1) S^1 (y_1) S^{1*} (z_1) \chi^2 (z_2) \chi^{2*} (y_2) \chi^1 (y_2) \chi^{1*} (z_2)$$

$$\times \exp(-\mathcal{L}) \, D\psi \, D\bar\psi$$

where ψ, $\bar\psi$ are eight-component supervectors similar to those in Chapter 4, and $\mathcal{L}$ is also similar to the Lagrangian $\mathcal{L}$ in Eq. (4.5) but with the Hamiltonian $\mathcal{H}_{\text{eff}}$ from Eq. (9.42). The variables z and y relate to the first and second leads, respectively. It is very important that the first pair of Green functions in Eq. (9.115) be written using conventional numbers whereas the second contains anticommuting variables. This excludes the possibility of an additional pairing when calculating the integral over supervectors ψ, $\bar\psi$, and prevents doubling of the number of components of the supervectors. Using the representation Eq. (9.115) one can execute the routine program averaging over impurities and decoupling the interaction term by integration over supermatrices Q. Then one has to integrate over the supervectors ψ using an effective Lagrangian $\mathcal{L}_{\text{eff}}$ like that in Eq. (4.19). Replacing $\mathcal{L}$ by $\mathcal{L}_{\text{eff}}$ one can still use Eq. (9.115).

Now, in principle, all pairings in this equation are possible. However, assuming that the distance between leads much exceeds the mean free path l, pairings of field variables with coordinates of different leads can be neglected. A further reduction of the number of possible pairings is due to the fact that the unitary ensemble is considered and averages like $\langle SS \rangle$, $\langle S^*S^* \rangle$, $\langle \chi\chi \rangle$, or $\langle \chi^*\chi^* \rangle$ are equal to zero. Integrating finally across the leads and using the functions f, Eqs. (9.58), one comes to the following integral for the average square of the conductance $\langle g^2 \rangle$:

$$\langle g^2 \rangle = - \int [N^4 f_{33}^{12} f_{33}^{21} f_{11}^{12} f_{11}^{21} + N^3 f_{11}^{12} f_{33}^{12} f_{13}^{21} f_{31}^{21} + N^3 f_{13}^{12} f_{31}^{12} f_{11}^{21} f_{33}^{21} \tag{9.116}$$

$$+ N^2 f_{31}^{12} f_{13}^{12} f_{31}^{21} f_{13}^{21}] \exp(-F[Q] \, dQ)$$

with $F[Q]$ and $f(Q)$ given by Eqs. (9.108, 9.109).

Calculation of the integral in Eq. (9.116) can be carried out in exactly the same way as when calculating average conductance $\langle g \rangle$. Using the representation for supermatrices f, Eqs. (9.109, 9.110), and integrating over supermatrices u and v one can reduce Eq. (9.116) to integrals over λ and λ_1. Integration of the product of the supermatrices f over u and v in the first term in the integrand gives

$$16\left(q^2 + q_1^2\right)^2 \tag{9.117}$$

After the same integration the sum of corresponding products in the second and third terms in the integrand gives

$$64q_1^2\left(q^2 + q_1^2\right) - 32\left(q^2 + q_1^2\right)^2 \tag{9.118}$$

whereas integration in the fourth term leads to

$$16\left(q^2 + q_1^2\right)^2 \tag{9.119}$$

Substituting Eqs. (9.117–9.119) into Eq. (9.116) and using Eqs. (9.108, 9.110) one obtains

$$\left\langle g^2 \right\rangle = 4\left(N^4 + N^2 - 2N^3\right)\int_1^\infty \int_{-1}^1 \frac{(1+\lambda)^{2N-2}}{(\lambda_1+1)^{2N+2}} d\lambda_1 d\lambda$$

$$+ 8N^3 \int_1^\infty \int_{-1}^1 \frac{(\lambda_1 - 1)(1+\lambda)^{2N-1}}{(\lambda_1+1)^{2N+2}} \frac{d\lambda_1 d\lambda}{\lambda_1 - \lambda} \tag{9.120}$$

Evaluation of the first integral in Eq. (9.120) is trivial. The second integral can be computed by using Eq. (9.112) and integrating each term of the resulting sum. Finally, one obtains

$$\left\langle g^2 \right\rangle = \frac{N^2 (N-1)^2}{4N^2 - 1} + \frac{N^2}{2N+1} = \frac{N^4}{4N^2 - 1} \tag{9.121}$$

Using Eq. (9.113) one arrives at the following expression:

$$\mathrm{Var}\,(g) = \frac{N^2}{4\left(4N^2 - 1\right)} \tag{9.122}$$

Eq. (9.122) holds for an arbitrary N. In the semiclassical limit $N \gg 1$ one obtains the value $1/16$, which agrees with the value obtained from Eq. (9.100). In the one-channel limit, $N = 1$, the variance is equal to $1/12$, which is not very far from the semiclassical limit. Eq. (9.122) agrees with the corresponding result obtained by Baranger and Mello (1994) from the theory of random scattering matrices.

Of course, similar calculations can be made for the orthogonal and symplectic ensembles but have not yet been done. As another extension of the calculation presented here one can try to obtain the entire distribution function of the conductances, and this will be done in the next section.

9.6 Conductance distribution function

9.6.1 The distribution function in terms of a definite integral

Having understood how to calculate the variance of the conductances one can try to compute higher moments of the conductances and the entire distribution function. In this section the distribution function of conductances, $P(g)$, of a quantum dot is calculated. This function is introduced as

$$P(g) = \langle \delta(g - g_{mn}) \rangle \tag{9.123}$$

where g_{mn} is the conductance between leads m and n.

As we have learned from Section 7.3, distribution functions of local quantities can be calculated. For example, the distribution function of the local density of states has been calculated exactly. Because any power of the local density of states can be expressed in terms of Green functions taken at one energy and one space point, one can apply the supersymmetry technique for any moment without increasing the size of the supermatrices.

Unfortunately, it is not simple to obtain the distribution function for an arbitrary number of channels in the leads. Although, according to Eq. (9.40), the conductance is expressed in terms of a product of two Green functions with the same energy, this equation contains integrals over coordinates across the leads. That means that the number of space points coming into play grows with an increase in the power of the conductance. The only exception is the case with point contacts because in this situation only two space points are involved. The possibility of obtaining the conductance distribution function for the model with point contacts was noticed by Prigodin, Efetov, and Iida (1993), and the discussion of the present section follows that work.

One can compute the distribution function $P(g)$ by expanding Eq. (9.123) in Taylor series. Any power n of the product of the Green functions $G_\varepsilon^R(z, y)$ and $G_\varepsilon^A(y, z)$ can be written in terms of the eight-component supervectors ψ as

$$(n!)^2 \left(G_\varepsilon^R(z, y)\, G_\varepsilon^A(y, z) \right)^n \tag{9.124}$$

$$= \int \left(S^2(z)\, S^{2*}(y)\, S^1(y)\, S^{1*}(z) \right)^n \exp(-\mathcal{L})\, D\psi$$

(All the notations here are the same as in Chapter 4.) Then, one repeats the procedure of averaging over impurities and decoupling the effective interaction by integration over 8×8 supermatrices Q. A further simplification occurs if the leads are separated by a distance larger than the mean free path l. Assuming also that the unitary ensemble is considered, one obtains only two different averages $\langle S^2(z)\, S^{1*}(z) \rangle$ and $\langle S^1(y)\, S^{2*}(y) \rangle$ with the effective Lagrangian $\mathcal{L}_{\text{eff}}$, Eq. (4.19). Performing the integration over the supervectors ψ one can thus reduce calculation of the average of any power of the conductance to an integral over the standard Q matrices (cf. Eqs. (6.42–6.47)). Summing up all the terms of the expansion of the distribution function $P(g)$ one obtains the following expression for this function:

$$P(g) = \int \delta\left(g + f_{33}^{12}(t_1, Q)\, f_{33}^{21}(t_2, Q)\right) \exp(-F[Q])\, dQ \tag{9.125}$$

where the function $f(t, Q)$ is defined in Eq. (9.58) and $F[Q]$ is the proper free energy function. If the leads are not very large we can use the zero-dimensional σ-model and, thus, the distribution function is reduced to a definite integral over supermatrix Q.

Later a model with two source and drain point contacts and a third extended lead is considered and it is assumed that the lead conductances t_1 and t_2 are arbitrary (but smaller than unity). Although different versions of the free energy $F[Q]$ for a quantum dot with leads have been given several times in the present chapter, let us present it once more for the model under consideration

$$F[Q] = -\frac{1}{4} \operatorname{str} \sum_{m=1}^{2} \ln\left(2 - t_m + t_m \Lambda Q^{\parallel}\right) - \frac{t}{8} \operatorname{str} \Lambda Q \qquad (9.126)$$

The first term in Eq. (9.126) describes the point contacts and was derived in Section 9.2 microscopically. The second term corresponds to the third extended lead. This term was also written in Section 7.3 to describe the possibility of electrons' leaving a metal particle into an environment. In contrast to the channel conductances t_1 and t_2, conductance t need not be smaller than unity. This corresponds to the assumption that the third lead has many channels. At the same time, because of the possibility of having a barrier at the contact, the parameter t may be smaller than unity as well.

Before starting calculation of the integral in Eq. (9.125) it is useful to understand qualitatively what one can expect from the model under consideration and what the physical difference between the two terms in Eq. (9.126) is. To do that, let us note that the free energy, Eq. (9.126), can also be derived from the following phenomenological Hamiltonian:

$$\mathcal{H} = \mathcal{H}_c \pm \frac{i\alpha\Delta}{\pi} \pm \frac{i}{\pi\nu}\left[\alpha_1\delta\left(\mathbf{r} - \mathbf{r}_1\right) + \alpha_2\delta\left(\mathbf{r} - \mathbf{r}_2\right)\right] \qquad (9.127)$$

where $\mathcal{H}_c$ stands for the Hamiltonian of the closed dot, and the upper (lower) sign should be taken when solving the Schrödinger equation for the advanced (retarded) Green function. The second term describes the extended contact and a dependence on coordinates can be neglected. The third term describes the point contacts at the points $\mathbf{r}_1$ and $\mathbf{r}_2$. Proceeding in the standard way one can obtain Eq. (9.126) with $t = 4\alpha$ and $t_{1,2}$ expressed through $\alpha_{1,2}$ by Eq. (9.60). The difference between the two terms in Eq. (9.126) is due to the different space dependence of the corresponding terms in Eq. (9.127).

To discuss properties of conductance qualitatively it is convenient to express the exact Green functions of the Hamiltonian, Eq. (9.127), in terms of the Green functions $G_0^{R,A}$ of the Hamiltonian without the point contacts

$$G_{0\varepsilon}^{R,A}\left(\mathbf{r}, \mathbf{r}'\right) = \sum_{\beta} \frac{\varphi_\beta\left(\mathbf{r}\right)\varphi_\beta^*\left(\mathbf{r}'\right)}{\varepsilon - \varepsilon_\beta \pm i\gamma/2}, \qquad \gamma = \frac{2\alpha\Delta}{\pi} \qquad (9.128)$$

where φ_β and ε_β are eigenfunctions and eigenenergies of the closed dot. Considering the last term in Eq. (9.127) as a perturbation and writing a corresponding Dyson equation that reduces in the case considered to a system of two algebraic equations one can express the exact functions $G_\varepsilon^{R,A}$ in terms of the functions $G_{0\varepsilon}^{R,A}$. The solution for the function $G_\varepsilon^A\left(\mathbf{r}_1, \mathbf{r}_2\right)$ can be written as

$$G_{12}^{R,A} = G_{012}^{R,A}\left(\left(1 \pm \frac{i\alpha_1}{\pi\nu}G_{011}^{R,A}\right)\left(1 \pm \frac{i\alpha_2}{\pi\nu}G_{022}^{R,A}\right) + \frac{\alpha_1\alpha_2}{(\pi\nu)^2}G_{012}^{R,A}G_{021}^{R,A}\right)^{-1} \qquad (9.129)$$

where $G_{in}^{R,A} \equiv G_\varepsilon^{R,A}\left(\mathbf{r}_i, \mathbf{r}_n\right)$.

A formula for conductance g between the point contacts has the form

$$g = \frac{4\alpha_1\alpha_2}{(\pi\nu)^2}G_\varepsilon^R\left(\mathbf{r}_1, \mathbf{r}_2\right)G_\varepsilon^A\left(\mathbf{r}_2, \mathbf{r}_1\right) \qquad (9.130)$$

With the help of Eq. (9.128) we see that the second term in Eq. (9.127) leads to a finite level width that is the same for all levels. The third term in Eq. (9.127) also leads to a smearing of the levels but gives a different contribution to the width of different levels. To see this explicitly let us consider the case when the level widths are much smaller than the distance between the levels. Taking the energy ε very close to a level k we may retain only one term in the sum, Eq. (9.128), corresponding to this level. This is a so-called resonance regime. In this approximation using Eq. (9.128) one has

$$G_{012}^{R,A} G_{021}^{R,A} = G_{011}^{R,A} G_{022}^{R,A}$$

and Eq. (9.129) takes the form

$$G_{\varepsilon}^{R,A}(\mathbf{r}_1, \mathbf{r}_2) = \frac{\varphi_n(\mathbf{r}_1) \varphi_n^*(\mathbf{r}_2)}{\varepsilon - \varepsilon_n \pm i\gamma/2 \pm i(\gamma_{1n} + \gamma_{2n})/2} \tag{9.131}$$

where the level widths γ_{an}, $a = 1, 2$ have the form

$$\gamma_{an} = \frac{2\alpha_a}{\pi \nu} |\varphi_n(\mathbf{r}_a)|^2 \tag{9.132}$$

Eqs. (9.131, 9.132) show that the level widths of the system with point contacts are very sensitive to the values of the wave functions at the contacts. The level width is proportional to the probability of tunneling through the contacts. To bring about this process one needs a state with a finite wave function at the point of the contact. Hence, it is quite natural that the level width given by Eq. (9.132) is proportional to the probability of being at the contact. Let us emphasize that the only assumption made when deriving Eqs. (9.131, 9.132) is that the level width is much smaller than the distance between the levels such that one can be in the resonance regime.

Using Eqs. (9.130, 9.131) one can write in the same approximation the conductance $g^{(n)}(\varepsilon)$

$$g^{(n)}(\varepsilon) = \frac{\gamma_{1n}\gamma_{2n}}{(\varepsilon - \varepsilon_n)^2 + (1/4)(\gamma + \gamma_{1n} + \gamma_{2n})^2} \tag{9.133}$$

We know from the discussion in Section 9.2 that the conductance for a system with a one-channel source and drain leads cannot exceed unity. This property can also be seen from Eq. (9.133). The maximum conductance $g(\varepsilon)$ is achieved if the Fermi energy is very close to a level n. It can become unity only if the Fermi energy coincides with the level, additional leads are absent ($\gamma = 0$), and $\gamma_{1n} = \gamma_{2n}$. It follows from this general argument that the conductance distribution function $P(g)$ must be zero for $g > 1$ and calculation with Eqs. (9.125, 9.126) must reproduce this property.

As concerns the opposite limit $\gamma \gg \Delta$ one can apply the semiclassical approximation of Section 9.3. An analogue of the Hauser–Feshbach formula, Eq. (9.76), can be written for the present case as

$$\bar{g} = \frac{t_1 t_2}{t} \tag{9.134}$$

with $t = 4\alpha$ and t_1, t_2 expressed through $\alpha_{1,2}$ by Eq. (9.60).

Now, we can start computing the integral in Eq. (9.125). Using the parametrization of supermatrix Q for the unitary ensemble, Eqs. (6.42–6.44, 6.50, 6.51), we see that the free energy $F[Q]$, Eq. (9.125), depends only on the variables θ and θ_1, and so one should integrate first over the supermatrices u and v in the preexponential. The calculation is, as usual, straightforward and I present only several steps. Using Eq. (9.109) and the first formula in Eq. (9.110) for f_0 one can write the product $f_{33}^{12}(t_1, Q) f_{33}^{21}(t_2, Q)$ as

$$f_{33}^{12}(t_1, Q) f_{33}^{21}(t_2, Q) = -[q_1(t_1) q_1(t_2)(1 + 4(\eta\eta^* - \kappa\kappa^*))$$
$$- 4\kappa\eta^* e^{i(\varphi-\chi)} q(t_1) q_1^*(t_2) - 4\eta\kappa^* e^{i(\chi-\varphi)} q_1(t_1) q^*(t_2) \qquad (9.135)$$
$$- 16\eta\eta^*\kappa\kappa^*(q_1(t_1) q_1(t_2) + q(t_1) q(t_2))]$$

where

$$q(t_a) = \frac{t_a \sin\theta}{2 + t_a(\cos\theta - 1)}, \qquad q_1(t_a) = \frac{t_a \sinh\theta_1}{2 + t_a(\cosh\theta_1 - 1)}, \qquad a = 1, 2 \quad (9.136)$$

Substituting Eq. (9.135) into Eq. (9.125) and integrating over the Grassmann variables one reduces the function $P(g)$ to the form

$$P(g) = \delta(g)$$

$$+16 \int_0^\infty \int_0^\pi (q_1(t_1) q_1(t_2) + q(t_1) q(t_2))[-\delta'(g - q_1(t_1) q_1(t_2))$$

$$+ \delta''(g - q_1(t_1) q_1(t_2)) q_1(t_1) q_1(t_2)] \exp[-F(\theta, \theta_1)] J(\theta, \theta_1) d\theta_1 d\theta \qquad (9.137)$$

where the Jacobian $J(\theta, \theta_1)$ is given by Eq. (6.67) and δ' and δ'' stand for the first and second derivatives of the δ-function, respectively. The first term in Eq. (9.137) originates from the term that did not contain the anticommuting variables. The function $q_1(t)$ is always smaller than unity and therefore the function $P(g)$ vanishes for $g > 1$ in agreement with the qualitative discussion. Further computation in Eq. (9.137) is rather simple. Integration over θ_1 can be done immediately because of the presence of the δ-function. The second integration can also be done analytically, and the result can be expressed in elementary functions for arbitrary coefficients t_1, t_2, and t.

9.6.2 The limit $t_{1,2} \ll t$: Random phases of eigenstates and independent fluctuations of amplitudes at different points

To understand the underlying physics better, let us discuss, first, the case when the third lead has a strong influence, such that $t_{1,2} \ll t$. It is clear that in this limit the first term in the free energy $F[Q]$ in Eq. (9.126), as well as the terms proportional to $t_{1,2}$ in Eq. (9.136), can be neglected. The distribution function $P(g)$ in this limit takes the form (Prigodin, Efetov, and Iida (1993))

$$P(g) = -\frac{1}{t\bar{g}y} \frac{d}{dy}\left[\left(\frac{2}{y}\cosh\left(\frac{t}{2}\right)\right) \qquad (9.138)$$

$$+ \left(1 - \frac{2}{ty} + \frac{1}{y^2} + \frac{2}{ty^3}\right) \sinh\left(\frac{t}{2}\right)\right) \exp\left(-\frac{ty}{2}\right)\right]$$

where

$$y = \sqrt{1 + \frac{4g}{t\bar{g}}}, \qquad \bar{g} = \frac{t_1 t_2}{t} \qquad (9.139)$$

Using Eq. (9.138) one can compute the average of any function of the conductance. For example, one can check easily that $\bar{g}$ in Eq. (9.139) is just the average conductance of the

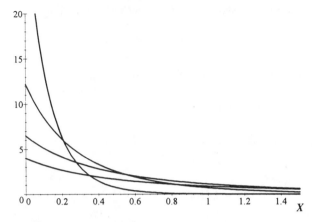

Fig. 9.7. Conductance distribution function $t\bar{g}P(g)$ as a function of $x = 4g/t_1t_2$ in the limit $t_{1,2} \ll t$ for coupling constants to the third lead $t = 0.2, 4, 10, 30$. The decay of the function is faster for larger values of t.

dot. The dependence of the average conductance $\bar{g}$ on the conductances of the leads looks very similar to the Hauser–Feshbach formula, Eqs. (9.76), and coincides with Eq. (9.134). This agreement is no surprise in the limit $t \gg 1$, where the semiclassical approximation can be used. What is surprising is that Eq. (9.139) also holds in the limit $t \ll 1$ provided $t_{1,2} \ll t$.

The function $P(g)$ is represented for several values of t in Fig. 9.7. This function decreases monotonously and $g = 0$ is the most probable value of the conductance. Such behavior is very different from that known for samples strongly connected with leads (Altshuler, Kravtsov, and Lerner (1986)), where the maximum of the distribution function corresponds to a finite value of the conductance. It is also different from the behavior of the distribution function of the density of states obtained in Section 7.3 that approaches the δ-function in the limit $1 \ll t \ll \mathcal{E}_c/\Delta$, whereas $P(g)$, Eqs. (9.138, 9.139), in this limit takes for $g \ll t\bar{g}$ the form

$$P(g) = \left(\frac{1}{\bar{g}}\right) \exp\left(-\frac{g}{\bar{g}}\right) \qquad (9.140)$$

The exponential behavior, Eq. (9.140), gives very interesting information about the phases $\chi_\beta(r_a)$ of the wave functions $\varphi_\beta(r_a)$ of a closed dot: namely, that the phases $\chi_\beta = \chi_\beta(r_1) - \chi_\beta(r_2)$ for different states β are completely uncorrelated. In order to see this, one can start from Eqs. (9.123, 9.128, 9.129) and write the Green function G_ε^R in the form

$$G_\varepsilon^R(\mathbf{r}_1, \mathbf{r}_2) = \sum_\beta \left(m_{\beta x} + i m_{\beta y}\right), \qquad (9.141)$$

$$m_{\beta x} + i m_{\beta y} = \frac{1}{(1+\alpha_1)(1+\alpha_2)} \frac{\left|\varphi_\beta(\mathbf{r}_1)\right| \left|\varphi_\beta(\mathbf{r}_2)\right|}{\varepsilon - \varepsilon_\beta + i\gamma/2} e^{i\chi_\beta} \qquad (9.142)$$

In the limit $t \gg 1$ ($\gamma \gg \Delta$) one can replace the Green functions $G_{011}^{R,A}$ and $G_{022}^{R,A}$ in Eq. (9.129) by their averages and neglect the last term in the brackets. This is the origin of the first multiplier in Eq. (9.142). The quantities $m_{\beta x}$ and $m_{\beta y}$ are the components of a two-dimensional vector $\mathbf{m}_\beta$. The phases χ_β determine the direction of this vector.

If the phases χ_β are really random, one may speak of a diffusion with the vector

$$\mathbf{m} = \sum_\beta \mathbf{m}_\beta \qquad (9.143)$$

playing the role of a total two-dimensional displacement vector. Each energy level in Eqs. (9.141–9.143) corresponds to an elementary random displacement. Using Eqs. (9.132–9.134) one can write the distribution function $P(g)$ in the form

$$P(g) = \left\langle \left(g - 4\alpha_1\alpha_2 \left(\sum_\beta \mathbf{m}_\beta \right)^2 \right) \right\rangle = \int \delta \left(g - \mathbf{m}^2 \right) \tilde{P}(m)\, d\mathbf{m}, \qquad (9.144)$$

$$\tilde{P}(m) = \left\langle \delta \left(\mathbf{m} - 2 \left(\alpha_1\alpha_2 \right)^{1/2} \sum_\beta \mathbf{m}_\beta \right) \right\rangle \qquad (9.145)$$

The averaging in Eqs. (9.144) has to be carried out over directions of each vector $\mathbf{m}_\beta$. Writing the function $\tilde{P}(m)$ as

$$\tilde{P}(m) = \int \exp\left(i\mathbf{k}\mathbf{m} \right) \left\langle \exp\left(-2i \left(\alpha_1\alpha_2 \right)^{1/2} \sum_\beta \mathbf{k}\mathbf{m}_\beta \right) \right\rangle \frac{d\mathbf{k}}{(2\pi)^2} \qquad (9.146)$$

and using the assumption that the directions of the vectors $\mathbf{m}_\beta$ are completely random and uncorrelated one can easily perform the averaging. As a result, the expression in the angle brackets in Eq. (9.146) equals

$$\exp\left(\sum_\beta \ln J_0\left(km_\beta \right) \right) \qquad (9.147)$$

where J_0 is the Bessel function.

The functions φ_β are normalized by volume and, hence, each elementary displacement $\mathbf{m}_\beta$ is small. This means that the main contribution is from small k. Expanding the expression in the exponential in Eq. (9.147), substituting the result into Eqs. (9.146, 9.144), and using Eqs. (9.142, 9.60) one obtains for the distribution function

$$P(g) = \left(\frac{1}{B} \right) \exp\left(-\frac{g}{B} \right), \qquad (9.148)$$

$$B = \sum_\beta m_\beta^2 = \frac{t_1 t_2}{(2\pi \nu)^2} \sum_\beta \frac{\left| \varphi_\beta(\mathbf{r}_1) \right|^2 \left| \varphi_\beta(\mathbf{r}_2) \right|^2}{\left(\varepsilon - \varepsilon_\beta \right)^2 + (t\Delta/4\pi)^2} \qquad (9.149)$$

Eq. (9.148) already has the desired form, Eq. (9.140), and we see that as soon as the phase differences χ_β for different states are uncorrelated one comes to the Poisson distribution, Eq. (9.148), for the conductances. The only thing that remains to be done is to evaluate the sum in Eq. (9.149) and this is also not difficult. In the limit under consideration $t \gg 1$ the sum in Eq. (9.149) can be replaced by the integral. Also replacing the absolute values of the wave functions by V^{-1}, one can perform the integration and obtain

$$B = \frac{t_1 t_2}{t} \qquad (9.150)$$

which coincides in the limit under consideration with the average conductance $\bar{g}$. Thus, we have derived Eq. (9.140) from the simple qualitative arguments. The high probability of having $g = 0$ is due to the phase fluctuations that reduce the conductance. At the same time, the local density of states does not depend on phases, and, therefore, the probability that this quantity turns to zero is small, in agreement with the result of Section 7.3.

In the limit $t \ll 1$ the asymptotics of the function $P(g)$ from Eqs. (9.138, 9.139) can be written in the form

$$
P(g) = \frac{t^2}{8\bar{g}}
\begin{cases}
\frac{32}{t^3}, & g \ll t\bar{g} \\
u^{-3/2}, & t\bar{g} \ll g \ll \frac{\bar{g}}{t} \\
u^{-1/2} \exp\left(-u^{1/2}\right), & \frac{\bar{g}}{t} \ll g
\end{cases}
\tag{9.151}
$$

where $u = tg/\bar{g}$.

Eq. (9.151) can be understood qualitatively by using Eqs. (9.123, 9.130, 9.141, 9.142) again. In order to reproduce $P(g)$ for $g \ll t\bar{g}$ one can repeat the argument about the destructive interference and come again to Eqs. (9.148, 9.149). Approximating $|\varphi_\beta|^2$ in Eq. (9.149) by V^{-1} and assuming that characteristic $|\varepsilon - \varepsilon_\beta|$ are of the order of Δ one can estimate the coefficient B as $B \sim t\bar{g}$, thus reproducing the first line of Eq. (9.151).

In the limit $t\bar{g} \ll g \ll \bar{g}/t$, a level k that is closest to energy ε gives the main contribution to the Green function, Eqs. (9.141, 9.142). Assuming that $\varepsilon_n - \varepsilon$ is uniformly distributed so long as $|\varepsilon_n - \varepsilon| \ll \Delta$, again replacing $|\varphi_n|^2$ by V^{-1}, and using Eqs. (9.123, 9.130) one can reproduce the second line in Eq. (9.151).

The nonzero probability of a finite g in the region $g \gg \bar{g}/t$ signals very strong fluctuations of the wave functions φ_β. To obtain large conductances one has to assume that, again, only one level k contributes essentially to the conductance. Then, the conductance g can be written in the form

$$
g = \frac{t_1 t_2}{(2\pi\nu)^2} \frac{|\varphi_n(\mathbf{r}_1)|^2 |\varphi_n(\mathbf{r}_2)|^2}{(\varepsilon - \varepsilon_n)^2 + (\gamma/2)^2}
\tag{9.152}
$$

The exponential dependence of $P(g)$ on $g^{1/2}$ in Eq. (9.151) can be reproduced (with the correct numerical coefficient) from simple arguments if one assumes that the quantities $v_a = V |\varphi_n(\mathbf{r}_a)|^2$, $a = 1, 2$ in Eq. (9.152) fluctuate independently and their distribution $W(v)$ is exponential (Gaussian for $|\varphi|$)

$$
W(v) = \exp(-v)
\tag{9.153}
$$

Taking $\varepsilon = \varepsilon_n$ in Eq. (9.152), substituting Eq. (9.152) into Eq. (9.123), and integrating over v_1 and v_2 with the distribution function $W(v)$, Eq. (9.153), one arrives at the third line of Eq. (9.151). This is a remarkable result because the assumption about the exponential behavior of the distribution function of the wave functions has in Section 7.3 allowed us to explain the asymptotics of the distribution function of the local density of states in the limit of large and low densities. Thus, we see that the results of the calculations of the distribution functions for both the conductance and the local density of states are consistent with each other and lead to Eq. (9.153).

The Gaussian distribution for $|\varphi_n|$ can also be confirmed by computation of the inverse participation ratio coefficients B_n

$$
B_n = \left\langle \sum_\beta |\varphi_\beta(\mathbf{r}_1) \varphi_\beta^*(\mathbf{r}_2)|^{2n} \delta(\varepsilon - \varepsilon_\beta) \right\rangle
\tag{9.154}
$$

in a procedure explained in Section 7.3. Using this approach, first suggested by Wegner

(1980), one can relate the coefficients B_n to the moments of conductance $\langle g^n \rangle$ by the following equation:

$$B_n = \frac{[(n-1)!]^2}{2^n \, (2n-2)!} \frac{\Delta}{V^{2n}} \lim_{t \to 0} t^{n-1} \left\langle \left(\frac{g}{\bar{g}} \right)^n \right\rangle \tag{9.155}$$

Finally from Eqs. (9.138, 9.139, 9.155) one obtains

$$B_n = \Delta \left(\frac{n!}{V^n} \right)^2 \tag{9.156}$$

Eq. (9.156) can be reproduced easily by using the distribution function $W(v)$, Eq. (9.153). This confirms a very important result that the absolute values of a wave function of a closed dot at different points fluctuate independently provided the distance between these points is larger than the mean free path l (in fact, the correlation of the wave functions already decays at distances larger than the wavelength). Although this result has a rather simple meaning, it was obtained after nontrivial calculations performed for the unitary ensemble. It is not clear in advance whether the same holds for the orthogonal ensemble or for the crossover region between the two ensembles.

The fluctuations in the unitary ensemble obey the Gaussian statistics, as is consistent with predictions of random matrix theory (Brody et al. (1981)). Together with the conclusion about the uncorrelated fluctuations of the phases of wave functions of different states, this result gives a good description of statistical properties of wave functions of quantum chaotic systems. Knowledge of statistical properties of wave functions is very important for the theory of Coulomb blockade oscillations.

9.6.3 General form of the distribution function

Postponing discussion of the Coulomb blockade problem to the next section, let us now express the conductance distribution function $P(g)$ without being limited by the inequality $t_{1,2} \ll t$ used when deriving Eqs. (9.138, 9.139). Eq. (9.137) is valid in the general case. Integrating in Eq. (9.137) over variables θ_1 and θ one reaches the result (Prigodin, Efetov, and Iida (1995)) valid for arbitrary $t_{1,2}$ and t

$$P(g) = \frac{dK(g)}{dg}, \tag{9.157}$$

$$K(g) = \frac{d_1 d_2}{ay + b} \left[\left(y \sinh \left(\frac{t}{2} \right) + \cosh \left(\frac{t}{2} \right) \right) \exp \left(-\frac{ty}{2} \right) \right. $$
$$\left. + \frac{2 \sinh (t/2)}{t} \left(y^2 - 1 \right) \frac{d}{dy} \frac{a}{ay + b} \exp \left(-\frac{ty}{2} \right) \right]$$

where

$$g = t_1 t_2 \left(y^2 - 1 \right) (d_1 d_2)^{-1}$$
$$d_{1,2} = 2 + (y - 1) t_{1,2}$$
$$a = -(y - 1)(t_1 + t_2 - t_1 t_2) - 2 \tag{9.158}$$
$$b = (y - 1)(t_1 + t_2 - t_1 t_2) - 2y$$

Eqs. (9.157, 9.158) give the most general expression for the quantum dot with two point

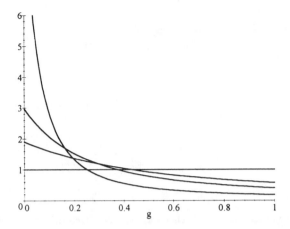

Fig. 9.8. Conductance distribution function for a quantum dot with two one-channel leads for $t_0 = 0.5, 0.75, 0.85, 1$. The decay becomes faster with decreasing t_0.

contacts and an additional extended lead. In the limit $t_{1,2} \ll t$ one may neglect in $d_{1,2}$, a and b all terms containing $t_{1,2}$ and then Eqs. (9.138, 9.139) are immediately reproduced.

The function $P(g)$, Eq. (9.157), monotonously decreases in the region $0 \leq g \leq 1$ (y varies from 1 to ∞). Outside this region the distribution function vanishes, $P(g) = 0$; therefore, the reduced conductance cannot be larger than 1. This is in agreement with the qualitative arguments presented at the beginning of this section. For $t = 0$ the dependence $P(g)$ is drawn in Fig. 9.8 for some values of $t_{1,2} = t_0$. In the limit of weak symmetric coupling to the leads $t_1 = t_2 = t_0 \ll 1$ and $t = 0$, the distribution function in the region $g \ll 1$ has a rather simple form

$$P(g) = 4t_0 \frac{t_0^2 + g}{\left(t_0^2 + 4g\right)^{5/2}} \tag{9.159}$$

This result agrees precisely with the random matrix calculation of Brouwer and Beenakker (1994). The function $P(g)$ has very interesting behavior at $t_1 = t_1 = 1$ ($t = 0$). Substituting these values into Eq. (9.157) one obtains

$$P(g) = 1, \qquad 0 \leq g \leq 1 \tag{9.160}$$

That means that any value of the conductance in the system with open channels is equally probable in the interval $(0, 1)$. The same result has recently been obtained by the random matrix approach for the scattering matrix by Baranger and Mello (1994) and Jalabert, Pichard, and Beenakker (1994). However, as shown by these authors, such a homogeneous distribution only holds for the unitary ensemble. Apparently, the unusual behavior, Eq. (9.160), is rather accidental and there is no deep reason for it.

Using Eq. (9.157) one can compute any moment of the conductance. In the absence of the third lead one arrives in the symmetric case at Eq. (9.114) (the parameter t in that equation has to be replaced by t_0 in the notations of this section).

All the results of this section were obtained for the unitary ensemble. This is because calculations are most compact within this ensemble. Extension to the orthogonal or symplectic ensembles is straightforward but has not yet been done. All electron interactions were neglected. Although one can argue that in many cases their influence can be taken into

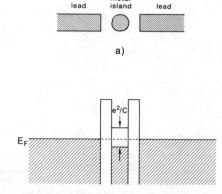

Fig. 9.9. Coulomb blockade system: a) scheme of the device b) corresponding potential and energy-level spectrum with filled (hatched) and empty levels. The quantum dot has a gap of width $e^2/(2C)$ in its tunneling density of states.

account within the Fermi liquid picture, the long-range part of the Coulomb interaction can lead to interesting new effects, which are considered in the next section.

9.7 Statistical theory of Coulomb blockade oscillations

9.7.1 Phenomenological approach

Up to now the theory of noninteracting particles has been considered. A short-range interaction can be included in the scheme by using the Landau theory of Fermi liquid. In this approach, one can replace the term *noninteracting particles* by the term *noninteracting quasi particles* and rewrite all results with renormalized parameters. A less trivial contribution comes from the long-range part of the Coulomb interaction, which is just the classical electrostatic energy. Adding N electrons to the quantum dot increases the energy by the amount $\mathcal{E}_N = e^2 N^2/(2C)$, where C is the capacitance of the dot. This capacitance is of the order of the size of the dot and, hence, electrostatic energy $\mathcal{E}_N$ can exceed mean level spacing Δ and even Thouless energy $\mathcal{E}_c$. If both the tunneling amplitude from a lead to the dot and the temperature are not very large, the transport is blocked by the charging. This is the well-known Coulomb blockade effect. A scheme of a device where the Coulomb blockade effect can be important is represented in Fig. 9.9.

In the quantum devices under experimental study one can change the Fermi energy in the dot by changing the gate voltage V_g. Then the energy $\mathcal{E}_N$ of N electrons added to the dot takes the form

$$\mathcal{E}_N = \frac{e^2 N^2}{2C} - eNV_g \tag{9.161}$$

Changing the gate voltage one can choose such a value that the energy $\mathcal{E}_N$ of N electrons becomes equal to the energy $\mathcal{E}_{N+1}$ of $N+1$ electrons. It is clear that, in this situation, the charging effect does not block transport through the dot and, in fact, the Coulomb interaction can be excluded from the consideration. It is easy to obtain values of V_g corresponding to the equality $\mathcal{E}_N = \mathcal{E}_{N+1}$

$$V_g = \frac{e}{C}\left(N + \frac{1}{2}\right) \tag{9.162}$$

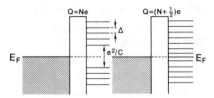

Fig. 9.10. Energy diagram for tunneling into the quantum dot when the average charge on the dot is Ne (left) and when it is $(N + 1/2)e$ (right). The energy levels for electronic excitations without changing the number of electrons in the dot are spaced in average by Δ. Adding an electron from the leads when the charge of the dot is Ne requires the energy $e^2/(2C)$.

Fig. 9.10 illustrates the spectrum of the single-particle density of states for the quantum dot. Experimentally, by changing the gate voltage one observes a sequence of peaks of the conductance at gate voltages determined by Eq. (9.162) (Kastner (1992)). These peaks are in the limit $e^2/C \gg \Delta$, when Eq. (9.162) holds, almost equidistant. The distance between them fluctuates slightly, but these fluctuations are of the order of Δ, which is much smaller than the distance between the peaks.

At first glance, one would expect that heights of the peaks in clean samples are also more or less equal. However, experimental study of transport through the dots has revealed order-of-magnitude fluctuations of the peak heights. This phenomenon is very similar to the Porter–Thomas level-width fluctuations in neutron scattering well known in nuclear physics (Porter (1965), Brody et al. (1981)). Using this analogy Jalabert, Stone, and Al-hassid (1992) suggested using phenomenological methods of nuclear physics to describe the fluctuations in quantum dots. The condensed matter problem is somewhat more reach by various physical phenomena than the nuclear physics problem. For example, one can have thermal effects. Applying a magnetic field one can study the entire crossover region between the orthogonal and unitary ensembles whereas in nuclear physics one has to deal with systems with unbroken time reversal symmetry.

Assuming that the level width and temperature are smaller than the mean-level spacing Δ (which is typically 0.5 K) one may consider only one single-particle level contribution to each resonance (Beenakker (1991), Meir, Wingreen, and Lee (1991)). This means that the Breit–Wigner formula can be used. Taking into account the temperature smearing of the Fermi distribution the dimensionless conductance g can be written in the form

$$g^{(n)} = - \int n'(\varepsilon) \, g^{(n)}(\varepsilon) \, d\varepsilon \qquad (9.163)$$

where $g^{(n)}$ is determined by Eqs. (9.133, 9.132) and $n(\varepsilon)$ is the Fermi function. Later, devices without an additional lead are considered and therefore one can put $\gamma = 0$. Eq. (9.163) describes the resonance transport, and conductance $g^{(n)}$ is the height of a peak in the dependence of the conductance of the dot on the gate voltage. Changing the gate voltage changes the number of electrons in the quantum dot. Therefore, different peaks correspond to different single-particle levels. But we know from previous sections that the level widths entering $g^{(n)}(\varepsilon)$, Eq. (9.133), fluctuate, and this naturally leads to the fluctuations of the peaks of the conductance. From Eq. (9.132), we see that in order to describe the fluctuations of the conductance peaks one has to know the distribution function of wave functions at points near the contacts. In the chaotic regime all space points of the dot are equivalent, so one needs to know the distribution function at an arbitrary point $\mathbf{r}$.

Experimentally, one can check a statistical description by averaging over different peaks rather than changing the shape of the dot (although the latter is also possible). Then, by making explicit calculations for the distribution function one should assume again that averaging over different parts of the one-particle energy spectrum is equivalent to averaging over impurities. This makes it possible to reduce the problem to computation of correlation functions with the zero-dimensional σ-model in the same way as used for numerous problems considered in preceding sections. However, before making the calculations let us show how the distribution function of wave functions was obtained in random matrix theory in nuclear physics (Brody et al. (1981)).

To start the calculation one should choose an arbitrary basis of eigenfunctions $\rho_m(\mathbf{r})$ and express the function $\varphi_n(\mathbf{r})$, with Eq. (9.132) determining the level width, as

$$\varphi_n(\mathbf{r}) = V^{-1/2} \sum_{m=1}^{\infty} a_{nm} \rho_m(\mathbf{r}) \tag{9.164}$$

where V is the volume. It is convenient to truncate the basis to a finite N-dimensional set and take the limit $N \to \infty$, as is usually done in random matrix theory. The main statistical hypothesis is that all coefficients a_{mn} are uniformly distributed. The only restriction on the coefficients $\{a_{nm}\}$ is imposed by normalization of the wave functions, and the probability density $\tilde{P}(\{a_{nm}\})$ can be written as

$$\tilde{P}(\{a_{nm}\}) = \frac{2}{\Omega_N} \delta \left(\sum_{m=1}^{N} |a_{nm}|^2 - 1 \right) \tag{9.165}$$

where Ω_N is the solid angle in N dimensions. Because of the truncation of the basis the condition of completeness of the basis $\{\rho_m\}$ should be written in the form

$$\sum_{m=1}^{N} \rho_m^2(\mathbf{r}) \equiv |\vec{\eta}|^2 = N \tag{9.166}$$

where $\vec{\eta}$ is an N-dimensional vector with components $\{\rho_m\}$. The distribution function $P(u)$ of the intensities at the point $\mathbf{r}$ is introduced as

$$W(v) = \frac{2}{\Omega_N} \int \delta \left(v - |\vec{a}\vec{\eta}|^2 \right) \delta \left(|\vec{a}|^2 - 1 \right) d\vec{a} \tag{9.167}$$

where the vector $\vec{a}$ is an N-dimensional vector with components $\{a_{nm}\}$. The distribution function $W(v)$, Eq. (9.167), is defined in such a way that

$$\int W(v)\, dv = 1 \tag{9.168}$$

In the unitary ensemble, one should integrate over complex vectors $\vec{a}$. Integrating first over the component of the vector $\vec{a}$ parallel to the vector $\vec{\eta}$ and using Eq. (9.166) one obtains

$$W(v) = \frac{2\pi}{N\Omega_N} \int \delta \left(|\vec{a}_\perp|^2 - \left(1 - \frac{v}{N} \right) \right) d\vec{a}_\perp \tag{9.169}$$

where $\vec{a}_\perp$ is the component perpendicular to $\vec{\eta}$. The remaining integration in Eq. (9.169) can be carried out easily. Taking the limit $N \to \infty$ one obtains for the unitary ensemble a simple formula

$$W(v) = \exp(-v) \tag{9.170}$$

Eq. (9.170) is already known to us. In the preceding section the same formula, Eq. (9.153), was obtained from the distribution function of conductances. The present derivation sheds light on the origin of the form of the distribution function, Eq. (9.170). We see that this form can be considered as a direct consequence of the uniform distribution of the coefficients $\{a_{nm}\}$ of the expansion, Eq. (9.164), of the wave function $\varphi_n(\mathbf{r})$ in functions $\{\rho_m\}$ forming a basis. Let us emphasize, however, that the microscopic calculation of the preceding section provided us with a stronger result, namely, the intensities at different points separated by a sufficiently large distance fluctuate independently. This result is very important for a study of fluctuations of peaks of conductance using Eqs. (9.132, 9.133) because the latter equation contains the wave functions at different points.

Computing the integral in Eq. (9.167) for real vectors $\vec{a}$ and $\vec{\eta}$ one can obtain the distribution function $W(v)$ for the orthogonal ensemble

$$W(v) = \frac{1}{\sqrt{2\pi v}}\exp\left(-\frac{v}{2}\right) \tag{9.171}$$

It is not difficult to see that both functions $W(v)$ determined by Eqs. (9.170, 9.171) satisfy the normalization condition, Eq. (9.168). The distribution functions $W(v)$ are universal and do not depend on details of models for the disorder.

9.7.2 Distribution function of conductance peaks

The statistical independence of fluctuations of the intensities in the unitary ensemble allows one immediately to calculate the distribution function $R(g_m)$ of the peaks g_m of the conductance (Prigodin, Efetov, Iida (1993)). This can be done by using the convolution

$$R(g_m) = \int\limits_0^\infty \int\limits_0^\infty \delta\left(g_m - g^{(n)}\right) W(v_1) W(v_2)\, dv_1 dv_2 \tag{9.172}$$

where $g^{(n)}$ is determined by Eqs. (9.133, 9.132, 9.163) and $\gamma_{1,2} = 2\Delta\alpha_{1,2}v_{1,2}/\pi$. The integral in Eq. (9.172) can be evaluated analytically in the regions $\alpha_{1,2}\Delta \ll T \ll \Delta$ and $T \ll \alpha_{1,2}\Delta$. In the former region for conductance $g^{(n)}$ one has

$$g^{(n)} = \frac{\pi}{2T}\frac{\gamma_{1n}\gamma_{2n}}{\gamma_{1n} + \gamma_{2n}} \tag{9.173}$$

whereas in the latter regime one obtains

$$g^{(n)} = \frac{4\gamma_{1n}\gamma_{2n}}{(\gamma_{1n} + \gamma_{2n})^2} \tag{9.174}$$

In the unitary ensemble, substituting Eq. (9.173) into Eq. (9.172) and calculating the integral with the function $W(v)$, Eq. (9.170), one can reduce the function $R(g_m)$ in the region $\alpha_{1,2}\Delta \ll T \ll \Delta$ to the form

$$g_m R(g_m) = x^2 \exp(-ax)\left[K_0(x) + aK_1(x)\right], \tag{9.175}$$

where $x = (2T/\Delta)\left(g_m/\sqrt{\alpha_1\alpha_2}\right)$,

$$a = 1 + \frac{1}{2}\left[\left(\frac{\alpha_1}{\alpha_2}\right)^{1/4} - \left(\frac{\alpha_2}{\alpha_1}\right)^{1/4}\right]^2 \tag{9.176}$$

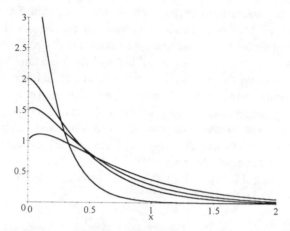

Fig. 9.11. Distribution of the resonance conductance $((\alpha_1\alpha_2)^{1/2}\Delta/2T)R(g_m)$ in the region $\alpha_{1,2}\nabla \ll T \ll \Delta$, Eq. (9.175), as a function of $x = (2T/\Delta)(g_m/(\alpha_1\alpha_2)^{1/2})$ for values $a = 1.0, 1.5, 2.0, 5.0$ characterizing the asymmetry of contacts a. The decay becomes faster with increasing a.

and $K_0(x)$, $K_1(x)$ are the modified Bessel functions. The function $R(g_m)$ is shown in Fig. 9.11. At $\alpha_1 = \alpha_2$ ($a = 0$) the distribution function $R(g_m)$ was first obtained through the phenomenological approach by Jalabert, Stone, and Alhassid (1992). As is seen from Fig. 9.11 the distribution is mainly concentrated in the region of small $g_m \sim \sqrt{\alpha_1\alpha_2}\Delta/2T \ll 1$. The probability of larger values of g_m decays exponentially with the rate dependent on the asymmetry of contacts a.

Experimentally, it is possible to change the heights of the barriers between the dot and the leads independently. Therefore, the function $R(g_m)$, Eq. (9.175), might be compared with a histogram of resonance conductance obtained for arbitrary α_1 and α_2. One can also see from Eq. (9.175) that if one of the barriers is much less penetrable than the other, such that $\alpha_1 \ll \alpha_2$, the conductance g_m is proportional to the square of the wave function at the weak contact 1. In this case the function $R(g_m)$, Eq. (9.175), reduces to the function W

$$R(g_m) = \frac{T}{\alpha_1\Delta}W\left(\frac{g_m T}{\alpha_1\Delta}\right) \tag{9.177}$$

which makes it possible to probe the distribution function of wave functions $W(v)$ directly.

In the limit of low temperature $T \ll \alpha_{1,2}\Delta$, by using Eqs. (9.172, 9.174) one can obtain the following expression for function $R(g_m)$:

$$R(g_m) = \frac{1}{2\sqrt{1-g_m}}\frac{1+(2-g_m)(a^2-1)}{\left[1+g_m(a^2-1)\right]^2} \tag{9.178}$$

We see from Eq. (9.178) that in this regime characteristic values of g_m have the order of unity. The function $R(g_m)$, Eq. (9.178), is represented in Fig. 9.12 for several values of a. One can see that now the distribution function $R(g_m)$ is shifted to $g_m = 1$. Additional discussion of the statistical properties of level widths and conductance peaks can be found in Mucciolo, Prigodin, and Altshuler (1995).

The preceding formulae were obtained for the unitary ensemble that corresponds to a sufficiently strong magnetic field. Experimentally, the magnetic field is an easily variable

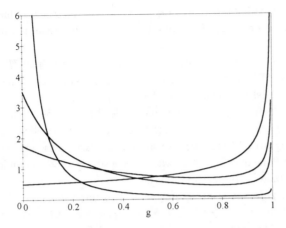

Fig. 9.12. Distribution of the resonance conductances $R(g_m)$ in the limit $T \ll \alpha_{1,2}\Delta$ for $a = 1.0,\ 1.5,\ 2.0,\ 5.0$. The value $R(0)$ grows with an increase in the degree of the asymmetry between the contacts.

parameter, and it is of interest to obtain the distribution functions of the resonance conductances for an arbitrary field. Technically, it is a more complicated task. Generally speaking, to describe the fluctuations of the resonance conductances, one should know the distribution function $W_2(v_1, v_2)$ of two variables

$$W_2(v_1, v_2) = \left\langle \delta\left(v_1 - V |\varphi_n(\mathbf{r}_1)|^2\right) \delta\left(v_2 - V |\varphi_n(\mathbf{r}_2)|^2\right) \right\rangle \qquad (9.179)$$

In the zero magnetic field one has to use the orthogonal ensemble. The function $W_2(v_1, v_2)$ has been calculated at arbitrary distance $|\mathbf{r}_1 - \mathbf{r}_2|$ for both the unitary (Prigodin (1995)) and orthogonal (Prigodin et al. (1995)) ensembles. The resulting formulae are rather cumbersome. However, if the points $\mathbf{r}_1$ and $\mathbf{r}_2$ are separated by distances much exceeding the electron wavelength, the wave functions at these points fluctuate independently and the distribution function $W_2(v_1, v_2)$ in this limit takes the form

$$W_2(v_1, v_2) \simeq W(v_1) W(v_2) \qquad (9.180)$$

with the functions $W(v)$ determined by Eqs. (9.170, 9.171). In the unitary ensemble this result has been extracted from Eq. (9.156).

To obtain the conductance peak distribution in the orthogonal ensemble one can start with Eq. (9.172) again. Using Eqs. (9.133, 9.132, 9.163) and calculating the integral with the function $W(v)$, Eq. (9.171), in the limit $\alpha_{1,2}\Delta \ll T \ll \Delta$ in the symmetric case ($\alpha_1 = \alpha_2 = \alpha_0$) one arrives at the following simple expression:

$$R(g_m) = \frac{2T}{\alpha_0 \Delta} \frac{\exp(-x)}{\sqrt{\pi x}} \qquad (9.181)$$

Eq. (9.181) was obtained for the first time by Jalabert, Stone, and Alhassid (1992). Changing the magnetic field one can obtain the entire crossover between Eqs. (9.181) and (9.175) taken at $a = 0$. However, such a function has not been computed yet. As we have seen (cf. Eq. (9.177)) the situation simplifies when one of the contacts is much weaker than the other.

In this case the distribution function of the resonance conductances reduces to the distribution function $W(v)$ of the wave functions at one point. This result is valid for an arbitrary magnetic field.

The function $W(v)$ in the presence of an arbitrary magnetic field H will be calculated later, but first I would like to demonstrate very recent results of an experimental study (Chang et al. (1996)) of the distribution of the conductance peaks in dots with point contacts where Coulomb blockade effects play an important role. Measurements of conductance were carried out at temperatures exceeding the level widths but smaller than the mean level spacing. Averaging over different peaks was done by varying the gate voltage. At $H = 0$, the averaging was done over 72 peaks, whereas at a strong $H \neq 0$ completely violating the time reversal symmetry, 216 peaks were used.

In the limits of zero and a strong magnetic field one can expect that the distribution function is described by Eqs. (9.181, 9.175), and these are the formulae that were checked in the experiment. The coefficients α_1, α_2 were not known and could be obtained by the best fit. At the same time, the contacts were symmetric such that one could put $a = 0$ in Eq. (9.175). Besides, it was assumed that the coefficients α_1, α_2 did not depend on such external parameters as the gate voltage and the magnetic field. Eqs. (9.175, 9.181) depend on only one parameter, $T/(\alpha_0 \Delta)$, and this parameter is the same in both equations. The histogram of conductance peak heights $G_{\max}$ extracted from the measurements by Chang et al. (1996) is depicted in Fig. 9.13. We see that the agreement of the theory and the experiment is quite reasonable. A similar experiment was carried out at the same time by Folk et al. (1996).

9.7.3 Distribution function of wave functions at an arbitrary magnetic field

Now, let us calculate the distribution function of wave functions at an arbitrary magnetic field. Of course, one can use the same method of extracting the distribution function of wave functions from the distribution function of conductances as in the preceding section. In this scheme one calculates all moments of the conductance, obtains coefficients of the participation ratio, and finally reconstructs the distribution function of wave functions. This is a rather indirect method and the computation for an arbitrary magnetic field can become very cumbersome.

Fortunately, there is another, more direct and simple scheme of calculation (Falko and Efetov (1994)). Following this method one should write the function $W(v)$ in the form

$$W(v) = \left\langle \Delta \sum_n \delta \left(v - V |\varphi_n(\mathbf{r})|^2 \right) \delta(\varepsilon - \varepsilon_n) \right\rangle \tag{9.182}$$

where the sum is extended over the full basis of states confined in a disordered box. For each configuration of disorder only one state contributes. Calculating moments of the distribution function $W(v)$, Eq. (9.182), one obtains the inverse participation ratio coefficients introduced in Eq. (9.154).

To use the supersymmetry technique, as usual one has to represent distribution function $W(v)$ in terms of the retarded and advanced Green functions. This can be done rather easily by using the discreteness of the energy levels in a finite volume. Using the Green functions $G_{\varepsilon\gamma}^{R,A}(\mathbf{r}, \mathbf{r}')$ in the form given by Eq. (7.78) (the dependence on γ is now explicitly

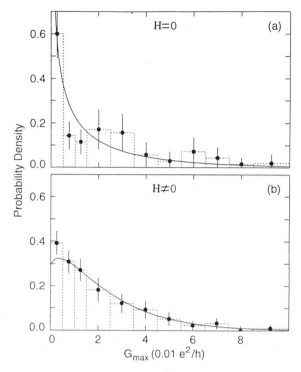

Fig. 9.13. Histograms of conductance peak heights G_{max} for different values of the magnetic field H: a) $H = 0$ and b) $H \neq 0$. Data are scaled to unit area as for a probability distribution. The solid lines correspond to a) Eq. (9.181) and b) Eq. (9.175) at $a = 0$. (From Chang et al. (1996).)

specified) one can rewrite the distribution function $W(v)$, Eq. (9.182), in the form

$$W(v) = \Delta \left\langle \lim_{\gamma \to 0} \sum_n \delta \left(v - \frac{i\gamma V}{2} G_{\varepsilon\gamma}^R (\mathbf{r}, \mathbf{r}) \right) \delta(\varepsilon - \varepsilon_n) \right\rangle \tag{9.183}$$

At first glance, the argument of the first δ-function in Eq. (9.183) is complex at finite γ, implying that this function is just a formal series. On the other hand, in the limit $\gamma \to 0$, the Green function is determined by the level n only and this argument becomes real. Eq. (9.183) can be rewritten further as

$$W(v) = \frac{\Delta}{2\pi} \lim_{\gamma \to 0} \lim_{\beta \to 0} < \int \delta \left(v - \frac{i\gamma V}{2} G_{\varepsilon\gamma}^R (\mathbf{r}, \mathbf{r}) \right)$$

$$\times \left[G_{\varepsilon\beta}^A (\mathbf{r}', \mathbf{r}') - G_{\varepsilon\beta}^R (\mathbf{r}', \mathbf{r}') \right] > d\mathbf{r}' \tag{9.184}$$

In Eq. (9.184) one should first take the limit $\beta \to 0$ and then $\gamma \to 0$. Since the distribution function $W(v)$ is represented in terms of a function of only two Green functions one can reduce the computation of this quantity to an integral over 8×8 supermatrices Q with the nonlinear σ-model. The reduction is similar to those used in Sections 7.3 and 9.6. One has to expand the δ-function in Eq. (9.184) and to write each term of the expansion in terms of an integral of the eight-component supervectors ψ, Eq. (4.6). This can be done by using the relation

$$n! i^{n+1} \left[G_{\varepsilon\gamma}^R \left(\mathbf{r}, \mathbf{r} \right) \right]^n G_{\varepsilon\beta}^A \left(\mathbf{r}', \mathbf{r}' \right) \tag{9.185}$$

$$= - \int \left| S^1 \left(\mathbf{r}', \mathbf{r}' \right) \right|^2 \left| S^2 \left(\mathbf{r}, \mathbf{r} \right) \right|^{2n} \exp \left(-\mathcal{L} \left[\psi \right] \right) D\psi$$

where $\mathcal{L}[\psi]$ is a proper Lagrangian. After averaging over impurities and decoupling by integration over supermatrix Q one should integrate over supervector ψ. When performing this Gaussian integration the main contribution is from pairing of S taken at the same space point. This approximation has been used in Chapter 6 for the two-level correlation function. It is justified provided the inequality $\tau \Delta \gg 1$ holds. Introducing the eight-component supervectors z_1 and z_2 such that $\bar{z}_1 = \left(0, \, 0, \, e^{i\zeta_1}, \, e^{-i\zeta_1}, \, 0, \, 0, \, 0, \, 0 \right)$ and $\bar{z}_2 = \left(0, \, 0, \, 0, \, 0, \, 0, \, 0, \, e^{i\zeta_2}, \, e^{-i\zeta_2} \right)$ and integrating over the supervectors ψ one finally arrives at

$$W \left(v \right) = \frac{1}{4} \int_0^{2\pi} \int_0^{2\pi} \frac{d\zeta_1 d\zeta_2}{(2\pi)^2} \lim_{\gamma \to 0} \int \bar{z}_1 \left(Q - \Lambda \right) z_1 \tag{9.186}$$

$$\times \delta \left(v - \frac{\gamma}{2\Delta} \left(\bar{z}_2 Q z_2 \right) \right) \exp \left(-F \left[Q \right] \right) dQ$$

If the parameter v is not very large, one may use the zero-dimensional σ-model, and Eq. (9.186) is written for this case. Taking the limit $\beta \to 0$ but keeping a finite γ one comes to the free energy F of the same form as in Eq. (8.42). The function F_1 is determined by Eq. (8.42), whereas the function F_0 takes the form

$$F_0 = -\frac{\pi \gamma}{8\Delta} \, \mathrm{str} \left(\Lambda Q \right) \tag{9.187}$$

Note that the expression for F_0, Eq. (9.187), is two times smaller than the corresponding expression written in Eq. (7.93). This is because Eq. (7.93) was written for the case when both the Green functions $G_{\varepsilon\gamma}^R$ and $G_{\varepsilon\beta}^A$ had equal absolute values of the imaginary parts, $\gamma = \beta$, whereas Eq. (9.187) is written for $\beta \to 0$.

To evaluate the integral in Eq. (9.186) it is convenient to use the magnetic field parametrization, Eqs. (8.45–8.51). As usual, because of the symmetry of free energy F one can integrate first over the anticommuting variables and phases, thus reducing the integral to an integral over the "eigenvalues" $\lambda_{1c,d}$ and $\lambda_{c,d}$. For an arbitrary γ one would obtain a rather complicated integral. However, we are interested in the limit $\gamma \to 0$ only and this simplifies all formulae. In this limit (taken at an arbitrary but nonvanishing magnetic field) the main contribution is from the range of $\lambda_{1d} \sim \Delta / (\pi \gamma \lambda_{1c}) \gg 1$, which infinitely increases at $\gamma \to 0$. For these large values of λ_{1d} one may keep in the matrix V_d, Eqs. (8.45, 8.47), terms of the highest order in λ_{1d}. The "scalar products" entering Eq. (9.186) can be reduced to the form

$$\left(\bar{z}_a Q z_a \right) = 4 \lambda_{1d} K_{a,d}^2 A_a, \qquad a = 1, \, 2 \tag{9.188}$$

where

$$K_{1,d} = 1 + 2\eta_d \eta_d^*, \qquad K_{2,d} = 1 - 2\kappa_d \kappa_d^*,$$

$$A_a = P + \frac{B}{2} e^{i\zeta_a} + \frac{B^*}{2} e^{-i\zeta_a}, \qquad P = \lambda_{1c} + 2 \left(\lambda_{1c} - \lambda_c \right) K_c,$$

$$B = \sqrt{\lambda_{1c}^2 - 1} \left(1 + 2K_c + 2K_c^2 \right) + 4 \sqrt{1 - \lambda_c^2} \, \eta_c \kappa_c, \qquad K_c = \eta_c \eta_c^* - \kappa_c \kappa_c^*$$

In the same limit for the energy F_0, Eq. (9.187), one obtains

$$F_0 = \frac{\pi \gamma}{\Delta} \lambda_{1d} P \tag{9.189}$$

The Jacobian J_d, Eq. (8.52), now becomes $J_d \to (8\pi \lambda_{1d})^{-2}$. All this makes it natural to change the variable $\gamma \lambda_{1d} = \lambda_1'$ and then the integration over all diffuson variables in the limit $\gamma \to 0$ becomes simple, especially because of the presence of the δ-function in the integrand in Eq. (9.186). As a result one obtains

$$W(v) = \int_0^{2\pi} \frac{d\zeta_2}{2\pi} \int_0^1 d\lambda_c \int_1^\infty d\lambda_{1c} J_c \frac{4\lambda_c^2}{(\lambda_{1c} + \lambda_c)^2} \frac{P^2}{A_2} \tag{9.190}$$

$$\times \exp\left(-\frac{vP}{A_2} - F_1\right) dR_c, \qquad dR_c = d\eta_c dn_c^* d\kappa_c^* d\kappa_c$$

Subsequent algebraic manipulations are straightforward, and the final expression for the distribution function $W(v)$ can be written as

$$W(v) = 2\int_1^\infty \left\{ \Phi_1(X) + \left[(tX)^2 - 1\right] \Phi_2(X) \right\} (tX)^2$$

$$\times I_0\left(vt\sqrt{t^2 - 1}\right) \exp\left[-vt^2 - X^2\left(t^2 - 1\right)\right] dt \tag{9.191}$$

In the preceding expression

$$\Phi_1(X) = \frac{\exp(-X^2)}{X} \int_0^X \exp(y^2)\, dy, \qquad \Phi_2(X) = \frac{1 - \Phi_1(X)}{X^2}$$

$I_0(x)$ is the Bessel function of the imaginary argument. The parameter X includes all the magnetic field dependence of $W(v)$. Its explicit dependence on the magnetic field is described by Eqs. (8.43, 8.44).

Using the general expression Eq. (9.191) for distribution function $W(v)$ one can obtain the asymptotics in the limits $X \to \infty$ and $X \to 0$, corresponding to the orthogonal and unitary ensembles. In these limits integration over t in Eq. (9.191) can be performed easily and one comes to Eqs. (9.170, 9.171). Thus, the results of the phenomenological approach used by Brody et al. (1981) and Jalabert, Stone, and Alhassid (1992) are confirmed by the microscopic calculation. The result for the unitary ensemble agrees with the corresponding result obtained in Sections 7.3 and 9.6. One can also check that $W(v)$ is properly normalized

$$\int_0^\infty W(v)\, dv = 1 \tag{9.192}$$

and that

$$\int_0^\infty vW(v)\, dv = 1 \tag{9.193}$$

This shows that the average density of states in the limit of weak disorder coincides with

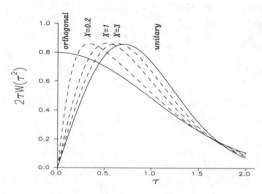

Fig. 9.14. The distribution function $2\tau W\left(\tau^2\right)$ of the local amplitude $\tau = |\varphi|\, V^{1/2}$ shown in the crossover regime for different values of parameter X proportional to the flux through the system.

the density of states v of the ideal electron gas. Eq. (9.193) demonstrates also that the wave functions are normalized by 1.

As usual, the crossover between the orthogonal and unitary ensembles occurs at relatively small magnetic fluxes ϕ through the area of the sample, $\phi \sim \phi_0\, (\Delta/\mathcal{E}_c)^{1/2}$. Fig. 9.14 illustrates the behavior of the distribution function $W\,(v)$ at different magnetic fields. To make the crossover between the orthogonal and unitary ensembles more spectacular the statistics of the single-electron wave functions is represented in terms of the function $2\tau W\left(\tau^2\right)$ of local amplitudes $\tau = |\varphi|\,\sqrt{V}$. As one can see from this figure and Eq. (9.191) the statistics of $|\varphi|^2$ undergoes the most pronounced changes in the asymptotical regions $v \to 0$ and $v \to \infty$. The probability of the zero amplitude $|\varphi|$ of a wave function immediately drops after any weak magnetic field is applied. One can see this fact from the asymptotical behavior at $v \to 0$

$$W\,(v) \approx \frac{2\sqrt{\pi}}{3X} \tag{9.194}$$

at small X.

The exponentially rare event of finding a pronounced splash of the single-electron density above the average level V^{-1} can be found from Eq. (9.191) using the saddle-point method of evaluating the integral over t. With exponential accuracy, it can be approximated as

$$W\,(v \gg 1) \propto \exp\left\{-\frac{v}{2}\left[1 + \left(\frac{X^2}{v}\right)\left(\sqrt{1 + \frac{2v}{X^2}} - 1\right)\right]\right\} \tag{9.195}$$

Eq. (9.195) describes the crossover between the exponents in Eqs. (9.170, 9.171) in a simple form.

Eqs. (9.191–9.195) were obtained in the approximation of the zero-dimensional σ-model and do not depend on details of disorder. The system can be very clean and the chaotic behavior can be due to the nonintegrability of the quantum billiard. However, such universality can hold only for amplitudes of the wave functions that are not very large when the wave function is still sensitive to the boundaries. The probability of very large amplitudes must be sensitive to disorder inside the quantum dot. Using Eq. (9.186) with free energy F expressed by taking into account the gradient term and the fact that essential $\gamma\lambda_{1d}$ is of the order of

v, one can obtain the criterion of the applicability of the zero-dimensional σ-model

$$|\varphi|^2 V \ll \frac{\mathcal{E}_c}{\Delta} \qquad (9.196)$$

Only amplitudes of the wave function satisfying the inequality Eq. (9.196) are described by Eqs. (9.191–9.195). For higher amplitudes one should use the σ-model in higher dimensions.

Recently Sommers and Iida (1994) calculated a distribution function of the amplitudes of wave functions directly using the random matrix approach. Their result is somewhat different from the distribution function, Eq. (9.191). At the same time, the asymptotics, Eqs. (9.170, 9.171), are correctly reproduced and the normalization conditions, Eqs. (9.192, 9.193), are fulfilled. The distribution function considered by Sommers and Iida (1994) is introduced in a different way; the difference in the definitions leads to the different results. Numerically, both distribution functions are very close to each other.

To finish the chapter I want to emphasize that transport through quantum devices is a very popular topic now and many interesting problems are being investigated both theoretically and experimentally. Some publications have appeared during the process of writing this book and I have had to insert new sections to update the material. At the same time some work clarifying important questions has to be done and as a result this chapter may appear somewhat incomplete. However, I hope that by reading it one can learn about important problems and methods to attack them.

10

Universal parametric correlations

10.1 Brownian motion model

10.1.1 Fokker–Planck equation for the Coulomb gas model

In the preceding chapters the relation between random matrix theory and the zero-dimensional supersymmetric σ-model was emphasized many times. This equivalence makes it possible to use the σ-model for a description of such nontrivial problems as transport through quantum dots in the regime of chaotic dynamics. By changing external parameters such as magnetic field or gate voltage one can study correlations of physical quantities at different values of these parameters. One example has been considered in Section 9.4. One can also study the dependence of average quantities on the external parameters, and the persistent currents considered in Chapter 8 are this type of problem. In many cases the final formulae are quite universal, depending only on the mean level spacing and several parameters characterizing changes in the initial Hamiltonian.

It is clear now that study of parametric correlations leads to new and very interesting and unexpected results. Again, much important work has been performed in the last 1–2 years, and therefore the content of this chapter reflects only the present state of the art.

I want to present first what was known about parametric correlations in random matrix theory in the past. The most important work in this area is that of Dyson (1962b), who proposed to use the idea of Brownian motion, well known in kinetic theory (Chandrasekhar ((1943), Uhlenbeck and Ornstein (1930), Wang and Uhlenbeck (1945), Isihara (1971)), to describe parametric variations of physical quantities. Using the ideas of a kinetic theory implies that time dependence was somehow included in the theory. What does it mean for quantum mechanical problems where one has to find energy levels and wave functions of the stationary Schrödinger equation?

To understand the logic of the Dyson model let us recall formulae of in Section 6.1. According to random matrix theory one can calculate physical quantities in a Gaussian ensemble by using the distribution function Eqs. (6.11, 6.12) with the variables r_i describing the energy levels. The same can be reformulated in terms of a one-dimensional gas of N particles moving in a potential W

$$W = \frac{1}{2\tilde{a}^2} \sum_i r_i^2 - \sum_{i<j} \ln |r_i - r_j| \qquad (10.1)$$

where $\tilde{a}^2 = \beta a^2$. In such an interpretation the variables r_i are the coordinates of the particles on an infinite straight line and calculation of averages with the distribution function given by Eqs. (6.11, 6.12) is equivalent to computation of thermodynamic quantities. Note that the particles repel each other and the confining potential in Eq. (10.1) is really necessary to keep the system in equilibrium. The partition function of the model can be found by integration over the coordinates of the particles of the function P

$$P(\{r_i\}) = A_{N\beta} \exp(-\beta W), \qquad i = 1, 2, \ldots, N \tag{10.2}$$

It is clear from Eq. (10.2) that β plays the role of inverse temperature. (The fact that β also enters the constant $\tilde{a}$ determining the confining potential is not important because after proper rescaling of the coordinates $\tilde{a}$ can be removed from the theory. Later, it is assumed that such rescaling has been done and so one may use Eq. (10.1) with $\tilde{a} = 1$.) This system of the particles in the thermodynamic equilibrium is often called the Coulomb gas model.

Having constructed the thermodynamics, why not extend the theory to include nonequilibrium processes? Of course, this cannot be done in a unique way and one should use such criteria as simplicity and beauty. It is also important to add that in the nonequilibrium processes for the Coulomb gas time is fictitious and can be related to an external parameter of the initial Hamiltonian. The validity of any kinetic picture can only be checked by using microscopic theories. According to Dyson (1962b) the most natural way to introduce motion of the particles is by using the assumption that each particle has no inertia and is subjected to action of a fluctuating force f_i and a friction force proportional to the velocity. This is exactly the description of a heavy Brownian particle. The motion of such particles obeys a system of Langevin equations

$$\gamma \frac{dr_i}{dt} = -\frac{\partial W}{\partial r_i} + f_i, \qquad i = 1, 2, \ldots, N \tag{10.3}$$

where the random force f is a Gaussian white noise of zero mean $\overline{f_i(t)} = 0$ and the variance

$$\overline{f_i(t) f_j(t')} = \frac{2\gamma}{\beta} \delta_{ij} \delta(t - t') \tag{10.4}$$

Using Eqs. (10.3, 10.4) one obtains

$$\gamma \overline{\delta r_i} = -\frac{\partial W}{\partial r_i} \delta t \tag{10.5}$$

$$\gamma \overline{(\delta r_i)^2} = \frac{2}{\beta} \delta t \tag{10.6}$$

Starting from Eqs. (10.3, 10.4) one can derive in a standard way the Fokker–Planck equation, which determines evolution of the distribution function $P(\{r_i\}, t)$

$$\gamma \frac{\partial P}{\partial t} = \sum_{i=1}^{N} \frac{\partial}{\partial r_i} \left(P \frac{\partial W}{\partial r_i} + \beta^{-1} \frac{\partial P}{\partial r_i} \right) \tag{10.7}$$

In the limit $t \to \infty$ one obtains the equilibrium solution P_{eq}, Eq. (10.2).

One important step to be taken is to relate the model given by Eqs. (10.3–10.7) to a physical system and, thus, express the fictitious time t through a physical parameter X characterizing the initial Hamiltonian. Furthermore, one needs a microscopic interpretation of the coefficient γ.

The random force f acting on the fictitious particles causes their coordinates to fluctuate. At the same time, because of repulsion the coordinates cannot be close to each other. This

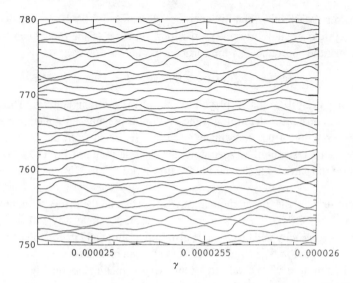

Fig. 10.1. Typical dependence of the energy levels $r_i(\gamma)$ on a parameter γ. This plot is based on a calculation of the spectrum of the hydrogen atom in a strong magnetic field by Goldberg et al. (1991). The parameter γ characterizes the magnetic field (cf. Eq. (6.16)).

is reminiscent of the dependence of the energy levels of a chaotic system on such external parameters as magnetic field. As an illustration of a dependence of the energy levels on the parameter γ in the model of the hydrogen atom in a strong magnetic field (cf. Eq. (6.16)) one may use Fig. 10.1.

A relation between the fictitious time t and a parameter X was proposed in recent works by Beenakker (1993) and Beenakker and Rejaei (1994). At $X = 0$ one has a set of energy levels $r_i^{(0)}$, so that the distribution function $P(\{r_i\}, 0)$ has the form

$$P(\{r_i\}, 0) = \prod_{i=1}^{N} \delta\left(r_i - r_i^{(0)}\right) \tag{10.8}$$

Choosing $t = 0$ at $X = 0$ one may identify

$$t = X^2 \tag{10.9}$$

This is the simplest relation between t and X that is consistent with the average initial rate of change of the energy levels. Eq. (10.9) can be understood with the help of the following relations:

$$\overline{\left(r_i(X) - r_i^{(0)}\right)^2} = X^2 \overline{\left(\frac{dr_i}{dX}\right)^2} + O\left(X^3\right) \tag{10.10}$$

This is the lowest order of the expansion in the small parameter X. On the other hand, the same can be calculated by using the Langevin equations, Eqs. (10.3, 10.4), to give

$$\overline{\left(r_i(t) - r_i^{(0)}\right)^2} = \frac{2t}{\beta\gamma} + O\left(t^2\right) \tag{10.11}$$

With the choice given by Eq. (10.9) one immediately obtains the relation

$$\frac{2}{\beta \gamma} = \overline{\left(\frac{dr_i}{dX}\right)^2}$$

(10.12)

between the friction coefficient and the mean square of the level velocity.

Correlations in the model under consideration can be characterized by the density-of-states $S\left(r, X, r', X'\right)$ and current $C\left(r, X, r', X'\right)$ correlation functions introduced as

$$S\left(r, X, r', X'\right) = \sum_{i,j} \overline{\delta\left(r - r_i\left(X\right)\right) \delta\left(r' - r_j\left(X'\right)\right)}$$

$$- \left(\sum_i \overline{\delta\left(r - r_i\left(X\right)\right)}\right) \left(\sum_i \overline{\delta\left(r' - r_j\left(X'\right)\right)}\right),$$

(10.13)

$$C\left(r, X, r', X'\right) = \sum_{i,j} \overline{\dot{r}_i\left(X\right) \dot{r}_j\left(X'\right) \delta\left(r - r_i\left(X\right)\right) \delta\left(r' - r_j\left(X'\right)\right)}$$

(10.14)

where

$$\dot{r}\left(X\right) = \frac{dr\left(X\right)}{dX}$$

is the level "velocity." One can check the following identity by direct differentiation:

$$\frac{\partial^2}{\partial r \partial r'} C\left(r, X, r', X'\right) = \frac{\partial^2}{\partial X \partial X'} S\left(r, X, r', X'\right)$$

(10.15)

Once Eq. (10.7) is solved and the function $P\left(\{r_i\}, t\right)$ is found, one can write correlation functions as an integral with this distribution function. For example, the density-of-states correlation function $S\left(r, X, r', X'\right)$ takes the form

$$S\left(r, 0, r', X\right) = \int_{-\infty}^{\infty} dr_1^{(0)} \cdots \int_{-\infty}^{\infty} dr_N^{(0)} \int_{-\infty}^{\infty} dr_1 \cdots \int_{-\infty}^{\infty} dr_N$$

$$\times \sum_{i,j} \delta\left(r - r_i^{(0)}\right) \delta\left(r' - r_j\right) P_{\text{eq}}\left(\{r_n^{(0)}\}\right) \left(P\left(\{r_n\}, X^2\right) - P_{\text{eq}}\left(\{r_n\}\right)\right)$$

(10.16)

Calculation of the functions C and S within the Brownian motion model is, generally speaking, not a simple task. In the next section of the chapter it will be demonstrated that the supersymmetry technique is very helpful for studying parametric correlations, but now let us discuss some results that can be obtained by older approaches.

10.1.2 Brownian motion of a matrix

The Brownian motion model can also be formulated for the matrix $\mathcal{H}$ depending on the fictitious time t, of which the variables r_i are eigenvalues. Brownian motion of $\mathcal{H}$ is defined by Langevin equations for each matrix element separately. The ensemble averages are written in the form

$$\gamma \overline{\delta \mathcal{H}_{ij}} = -\mathcal{H}_{ij} \delta t$$

(10.17)

$$\gamma \overline{\delta \mathcal{H}_{ij} \delta \mathcal{H}_{ji}} = 2 g_{ij} \delta t$$

(10.18)

where $g_{ij} = \delta_{ij} + (1 - \delta_{ij})(\beta/2)$. The corresponding Fokker–Planck equation for distribution $P(\{\mathcal{H}_{ij}\}, t)$ of the matrix elements $\mathcal{H}_{ij}$ can be written as

$$\gamma \frac{\partial P}{\partial t} = \sum_{i,j} \left[g_{ij} \frac{\partial^2 P}{\partial \mathcal{H}_{ij} \partial \mathcal{H}_{ji}} + \frac{\partial}{\partial \mathcal{H}_{ij}} (\mathcal{H}_{ij} P) \right] \qquad (10.19)$$

The solution of Eq. (10.19) with the initial condition $\mathcal{H} = \mathcal{H}^{(0)}$ at $t = 0$ is known explicitly (Uhlenbeck and Ornstein (1930), Mehta (1991)) and can be written as

$$P(\mathcal{H}, t) = C \left(1 - q^2\right)^{-p/2} \exp\left(-\frac{\beta \left(\mathcal{H} - q\mathcal{H}^{(0)}\right)^2}{2 \left(1 - q^2\right)}\right) \qquad (10.20)$$

where $q = \exp(-t/\gamma)$. The function $P(\mathcal{H}, t)$ is invariant under simultaneous unitary transformations of the matrices $\mathcal{H}$ and $\mathcal{H}^{(0)}$. In the limit $t \to 0$ the solution $P(\mathcal{H}, t)$ is the product of δ-functions of all matrix elements of $(\mathcal{H} - \mathcal{H}^{(0)})$. In the opposite limit the probability density Eq. (10.20) tends to the stationary form $P_{eq}(\mathcal{H})$

$$P_{eq}(\mathcal{H}) = C \exp\left(-\frac{\beta \, \mathrm{tr} \, \mathcal{H}^2}{2}\right) \qquad (10.21)$$

Within this formulation the density-of-states correlation function $S(\lambda, 0, \lambda', X)$ can be written as

$$S(r, 0, r', X) = \int d\mathcal{H}^{(0)} d\mathcal{H} \, \mathrm{tr} \, \delta \left(r - \mathcal{H}^{(0)}\right) \mathrm{tr} \, \delta \left(r' - \mathcal{H}\right)$$

$$\times P_{eq}\left(\mathcal{H}^{(0)}\right) \left(P\left(\mathcal{H}, X^2\right) - P_{eq}\left(\mathcal{H}^{(0)}\right)\right) \qquad (10.22)$$

We see that correlation functions of quantities taken at different values of physical parameters X can be represented in a form of Gaussian integral. This can be convenient because one does not need to solve equations and the integral that arises is similar to the corresponding integral in the stationary regime. The only difference is that now, instead of integrating over one matrix $\mathcal{H}$, one should integrate over two coupled matrices $\mathcal{H}$ and $\mathcal{H}^{(0)}$.

Are the approaches based on the different Fokker–Planck equations, Eqs. (10.7, 10.19), equivalent? The answer is yes and this can be proved by using the independence of Eqs. (10.17, 10.18) of the representation of $\mathcal{H}$. With this property one may choose such a representation that $\mathcal{H}$ is diagonal at time t and its components are

$$\mathcal{H}_{ii}(t = 0) = r_i \qquad (10.23)$$

At a later time $t + \delta t$ the matrix $\mathcal{H} + \delta \mathcal{H}$ is no longer diagonal. Its eigenvalues $r_i + \delta r_i$ can be found by the standard quantum mechanical perturbation theory. In the second order the result reads

$$\delta r_i = \delta \mathcal{H}_{ii} + \sum_{j \neq i} \frac{\delta \mathcal{H}_{ij} \delta \mathcal{H}_{ji}}{r_j - r_i} \qquad (10.24)$$

Higher terms of the perturbation theory do not contribute to the first order in δt, which is necessary to write the Fokker–Planck equation. Taking the ensemble average of both sides of Eq. (10.24) and using Eqs. (10.17, 10.18) one comes to Eq. (10.5). Taking the ensemble average of the squares of both sides of Eq. (10.24) with the help of Eq. (10.18) one arrives at Eq. (10.6). Eqs. (10.5, 10.6) supplemented with the assumption that the process is Markovian are sufficient to obtain Eq. (10.7), thus proving the equivalence of the approaches.

10.1.3 Relation between the Brownian motion and Calogero–Sutherland models

The Fokker–Planck equation, Eq. (10.7), can be reduced to a multiparticle Schrödinger equation (Sutherland (1972), Beenakker and Rejaei (1994)) that helps to solve it. This can be done by using the substitution

$$P\left(\{r_i\},t\right) = \exp\left(-\frac{\beta}{2}W\left(\{r_i\}\right)\right)\Psi\left(\{r_i\},t\right) \tag{10.25}$$

Substituting Eq. (10.25) into Eq. (10.7) one obtains the following equation for the function Ψ:

$$-\frac{\partial\Psi}{\partial t} = -\frac{1}{\beta\gamma}\sum_{i=1}^{N}\frac{\partial^2\Psi}{\partial r_i^2} + \frac{1}{2\gamma}\Psi\sum_{i=1}^{N}\left[\frac{\beta}{2}\left(\frac{\partial W}{\partial r_i}\right)^2 - \frac{\partial^2 W}{\partial r_i^2}\right] \tag{10.26}$$

Calculation of the derivatives of W in Eq. (10.26) with W determined by Eq. (10.1) is straightforward (remember that $\tilde{a} = 1$). As a result of the identity

$$\sum_i\sum_{j\neq i}\sum_{k\neq i}\frac{1}{\left(r_i - r_j\right)\left(r_i - r_k\right)} = 0 \tag{10.27}$$

all three-body terms are canceled and one obtains the final form of the Schrödinger equation

$$-\frac{\partial\Psi}{\partial t} = \left(\mathcal{H}_C - \mathcal{E}_0\right)\Psi, \tag{10.28}$$

$$\mathcal{H}_C = \frac{1}{\beta\gamma}\left[-\sum_i\frac{\partial^2}{\partial r_i^2} + \beta\left(\frac{\beta}{2} - 1\right)\sum_{i<j}\frac{1}{\left(r_i - r_j\right)^2} + \frac{\beta^2}{4}\sum_i r_i^2\right], \tag{10.29}$$

$$\mathcal{E}_0 = \frac{N}{2\gamma} + N\left(N - 1\right)\frac{\beta}{4\gamma} \tag{10.30}$$

Eq. (10.29) describes a model of interacting particles with the interaction inversely proportional to the square of the distance between the particles. The interaction is attractive for $\beta = 1$, repulsive for $\beta = 4$, and equal to zero at $\beta = 0$. The system is confined by the parabolic potential. The model with the Hamiltonian, Eq. (10.29), was considered by Calogero (1969) long ago. Sutherland (1971, 1972) considered a somewhat different model of particles not confined by the parabolic potential but moving on a ring. However, in the limit $N \to \infty$ both models coincide. Calogero has shown that the model described by Eq. (10.29) is completely separable and the spectrum is integrable. One can rather easily obtain a time-independent solution of Eqs. (10.28–10.30) by using the fact that $\exp\left(-\beta W\right)$ is a time-independent solution of the initial Fokker–Planck equation, Eq. (10.7). Using the substitution Eq. (10.25) one comes to the eigenvalue equation

$$\mathcal{H}_C \exp\left(-\frac{\beta}{2}W\right) = E_0 \exp\left(-\frac{\beta}{2}W\right) \tag{10.31}$$

For a particular ordering of the "coordinates" $r_1, r_2, \ldots, r_N$, the function $\Psi_0 \propto \exp\left(-\beta W/2\right)$ is an eigenfunction of the N-fermion Hamiltonian $\mathcal{H}_C$. Since it has no nodes, it is the ground state. Antisymmetrization yields the fermion wave function Ψ_0 for the ground state

$$\Psi_0\left(\{r_i\}\right) = \text{const} \times \exp\left(-\frac{\beta}{2}W\left(\{r_i\}\right)\right)\prod_{i<j}\frac{r_i - r_j}{|r_i - r_j|} \tag{10.32}$$

Calculation of correlation functions at different times is more difficult because one has to know all matrix elements. Drastic simplifications occur only in the unitary ensemble because at $\beta = 2$ the interparticle interaction in Eq. (10.29) vanishes. In this case, the correlation functions C and S, Eqs. (10.13) and (10.14), are just the current and density-of-states correlation functions of an ideal one-dimensional fermion gas. This can be shown by using the relation

$$P_{eq}\left(\{r_i\}\right) = \Psi_0^2\left(\{r_i\}\right) \tag{10.33}$$

The solution of Eq. (10.28) can be written in the operator form as

$$\Psi\left(\{r_i\},t\right) = \exp\left(-\left(\mathcal{H}_C - \mathcal{E}_0\right)t\right)\Psi\left(\{r_i\},0\right) \tag{10.34}$$

The wave function at zero time, $\Psi\left(\{r_i\},0\right)$, can be obtained from Eqs. (10.8, 10.25, 10.31)

$$\Psi\left(\{r_i\},0\right) = \Psi_0^{-1}\left(\{r_i\}\right)\prod_i \delta\left(r_i - r_i^{(0)}\right) \tag{10.35}$$

Introducing the density operators as

$$n\left(r\right) = \sum_{i=1}^{N}\delta\left(r - r_i\right), \tag{10.36}$$

$$n\left(r,t\right) = \exp\left(\mathcal{H}_C t\right)n\left(r\right)\exp\left(-\mathcal{H}_C t\right) \tag{10.37}$$

and using Eqs. (10.25, 10.32, 10.33, 10.35) one can reduce Eq. (10.16) to the form

$$S\left(r,0,r',X\right) = \left\langle n\left(r,0\right)n\left(r',X^2\right)\right\rangle_0 - \left\langle n\left(r\right)\right\rangle_0\left\langle n\left(r'\right)\right\rangle_0 \tag{10.38}$$

where $\langle A\rangle_0$ represents the following integral:

$$\langle A\rangle_0 = \int_{-\infty}^{\infty} dr_1 \cdots \int_{-\infty}^{\infty} dr_N \Psi_0^*\left(\{r_i\}\right)A\Psi_0\left(\{r_i\}\right)$$

for an arbitrary operator A. Eq. (10.38) is exactly the density-of-states correlation function of the ideal fermion gas with the parameter X^2 playing the role of imaginary time (cf. Eq. (10.9)). The same can be done for the current correlation function C. One can compute the correlation function given by Eq. (10.38) by writing the density operator, for example, in second quantization. After standard manipulations the density-of-states correlation function S can be written as

$$S\left(r,0,r',X\right) = \sum_{m=N}^{\infty}\sum_{k=0}^{N-1}\phi_k\left(r\right)\phi_k\left(r'\right)\phi_m\left(r\right)\phi_m\left(r'\right)\exp\left(\left(\epsilon_k - \epsilon_m\right)X^2\right) \tag{10.39}$$

where ϕ_k and ϵ_k are single-particle eigenfunctions and eigenenergies of the Hamiltonian of noninteracting particles with the quadratic potential (Eq. (10.29) with $\beta = 2$). In the limit $N \to \infty$ the sum in Eq. (10.39) can be simplified. The main contribution is from large k and m, and one may use the quasi-classical approximation. The sum can be replaced by the integral and the final expression for the density-of-states correlation function can be written in the form (Beenakker and Rejaei (1994))

$$S\left(r,X\right) = \rho_0^2 \int_0^1 ds \int_1^{\infty} ds' \exp\left[\alpha X^2\left(s^2 - s'^2\right)\right]\cos\left(\pi\rho_0 rs\right)\cos\left(\pi\rho_0 rs'\right) \tag{10.40}$$

where $S\left(r'-r,\,X\right) \equiv S\left(r,\,0,\,r',\,X\right)$, $\rho_0 = (2N)^{1/2}/\pi$, $\alpha = N/\gamma$. All calculations in the present section were performed for the model with the quadratic confining potential, Eq. (10.1). However, in the limit $N \to \infty$ the form of confining potential is not essential and everything depends only on the mean interparticle spacing ρ_0^{-1}. For example, one can obtain similar results by studying the periodic Sutherland model with the Hamiltonian $\mathcal{H}_S$

$$\mathcal{H}_S = -\sum_{i=1}^{N} \frac{\partial^2}{\partial r_i^2} + \beta \left(\frac{\beta}{2} - 1\right) \left(\frac{\pi}{N}\right)^2 \sum_{i<j} \frac{1}{\sin^2\left[\pi\left(r_i - r_j\right)/N\right]} \tag{10.41}$$

The pair interaction in Eq. (10.41) scales as the inverse square of the chord length between the particles on the ring of circumference N. The mean interparticle spacing is chosen to be 1. The Sutherland model can be obtained from the Brownian motion model for the Dyson circular ensembles (Mehta (1991)) in the same way as the Calogero model has been obtained for the Gaussian ensembles. Although in the limit $N \to \infty$ the Sutherland model must give the same results as the Calogero model, the periodicity of the Hamiltonian $\mathcal{H}_S$, Eq. (10.41), leads to a simplification of mathematical manipulations because there is no need to introduce a confining potential. In the unitary ensemble, the Hamiltonian $\mathcal{H}_S$ describes free noninteracting fermions. The single-particle wave functions ϕ_p take the simple form

$$\phi_p(r) = N^{-1/2} \exp\left(irp\right), \qquad p = \frac{2\pi k}{N} \tag{10.42}$$

where $k = 0, \pm 1, \pm 2, \ldots$ This leads immediately to the standard expression for the density-of-states correlation function S, Eq. (10.38),

$$S(r, t) = \frac{2}{N^2} \sum_{|p| \leq p_0} \sum_{|p_1| > p_0} \exp\left[-t\left(p_1^2 - p^2\right) + i\left(p_1 - p\right)r\right] \tag{10.43}$$

where p_0 is the Fermi momentum $p_0 = \pi$. In the limit $N \to \infty$ the sum in Eq. (10.43) has to be replaced by the integral and it is clear that after proper rescaling Eqs. (10.40–10.43) coincide.

The material presented in this section demonstrates that one can generalize the random matrix theory by introducing a fictitious time and writing equations of Brownian motion for the matrices. The Brownian motion models are equivalent to one-dimensional models of interacting fermions; that equivalence helps in some cases to solve the corresponding Fokker–Planck equations. Using the Brownian motion one can describe parametric correlations of different physical quantities. However, the theory developed in the present section is purely phenomenological and it is not clear whether it is related to realistic models or not. Besides, the external parameters were such that they did not change the symmetry of the ensemble.

The preceding results have only recently been confirmed by the supersymmetry technique. In the next section, important new properties of parametric correlations for mesoscopic systems and quantum chaos are obtained directly from first principles using a proper supersymmetric σ-model. The connection between the phenomenological theories of the present section and the microscopic consideration of the next is established in Section 10.3.

10.2 Universalities in disordered and chaotic systems spectra

10.2.1 General formulae

Although the Brownian motion model was proposed long ago, its applicability to a description of real systems remained unclear and less attention was paid to the problem of
the dependence of the energy levels on external parameters. Only recently was interest in
studying spectral correlations in disordered and chaotic systems aroused by works by Szafer
and Altshuler (1993), Simons and Altshuler (1993), Simons et al. (1993), Faas et al. (1993).

The main issue of these works is that some correlation functions of physical quantities
taken at different parameters can be as universal as the two-level correlation functions in the
random matrix theory of Wigner and Dyson. In other words, after a proper rescaling such
correlation functions have a universal form that does not depend on details of the system
under consideration. The formulae obtained were suggested to be valid for such different
models as chaotic billiards, a hydrogen atom in a strong magnetic field, and disordered
conductors and were checked by computer simulations. The basic analytical results were
derived by using the supersymmetry technique, and in this section I want to present the
main ideas and formulae of the works.

In preceding chapters, a parametric dependence of several physical quantities has already
been considered. The persistent current as a function of a magnetic flux through mesoscopic
rings discussed in Chapter 8 is a typical example. In Section 9.4 a correlation function
of conductances at different magnetic fields and energies was calculated. The discussion
presented later is of a more general character. Rather general autocorrelators as a function
of both the energy and some arbitrary external perturbation controlled by a parameter Y
will be examined. The variable Y can denote the strength of some field, such as magnetic
flux through a ring or the position of some impurity. The Schrödinger equation describing
the model can be written in the form

$$\mathcal{H}(Y)\,\varphi_i(Y) = \varepsilon_i(Y)\,\varphi_i(Y) \tag{10.44}$$

where $\varphi_i(Y)$ and $\varepsilon_i(Y)$ are eigenfunctions and eigenenergies depending on the parameter
Y. An interesting quantity to be studied is the correlation function of level "velocities"

$$C(Y) = \Delta^{-2}\langle \partial_{\bar{Y}}\varepsilon_i(\bar{Y})\,\partial_{\bar{Y}}\varepsilon_i(\bar{Y}+Y)\rangle \tag{10.45}$$

where $\partial_Y \equiv \partial/\partial Y$, and the statistical averaging, denoted by $\langle\cdots\rangle$, may be performed
over realizations of impurities, or over a range of energy levels, or over the parameter $\bar{Y}$,
provided that by changing the parameter one does not change the symmetry of the system.
For example, if $\bar{Y}$ corresponds to a magnetic field, this field must be strong enough that in
varying this parameter one always remains within the unitary ensemble. It is also convenient
to assume that

$$\langle \partial_Y \varepsilon_i(Y)\rangle = 0 \tag{10.46}$$

The statistical property of the spectrum $C(Y)$ can be related to the conductance of the
sample. According to Thouless (1974) the conductance depends on the sensitivity of the
energy levels to a change in the boundary conditions. If the parameter Y describes the phase
of a quasi-periodic boundary condition, the dimensionless conductance g_c is proportional to
the typical level curvature $\Delta^{-1}\partial^2\varepsilon_n/\partial Y^2$. In a broader context the meaning of this relation
was discussed by Akkermans and Montambaux (1992) (see also the detailed discussion by
Kamenev and Braun (1994) and by Braun and Montambaux (1994)). Using random matrix

theory it was shown that up to a constant prefactor, the conductance can be reduced to the average square of the level velocity. Therefore, it is convenient to define a generalized conductance of the sample as

$$C\left(0\right) = \Delta^{-2} \left\langle \left(\partial_Y \varepsilon_i\right)^2 \right\rangle \tag{10.47}$$

for an arbitrary perturbation Y. The term *conductance* used in this context originates from the Kubo linear response theory, where the conductance is expressed in terms of velocity–velocity correlators. For some perturbations such as magnetic flux perturbation the quantity $C\left(0\right)$ is proportional to the real conductance. This quantity is very important for discussing the universal properties of the parametric correlations because its square root can serve as a natural unit of measuring the parameter Y. After the reparametrization

$$\epsilon_i\left(Y\right) = \frac{\varepsilon_i\left(Y\right)}{\Delta}, \qquad y = Y\sqrt{C\left(0\right)} \tag{10.48}$$

one can eliminate from, for example, the correlation function of the density of states any dependence on detailed properties of the system. This function allows the possibility of obtaining numerous correlators of energy derivatives, which, expressed in terms of the rescaled variables, also become universal. Applying the rescaling of Eq. (10.48) to the correlation function $C\left(Y\right)$ and defining $c\left(y\right) = C\left(Y\right)/C\left(0\right)$ one obtains

$$c\left(y\right) = \left\langle \partial_{\bar{y}} \epsilon_i\left(\bar{y}\right) \partial_{\bar{y}} \epsilon_i\left(\bar{y} + y\right) \right\rangle \tag{10.49}$$

It follows from Eqs. (10.47–10.49) that

$$c\left(0\right) = 1 \tag{10.50}$$

Assuming the vanishing of the boundary terms at large enough y one comes to the sum rule

$$\int c\left(y\right) dy = 0 \tag{10.51}$$

Alternatively one could examine the following correlation function of velocities:

$$\tilde{c}\left(x, y\right) = \frac{\sum_{ij} \left\langle \partial_{\bar{y}} \epsilon_i\left(\bar{y}\right) \partial_{\bar{y}}\left(\bar{y} + y\right) \delta\left(\epsilon_i\left(\bar{y} + y\right) - \epsilon_j\left(\bar{y}\right) - x/\pi\right)\right\rangle}{\sum_{kl} \left\langle \delta\left(\epsilon_k\left(\bar{y} + y\right) - \epsilon_l\left(\bar{y}\right) - x/\pi\right)\right\rangle} \tag{10.52}$$

where $x = \pi \omega / \Delta$.

In contrast to $c\left(y\right)$ containing the level velocities of the same level, the correlation function $\tilde{c}\left(\omega, y\right)$ measures the averaged correlation of the velocities at fixed energies. Although one can argue that both correlation functions $c\left(y\right)$ and $\tilde{c}\left(\omega, y\right)$ are universal for all perturbations X and depend only on the Dyson ensemble of the system, only the function $\tilde{c}\left(\omega, y\right)$ can be obtained in an explicit analytical form.

In order to compute the correlation functions and to check the universalities one can start with a usual microscopic Hamiltonian $\mathcal{H}$ containing disorder and a magnetic field. A nonrandom external potential $U_0 w\left(\mathbf{r}\right)$ can also be added

$$\mathcal{H} \to \mathcal{H} + U_0 w\left(\mathbf{r}\right) \tag{10.53}$$

to make the situation more general. The profile of the background potential $U_0 w\left(\mathbf{r}\right)$ is assumed to be arbitrary, but with the zero spatial average

$$\int w\left(\mathbf{r}\right) d\mathbf{r} = 0 \tag{10.54}$$

The zero space harmonics of the potential can be included in the ground energy. With the Hamiltonian $\mathcal{H}$, Eq. (10.53), one can study correlations of quantities taken both at different magnetic fields and at the parameters U_0. This can lead to different forms of the correlation functions because the potential $U_0 w(\mathbf{r})$ does not violate the time-reversal symmetry whereas the magnetic field does. The correlation function of the density of states $R(\omega, U_0, \phi)$ calculated later is defined as

$$R\left(\omega, U_0, \bar{U}_0, \phi, \bar{\phi}\right) = \Delta^2 \left\langle \rho\left(\varepsilon - \omega, \bar{U}_0 - U_0, \bar{\phi} - \phi\right) \rho\left(\varepsilon, \bar{U}_0, \bar{\phi}\right)\right\rangle \qquad (10.55)$$

with the density of states $\rho(\varepsilon, U_0, \phi)$ given by the standard formula

$$\rho(\varepsilon, U_0, \phi) = \sum_i \delta\left(\varepsilon - \varepsilon_i(U_0, \phi)\right) \qquad (10.56)$$

In Eqs. (10.55, 10.56) ϕ is the magnetic flux through the system, ν is the average density of states, V is the volume, Δ is the mean level spacing, and $\langle...\rangle$ stands for the statistical average over a typical range of energies and/or over realizations of the disorder potential. The function $R\left(\omega, U_0, \bar{U}_0, \phi, \bar{\phi}\right)$ is a direct generalization of the two-level correlation function $R(\omega)$ considered in Chapter 6 and can be calculated by using the supersymmetry technique in the standard way. First, one should express the function R in terms of Green functions G^R and G^A

$$R\left(\omega, U_0, \bar{U}_0, \phi, \bar{\phi}\right) = \pi^{-2} \int \left\langle \left(\operatorname{Im} G^R_{\varepsilon - \omega/2, \bar{U}_0 - U_0/2, \bar{\phi} - \phi/2}(\mathbf{r}, \mathbf{r})\right.\right.$$

$$\left.\left. \times \operatorname{Im} G^R_{\varepsilon + \omega/2, \bar{U}_0 + U_0/2, \bar{\phi} + \phi/2}(\mathbf{r}', \mathbf{r}')\right\rangle d\mathbf{r} d\mathbf{r}' \qquad (10.57)$$

where the Green functions are expressed in terms of the eigenfunctions $\varphi_k(\mathbf{r})$ and eigenenergies ε_k in the usual way

$$G^{R,A}_{\varepsilon, U_0, \phi}(\mathbf{r}, \mathbf{r}') = \sum_k \frac{\varphi_k(\mathbf{r}) \varphi_k^*(\mathbf{r}')}{\varepsilon - \varepsilon_k(U_0, \phi) \pm i\delta} \qquad (10.58)$$

The subsequent derivation of the σ-model is performed in the same way as in Chapters 4, 6, and 8. For example, equations like Eqs. (4.42, 4.45) can be written again provided one substitutes $\omega \rightarrow \omega + U_0 w(\mathbf{r})$. The magnetic flux and the potential are assumed to act as a weak perturbation and therefore one may make an expansion in these terms. After standard manipulations one comes to the following expression for the free energy

$$F[Q] = \frac{\pi \nu D_0}{8} \int \operatorname{str}\left(\nabla Q - \frac{ie}{c}\left[Q, \left(\bar{\mathbf{A}}\tau_3 + \frac{\Lambda}{2}\mathbf{A}\tau_3\right)\right]\right)^2 d\mathbf{r}$$

$$+ \frac{\pi \nu i}{4} \int \operatorname{str}\left[(\omega + U_0 w(\mathbf{r})) \Lambda Q\right] d\mathbf{r} \qquad (10.59)$$

where the vector potentials $\bar{\mathbf{A}}$ and $\mathbf{A}$ correspond to the fluxes $\bar{\phi}$ and ϕ. The term with $\bar{\mathbf{A}}$ cuts only the cooperon degrees of freedom, whereas the term with $\mathbf{A}\tau_3\Lambda$ cuts the diffusons too.

To calculate the correlation function R one should write an integral over supermatrices Q with free energy $F[Q]$. This can be done in the same way as when deriving Eq. (6.41) for the function $R(\omega)$ and the result can be written as

$$R\left(\omega, U_0, \bar{U}_0, \phi, \bar{\phi}\right) = -\frac{1}{2}\left[\frac{1}{V^2} \int \left\langle Q^{11}_{33}(\mathbf{r}) Q^{22}_{33}(\mathbf{r}')\right\rangle_Q d\mathbf{r} d\mathbf{r}' - 1\right] \qquad (10.60)$$

where the symbol $\langle \cdots \rangle_Q$ stands for integration with free energy $F[Q]$, Eq. (10.59), and V is the volume.

10.2.2 Correlation function of density of states at different values of external fields

We have learned from the preceding chapters that random matrix theory and universality are equivalent to the zero-dimensional σ-model. With the purpose of obtaining new universalities it is natural to use the zero-mode approximation again. For a disordered metal, this is possible provided the size of the sample is smaller than the localization length. In the case when $U_0 = 0$ one can arrive at the zero-dimensional σ-model simply by assuming that supermatrix Q does not depend on coordinates. (For the ring geometry it is not quite correct because physical quantities must be periodic in flux. Nevertheless, a proper modification is not difficult and this question has been discussed in Section 8.3.) However, a nonzero value of the potential U_0 cannot be considered in the same way. Using the convention Eq. (10.54) one can see that if Q does not depend on the coordinates, the last term in the free energy $F[Q]$, Eq. (10.59), vanishes.

To take the contribution of the potential $U_0 w(\mathbf{r})$ into account properly it is convenient to represent supermatrix Q in the form

$$Q = V_0 \tilde{Q} \bar{V}_0 \qquad (10.61)$$

where V_0 is an arbitrary unitary matrix, $V_0 \bar{V}_0 = 1$ (cf. Eq. (4.25)). Let us assume that V_0 does not depend on the coordinates and all coordinate dependence that has the same structure as Q is included in $\tilde{Q}$. If the amplitude of the potential U_0 is not large, essential deviations of $\tilde{Q}$ from Λ are small and one can make an expansion in these deviations. For that purpose it is convenient to use the parametrization for $\tilde{Q}$ written in Eq. (5.1). If all external perturbations are small, their contributions can be calculated independently. Therefore, deriving the term originating from the potential $U_0 w(\mathbf{r})$ one may omit the terms with $\bar{A}$, A, and ω in Eq. (10.59). Using the parametrization of Eq. (5.1) for $\tilde{Q}$, making an expansion in P, and substituting Eq. (10.61) into Eq. (10.59) one obtains the free energy functional $\tilde{F}$ in the lowest nontrivial order

$$\tilde{F} = \frac{\pi \nu}{2} \int \mathrm{str} \left[D_0 \left(\nabla P(\mathbf{r}) \right)^2 - U_0 w(\mathbf{r}) \left(\bar{V}_0 \Lambda V_0 \right)^{\perp} \Lambda P(\mathbf{r}) \right] d\mathbf{r} \qquad (10.62)$$

with the superscript $\perp$ denoting the anticommuting with Λ part of a supermatrix.

With the same accuracy, one can substitute supermatrix $\tilde{Q}$ in the preexponential in Eq. (10.60) by Λ. So what remains to do is to compute the integral

$$I(V_0) = \int \exp\left(-\tilde{F}\right) dP \qquad (10.63)$$

The exponent in the integrand in Eq. (10.62) is quadratic with a linear term and the integration is simple. As a result, one obtains

$$-\ln I = -\frac{1}{32} \frac{\pi U_0^2}{V D_0 \Delta} \, \mathrm{str} \, [Q_0, \Lambda]^2 \int \tilde{g}(\mathbf{r}, \mathbf{r}') \, w(\mathbf{r}) w(\mathbf{r}') \, d\mathbf{r} d\mathbf{r}' \qquad (10.64)$$

where $Q_0 = V_0 \Lambda \bar{V}_0$ and $\tilde{g}(\mathbf{r}, \mathbf{r}')$ is the Green function of the Laplace equation

$$-\Delta_{\mathbf{r}} g(\mathbf{r}, \mathbf{r}') = \delta(\mathbf{r} - \mathbf{r}') \qquad (10.65)$$

As the boundary condition for the function $\tilde{g}\left(\mathbf{r},\ \mathbf{r}'\right)$ one should require that normal gradients at the boundary of the sample vanish. Thus, we see that both the preexponential in Eq. (10.60) and the free energy are expressed through the supermatrix Q_0, which does not depend on coordinates. This means that we return again to the zero-dimensional σ-model. The quantity $-\ln I$ is the additional term in the free energy due to the potential. Omitting the subscript of supermatrix Q_0 one can express the free energy F of the zero-dimensional σ-model with all the perturbations in the form

$$F = \operatorname{str}\left(\frac{1}{4}ix\Lambda Q - \frac{1}{16}\left[Q,\left(\bar{X}\tau_3 + \frac{1}{2}X\tau_3\Lambda\right)\right]^2 - \frac{X_0^2}{16}[Q,\ \Lambda]^2\right) \qquad (10.66)$$

For computation of the correlation function of the density of states one can use Eq. (10.60), implying that supermatrices Q entering this equation do not depend on the coordinates. Then integration over $\mathbf{r}$ and $\mathbf{r}'$ cancels the factor V^{-2}.

The variables x and $X\left(\bar{X}\right)$ are related to the frequency ω and the vector-potential $\mathbf{A}\left(\bar{\mathbf{A}}\right)$, respectively, by Eq. (8.43). The last term in Eq. (10.66) originates from the potential $U_0 w\left(\mathbf{r}\right)$ and equals the quantity $-\ln I$, Eq. (10.64). The parameter X_0 can be represented in a more explicit form using a Fourier expansion in eigenfunctions of the Laplace equation, Eq. (10.65). The normal gradients of the functions at the boundary must be zero. The result takes the form

$$X_0^2 = \frac{\pi U_0^2}{2D_0 V\Delta}\sum_{k\neq 0}\frac{w_k^2}{\mathcal{E}_k} \qquad (10.67)$$

where $\mathcal{E}_k$ are eigenenergies of the Laplace equation and w_k are coefficients of expansion of the function $w\left(\mathbf{r}\right)$ in the eigenfunctions of the Laplace equation. In Eq. (10.67), one has to sum over all states with nonzero eigenenergies $\mathcal{E}_k$. For a simple geometry, the sum in Eq. (10.67) can be calculated analytically. In the case of a cube with size L one obtains

$$X_0 = \pi\left(\frac{U_0}{\mathcal{E}_c}\right)\left(\frac{\Gamma\mathcal{E}_c}{\Delta}\right)^{1/2} \qquad (10.68)$$

where $\mathcal{E}_c = \pi^2 D_0/L^2$ is the Thouless energy and the dimensionless parameter Γ is equal to

$$\Gamma = \sum_{\mathbf{n}\neq 0}\frac{1}{2\pi\mathbf{n}^2}\left|\int v_0\left(\mathbf{r}\right)w\left(\mathbf{r}\right)v_{\mathbf{n}}\left(\mathbf{r}\right)d\mathbf{r}\right|^2 \qquad (10.69)$$

where $\mathbf{n} = (n_1,\ n_2,\ n_3)$ is a vector with n_1, n_2, n_3 positive integers, and $v_{\mathbf{n}}$ are corresponding eigenfunctions. If $w\left(\mathbf{r}\right)$ represents a step potential that acts along one spatial direction, one can find that $\Gamma = \pi/24$.

Eqs. (10.67–10.69) were obtained by Simons and Altshuler (1993) from Eq. (10.59) using a somewhat different procedure. Using Eqs. (10.60–10.66) one can compute correlation function R. However, the main issue of the present section is some new universalities and so before performing explicit evaluation of the integral in Eq. (10.60) let us discuss this question.

10.2.3 Universal form of parametric correlations

Eq. (10.66) contains the dimensionless variable x, which is just the parameter entering the Wigner–Dyson correlation functions, Eq. (6.13–6.15). The dependence on x does not contain any detail about the model involved and is thus universal. In order to change from the physical variable ω to x one has to find the mean level spacing Δ for a given model and use Eq. (8.43). At the same time, the dependence of the correlation function R on the parameters $\bar{\phi}$, ϕ, and U_0 contains some parameters of the model. Fortunately, these parameters enter the result in combinations that can be considered as "generalized conductances." A proper scaling of the variables ϕ, $\bar{\phi}$, U_0 in units of these conductances leads to universal dependencies on the new variables that are not sensitive to microscopic details of the model.

In fact, the conductances can be extracted from the correlation function $R(\omega, U_0, \bar{U}_0,$ $\phi, \bar{\phi})$, Eqs. (10.55, 10.56). Let us demonstrate this relation for an arbitrary perturbation Y. The function $R(\omega, Y)$ is introduced as

$$R\left(\omega, Y, \bar{Y}\right) = \Delta^2 \left\langle \sum_{i,j} \delta\left(\varepsilon - \omega - \varepsilon_i\left(\bar{Y} - Y\right)\right) \delta\left(\varepsilon - \varepsilon_j\left(\bar{Y}\right)\right) \right\rangle \tag{10.70}$$

$$= \Delta^2 \left\langle \sum_{i,j} \delta\left(\omega - \varepsilon_j\left(\bar{Y}\right) + \varepsilon_i\left(\bar{Y} - Y\right)\right) \delta\left(\varepsilon - \varepsilon_j\left(\bar{Y}\right)\right) \right\rangle$$

Now let us take the limit $\omega \to 0$, $Y \to 0$ but $\omega/Y = \text{const}$. Because of the discreteness of the energy spectrum only terms with $i = j$ contribute in this limit to the sum in Eq. (10.70). Expanding the eigenenergies $\varepsilon_i\left(\bar{Y} - Y\right)$ in Y one obtains

$$\lim_{\omega \to 0, Y \to 0} R\left(\omega, Y, \bar{Y}\right) = \frac{\Delta}{Y} P\left(\frac{\omega}{Y}\right) \tag{10.71}$$

where the function $P(t)$ is the distribution function of the level velocities

$$P(t) = \Delta \left\langle \sum_i \delta\left(t - \partial_{\bar{Y}} \varepsilon_i\left(\bar{Y}\right)\right) \delta\left(\varepsilon - \varepsilon_i\left(\bar{Y}\right)\right) \right\rangle \tag{10.72}$$

The conductance $C(0)$, Eq. (10.47), corresponding to the perturbation Y can be written as

$$C(0) = \Delta^{-2} \int_{-\infty}^{\infty} t^2 P(t)\, dt \tag{10.73}$$

Eqs. (10.47) and (10.73) coincide exactly with each other if, in Eq. (10.72), averaging over an interval of energies is performed together with averaging over realizations of a random potential. For a weakly disordered metal in the diffusive regime the former averaging does not add anything to the latter. Eqs. (10.71–10.73) show that the generalized conductances that will be used for scaling the perturbation parameters are completely determined by the combinations of the microscopic parameters entering the correlation function R and therefore the σ-model, Eq. (10.66). Hence, the combinations can be removed by proper scaling in the units of the conductances.

This is very important because, for example, Eqs. (10.68, 10.69) were derived from Eq. (10.59), which contains only a linear term in the potential $U_0 w(\mathbf{r})$. Terms with higher powers of this potential can really be neglected in the limit $l \ll L$, where l is the mean free path. But can one apply the same formulae to chaotic billiards where these quantities are of the same order of magnitude? In this case Eqs. (10.68, 10.69) are no longer valid but the symmetry of the last term in Eq. (10.66) should be the same. This fact is crucial for the universality because whatever the coefficient in front of this term is it can be removed by the scaling procedure. In the language of the σ-model, all results after rescaling are completely determined by the symmetry of the terms in the σ-model. The only requirement is that the σ-model be zero-dimensional.

Although evaluation of the integral in Eq. (10.60) with the free energy F, Eq. (10.66), can, in principle, be performed at arbitrary $\bar{\phi}$, let us restrict ourselves with the limiting cases $\bar{\phi} = \phi = 0$ and $\bar{\phi} \to \infty$ that correspond to the orthogonal and unitary ensembles. In the unitary ensemble supermatrix Q commutes with τ_3 and the last two terms in the free energy F, Eq. (10.66), are equivalent. Using the parametrization of Section 6.3 for supermatrix Q one can see that the free energy F in both cases contains only the eigenvalues λ, λ_1 in the unitary ensemble and λ, λ_1, and λ_2 in the orthogonal one. The corresponding expressions can be written as

$$F_{\text{orth}} = X_0^2 \left(2\lambda_1^2 \lambda_2^2 - \lambda_1^2 - \lambda_2^2 - \lambda^2 + 1 \right) + i\,(x + i\delta)\,(\lambda - \lambda_1 \lambda_2) \tag{10.74}$$

$$F_{\text{unit}} = \left(X_0^2 + X^2/4 \right) \left(\lambda_1^2 - \lambda^2 \right) + i\,(x + i\delta)\,(\lambda - \lambda_1) \tag{10.75}$$

where F_{orth} and F_{unit} are the free energies for the orthogonal and unitary ensembles, respectively. The fact that these quantities depend only on the eigenvalues enables us just to use Eqs. (6.74, 6.75) with the corresponding substitution for the exponents. The two-level correlation function $R(\omega, \phi, U_0)$ can be written in the form

$$R(\omega, \phi, U_0) = R^{(0)}(\omega, \phi, U_0), \tag{10.76}$$

$$R_{\text{orth}}^{(m)}(\omega, 0, U_0) = \tag{10.77}$$

$$1 + \pi^{-m} \operatorname{Re} \int\limits_{-1}^{1} d\lambda \int\limits_{1}^{\infty} d\lambda_1 \int\limits_{1}^{\infty} d\lambda_2 \frac{(1 - \lambda^2)(\lambda_1\lambda_2 - \lambda)^{2-m} \exp(-F_{\text{orth}})}{\left(\lambda_1^2 + \lambda_2^2 + \lambda^2 - 2\lambda\lambda_1\lambda_2 - 1 \right)^2},$$

$$R_{\text{unit}}^{(m)}(\omega, \phi, U_0) = 1 + \frac{1}{2}\pi^{-m} \operatorname{Re} \int\limits_{-1}^{1} d\lambda \int\limits_{1}^{\infty} d\lambda_1 (\lambda_1 - \lambda)^{-m} \exp(-F_{\text{unit}}) \tag{10.78}$$

In Eqs. (10.77, 10.78), the free energies F_{orth} and F_{unit} are determined by Eqs. (10.74, 10.75). Although for the correlation function of the density of states one should calculate the integrals, Eqs. (10.77, 10.78), only for $m = 0$, another value of this parameter is important for calculation of the correlation function $\tilde{c}(x, y)$, Eq. (10.52).

In order to obtain the distribution function of the level velocities $P(t)$, Eq. (10.72), one has to evaluate the integrals, Eqs. (10.77, 10.78), in the limit X, X_0, $x \to 0$. Let us choose as the variable Y the following combinations of physical parameters:

$$Y = \begin{cases} \dfrac{U_0}{\mathcal{E}_c}, & \text{potential perturbation} \\[2ex] \dfrac{\phi}{\phi_0}, & \text{magnetic flux} \end{cases} \tag{10.79}$$

Calculating the integral in Eq. (10.78) and using Eqs. (10.68, 10.71, 8.44) in the unitary case one arrives at a Gaussian distribution function

$$P(t) = \frac{1}{\sqrt{2\pi C(0)}} \exp\left(-\frac{t^2}{2C(0)} \right) \tag{10.80}$$

where

$$
C(0) = \begin{cases} \frac{4\mathcal{E}_c}{\pi\Delta}, & \text{flux} \\[2mm] \frac{2\Gamma\mathcal{E}_c}{\Delta}, & \text{potential} \end{cases}
$$
(10.81)

The quantity $C(0)$ is the generalized conductance introduced in Eq. (10.47). This follows from the fact that the second moment of distribution $P(t)$, which is just the average square of the level velocity, equals $C(0)$. The generalized conductance $C(0)$ is proportional to the dimensionless conductance $\mathcal{E}_c/\Delta$ introduced by Thouless (1974).

Calculation of the integral in Eq. (10.77) for the orthogonal case in the limit X_0, $x \to 0$ leads again to the Gaussian distribution function, Eq. (10.80). Now, only a potential perturbation does not violate the symmetry of the system. Using the same variable Y, Eq. (10.79), one comes to a different value of the conductance $C(0)$

$$
C(0) = \frac{4\Gamma\mathcal{E}_c}{\Delta}
$$
(10.82)

We see from Eqs. (10.81, 10.82) that the conductance in the orthogonal ensemble is twice as large as that in the unitary ensemble.

The Gaussian form of the level velocity distribution function is not general although it is valid in both the orthogonal and unitary ensembles. Recently, Taniguchi et al. (1994) have calculated the correlation function of the density of states as a function of two external perturbations, one of which preserved time reversal symmetry, whereas the other violated it. A general expression they obtained enabled them also to deduce the level velocity distribution function. In the crossover region between these two limiting cases the distribution function has a rather complicated form.

Using Eqs. (10.81, 10.82) one can express the correlation functions of the density of states in terms of the dimensionless variables y, Eq. (10.48). The parameters X and X_0 entering free energy F, Eqs. (10.74, 10.75), differ from these variables in numerical coefficients only. In the unitary ensemble, in Eq. (10.75) one has to make the substitution

$$
X_0^2 + \frac{X^2}{4} \to \frac{\pi^2}{2} y^2
$$
(10.83)

whereas in the orthogonal case one has for X_0 in Eq. (10.74)

$$
X_0^2 = \frac{\pi^2}{4} y^2
$$
(10.84)

The representation of the correlation function R for the unitary ensemble, Eqs. (10.75, 10.78), can be somewhat simplified by using the variables $v = \lambda_1 - \lambda$ and $u = \lambda_1 + \lambda$. The integration over u can be performed and one obtains for the function R_{unit}

$$
R_{\text{unit}}(x, y) = 1 + \frac{1}{2\pi^2 y^2}
$$
(10.85)

$$
\times \int_0^{\infty} \left\{ \exp\left[-\frac{\pi^2 y^2}{2} \left| 2v - v^2 \right| \right] - \exp\left[-\frac{\pi^2 y^2}{2} \left| 2v + v^2 \right| \right] \right\} \cos(xv) \frac{dv}{v}
$$

As concerns the orthogonal ensemble, the integral, Eqs. (10.74, 10.77), cannot be further simplified. It is not difficult to complement Eqs. (10.74–10.78) by corresponding formulae for the symplectic ensemble. To derive the correlation function $R_{\text{sympl}}(x, y)$ in the universal variables x and y it is convenient, when doing the rescaling according to Eq. (10.48), to use instead of Δ the parameter $\bar{\Delta}$, Eq. (6.21), which is the physical mean level spacing in a situation when the spin degeneracy is lifted by spin interactions. (With such rescaling the universal form of the correlation function R_{unit} for the unitary ensemble, Eqs. (10.75, 10.83), is the same for both the cases with magnetic field and magnetic impurities.) For a potential perturbation, repeating the same procedure as when deriving Eqs. (10.74, 10.77, 10.84) one obtains the following integral (Simons, Lee, and Altshuler (1993a)):

$$R_{\text{sympl}}(x, y) = \tag{10.86}$$

$$1 + \text{Re} \int_1^\infty d\lambda \int_{-1}^1 d\lambda_1 \int_0^1 d\lambda_2 \frac{(\lambda^2 - 1)(\lambda - \lambda_1\lambda_2)^2 \exp\left(-F_{\text{sympl}}\right)}{\left(2\lambda\lambda_1\lambda_2 - \lambda^2 - \lambda_1^2 - \lambda_2^2 + 1\right)^2},$$

$$F_{\text{sympl}} = \frac{\pi^2}{2} y^2 \left(\lambda_1^2 + \lambda_2^2 + \lambda^2 - 2\lambda_1^2\lambda_2^2 - 1\right) + i(x + i\delta)(\lambda_1\lambda_2 - \lambda) \tag{10.87}$$

When doing numerical computation of spectra for different models such as chaotic billiards, the velocity correlation functions $c(y)$, Eq. (10.49) and $\tilde{c}(x, y)$, Eq. (10.52), provide a more exacting test of universality than $R(x, y)$ since they display more features and are less sensitive to the tilt of the spectrum. Unfortunately, as a result of technical difficulties the function $c(y)$ cannot be calculated analytically. Function $\tilde{c}(x, y)$ contains the correlation function $R(x, y)$ in the denominator and a velocity–velocity correlation function in the numerator. Correlation functions of density of states and level velocities are related to each other by Eq. (10.15), which makes it possible to derive the function $c(x, y)$ from the functions $R^{(0)}$ and $R^{(2)}$ given by Eqs. (10.77, 10.78). The function $R^{(2)}$ is obtained from the function $R^{(0)}$ by double integration over energies. The formula for $\tilde{c}(x, y)$ reads

$$\tilde{c}(x, y) = -\frac{\partial_y^2 R^{(2)}(x, y)}{R^{(0)}(x, y)} \tag{10.88}$$

The integral representation of the functions $R^{(0)}$ and $R^{(2)}$, Eqs. (10.77, 10.78), enables us to compute the function $\tilde{c}(x, y)$ at least numerically and compare it with numerical simulations. The function $\tilde{c}(x, y)$ in the dimensionless units x and y is as universal as $R(x, y)$. Because of the universality one can hope that the correlation functions obtained in the present section are applicable whenever one is in the regime of quantum chaos.

The correlation functions calculated in this section by the supersymmetry technique were compared with results of numerical simulations for different models. Simons and Altshuler (1993) made the comparison with numerical simulations for disordered systems and chaotic billiards. Faas et al. (1993) considered a one-dimensional model of strongly interacting spinless fermions, focusing on a range of parameters for which the Hamiltonian is nonintegrable. In this situation one expects chaos and a check of the universalities becomes possible.

Simons et al. (1993) studied the universal correlations in the spectra of hydrogen in a magnetic field. This model has been briefly mentioned in Section 6.1. For some values of the parameter γ in Eq. (6.16), which is a reduced magnetic field, one obtains the regime of quantum chaos. Changing the magnetic field can be considered as applying a perturbation Y. Computing spectra as functions of the magnetic field one can extract the universal correlation functions.

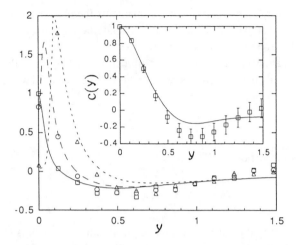

Fig. 10.2. The velocity–velocity correlation function $\tilde{c}(x,\ y)$ extracted from the spectra of hydrogen in a magnetic field for three values of x and compared with the results of analytical calculation (Simons et al. (1993)). The curves are theoretical with no adjustable parameters. The results are shown for $x = 0$ (squares, continuous), $x/\pi = 0.1$ (circular, dashed), and $x/\pi = 0.25$ (triangles, dotted). The inset shows the second correlation function, $c(y)$. The results for the hydrogen model (squares) are compared to those from numerical simulation of an Anderson model of disorder for an orthogonal ensemble taken from Simons and Altshuler (1993).

The preceding publications contain numerous figures demonstrating very good agreement of the numerical results with the correlation functions calculated analytically by the supersymmetry technique. As an illustration, the results of both analytical and numerical computation of the velocity–velocity correlation functions $\tilde{c}\,(x,\ y)$ and $c\,(y)$, Eqs. (10.49, 10.52), are represented in Fig. 10.2. When comparing the numerical results for the hydrogen model with the results of the analytical theory one should keep in mind that, in spite of the presence of the magnetic field, the Hamiltonian $\mathcal{H}$, Eq. (6.16) at $L_z = 0$ obeys time-reversal symmetry and the appropriate Dyson ensemble for the chaotic region is orthogonal. The theoretical curves in Fig. 10.2 correspond to this ensemble.

10.2.4 Some other parametric correlations

Parametric correlations were further studied by the supersymmetry technique in a number of subsequent works. Taniguchi, Andreev, and Altshuler (1994) calculated the correlation function $S\,(\omega,\ B)$ of oscillator strengths $W\,(\bar{\omega},\ \bar{B})$ introduced as

$$S\,(\omega,\ B) = \left\langle W\,(\bar{\omega},\ \bar{B})\,W\,(\bar{\omega} + \omega,\ \bar{B} + B)\right\rangle \tag{10.89}$$

where

$$W\,(\bar{\omega},\ \bar{B}) = 2\bar{\omega}\sum_i \left|\langle 0\,|\hat{z}|\,i\rangle\right|^2 \delta\left(\varepsilon_i\,(\bar{B}) - \varepsilon_0\,(\bar{B}) - \bar{\omega}\right) \tag{10.90}$$

In Eq. (10.90), $\langle 0\,|\hat{z}|\,i\rangle$ is the dipole matrix element between a certain initial state, such as the ground state, $|0>$, and a highly excited state, $|i>$, with their energies equal, respectively, to ε_0 and ε_i. The parameter B can stand for any external parameter of the system. The correlator $S\,(\omega,\ B)$ contains not only the energy levels but also matrix elements and therefore provides a characterization of quantum chaos that complements the usual energy level correlation. Again, it was found that the correlator is universal and one can hope to apply the theory to, for example, the hydrogen atom in a magnetic filed.

Recently, a work appeared (Fyodorov and Sommers (1995)) in which the supersymmetric σ-model was used to study the distribution function $\mathcal{P}(K)$ of the level curvature K. The curvature of a level i is introduced as

$$K_i = \frac{\partial^2 \varepsilon_i}{\partial Y^2} \tag{10.91}$$

and the distribution function $\mathcal{P}(K)$ equals

$$\mathcal{P}(K) = \left\langle \Delta \sum_i \delta(K - K_i)\, \delta(\varepsilon - \varepsilon_i) \right\rangle \tag{10.92}$$

The interest in studying this quantity originates from numerical works in which universal behavior was discovered. Gaspard et al. (1990) and Saher, Haake, and Gaspard (1991) found that the asymptotic behavior in the limit $K \to \infty$ obeyed a power law $\mathcal{P}(K) \propto K^{-(\beta+2)}$, $\beta = 1, 2, 4$ depending on the symmetry of the ensemble. Later the analytical form of the entire distribution function $\mathcal{P}(K)$ was guessed by Zakrzewski and Delande (1993) on the basis of numerical results; they proposed the following formula

$$\mathcal{P}(K) = C_\beta \left(1 + k^2\right)^{-(1+\beta/2)} \tag{10.93}$$

where C_β is a normalization constant and k is the dimensionless level curvature. Eq. (10.93) was confirmed for the unitary ensemble by von Oppen (1994) by random matrix theory. With the supersymmetry technique, Fyodorov and Sommers (1995) managed to derive Eq. (10.93) from a standard correlation function similar to function R, Eq. (10.70). They considered the case when a time-reversal invariant system is perturbed by a magnetic flux ϕ and calculated the distribution function at zero flux. In this situation the level velocity is zero at $\phi = 0$ for all levels. This makes it possible to get the distribution of the curvatures in a procedure similar to that used above for the derivation of the distribution function of the level velocities, Eq. (10.80). It turns out that even taking into account the lowest corrections to the zero-dimensional σ-model does not change Eq. (10.93).

Universal parametric correlations of eigenfunctions in chaotic and disordered systems were studied in a recent publication by Alhassid and Attias (1995). These authors introduced a Gaussian process that allows one to consider the parametric correlations in a general way. According to their suggestion a chaotic system is to be described by a random $N \times N$ matrix $\mathcal{H}(x)$ that depends on a parameter x. The probability measure for matrix elements $\mathcal{D}(\mathcal{H})$ is introduced as

$$\mathcal{D}(\mathcal{H}) = A \exp\left(-\frac{1}{2a^2} \operatorname{tr} \int \mathcal{H}(x)\, f^{-1}(x, x')\, \mathcal{H}(x')\, dx\, dx'\right) \tag{10.94}$$

which is the most natural extension of Eq. (6.1). The Brownian motion model is a particular case of Eq. (10.94). Putting

$$f(x, x') = \exp\left(-\frac{|x - x'|}{\gamma}\right) \tag{10.95}$$

one comes to the Fokker–Planck equation for the matrix distribution, Eq. (10.19).

Using the supersymmetry technique Alhassid and Attias (1995) showed that the Gaussian process in its general form, Eq. (10.94), leads to the universalities discussed in the preceding sections and calculated some universal correlation functions of eigenfunctions.

Some universal correlation functions have been examined experimentally. Sivan et al. (1994) studied the spectrum of heavily doped quantum dots. They designed a device in such a way that it consisted of two dots, one of which had a small volume and served as a spectrometer. Electrons could tunnel to the dots from the top and not through a rather narrow contact as in the experiments discussed in Chapter 9. In such a geometry one directly probes the density of states in the dots and the differential conductance $g(\Omega, \phi)$, where Ω is the applied voltage and ϕ is the magnetic flux through the studied large dot, can be written in the form

$$g(\Omega, \phi) \propto \partial_\Omega \int \rho(\Omega - \Omega', \phi) f(\Omega') d\Omega' \tag{10.96}$$

In Eq. (10.96), $\rho(\Omega, \phi)$ is the density of states in the large dot and $f(\Omega)$ is the line shape of a spectrometer level. The partial derivative with respect to Ω results from the fact that the differential conductance is measured rather than the conductance itself.

From the experimental measurement of the the the quantity $g(\Omega, \phi)$ Sivan et al. (1994) extracted the correlation function $K(\bar{\phi}, \phi)$

$$K(\bar{\phi}, \phi) = \int [g(\Omega, \phi) - \langle g(\phi) \rangle][g(\Omega, \bar{\phi} + \phi) - \langle g(\bar{\phi} + \phi) \rangle] d\Omega \tag{10.97}$$

where $\langle g \rangle$ is the average differential conductance.

To compare this function with a theoretical curve one can use the following equations:

$$K(\bar{\phi}, \phi) \propto \int f(\varepsilon) f(\varepsilon') \tilde{K}(\varepsilon - \varepsilon', \bar{\phi}, \phi) d\varepsilon d\varepsilon', \tag{10.98}$$

$$\tilde{K}(\varepsilon - \varepsilon', \bar{\phi}, \phi) = \langle \partial_\varepsilon \rho(\varepsilon, \bar{\phi}) \partial_{\varepsilon'} \rho(\varepsilon', \bar{\phi} + \phi) \rangle - \langle \partial_\varepsilon \rho(\varepsilon, \bar{\phi}) \rangle^2$$

The physics is contained in the function $\tilde{K}(\varepsilon - \varepsilon', \bar{\phi}, \phi)$, which measures the correlation between the derivative of the density of states with respect to energy at two different energies and fluxes. This function is obtained from the function R, Eq. (10.70), by differentiations with respect to the energies, and one can use Eqs. (10.75, 10.76, 10.78) to get the desired theoretical curves for the unitary ensemble. Of course, to make a comparison at weak magnetic fields one should use the correlation function R at arbitrary flux $\bar{\phi}$ (this has been computed by Taniguchi et al. (1994)). Using a Lorentzian form for the function $f(\varepsilon)$ with a width determined from the experiment, Sivan et al. (1994) compared the experimental function $K(\bar{\phi}, \phi)$, Eq. (10.97), normalized to its value at $\bar{\phi} = 0$ with the corresponding theoretical curve. The result for the normalized function is depicted in Fig. 10.3.

The dashed line corresponds to the theoretical curve with the parameter $\mathcal{E}_c/\pi^2 \Delta \approx 10.6$, which was estimated from the doping level and lithography. The best fit corresponds to a slightly different value of this parameter, but this discrepancy can easily result from deviations in geometry and/or doping level. In fact, the agreement is good enough to suggest a new method for measuring the conductance $\mathcal{E}_c/\Delta$ of a microscopic object that cannot be otherwise measured. The experimental curves at the magnetic fields $H = 0$ and $H = 30G$ are different from those at higher fields. Apparently, at these values of the magnetic fields one should use formulae for the crossover region between the orthogonal and unitary ensembles.

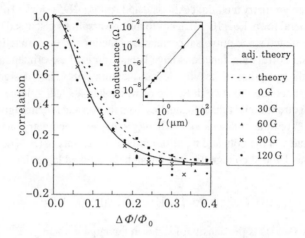

Fig. 10.3. Level correlation function $K(\bar{\phi}, \phi)$ versus flux separation. Various symbols correspond to different base magnetic fields. Dashed line: theoretical curve with no adjustable parameters ($\mathcal{E}_c/\Delta = 10.6$). Solid line: theoretical curve with flux scale adjusted by a factor of 1.2 for best fit. (From Sivan et al. (1994).)

Thus, we see from the results of this section that study of parametric correlations can be interesting from the point of view of both theory and experiment. The analytical formulae obtained by the supersymmetry technique are confirmed by numerical investigation of corresponding models and agree with the results of experiments. Now a relevant question can be asked: What does the theory presented in this section have in common with the phenomenological Brownian motion model considered in the preceding one?

One can already guess that there is some similarity between the correlation function of the density of states $R(\omega, \phi, U_0)$ for the unitary ensemble, Eqs. (10.75, 10.76, 10.78), and the corresponding Eqs. (10.40, 10.43). After proper rescaling of the variables all these equations are identical. It is more difficult to compare the corresponding results obtained by the supersymmetry technique for the orthogonal and symplectic ensembles to those in the Brownian motion model because the latter have not been available. However, it turns out that one can rigorously derive the σ-model from the Brownian motion model. This remarkable equivalence will be discussed in the next section.

10.3 "Grand unification"

10.3.1 Parametric correlations and one-dimensional fermions

In this section the recently discovered equivalence of models that originated in completely different fields will be discussed. We have seen in Chapter 6 that the random matrix models can be mapped on the zero-dimensional σ-model. This shows that, for example, computation of the level–level correlation function for models with disorder is equivalent to study of the thermodynamics of the Coulomb gas. In Section 10.1, it was demonstrated that calculation of thermodynamic quantities for the Coulomb gas model could be considered as calculation of ground state averages for the one-dimensional Calogero–Sutherland models. The latter models are equivalent to the Brownian motion model that was designed originally for studying parametric correlations in spectra of chaotic systems. Time-dependent correlations for the one-dimensional fermionic models can be equivalent to parametric correlations in the chaotic models described by the σ-model and several arguments in favor of this suggestion have been presented. The clearest evidence comes from the coincidence of Eqs. (10.75, 10.76, 10.78) and the corresponding Eqs. (10.40, 10.43) written for the unitary ensemble. This fact has been noted recently by Simons, Lee, and Altshuler (1993).

Is this coincidence occasional? Of course, the first attempt to check this fact would be to compare the same correlation functions for the orthogonal and symplectic ensembles. However, although the level–level correlation function can be obtained rather easily by using the σ-model, Eqs. (10.74, 10.76, 10.77) (the corresponding formula for the symplectic ensemble is only slightly different), direct calculation of the correlation function for the one-dimensional models is a difficult task. This is because the Hamiltonians $\mathcal{H}_C$ and $\mathcal{H}_S$, Eqs. (10.29, 10.41), describe strongly interacting particles at $\beta = 1$ and $\beta = 4$. In the orthogonal ensemble the interaction is attractive, whereas in the symplectic ensemble it is repulsive. The unitary ensemble is exceptional because in this case the interaction vanishes, thus making it possible to obtain all correlation functions immediately.

Nevertheless, the one-dimensional fermionic models prove to be really equivalent in the limit $N \to \infty$ to the σ-model for all ensembles and a general proof exists (Simons, Lee, and Altshuler (1993a, 1994)). This proof will be presented later. Let us first check the agreement of the density–density correlation function S, Eq. (10.38), for the one-dimensional fermionic models with the correlation function $R - 1$, Eq. (10.76), in the limits of large and small t. The variable t plays the role of the imaginary time for the one-dimensional models. Its relation to the dimensionless parameter y, Eq. (10.83, 10.84), as well as the relation between x and r can be extracted from Eqs. (10.75, 10.76, 10.78, 10.40, 10.43)

$$y^2 = 2t, \qquad x = \pi r \tag{10.99}$$

In fact, Eqs. (10.99) holds for all three ensembles, as will be seen shortly.

For an arbitrary pair interaction between particles one has a general f-sum rule (see, e.g., Pines and Nozières (1992)), which can be written for the function S defined in Eq. (10.38) as

$$\int [S(r, t) - S(r, 0)] \exp(iqr)\, dr = -2tq^2 \theta(q) + O\left(t^2\right) \tag{10.100}$$

where $\theta(q)$ is the step function. The same relation holds for the function $R(x, y)$. This can be proved by calculating integrals in Eqs. (10.74–10.78). Another way to get this general relation is to start from Eqs. (10.55, 10.56). The correlation function $R(x, y)$ contains the double sum over the energy levels. Expanding each term of the sum in y one obtains a sum that gives $R(x, 0)$ and a term quadratic in y. In the latter term, the contribution is from the same level. Using Eqs. (10.49, 10.50) one obtains

$$R(x, y) - R(x, 0) = \pi^3 y^2 \frac{\partial^2}{\partial x^2} \delta(x) + O\left(y^4\right) \tag{10.101}$$

With the relation given by Eq. (10.99), Eqs. (10.100) and (10.101) agree. The limit of large t for the one-dimensional models can be obtained by using an idea suggested long ago (Efetov and Larkin (1975)). This approach can be applied to one-dimensional systems that have a mode with a linear energy spectrum. Later this method was developed further by Haldane (1981), who called the state obtained in the one-dimensional models with a linear spectrum

"Luttinger liquid" (see also Korepin, Izergin, and Bogoliubov (1992)). The density–density correlation function is very sensitive to sound excitations that have the linear dispersion because this leads to fluctuations of the local Fermi momentum.

By this method the correlation function can be computed easily. At zero temperature it obeys a power law and contains only parameters characterizing linear excitations. Sutherland has shown (1971, 1972) that there are sound excitations in the models involved and has calculated the sound velocity v_s. Proceeding in the standard way one obtains for the correlation function $S(r, t)$

$$S(r, t) = -\frac{1}{2\pi^2 \beta} \sum_{\sigma=\pm 1} \frac{1}{(r + i\sigma v_s t)^2} \qquad (10.102)$$

where the sum is over right and left movers. The sound wave velocity is given by $v_s = \pi\beta$. At the same time the large x asymptotics of the correlation function $R(x, y) - 1$, Eqs. (10.74–10.78, 10.86, 10.87), can be found to be

$$R(x, y) - 1 = \frac{1}{\beta\pi^2} \text{Re} \frac{1}{\left(ix/\pi + \pi\beta u^2/2\right)^2} \qquad (10.103)$$

and, with Eq. (10.99), the correspondence of the functions $S(r, t)$ and $R(x, y) - 1$ is verified. If the correspondence is really exact, it enables us to obtain the density–density correlation function for the Sutherland model by using Eqs. (10.74–10.78) obtained from the σ-model. In fact, using the supersymmetry technique, Simons, Lee, and Altshuler (1993) were first to calculate the density–density correlation function for the Sutherland model. Eq. (10.103) can also be obtained by using the diagrammatic expansion in the diffusion modes, and this contribution is from the two-diffuson approximation. Thus, the lowest order of the diagrammatic perturbation theory is equivalent to the hydrodynamic limit of the Sutherland model.

10.3.2 Continuous matrix model and one-dimensional fermionic systems

Now, having presented semiqualitative arguments, let us prove the equivalence of the models using a more general consideration suggested by Simons, Lee, and Altshuler (1993a, 1994). The proof is correct for m-point correlation functions for an arbitrary m, but later it will be presented for two-point correlations functions, $m = 2$. The generalization requires introduction of supermatrices of a size proportional to m and is straightforward.

To begin the proof it is convenient to discuss first a connection between the one-dimensional models considered previously, which belong to the continuum Sutherland–Calogero–Moser class of fermionic systems (see also Moser (1975)) and a continuous matrix model. The connection was originally discussed by Brezin et al. (1978). The partition function Z of the continuous matrix model is introduced as

$$Z = \int \exp\left(-S[\Phi]\right) D\Phi \qquad (10.104)$$

where the Euclidean action $S[\Phi(t)]$ describes the free propagation of $N \times N$ random matrices in a potential $V(\Phi)$ and has the form

$$S[\Phi] = \int_{-\infty}^{\infty} \text{tr} \left[\frac{m}{2} \left(\frac{\partial \Phi(t)}{\partial t} \right)^2 + V(\Phi) \right] dt \qquad (10.105)$$

The potential $V(\Phi)$ together with the symmetry of Φ is left unspecified. The matrices Φ may be real symmetric, general Hermitian, or composed of real quaternions. These three cases correspond to the Dyson ensembles with $\beta = 1, 2, 4$. The matrix models of the type defined in Eqs. (10.104, 10.105) are under intensive study and the limit $N \to \infty$ seems to be most important. Making an expansion of the partition function in $V(\Phi)$ one obtains a series, each term of which is described by a diagram. In the limit $N \to \infty$ only planar diagrams contribute (t'Hooft (1974)), and for certain choices of $V(\Phi)$ the logarithm of Z generates the sum of connected surfaces (David (1985), Kazakov (1985)). Study of the models given by Eqs. (10.104, 10.105) is interesting in itself but is not reviewed here (see for a review Gross, Piran, and Weinberg (1991), Di Francesco, Ginsparg, and Zinn-Justin (1995)). I want to mention only that the continuous matrix models are believed to be useful for constructing string theories and theories of quantum gravity (see, e.g., Gross and Klebanov (1991)).

Instead of calculating the functional integral with classical action S, Eq. (10.105), one can study the corresponding quantum model (Feynman and Hibbs (1965)) with the Hamiltonian

$$\mathcal{H} = \mathrm{tr}\left[-\frac{1}{2m}\frac{\partial^2}{\partial\Phi\partial\Phi^+} + V(\Phi)\right] \qquad (10.106)$$

According to the symmetry of Φ, the Hamiltonian is invariant under global $SO(N)$, $SU(N)$, or $SU(N) \times SU(N)$ rotations for orthogonal, unitary, or symplectic ensembles, respectively. The symmetry makes it convenient to use the coordinates arising from the diagonalization of the matrix Φ

$$\Phi = \Omega R \Omega^+, \qquad \Omega^+\Omega = 1 \qquad (10.107)$$

where $R = \mathrm{diag}(r_1, \ldots r_N)$ is the diagonal matrix of eigenvalues. The variables r_i can be considered as new "radial" coordinates, with the "angular" coordinates arising from the unitary matrix Ω. The interaction term in the Hamiltonian $\mathcal{H}$, Eq. (10.106), which is assumed to be invariant under unitary transformations of Φ contains only r_i, whereas the kinetic energy contains derivatives with respect to both types of variables. The Jacobian $J(\{r_i\})$ of the transformation to the new variables has been discussed in Chapter 6; it can be written as

$$J(\{r_i\}) = \prod_{i<j}|r_i - r_j|^\beta \qquad (10.108)$$

The Laplacian entering Eq. (10.106) consists of a radial part containing derivatives with respect to the variables r_i only and an angular part containing derivatives over the elements of the matrix Ω. The general expression for the Laplacian reads

$$\mathrm{tr}\frac{\partial^2}{\partial\Phi\partial\Phi^+} = \sum_{i=1}^{N}\frac{1}{J}\frac{\partial}{\partial r_i}J\frac{\partial}{\partial r_i} + Y(\Omega, \{r_i\}) \qquad (10.109)$$

where $Y(\Omega, \{r_i\})$ is the angular part, the explicit form of which is not specified. It is clear that the Schrödinger equation with the Hamiltonian $\mathcal{H}$, Eqs. (10.106–10.108), has singlet wave functions $\Psi(\{r_i\})$ that depend only on the radial coordinates r_i. Of course, wave functions with higher "angular momenta" exist too, and the continuous matrix model is definitely more reach by a larger number of variables at the same N than the fermionic models.

A drastic simplification occurs for large $N \to \infty$, because in this limit the contribution of the excitations with nonzero angular momenta to the partition function Z is exponentially small (Brezin et al. (1978), Gross, Piran, and Weinberg (1991)) and attention can be restricted to the singlet component. Making the substitution for the radial part of the Hamiltonian

$$\mathcal{H} = \prod_{i<j} |r_i - r_j|^{-\beta/2} \, \tilde{\mathcal{H}}_C \prod_{i<j} |r_i - r_j|^{\beta/2} \tag{10.110}$$

and using the identity Eq. (10.27) one obtains for $\tilde{\mathcal{H}}_C$

$$\tilde{\mathcal{H}}_C = -\sum_{i=1}^{N} \frac{\partial^2}{\partial r_i^2} + \beta \left(\frac{\beta}{2} - 1\right) \sum_{i<j} \frac{1}{(r_i - r_j)^2} + \sum_{i=1}^{N} V(\{r_i\}) \tag{10.111}$$

For a quadratic potential $V \{r_i\}$ the Hamiltonian $\tilde{\mathcal{H}}_C$ from Eq. (10.111) coincides, after rescaling of the coordinates, with the Hamiltonian $\mathcal{H}_C$, Eq. (10.29). One assigns fermionic or bosonic statistics to the particles using a Jordan–Wigner transformation. Since only density correlations are considered, the phase associated with the gauge transformation does not enter, and without loss of generality the statistics of the particles can be taken to be fermionic. The superiority of the interacting particle representation to the matrix model is clear because now one has to deal with a smaller number of independent variables. The reduction is especially useful for the unitary ensemble when one comes to the model of noninteracting particles.

Although the Hamiltonian depends explicitly on the form of $V(\{r_i\})$, various properties become universal in the thermodynamic limit $N \to \infty$. One can also omit the confining potential and start, instead of from Eq. (10.105) containing the Hermitian matrices, from an action containing unitary ones. Proceeding in the same way one then arrives at the Sutherland model, Eq. (10.41).

10.3.3 Continuous matrix model and a supermatrix σ-model

Thus, the one-dimensional fermionic models are equivalent in the limit $N \to \infty$ to the continuous models and now we are to prove the equivalence of these models and the super-matrix σ-model. The derivation of the σ-model from Eq. (10.105) can be done in the same way as the derivation of the σ-model without external perturbations, as in Section 6.5. Only a slight modification is necessary.

The density–density correlation function for the Calogero model contains densities $\sum_{i=1}^{N} \delta(r - r_i)$ and, thus, corresponds to the following correlator $k(\omega, t)$ for the matrix model:

$$k(\omega, t) = \Delta^2(\varepsilon) \left\langle \mathrm{tr}\, \delta \left(\varepsilon - \omega - \Phi(\bar{t} - t)\right) \mathrm{tr}\left(\varepsilon - \Phi(\bar{t})\right)\right\rangle_S \tag{10.112}$$

The averaging in Eq. (10.112) has to be performed with action S, Eq. (10.105), and the parameter $\Delta(\varepsilon)$ is introduced as

$$\Delta^{-1} = \langle \mathrm{tr}\, \delta(\varepsilon - \Phi(t))\rangle_S \tag{10.113}$$

Following the procedure of Section 6.5 one should introduce the Green functions $G_\varepsilon^{R,A}(t)$ as

$$G_\varepsilon^{R,A}(t) = (\varepsilon - \Phi(t) \pm i\delta)^{-1} \tag{10.114}$$

which depend now on the "time" t. The equation analogous to Eq. (6.94) takes the form

$$k(\omega, t) = \frac{\Delta^2(\varepsilon)}{4\pi^2} \left\langle \mathrm{tr}\left(G_{\varepsilon-\omega}^{A}(\bar{t} - t) - G_{\varepsilon-\omega}^{R}(\bar{t} - t)\right) \mathrm{tr}\left(G_\varepsilon^{R}(\bar{t}) - G_\varepsilon^{A}(\bar{t})\right)\right\rangle_S \tag{10.115}$$

Writing the Green functions in terms of an integral over supervectors one obtains an equation analogous to Eq. (6.96). The difference is that now the pairs of the vectors ψ, $\bar{\psi}$ should be taken at the different times $\bar{t} - t$ and $\bar{t}$; therefore, the Lagrangian $\mathcal{L}$ has the form

$$\mathcal{L} = \mathcal{L}_1 + \mathcal{L}_2, \tag{10.116}$$

$$\mathcal{L}_1 = i \left[\sum_{m=1}^{N} \bar{\psi}_m \left(\tilde{\varepsilon} - \frac{\omega + i\delta}{2} \Lambda \right) \psi_m \right.$$
$$\left. - \frac{1}{2} \sum_{m,n=1}^{N} \bar{\psi}_m \left(\Phi_{mn} \left(\bar{t} - t \right) + \Phi_{mn} \left(t \right) \right) \psi_n \right],$$

$$\mathcal{L}_2 = -\frac{i}{2} \sum_{m,n=1}^{N} \bar{\psi}_m \Lambda \left(\Phi_{mn} \left(\bar{t} - t \right) - \Phi_{mn} \left(\bar{t} \right) \right) \psi_n$$

The next step is to average $\exp(-\mathcal{L})$ over the matrices Φ with S, Eq. (10.105). As usual, the parameters ω and t are assumed not to be very large. Therefore, t in $\mathcal{L}_1$ can be neglected and the first part, $\mathcal{L}_1$, then coincides with the Lagrangian from Eq. (6.96). The second part of the Lagrangian, $\mathcal{L}_2$, can be considered as a perturbation. Neglecting $\mathcal{L}_2$ in the zero-order approximation one can repeat the manipulations of Section 6.5 and obtain the conventional zero-dimensional σ-model, Eq. (6.110), with $\Delta(\varepsilon)$ given by Eq. (6.109).

The term $\mathcal{L}_2$ after averaging over the matrix Φ gives an additional term to the Lagrangian written in the supervectors ψ. This term should be averaged over ψ with the Lagrangian $\mathcal{L}_{\text{eff}}$, Eq. (6.99). With the function $g(Q)$, Eq. (4.41), for the matrix models with real symmetric ($\beta = 1$) or Hermitian ($\beta = 2$) matrices one obtains the following additional term F_2 in the free energy:

$$F_2[Q] = -\frac{N\beta}{32\pi} \operatorname{str}(\Lambda g(Q) \Lambda g(Q)) \int_{-\infty}^{\infty} \frac{1 - \cos qt}{q^2 + a^{-2}} dq \tag{10.117}$$

Using Eqs. (6.103, 6.105–6.108) one obtains for $t \ll a$ the final result

$$F[Q] = \frac{i\pi\omega}{4\Delta} \operatorname{str}(\Lambda Q) - \frac{t\pi^2\beta}{64m\Delta^2} \operatorname{str}[Q, \Lambda]^2 \tag{10.118}$$

which agrees with the corresponding equations derived in Section 10.2 for chaotic systems with external perturbations. For the orthogonal and unitary ensembles one obtains from Eq. (10.118) the corresponding Eqs. (10.74, 10.75), provided the dimensionless variables x, y are related to energy ω and time t as

$$x = \frac{\pi\omega}{\Delta}, \qquad y^2 = \frac{t}{m\Delta^2} \tag{10.119}$$

The correlation function k, Eq. (10.112), coincides with the correlation function R from the preceding section. Of course, one can establish the equivalence for the symplectic ensemble, too. The same mapping can be done for many-point correlation functions depending on many variables t_α. The extension is straightforward; for details see Simons, Lee, and Altshuler (1993a). The parameters y_α and t_α entering the models are related to each other as

$$\left(y_\alpha - y_\beta \right)^2 = \frac{1}{m\Delta^2} \left| t_\alpha - t_\beta \right| \tag{10.120}$$

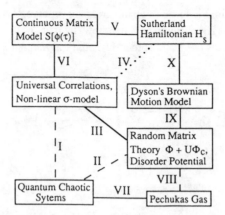

Fig. 10.4. A schematic diagram showing how the continuous matrix model draws together different branches of physics and relates them through the nonlinear supermatrix σ-model. (From Simons, Lee, and Altshuler (1993a, 1994).)

Thus, the matrix model with action S, Eq. (10.105), is equivalent in the limit N to the supersymmetric σ-model, and, hence, the latter is equivalent to the Calogero–Sutherland–Moser one-dimensional systems. In their turn the one-dimensional models are equivalent, as discussed in Section 10.1, to the Brownian motion model. The use of the relation between the models is evident because the density–density correlation function for the one-dimensional models for $\beta = 1$ and $\beta = 4$ has been computed for the first time using the σ-model, and Eqs. (10.74, 10.76, 10.77, 10.86, 10.87) give the final result. The possibility that the σ-model can turn out to be useful for string theories and the theory of quantum gravity cannot be excluded. Anyway, the unification of all these, at first glance, completely different theories from different branches of physics looks very exciting.

10.3.4 Discussion of the connections

To visualize all the connections discussed in the present chapter a schematic diagram summarizing them is represented in Fig. 10.4. The solid lines correspond to the links established analytically without using additional hypotheses.

The link V is the equivalence in the limit $N \to \infty$ of the continuous matrix models with the action S, Eq. (10.105), containing Hermitian matrices Φ (or containing unitary matrices instead of introducing a confining potential) to the one-dimensional Calogero–Sutherland models determined by Eqs. (10.29, 10.41). The parameter N is the size of the matrices in the matrix models and the number of particles in the one-dimensional models. This connection is well known in superstring theory and quantum gravity theory (Brezin et al. (1978), Gross, Piran, and Weinberg (1991), Di Francesco, Ginsparg, and Zinn-Justin (1995)).

The link VI is the mapping of the continuous matrix models on the zero-dimensional nonlinear σ-model with additional terms describing external perturbations, Eq. (10.118). This mapping has been discussed and was established by Simons, Lee, and Altshuler (1993a, 1994).

Link X is the equivalence of the Calogero–Sutherland models and the Dyson Brownian motion model (Dyson (1962a, b), Mehta (1991)), Eqs. (10.1–10.7), first discovered by Sutherland. The equivalence was discussed in Section 10.1.

Line IX is the connection between the nonstationary Dyson Coulomb gas model and the Brownian motion of the Hermitian matrix H, eigenvalues of which correspond to co-ordinates of particles in the Coulomb gas model (Dyson (1962a, b), Mehta (1991)). The solution of the Fokker–Planck equation is a Gaussian that allows consideration of para-metric correlations using a random matrix theory with several random matrices originating from a perturbation by an external parameter. This leads to many-matrix Gaussian integrals (Narayan and Shastry (1993)).

Link III describes the possibility of deriving the zero-dimensional nonlinear supersym-metric σ-model from microscopic models with disorder potential and from random matrix theory. The derivation has been discussed in Chapter 6 and in the present chapter. The σ-model was first obtained from models with random potential by Efetov (1982a) and then from random matrix theory by Verbaarschot, Weidenmüller, and Zirnbauer (1985). Study of universal parametric correlations with the σ-model was begun by Simons and Altshuler (1993).

Connection VII stands for a description of general quantum chaotic systems in an al-ternative way suggested by Pechukas (1983). Pechukas has shown that the dispersion of the energy levels of a nonintegrable Hamiltonian in response to an external perturbation can be expressed as a set of first-order differential equations, which were later shown to be integrable (Nakamura and Lakshmanan (1986)).

It is seen that the diagram in Fig. 10.4 is not "irreducible." One cannot get into each block by following only the solid lines. The dashed lines stand for relations that are established by using hypotheses that can be checked at the present time only numerically.

Connection II between quantum chaotic systems such as quantum billiards and a hydrogen atom in a magnetic field and random matrix theory has been established in a large number of numerical works (for a review see Mehta (1991)). Of course, models with disorder potential are a particular class of quantum chaotic systems and for these models the σ-model has been derived and the applicability of the random matrix model thus has been demonstrated.

A direct connection between the σ-model and quantum chaotic systems, and a check of the universality of the parametric correlations was done recently in a number of numerical works by Simons and Altshuler (1993), Simons et al. (1993), Faas et al. (1993). The equivalence of chaotic billiards to the σ-model can be proved by using an ergodic hypothesis that enables us to replace averaging over energies by averaging over a random potential. These relations correspond to link I.

The differential equations derived by Pechukas can be interpreted as equations of motion of classical particles in one dimension confined in a box (Pechukas gas). Assuming that the gas of these particles obeys principles of conventional statistical mechanics, Yukawa (1985) arrived at the random matrix theory (for a review see Haake (1992)). This relation is represented by line VIII.

Dotted line IV is established in limiting cases by Simons, Lee, and Altshuler (1993); this point was discussed at the beginning of this section.

Analytical confirmation of the possibility of describing quantum chaotic systems by random matrix theory or the nonlinear σ-model is highly desirable. This would allow one to replace the dashed lines in Fig. 10.4 by solid lines and (more important) would provide a unified theory for the variety of all the systems considered. My opinion is that one should try to derive the σ-model by starting from some general conditions making the system chaotic. In one particular case, for models with disorder, the derivation has proved to be possible.

11

Localization in systems with one-dimensional geometry

11.1 One-dimensional σ-model

11.1.1 One-dimensional σ-model for disorder problems

Most of the results of Chapters 6–10 have been obtained by using the zero-dimensional (0D) σ-model and we have seen that many rather different physical problems can be treated within this approximation successfully. Nevertheless, for some problems that can be described by the supermatrix σ-model the 0D version is not applicable. For example, localization in two-dimensional disordered metals was discussed in Chapter 5 and it had to be described by the two-dimensional (2D) σ-model. The problem is rather complicated and the solution presented there is not complete.

There are many problems of disordered metals and quantum chaos that can be reduced to a study of the one-dimensional (1D) σ-model. In contrast to those with higher dimensions, the 1D σ-model can be studied exactly by the transfer matrix technique, and very often it is possible to get explicit final results although the calculations are somewhat more difficult than in the 0D case. By now, the procedure of computation of different correlation functions with the 1D σ-model is well worked out, but before presenting it in detail let us discuss physical problems that can be treated in this way.

It is natural to suppose that the 1D σ-model describes one-dimensional disordered metals. However, when using the term *one-dimensional* one should distinguish between truly one-dimensional chains and microscopically three-dimensional metals with a one-dimensional geometry of the sample. The latter system will be called a *wire,* whereas for truly one-dimensional systems, the word *chain* will be used.

Since the work of Mott and Twose (1961) the kinetic properties of disordered chains have been attracting a lot of attention. A very interesting and important conclusion of this work is that for arbitrarily weak disorder all states of the system are localized. Because of the localization of the wave functions the system is an insulator for any disorder. The localization of a wave function near a center of the localization is a consequence of quantum interference, and therefore the insulating properties for any weak disorder are of essentially quantum origin. Later the one-dimensional chains were studied exactly by different methods by Berezinsky (1973), Abrikosov and Ryzhkin (1978), and Berezinsky and Gorkov (1979), who managed to calculate the frequency dependence of the conductivity and demonstrate the localization at arbitrary disorder. A mathematically rigorous proof of the localization can be found in Goldsheid, Molchanov, and Pastur (1977) and Pastur and Figotin (1978). The characteristic localization length L_c in one-dimensional chains has the same order of magnitude as the mean free path l. Therefore, if length L of the chain is finite, one can have either a regime of localization ($L > l$) or a ballistic regime ($L < l$). Classical diffusion does not occur in the chains.

The physics of thick wires is more complicated. At distances larger than the mean free path l electrons can diffuse, and, at first glance, nothing prevents them from moving classically around the sample. Thouless (1974, 1975, 1977) was first to predict that wires were similar to chains in the sense that all wave functions had to be localized with localization length L_c. In order to have classical diffusion at short distances, localization length L_c must exceed mean free path l. The characteristic energy corresponding to diffusion time τ_c through a sample of length L has the order of magnitude of the Thouless energy $\mathcal{E}_c = \pi^2 D_0/L^2$, where D_0 is the classical diffusion coefficient. If this energy is larger than the mean level spacing $\Delta = (\nu L S)^{-1}$, where ν is the density of states and S is the cross section of the wire, the discreteness of the levels is not important and classical mechanics provides a good description. The condition $\mathcal{E}_c \gg \Delta$ corresponds to the inequality $L \ll L_c \sim l\left(p_0^2 S\right)$, where p_0 is the Fermi momentum. Only in the opposite limit $L \gg L_c$ are the interference effects important and one can obtain localization. The quantity $p_0^2 S$ is often referred to as the number of transverse channels $N_\perp$, which can be very large, and hence, the localization length is proportional to the mean free path multiplied by the number of transverse channels. The case $N_\perp = 1$ corresponds to the one-dimensional chain.

The first work that presented an analytical treatment of a wire with N conducting channels was that of Anderson et al. (1980) using the Landauer transmission formula. Later, a model of N coupled chains was considered by Weller, Prigodin, and Firsov (1982) and by Dorokhov (1982, 1983), who used the Berezinsky technique to treat the problem. These authors also reached the same conclusion about the localization and Dorokhov showed that localization length was proportional to the number of chains. In principle, the models used in these works are valid for a description of wires. However, the number of chains or channels cannot be large (or alternatively, the mean free path must be large); otherwise one cannot begin with the description in terms of separate chains. This limit corresponds to sufficiently thin or sufficiently pure wires.

Treating the kinetic properties of wires with the one-dimensional σ-model was proposed by Efetov and Larkin (1983). This version of the σ-model can be obtained taking into account only the zero transversal space harmonics of the space dependence of supermatrix Q. In this approximation the free energy functional $F[Q]$ has the form

$$F[Q] = \frac{\pi \tilde{\nu}}{8} \operatorname{str} \int \left[D_0 \left(\frac{dQ}{dx} \right)^2 + 2i\omega\Lambda Q \right] dx \tag{11.1}$$

where

$$\tilde{\nu} = \begin{cases} \nu S, & \text{Models I and IIa} \\ 2\nu S, & \text{Models IIb and III} \end{cases} \tag{11.2}$$

and S is the cross section of the wire. Depending on the presence of magnetic and/or spin-orbit interactions the model has different symmetries and the proper classification was introduced in Chapter 5.

It is not difficult to obtain a criterion for the applicability of the 1D σ-model. The constraint $Q^2 = 1$ follows from the condition that fluctuations of the eigenvalues of the supermatrix are small. The average square of the deviation of the eigenvalue from unity $\delta\Lambda$ can be

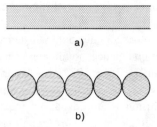

a)

b)

Fig. 11.1. a) A typical wire with a disorder; at short distances electron motion is diffusive. b) A granulated metal; macroscopic motion is possible if electrons can tunnel from granule to granule.

written as

$$\langle (\delta \Lambda)^2 \rangle \sim \frac{1}{\nu S} \int \frac{dk}{D_0 k^2 + 1/\tau} \qquad (11.3)$$

The propagator in Eq. (11.3) contains the energy $1/\tau$ instead of ω because modes corresponding to the fluctuations of the eigenvalues are "massive." The free energy functional has the form of the one-dimensional σ-model, Eq. (11.1), if $\langle (\delta \Lambda)^2 \rangle \ll 1$. Calculating the integral in Eq. (11.3) we come to the inequality

$$p_0^2 S \gg 1 \qquad (11.4)$$

This means that the number of transversal channels must be large. Of course, the inequality $\tau \varepsilon_0 \gg 1$ is implied. If an additional inequality $\tau \ll Sm$, where $(Sm)^{-1}$ is the characteristic energy of the transversal quantization, is fulfilled, electron motion at short distances is 3D diffusion.

The transversal nonzero space harmonics of fluctuations of Q can be neglected if the frequency ω is not very high and the corresponding inequality can be written as $\omega \ll \min \{ D_0/S, \ 1/\tau \}$. At higher frequencies corrections to the classical diffusion become three-dimensional and their contribution is small. A typical wire is represented in Fig. 11.1a.

Recently, a random scattering matrix approach for description of wires has been discussed in a number of works (Mello, Pereyra, and Kumar (1988), Mello (1988), Mello and Stone (1991), Macedo and Chalker (1992), Beenakker and Rejaei (1993, 1994)). Following the scheme suggested in these works a transfer matrix is written as a product of statistical independent building blocks modeling the quasi-one-dimensional structure of the wire. If the wire has N channels, a Fokker–Planck equation depending on N variables can be derived and the equation coincides with an equation derived by Dorokhov (1982, 1983). This is an alternative approach that does not depend on the details of the systems. In the limit of a large number of channels the scattering matrix approach is equivalent to the supersymmetry method based on Eq. (11.1) (Frahm (1995)).

The continuous σ-model, Eq. (11.1), is written for weak disorder. In many cases this limit is most interesting because the continuous σ-model contains only macroscopic parameters, often leading to universal properties of physical quantities and different scaling relations. However, the strong disorder limit can also be of interest. How can one work in this situation?

In principle, one can introduce strong disorder in different ways. A very convenient model of a disordered metal that can be treated with a slight modification of the σ-model is the ' model of a granulated metal suggested in the review (Efetov (1983)). The system consists of metallic granules that are in contact. In Fig. 11.1b a 1D chain of the granules is represented, but generally this can be a system of an arbitrary dimensionality. Let us assume that there are impurities in each granule or the shape of each granule is not simple and electrons are randomly scattered by the boundaries. In other words, it is assumed that the electron motion in an isolated granule is chaotic. So, if the granules were isolated, the 0D σ-model would be a good description. The Hamiltonian $\mathcal{H}$ of the system can be written in the form

$$\mathcal{H} = \sum_{i,j,p,q} T_{ij}^{pq} a_{ip}^+ a_{jq} + \sum_i \mathcal{H}_i, \tag{11.5}$$

$$\mathcal{H}_i = \sum_p \varepsilon_p^{(i)} a_{ip}^+ a_{ip} + \mathcal{H}_i'$$

where a_{ip}^+ (a_{ip}) are operators of creation (annihilation) of electrons in granule i in state p and T_{ij}^{pq} stands for the amplitudes of electron hops from the state p of the granule i to the state q of the granule j. The energies $\varepsilon^{(i)}(p)$ denote eigenenergies of isolated granules. The terms $\mathcal{H}_i'$ stand for a random potential in the granules that can be due to impurities inside the bulk or to irregular shape of each granule. With the ergodic hypothesis averaging over energy is equivalent to averaging over the random potential, and therefore the model described by Eqs. (11.5) is rather general. The granules can contain magnetic or spin-orbit impurities and a magnetic field can be applied.

As we know from the preceding chapters each isolated granule can be characterized by the mean level spacing Δ and the class of symmetry depending on the presence of magnetic and spin-orbit interactions. Provided the hopping amplitudes are not very large, macroscopic transport in the system of the granules is determined by the ratio of the hopping amplitudes T_{ij} to Δ. It is intuitively clear that in the limit $T_{ij} \gg \Delta$ the discreteness of the spectrum in a single granule is not perceived and the electron motion is diffusive through many granules. This limit corresponds to macroscopically weak disorder. In the opposite limit, $T_{ij} \ll \Delta$, electrons are almost completely localized in granules and this is the strong disorder limit.

The hopping term in the Hamiltonian $\mathcal{H}$ in Eq. (11.5) leads to an additional term in the σ-model describing a coupling of the supermatrices Q_i, corresponding to different granules. The derivation of such a term is completely analogous to the derivation of the Josephson energy in superconductors. Assuming that the dependence of T_{ij}^{pq} on p and q near the Fermi surface is weak, one comes to the following free energy:

$$F[Q] = \text{str} \left(-q \sum_{i,j} J_{ij} Q_i Q_j + \frac{i}{4} (\omega + i\delta) \sum_i \Delta_i^{-1} \Lambda Q \right), \tag{11.6}$$

$$J_{ij} = T_{ij}^2 v^2 V_i V_j = \frac{T_{ij}^2}{\Delta_i \Delta_j}$$

Here, $q = 1$ for Models I and IIa and $q = 2$ for Models IIb and III. The second term in Eq. (11.6) is the free energy of isolated granules; the first describes coupling. The coupling between nearest neighbors is largest, but, in principle, one may consider arbitrary J_{ij}. Different volumes of granules are possible, and so, the mean level spacing Δ_i and J_{ij} can depend on i. However, in later discussion this dependence will be neglected. The model described by Eq. (11.6) is a σ-model on a lattice and is analogous to the Heisenberg spin model.

In the limit $T_{ij} \gg \Delta$ ($J_{ij} \gg 1$) only small variations of supermatrix Q in space are important. Assuming that Δ is the same for all granules and that J_{ij} depend only on $|j - i|$, in this limit one comes to the continuous model, Eq. (11.1). The diffusion coefficient D_0 entering the latter equation in the granulated model has the form

$$D_0 = \frac{4\Delta}{\pi} \sum_j J_{ij} \left(r_i - r_j\right)^2 \tag{11.7}$$

In the limit of strong disorder, $J_{ij} \ll 1$, no further simplification of Eq. (11.6) is possible. A very important advantage of the lattice σ-model, Eq. (11.6), with respect to the continuous model, Eq. (11.1), is that it is well defined at short distances; this characteristic is very important for studying the Andersom metal–insulator transition. In one dimension, many quantities are highly universal at weak disorder and study of the lattice model can help in understanding the limits of the universalities.

A model equivalent to the model of granulated metal was used by Iida, Weidenmüller, and Zuk (1990) (the granules were called "slices"). In a recent paper Dittrich, Doron, and Smilansky (1994) considered a chain of chaotic billiards (domino model). So, the lattice model, Eq. (11.6), can give a proper description of both disorder problems and problems of quantum chaos.

11.1.2 One-dimensional σ-model and some problems of quantum chaos

Recently one more class of quantum chaotic models, namely, the models of random band matrices, was mapped by Fyodorov and Mirlin (1991) onto the 1D supermatrix nonlinear σ-model. The random band matrices are believed to give a proper description of such a chaotic model as the model of a quantum kicked rotator but also are used for a general modeling of the crossover from regular to chaotic behavior (Seligman, Verbaarschot and Zirnbauer (1984)).

The kicked rotator model was introduced by Casati et al. (1979) and since then its properties have been discussed in many publications (for a review see Izrailev (1990)). The Schrödinger equation of the rotator is written as

$$i\frac{\partial \psi(\theta, t)}{\partial t} = \left[-\frac{1}{2I}\frac{\partial^2}{\partial \theta^2} + kV(\theta)\delta_T(t)\right]\psi(\theta, t), \tag{11.8}$$

$$\delta_T(t) = \sum_{n=-\infty}^{\infty} \delta(t - nT)$$

Eq. (11.8) corresponds to a quantum rotator kicked periodically in time with the period T. The potential $V(\theta)$ is assumed to satisfy the relation $V(\theta) = V(-\theta) = V(\theta + 2\pi)$. In the classical limit, equations for p_n and θ_n, the classical angular momentum and angle just after the nth kick, are

$$p_{n+1} - p_n = -kV'(\theta_{n+1}), \qquad \theta_{n+1} - \theta_n = \frac{Tp_n}{I} \tag{11.9}$$

with $V'(\theta) = dV/d\theta$. If the quantity $K = kT/I$ is large enough, for many functions V, such as $V = \cos\theta$, the function V' generates random numbers for successive n values (Ott (1981)). Then, the momentum makes a random walk and the energy increases as $p_n^2 \sim k^2 \langle V'^2 \rangle n$. This relation corresponds to diffusion in momentum space.

However, in the quantum limit of small $k \ll 1$ the energy diffusion is suppressed and one obtains a localization of the energy (which is called the *dynamic localization*). The crossover from diffusion to localization motivated Fishman, Grempel, and Prange (1982) to try to map the kicked rotator model onto an electron model with disorder. Since Eq. (11.8) is invariant under $t \to t + 1$, the eigenstates are characterized by the quasi energy ε and can be written as

$$\psi_\varepsilon = \exp\left(-i\varepsilon t\right) u\left(\theta, t\right), \qquad u\left(\theta, t\right) = u\left(\theta, t+1\right) \tag{11.10}$$

Introducing the quantities $u_\pm$, which are the values of u just after (before) a jump, and the quantity $\bar{u} = (u_+ + u_-)/2$ Fishman, Grempel, and Prange (1982) arrived at the following equation:

$$T_m \bar{u}_m + \sum_{r \neq 0} U_r \bar{u}_{m+r} = -U_0 \bar{u}_m \tag{11.11}$$

where $U_r = U_{-r}$ is the Fourier coefficient of $U\left(\theta\right) = -\tan\left[kV\left(\theta\right)/2\right]$, $T_m = \tan(\varepsilon - Tm^2/2I)$. This is the equation for a one-dimensional tight binding model with elements U_r of hopping to the rth neighbor. Of course, the parameters T_m are values of a well-defined regular function of m. However, the function $\tan x$ oscillates in a complicated way and numerical tests show that, with good accuracy, one can consider T_m random. Then, Eq. (11.11) specifies the Anderson model with disorder. The function U_r can decay slowly provided k is not small.

We know already that for the "usual" 1D disordered chains the σ-model is not applicable because the condition (11.4) corresponds to a large number of transversal channels. However, if U_r decays slowly, Eq. (11.11) describes a long-range hopping model. It turns out that the long-range condition of the hopping is sufficient to derive the 1D σ-model again and, therefore, the kicked rotator model can be mapped onto it. The Hamiltonian corresponding to Eq. (11.11) does not contain any magnetic or spin-orbit interactions and so one comes to the orthogonal ensemble.

This Hamiltonian can be considered as a matrix whose elements are essentially nonzero only in a strip near the diagonal. One can introduce the ensemble of random band $N \times N$ matrices assuming that the matrix elements $\mathcal{H}_{ij} = \mathcal{H}_{ji}$ are statistically independent and distributed according to the Gaussian law with the zero mean value and the variance

$$\left\langle \mathcal{H}_{ij}^2 \right\rangle = \frac{1}{2} A_{ij} \left[1 + \delta_{ij}\right] \tag{11.12}$$

where $A_{ij} = a(i - j)$ depends only on the distance $r = |i - j|$. The function $a(r)$ is assumed to decrease sufficiently fast at $r > b$, with b the effective bandwidth. The band matrices have been used to study the crossover from diffusion to localization in many publications (Seligman, Verbaarschot, and Zirnbauer (1984), Casati, Molinary, and Izrailev (1990), Evangelou and Economou (1990), Feingold, Leitner, and Wilkinson (1991)), and the results for the kicked rotator and the random band matrix models are in good agreement with each other. The effect of dynamic localization was shown to occur in a number of other systems (e.g., the hydrogen atom in a monochromatic field) and was claimed to be observed in ionization experiments in a monochromatic field (Casati et al. (1987)). Deriving the 1D σ-model from the band matrix models Fyodorov and Mirlin (1991) showed that the inequalities $N \gg 1$, $b \gg 1$ are sufficient for the validity of Eq. (11.1). For the parameter Δ^{-1} that is proportional to density of states, one obtains the Wigner semicircle law; the diffusion coefficient D_0 is proportional to b^2.

Thus, the 1D σ-model can describe many interesting problems of condensed matter physics and quantum chaos and what remains to be done is to solve the model. In the preceding chapters many problems reduced to the 0D σ-models have been successfully solved only because, in this case, all one had to do was compute definite integrals. Of course, the 1D case is more difficult, but nevertheless the most interesting problems can still be studied exactly by using the transfer matrix technique. This will be done in the next sections of this chapter.

11.2 Transfer matrix technique

11.2.1 General equations

Nowdays the transfer matrix technique is well known to everybody working in theoretical physics. However, it is not so easy to identify who was the first to use this method. Apparently, the idea originated from the Feynman representation of quantum mechanics in terms of path integrals (Feynman (1948), Feynman and Hibbs (1965)). According to this approach one has to compute a functional integral instead of solving the Schrödinger equation. In statistical physics one usually has the opposite situation. The main quantity to be calculated is a partition function that is either a manifold integral for lattice models or a functional integral for a continuous model. If a 1D classical model is considered, the only space coordinate appears to be very similar to the time variable in the action in the Feynman representation. This leads to the idea of reducing the computation of the functional integral for the partition function to solving a corresponding Schrödinger equation, which is usually a simpler task. This approach was used by Vaks, Larkin, and Pikin (1966) to reduce a model of 1D classical interacting bosons to the Schrödinger equation for an anharmonic oscillator.

The same idea can be used to study the 1D σ-model. The only difficulty is that the Schrödinger equation obtained contains as coordinates the matrix elements of Q. The number of these elements is rather large, and supermatrix Q obeys the constraint $Q^2 = 1$. Kinetic properties of the model can be extracted from the density–density correlation function $Y^{00}(\mathbf{r}, \omega)$, Eqs. (3.24, 3.49, 4.57). It is more convenient to start from the model of granulated metal, Eq. (11.6). We know how to go to the continuous limit, Eq. (11.7), and, if necessary, this limit can be obtained from general formulae without difficulty. The correlation function $Y^{00}(\mathbf{r}, \omega)$ can be written for the model involved as

$$ Y^{00}(r_1, r_2, \omega) = -\frac{2\pi^2 \nu \tilde{\nu}}{S} \int Q^{12}_{33}(r_1) \, Q^{21}_{33}(r_2) \exp\left(-F[Q]\right) DQ \qquad (11.13) $$

with $F[Q]$ and $\tilde{\nu}$ given by Eqs. (11.6, 11.1). It will be assumed that the granules have macroscopically equal sizes and are enumerated by integer r. In all later discussions only the limiting cases of the orthogonal, unitary, and symplectic ensembles are considered and the standard parametrization for the supermatrices Q described in Section 6.3 is used.

The correlation function Y^{00} is an averaged quantity, and it is relevant to say a few words about the applicability of averaged quantities to describing transport in disordered systems. In several preceding chapters mesoscopic fluctuations were discussed and one could see that the fluctuations were very important. To obtain an averaged physical quantity like conductance one has to carry out the additional procedure of averaging over energy or different samples. The conductance of a given sample at a given energy is not a quantity considered by mesoscopic theory.

Now an averaged quantity $Y^{00}(r_1, r_2, \omega)$ for an infinitely long wire will be computed. One can calculate the average conductivity in the same way. Can these quantities be relevant to a measurement on a single sample? That would mean that they are self-averaging. The answer depends on the value of $\eta = \omega L^2/D_0$. If $\eta \to 0$, one has to think about the fluctuations and study statistical properties of the fluctuations. In the opposite limit $\eta \to \infty$, the wire is effectively subdivided into $\sqrt{\eta}$ independent pieces. Each piece has its own configuration of impurities and their contributions to the total conductance add together. This leads to averaging of the specific conductivity. So working with a finite frequency in the thermodynamic limit $L \to \infty$ one needs averaged quantities only. But at the same time, one should remember that the frequency should be kept finite in all calculations. Only in the final result may one take the limit $\omega \to 0$.

This requirement following from the intuitive arguments manifests itself in the mathematical structure of the σ-model. The matrix $\hat{\theta}$ parametrizing supermatrix Q, Eqs. (6.42–6.45), consists of blocks θ_{11} and θ_{22}, where the block θ_{11} is real and θ_{22} imaginary. Therefore, supermatrix Q contains $\cosh\theta_{1,2}$ and $\sinh\theta_{1,2}$ and one has to integrate over the hyperbolic noncompact group. As a consequence of the noncompactness the integral over Q in, for example, Eq. (11.13) diverges at $\omega = 0$ because the free energy functional $F[Q]$ is invariant under simultaneous rotation of all Q_i and nothing cuts the integral at large $\theta_{1,2}$. It is the finite frequency ω that makes the integrals convergent. Without the noncompact symmetry of the supermatrices it would not be possible to obtain localization. The general form of the correlation function Y^{00}, Eq. (3.32), contains ω in the denominator and, as will be seen, this arises from noncompactness. Cutting integrals over $\theta_{1,2}$ by finite ω leads to the appearance of ω in the denominator.

The function $Y^{00}(r, \omega)$ was calculated exactly in the work of Efetov and Larkin (1983) and the presentation of this section follows this work with some modifications. First, one has to rewrite Eq. (11.13) as follows:

$$Y^{00}(r_1, r_2, \omega) = -2\pi^2 \nu \tilde{\nu} \int \Psi(Q_1)(Q_1)_{33}^{12} \tag{11.14}$$

$$\times \Gamma(r_1, r_2; Q_1, Q_2)(Q_2)_{33}^{21} \Psi(Q_2) \, dQ_1 dQ_2$$

where the kernel $\Gamma(r_1, r_2; Q_1, Q_2)$ is the partition function of the segment between the points r_1 and r_2. It is assumed that integration for this kernel is performed over all Q except Q_1 and Q_2 at the points r_1 and r_2. So the kernel $\Gamma(r_1, r_2; Q_1, Q_2)$ depends on supermatrices Q_1, Q_2 and distance $r_2 - r_1$ (the point r_2 is assumed to be to the right of point r_1). The function $\Psi(Q)$ is the partition function of the parts of the wire located to the right of point r_2 and to the left of point r_1. This function depends only on supermatrix Q at the end point r_1 or r_2. Although these parts are infinitely long, function Ψ is finite. This is because this function must satisfy the following condition:

$$\int \Psi(Q) \, dQ = 1 \tag{11.15}$$

The integral in Eq. (11.15) is the total partition function of a wire and must be unity because of the supersymmetric structure, which leads to cancellation of the bosonic and fermionic contributions. The finiteness of Ψ implies that this function does not depend on the coordinate. With this property, comparing the function Ψ on neighboring sites one immediately arrives at the following relation:

$$\Psi(Q) = \int N(Q, Q') Z_0(Q') \Psi(Q') dQ' \qquad (11.16)$$

In Eq. (11.16), functions $N(Q, Q')$ and $Z_0(Q)$ equal

$$N(Q, Q') = \exp\left(\frac{\alpha}{4} \operatorname{str} QQ'\right), \qquad \alpha = 8J_{12} \qquad (11.17)$$

$$Z_0(Q) = \exp\left(\frac{\beta}{4} \operatorname{str} \Lambda Q\right), \qquad \beta = \frac{-i(\omega + i\delta)}{\Delta}$$

A similar equation can be written for the central part $\Gamma(r, r'; Q, Q')$. Comparing this function at the neighboring points r and $r + 1$ one obtains the recurrence equation

$$\Gamma(r, r'; Q, Q') - \int N(Q, Q'') Z_0(Q'') \Gamma(r+1, r'; Q'', Q') dQ''$$

$$= \delta_{rr'} \delta(Q - Q') \qquad (11.18)$$

where J_{12} is the coupling constant for the nearest neighbors. Eq. (11.18) should be complemented by the requirement

$$\Gamma(r, r'; Q, Q') = 0, \quad \text{for } r < r' \qquad (11.19)$$

The δ-function entering Eq. (11.18) satisfies the usual equality

$$\int f(Q') \delta(Q - Q') dQ' = f(Q) \qquad (11.20)$$

for an arbitrary function $f(Q)$. This δ-function can be written explicitly as

$$\delta(Q - Q') = \lim_{\gamma \to 0} \exp\left(-\frac{\operatorname{str}(Q - Q')^2}{\gamma}\right) \qquad (11.21)$$

Thus, the computation of the integrals for $\Psi(Q)$ and $\Gamma(r, r'; Q, Q')$ has been reduced to solving the integral equations, Eqs. (11.16, 11.18). Solving the equations one should substitute the solutions into Eq. (11.14) and calculate the integrals over Q and Q'. This procedure for obtaining recurrence equations is known as the *transfer matrix method*. Although Eqs. (11.16, 11.18) have simple structure, their solution is not trivial and one has to use several tricks.

A general solution of Eq. (11.18) is not necessary because the function Γ enters only one specific integral, Eq. (11.14). It is simpler to write a reduced equation for the matrix function $P_k(Q)$ introduced as

$$P_k(Q) = \sum_r \exp\left(i(r - r')k\right) \int \Gamma(r, r'; Q, Q') (Q')^{21} \Psi(Q') dQ' \qquad (11.22)$$

Such an equation follows easily from Eq. (11.18) and has the form

$$P_k(Q) - \exp(-ik) \int N(Q, Q') Z_0(Q') P_k(Q') dQ' = Q^{21} \Psi(Q) \qquad (11.23)$$

Using Eqs. (11.14–11.22) we can write the Fourier-transformed density–density correlation function $Y^{00}(k, \omega)$ as

$$Y^{00}(k, \omega) = -2\pi^2 v \tilde{v} \int \Psi(Q) Q_{33}^{12} (P_{k33}(Q) + P_{-k33}(Q)) dQ \qquad (11.24)$$

Eqs. (11.16, 11.23, 11.24) are sufficient to compute the density–density correlation function $Y^{00}(k, \omega)$. In this section only the weak disorder limit described by the continuous σ-model will be considered. In this limit the integral equations, Eqs. (11.16, 11.23), can be approximated by partial differential equations to simplify their solution. However, before carrying out the reduction let us rewrite Eqs. (11.16, 11.23) in a simpler form, taking into account the symmetry properties of the functions $N(Q, Q')$ and $Z_0(Q)$. It follows from Eq. (4.25) that the kernel $N(Q, Q')$ is invariant with respect to simultaneous rotation of supermatrices Q and Q' by a unitary supermatrix V. Although function Z_0 does not possess this general property for all V, it follows from Eq. (6.47) that with respect to rotations by the block-diagonal matrices U it is invariant. Because of this property $Z_0(Q)$ contains only the variables $\hat{\theta}$.

It is clear from these properties that one should seek a solution Ψ of Eq. (11.16) depending only on the variables $\hat{\theta}$. By using this assumption, using parametrization Eq. (6.47), and changing the variables of integration from Q' to $\tilde{Q} = \bar{U} Q' U$, one can reduce Eq. (11.16) to the form

$$\int \bar{N}(Q_0, \tilde{Q}_0) \left[Z_0(\tilde{Q}_0) \Psi(\tilde{Q}_0) - \Psi(Q_0) \right] d\tilde{Q}_0 = 0 \qquad (11.25)$$

where Q_0, $\tilde{Q}_0$ are the central parts of supermatrices Q and $\tilde{Q}$ in the parametrization Eq. (6.47). The function $\bar{N}(Q_0, \tilde{Q}_0)$ arises after integration over the unitary supermatrices $\tilde{u}$ and $\tilde{v}$ used for the parametrization of $\tilde{Q}$ and equals

$$\bar{N}(Q_0, \tilde{Q}_0) = \int N(Q_0, \tilde{Q}) d\tilde{u} d\tilde{v} \qquad (11.26)$$

Eq. (11.25) demonstrates explicitly that the assumption about the form of the solution Ψ does not contradict Eq. (11.16). Eq. (11.25) is linear and one should supplement it with a boundary condition. Such a condition can be obtained from Eq. (11.15) and from the fact that the function Ψ does not depend on the unitary supermatrices u and v. At the same time the Jacobian of the transformation to the variables $\hat{\theta}$, u, v is singular at $\hat{\theta} = 0$. General methods of calculating such integrals were discussed in Section 2.5 and have been used in all calculations starting from Chapter 6 (Section 6.3). The result of integration in Eq. (11.15) can be written as

$$\Psi(\hat{\theta} = 0) = 1 \qquad (11.27)$$

and this is just the required boundary condition. When deriving Eq. (11.25) the identity

$$\int N(Q_0, \tilde{Q}) d\tilde{Q} = 1 \qquad (11.28)$$

was used. This identity has a similar origin to Eq. (11.27) and can be proved in the same way.

As concerns Eq. (11.23), one can again use the parametrization Eq. (6.47) and represent the solution $P_k(Q)$ in the form

$$P_k(Q) = v R_k(Q_0) \bar{u} \qquad (11.29)$$

Substituting Eq. (11.29) into Eq. (11.23) and integrating over $\tilde{Q}$ one obtains the following equation for $R_k(Q_0)$:

$$R_{k;mn}(Q_0) - \exp(-ik) \sum_{i,j} \int N_{mijn}\left(Q_0, \tilde{Q}_0\right) Z_0\left(\tilde{Q}_0\right) R_{k;ij}\left(\tilde{Q}_0\right) d\tilde{Q}$$

$$= \left(Q_0^{21}\right)_{mn} \Psi(Q_0) \tag{11.30}$$

where

$$N_{mijn}\left(Q_0, \tilde{Q}_0\right) = \int N\left(Q_0, \tilde{Q}\right) \tilde{v}_{mi} \bar{\tilde{u}}_{jn} du\,dv \tag{11.31}$$

Although Eqs. (11.25, 11.30) occupy no less space than Eqs. (11.16, 11.23), they are considerably simpler because only the variables θ, $\theta_{1,2}$ enter these equations. So, depending on the ensemble, one has to solve equations containing two or three variables only.

11.2.2 Partial differential equations

This "only" does not indicate a simple task because the functions $\bar{N}$ and N_{mjin} are very complicated. Now, Eqs. (11.25, 11.30) will be simplified in the limit $\alpha \gg 1$ corresponding to weak disorder. In this limit, the main contribution to the integrals in Eqs. (11.25, 11.26, 11.30, 11.31) is from $\tilde{Q}$ close to Q_0. This means that the functions $\Psi(\tilde{Q}_0)$ and $R_{k;mn}(\tilde{Q}_0)$ vary with changing $\tilde{Q}_0$ much more slowly than the kernels $\bar{N}$ and N_{mijn} and therefore can be expanded near Q_0. For small deviations of $\tilde{Q}$ from Q_0 the function $N(Q_0, \tilde{Q})$, Eq. (11.17), can be written in the form

$$N\left(Q_0, \tilde{Q}\right) = \exp\left(-\Delta F_0 - \Delta F_1\right) \tag{11.32}$$

where

$$\Delta F_0 = \frac{\alpha}{4} \operatorname{str}\left(\Delta\hat{\theta}\right)^2, \tag{11.33}$$

$$\Delta F_1 = \frac{\alpha}{4} \operatorname{str}\left[\left(\Delta u \cos\hat{\theta}_m\right)^2 + \left(\Delta v \cos\hat{\theta}_m\right)^2 \right.$$
$$\left. + 2\Delta u \sin\hat{\theta}_m \Delta v \sin\hat{\theta}_m - \left(\Delta u\right)^2 - (\Delta v)^2\right] \tag{11.34}$$

In Eqs. (11.33, 11.34), $\hat{\theta}_m = (1/2)\left(\widehat{\tilde{\theta}} + \hat{\theta}\right)$, and the variables $\Delta\hat{\theta}$, Δu, and Δv are small deviations of the matrices $\widehat{\tilde{\theta}}$, $\tilde{u}$, $\tilde{v}$ from $\hat{\theta}$ and 1, respectively:

$$\Delta\hat{\theta} = \widehat{\tilde{\theta}} - \hat{\theta}, \qquad u = 1 - \Delta u + \frac{1}{2}\left(\Delta u\right)^2, \qquad v = 1 - \Delta v + \frac{1}{2}\left(\Delta v\right)^2 \tag{11.35}$$

The supermatrices Δu and Δv satisfy the conditions $\overline{(\Delta u)} = -\Delta u$, $\overline{(\Delta v)} = -\Delta v$, and their explicit form can be obtained from Eqs. (6.50–6.61). The quadratic form $\Delta F_0 + \Delta F_1$, Eqs. (11.34, 11.35), is written with the precision necessary for obtaining second derivatives in the effective Schrödinger equation for the functions Ψ and R_k. With this accuracy the variables $\Delta\hat{\theta}$, Δu, Δv can be considered as free variables of integration, and one is left with Gaussian integrations in Eqs. (11.25, 11.30). Let us note that the quadratic form $\Delta F_1 + \Delta F_2$ is the same as the form, Eq. (1.2), used in Appendix 1 for calculation of the Jacobian $J\left(\hat{\theta}\right)$, Eqs. (6.65–6.68), of the transformation to the variables $\hat{\theta}$, u, v. This means that the integration over Δu and Δv must have a contribution proportional to $J^{-1}\left(\hat{\theta}_m\right)$. As a result, Eq. (11.25) takes the form

$$\int \exp\left[-\frac{\alpha}{4}\operatorname{str}\left(\Delta\hat{\theta}\right)^2\right] J\left(\tilde{\theta}\right) J^{-1}\left(\hat{\theta}_m\right)\left(\Psi\left(\tilde{\theta}\right) Z_0\left(\tilde{\theta}\right) - \Psi\left(\hat{\theta}\right)\right) d\tilde{\theta} = 0 \quad (11.36)$$

Writing $\tilde{\theta} = \hat{\theta} + \Delta\hat{\theta}$, $\hat{\theta}_m = \hat{\theta} + (1/2)\Delta\hat{\theta}$, expanding all the functions depending on these variables in $\Delta\hat{\theta}$ up to the second order in $\Delta\hat{\theta}$, and assuming that $\omega \ll \Delta$ after calculating Gaussian integrals over $\Delta\hat{\theta}$ one obtains the final Schrödinger equation for function Ψ

$$\mathcal{H}_0\Psi = 0 \tag{11.37}$$

With Eqs. (11.7, 11.17), the effective Hamiltonian $\mathcal{H}_0$ can be written as

$$\mathcal{H}_0 = -\frac{1}{2\pi\tilde{v}D_0}\left[\frac{1}{J_\lambda}\frac{\partial}{\partial\lambda}\left(J_\lambda\left|1-\lambda^2\right|\frac{\partial}{\partial\lambda}\right)\right. \tag{11.38}$$

$$\left. + \sum_n \frac{1}{J_\lambda}\frac{\partial}{\partial\lambda_n}\left(J_\lambda\left|1-\lambda_n^2\right|\frac{\partial}{\partial\lambda_n}\right)\right] - i\left(\omega + i\delta\right)\pi\tilde{v}\left|\prod_n\lambda_n - \lambda\right|$$

The Hamiltonian $\mathcal{H}_0$ is written in the variables λ, λ_1, and λ_2, Eqs. (6.73), in the form valid for all ensembles. The subscript n in Eq. (11.38) can be 1 and 2 in Models I and III and is equal to 1 in Model II. The Jacobian J_λ corresponds to the use of the variables λ, λ_1, and λ_2 and is equal to

$$J_\lambda = \frac{\left|1-\lambda^2\right|}{\left(\lambda_1^2 + \lambda_2^2 + \lambda^2 - 2\lambda\lambda_1\lambda_2 - 1\right)^2}, \quad \text{Models I and III} \tag{11.39}$$

$$J_\lambda = \frac{1}{\left(\lambda_1 - \lambda\right)^2}, \quad \text{Model II} \tag{11.40}$$

The Hamiltonian $\mathcal{H}_0$ can be considered as the "radial" part of the total Hamiltonian. The corresponding "angular" part contains derivatives with respect to the elements of the unitary matrices u and v. This part is not important to the computation of function Ψ, which is assumed to be dependent on the variables λ, λ_1, λ_2 only. The way the Jacobian J_λ enters the Laplacian in Eq. (11.38) is usual for a situation when some "radial" and "angular" coordinates are used. The function Ψ corresponds to a state with zero "angular momentum." It is, in fact, the ground state wave function, where the energy of the ground state is zero. The zero ground state energy is the consequence of the zero free energy of the initial supermatrix σ-model.

Seeking the solution of Eq. (11.23) in the form of Eq. (11.29) means that this solution is in the subspace of wave functions with "angular momentum" 1. Using Eqs. (11.32–11.35) again in the main nonvanishing approximation in $\Delta\hat{\theta}$, Δu, Δv one can write

$$\sum_{i,j}\int N_{mijn}\left(Q_0, \tilde{Q}_0\right)\left(R_{k;ij}\left(\tilde{Q}_0\right) - R_{k;ij}\left(Q_0\right)\right)d\tilde{Q} \tag{11.41}$$

$$= \left[\mathcal{H}_0 R_k - \frac{1}{2}\left\langle(\Delta v)^2\right\rangle_1 R_k - \frac{1}{2}R_k\left\langle(\Delta u)^2\right\rangle_1 + \left\langle\Delta v R_k \Delta u\right\rangle_1\right]_{mn}$$

The symbol $\langle \ldots \rangle_1$ stands for the following averaging:

$$\langle \ldots \rangle_1 = \frac{\int (\ldots) \exp(-\Delta F_1) \, d(\Delta u) \, d(\Delta v)}{\int \exp(-\Delta F_1) \, d(\Delta u) \, d(\Delta v)}$$

Using Eq. (11.41) one can write the final differential equation for R_k. This equation, like Eq. (11.38), contains only the variables λ, λ_1, and λ_2. However, because of the presence of the last three terms on the right-hand side of Eq. (11.41) the corresponding effective Hamiltonian differs from $\mathcal{H}_0$. This is usual for quantum mechanics problems with a central potential. In such problems angular variables decouple from radial variables but for nonzero angular momentum give additional terms in the radial part of the Hamiltonian.

For further manipulations it is convenient to make the substitution

$$R_k = -2i\pi\tilde{\nu}D_0 \begin{pmatrix} \tilde{R}_{k;11} & 0 \\ 0 & \tilde{R}_{k;22} \end{pmatrix} \tag{11.42}$$

The matrices $\tilde{R}_{k;11}$ and $\tilde{R}_{k;22}$ should be written for each model separately

$$\tilde{R}_{k;11}^{(I)} = \left(1 - \lambda^2\right)^{1/2} f_k \mathbf{1}, \tag{11.43}$$

$$\tilde{R}_{k;22}^{(I)} = i \begin{pmatrix} \left(\lambda_1^2 - 1\right)^{1/2} \lambda_2 f_{1k} & \left(\lambda_2^2 - 1\right)^{1/2} \lambda_1 f_{2k} \\ \left(\lambda_2^2 - 1\right)^{1/2} \lambda_1 f_{2k} & \left(\lambda_1^2 - 1\right)^{1/2} \lambda_2 f_{1k} \end{pmatrix}$$

$$\tilde{R}_{k;11}^{(II)} = \left(1 - \lambda^2\right)^{1/2} f_k \mathbf{1}, \qquad \tilde{R}_{k;22}^{(II)} = i \left(\lambda_1^2 - 1\right)^{1/2} f_{1k} \mathbf{1} \tag{11.44}$$

$$\tilde{R}_{k;11}^{(III)} = \begin{pmatrix} \left(1 - \lambda_1^2\right)^{1/2} \lambda_2 f_{1k} & \left(1 - \lambda_2^2\right)^{1/2} \lambda_1 f_{2k} \\ \left(1 - \lambda_2^2\right)^{1/2} \lambda_1 f_{2k} & \left(1 - \lambda_1^2\right)^{1/2} \lambda_2 f_{1k} \end{pmatrix} \tag{11.45}$$

$$\tilde{R}_{k;22}^{(III)} = i \left(\lambda^2 - 1\right)^{1/2} f_k \mathbf{1}$$

where $\mathbf{1}$ stands for the 2×2 unity matrix, and the superscripts I, II, and III denote Models I, II, and III. Substituting Eqs. (11.42–11.45) into Eqs. (11.41, 11.30, 11.31); performing integration over $\Delta\hat{\theta}$, Δu, and Δv; and expanding in k and ω one arrives at the final equations. In Model II, there are only two independent variables, λ and λ_1, and two unknown functions, f_k and f_{1k}. The corresponding equations can be written as

$$2\pi\tilde{\nu}D_0 (ik + \mathcal{H}_0) f_k + 2\lambda \frac{\partial f_k}{\partial \lambda} + 2\left(\lambda_1^2 - 1\right) J_\lambda (f_k - f_{1k}) = \Psi, \tag{11.46}$$

$$2\pi\tilde{\nu}D_0 (ik + \mathcal{H}_0) f_{1k} - 2\lambda_1 \frac{\partial f_{1k}}{\partial \lambda_1} + 2\left(1 - \lambda^2\right) J_\lambda (f_{1k} - f_k) = \Psi \tag{11.47}$$

The corresponding equations for Models I and III are considerably more complicated. After rather cumbersome manipulations one obtains

$$2\pi\tilde{\nu}D_0 (ik + \mathcal{H}_0) f_k + 2(-1)^n \lambda \frac{\partial f_k}{\partial \lambda} \tag{11.48}$$

$$+ 4[\lambda_2 |\lambda_1^2 - 1| B(\lambda_1, \lambda, \lambda_2) (f_k - f_{1k}) + \lambda_1 |\lambda_2^2 - 1| B(\lambda_2, \lambda, \lambda_1) (f_k - f_{2k})] = \Psi$$

$$2\pi \tilde{v} D_0 (ik + \mathcal{H}_0) f_{ak} - 2(-1)^n \left(\lambda_a \frac{\partial f_{ak}}{\partial \lambda_a} + \frac{|\lambda_b^2 - 1|}{\lambda_b} \frac{\partial f_{ak}}{\partial \lambda_b} \right) \tag{11.49}$$

$$+ 4 \left[\frac{\lambda_a |\lambda_b^2 - 1|}{\lambda_b} B(\lambda_a, \lambda_b, \lambda)(f_{bk} - f_{ak}) + \frac{|1 - \lambda^2|}{\lambda_b} B(\lambda_a, \lambda, \lambda_b)(f_{ak} - f_k) \right] = \Psi$$

In Eqs. (11.48, 11.49) $n = 0$ for Model I and $n = 1$ for Model III. The subscripts a and b can take the values 1 and 2, $a \neq b$. The three-variable function $B(x, y, z)$ is equal to

$$B(x, y, z) = \frac{-2xy + z - z^3 + zx^2 + zy^2}{\left(x^2 + y^2 + z^2 - 2xyz - 1\right)^2} \tag{11.50}$$

The last step is to express the density–density correlation function $Y^{00}(k, \omega)$ in terms of an integral of the functions f, $f_{1,2}$, and Ψ. Substituting Eqs. (11.29, 11.42–11.45) into Eq. (11.24) we obtain

$$Y^{00}(k, \omega) = \tilde{Y}^{00}(k, \omega) + \tilde{Y}^{00}(-k, \omega) \tag{11.51}$$

The function $\tilde{Y}^{00}(k, \omega)$ has the following form for Model II:

$$\tilde{Y}^{00}(k, \omega) = 4\pi^3 v \tilde{v} D_0 \int_1^\infty \left[\left(\lambda_1^2 - 1 \right) f_{1k} + \left(1 - \lambda^2 \right) f_k \right] \Psi J_\lambda d\lambda_1 d\lambda \tag{11.52}$$

For Models I and III, it can be written as

$$\tilde{Y}^{00}(k, \omega) = 4\pi^3 v \tilde{v} D_0 \tag{11.53}$$

$$\times \iiint \left[|\lambda_1^2 - 1| \lambda_2^2 f_{1k} + |\lambda_2^2 - 1| \lambda_1^2 f_{2k} + |1 - \lambda^2| f_k \right] \Psi J_\lambda d\lambda_1 d\lambda_2 d\lambda$$

In Eq. (11.53), integration is performed over the domain $1 < \lambda_1 < \infty$, $1 < \lambda_2 < \infty$, and $-1 < \lambda < 1$ for Model I and $-1 < \lambda_1 < 1$, $0 < \lambda_2 < 1$, and $1 < \lambda < \infty$ for Model III.

Eqs. (11.27, 11.37–11.40, 11.46–11.53) completely solve the problem of electron motion in a long metallic wire for all types of symmetry. Eqs. (11.37, 11.46–11.49) are very complicated and can be solved for arbitrary ω only numerically. Fortunately, in the most interesting cases of high, $\omega \gg 1/(\tilde{v}^2 D_0)$, and low, $\omega \ll 1/(\tilde{v}^2 D_0)$, frequencies the result can be obtained analytically and written in a rather simple form. This will be done in the next section.

11.3 Density–density correlator and dielectric permeability

11.3.1 High-frequency limit

Let us consider first the limit of high frequencies, $\omega \gg (\tilde{v}^2 D_0)^{-1}$. This is the weak localization limit and for calculations in this region the transfer-matrix technique is not necessary. The perturbation theory presented in Section 5.1 is more convenient for this purpose and, besides, it is applicable in any dimension. However, the calculation presented later for high frequencies allows one to check the equations obtained and demonstrates in a simple example how the transfer-matrix method works. In the preceding section, when deriving the partial differential equations from the granulated model, Eq. (11.6), it was assumed not only that $J_{ij} \gg 1$ but also $\omega \ll \Delta$, where Δ is the mean level spacing in a single granule. Using Eqs. (11.7) the condition of high frequencies together with this restriction can be written as

$$\Delta / J_{ij} \ll \omega \ll \Delta \qquad (11.54)$$

At the same time, in the model of a homogeneous weakly disordered metal, the frequency ω is restricted from above by τ^{-1} and the weak localization region can be written as

$$\tau^{-1} \left(p_0^2 S \right)^{-2} \ll \omega \ll \tau^{-1} \qquad (11.55)$$

In other words, Eqs. (11.54, 11.55) determine the regions where electrons diffuse almost classically. We see that the classical diffusion region for the weakly disordered homogeneous metal exists when the number of transversal channels $p_0^2 S$ is large. For the granulated system, the coupling constant J_{ij} must be large.

At high frequencies, the essential values λ, λ_1, and λ_2 are close to unity in all three models. Therefore, it is convenient to use the following variables:

$$\begin{cases} \lambda_1 = 1 + x_1, & \lambda_2 = 1 + x_2, & \lambda = 1 - x, & \text{Model I} \\ \lambda_1 = 1 + x_1, & \lambda = 1 - x, & & \text{Model II} \\ \lambda_1 = 1 - x_1, & \lambda_2 = 1 - x_2, & \lambda = 1 + x, & \text{Model III} \end{cases} \qquad (11.56)$$

When integrating over x_1, x_2, and x in Eqs. (11.52, 11.53), the main contribution is from $x \sim x_1 \sim x_2 \sim \left(D_0 \tilde{v}^2 \omega \right)^{-1/2} \ll 1$. In the principal approximation one may integrate in Eqs. (11.52, 11.53) over all positive x_1, x_2, and x. Expanding in Eqs. (11.37–11.40, 11.46–11.53) in small x_1, x_2, and x and then solving the equations in this limit we find for Model II

$$\Psi = \exp\left(-\tilde{\gamma}\,(x_1 + x)\right), \qquad f_k \approx f_{1k} \approx \frac{1}{2\pi\tilde{v}D_0} \frac{\Psi}{ik + (-i\omega/D_0)^{1/2}} \qquad (11.57)$$

The behavior of the corresponding functions is the same for both Models I and III and is described by the following formula:

$$\Psi = \exp\left(-\tilde{\gamma}\,(x_1 + x_2 + x)\right), \qquad f_k \approx f_{1k} \approx f_{2k} \approx \frac{1}{2\pi\tilde{v}D_0} \frac{\Psi}{ik + (-i\omega/D_0)^{1/2}} \qquad (11.58)$$

where $\tilde{\gamma}^2 = -\left(\pi\tilde{v}\right)^2 D_0 i \left(\omega + i\delta\right)$.

The ground state wave function Ψ given by Eqs. (11.57, 11.58) obeys the boundary condition, Eq. (11.27). Inserting Eqs. (11.57, 11.58) into integrals in Eqs. (11.52, 11.53) and integrating over x, x_1, and x_2 we find in the principal approximation for all the models

$$Y^{00}(k, \omega) = \frac{4\pi v}{-i\omega + D_0 k^2} \qquad (11.59)$$

Eq. (11.59) describes the classical diffusion and agrees with Eq. (3.52). We see that for the description of the weak localization limit the noncompactness of the symmetry group of the supermatrices Q is not important. The main contribution is from the region $\lambda_1 \approx \lambda_2 \approx \lambda \sim 1$ and the difference between these variables is not important.

11.3.2 Low-frequency limit

The region $\omega \ll \left(\tilde{v}^2 D_0\right)^{-1}$ is more interesting because the phenomenon of localization manifests itself just in this limit. From the mathematical point of view this region is very interesting because here noncompactness plays a decisive role. In the low-frequency limit, the main contribution to the integrals in Eqs. (11.52, 11.53) is from the domain $\lambda_1 \gg 1$, $\lambda_2 \gg 1$, and $\lambda \sim 1$ in Model I; $\lambda_1 \gg 1$ and $\lambda \sim 1$ in Model II; and $\lambda_1 \sim 1$, $\lambda_2 \sim 1$, and $\lambda \gg 1$ in Model III. In these domains the partial differential equations, Eqs. (11.37, 11.38, 11.46–11.50), as well as the integrals, Eqs. (11.52, 11.53), can be considerably simplified.

For Model II, in Eqs. (11.46, 11.47) one may keep only the function f_{1k} and neglect the function f_k. As concerns Models I and III, one may retain only the functions f_{1k} and f_{2k} for Model I and f_k for Model III. In the principal approximation, one can seek solutions of the equations in the form

$$
\begin{cases}
\Psi = \Psi\left(\lambda_1 \lambda_2\right), & f_{1k} = f_{2k} = \Phi_k\left(\lambda_1 \lambda_2\right), & \text{Model I} \\[2mm]
\Psi = \Psi\left(\lambda_1\right), & f_{1k} = \Phi_k\left(\lambda_1\right), & \text{Model II} \\[2mm]
\Psi = \Psi\left(\lambda\right), & f_k = \Phi_k\left(\lambda\right), & \text{Model III}
\end{cases}
\tag{11.60}
$$

Using Eq. (11.60) one can reduce Eqs. (11.37, 11.40, 11.46–11.50) in the limit involved to surprisingly universal equations valid for all three models. It is convenient first to remove the frequency ω, changing the variables as

$$
\begin{cases}
z = -i\omega\pi^2\tilde{v}^2 D_0 \lambda_1 \lambda_2, & \text{Model I} \\[2mm]
z = -i\omega 2\pi^2\tilde{v}^2 D_0 \lambda_1, & \text{Model II} \\[2mm]
z = -i\omega 2\pi^2\tilde{v}^2 D_0 \lambda, & \text{Model III}
\end{cases}
\tag{11.61}
$$

Calculating integrals in Eqs. (11.52, 11.53) over the remaining variables one can reduce the density–density correlation function $Y^{00}(k, \omega)$, Eq. (11.51), to the form

$$
Y^{00}(k, \omega) = \frac{4\pi v A(k)}{-i\omega}, \qquad A(k) = \int_0^\infty \left(\Phi_k(z) + \Phi_{-k}(z)\right) \Psi(z)\, dz
\tag{11.62}
$$

The functions $\Phi_k(z)$ and $\Psi(z)$ satisfy the following equations:

$$
-z\frac{d^2\Psi}{dz^2} + \Psi = 0
\tag{11.63}
$$

$$
-\frac{d}{dz}\left(z^2\frac{d\Phi_k}{dz}\right) + ikL_c\Phi_k + z\Phi_k = \Psi(z)
\tag{11.64}
$$

where length L_c equals

$$
L_c = \begin{cases}
\pi\tilde{v}D_0, & \text{Model I} \\[2mm]
2\pi\tilde{v}D_0, & \text{Models II and III}
\end{cases}
\tag{11.65}
$$

Recalling Eq. (11.2) for $\tilde{v}$ we can rewrite length L_c in the explicit form valid for all the models

$$
L_c = \begin{cases} \pi \nu S D_0, & \text{Model I} \\ 2\pi \nu S D_0, & \text{Model IIa} \\ 4\pi \nu S D_0, & \text{Model IIb and III} \end{cases} \tag{11.66}
$$

Thus, in all the models considered the density–density correlator is determined by the same equations. Only the characteristic length L_c, which will be shown later to be the localization length, is different. The length L_c, Eq. (11.66), is proportional to $l\left(p_0^2 S\right)$; that means that this length is $N_\perp$ times, where $N_\perp$ is the number of transversal channels, larger than the mean free path l. The solutions $\Psi(z)$ and $\Phi_k(z)$ have to decrease at infinity. Furthemore, function $\Psi(z)$ must satisfy the boundary condition

$$
\Psi(0) = 1 \tag{11.67}
$$

The condition Eq. (11.67) is a consequence of the fact that at $\omega = 0$ Eq. (11.37) has the exact solution $\Psi(\lambda_1, \lambda_2, \lambda) = 1$ obeying the normalization condition, Eq. (11.27). Then the limit $\omega \to 0$ corresponds in terms of the variables z, Eq. (11.61), to Eq. (11.67). Besides, the assumption that the main contribution to the integral for $A(k)$ in Eq. (11.62) is from $z \sim 1$ implies that the integral must be convergent. A divergence of the integral at small z would imply that the region $\lambda, \lambda_1, \lambda_2 \sim 1$ is most essential, but this would not give ω in the denominator in the expression for $Y^{00}(k, \omega)$, Eq. (11.62). So, the function $\Phi_k(z)$ must be integrable at small z.

Eqs. (11.62–11.64) with the specified boundary conditions exactly coincide with the low-frequency limit of the Berezinsky equations written for one-dimensional disordered chains (Berezinsky (1973)) provided the localization length L_c is replaced by the mean free path l. This means that the equations are remarkably universal, a characteristic that is far from trivial. As discussed in Section 11.1, the σ-model is not applicable to description of the chains because fluctuations of eigenvalues of supermatrices Q can be important. It is clear that at arbitrary frequency the chain and wire models give different results. For example, the diffusive regime does not exist in the chain model at all.

The form of the correlator $Y^{00}(k, \omega)$ given by Eq. (11.62) is typical for localized states (cf. Eq. (3.32)) and can even be computed exactly. It is not difficult to find the solution of Eqs. (11.63–11.64) satisfying the boundary conditions discussed. As concerns Eq. (11.63) the solution $\Psi(z)$ can be written immediately as

$$
\Psi(z) = 2\sqrt{z} K_1\left(2\sqrt{z}\right) \tag{11.68}
$$

To find the solution of Eq. (11.64) one first has to solve the corresponding homogeneous equation and then obtain the final solution in a standard way. The result has the form

$$
\Phi_k(z) = \frac{2}{\sqrt{z}}\left[I_{\sqrt{1+4i\kappa}}\left(2\sqrt{z}\right) \int_{2\sqrt{z}}^{\infty} \xi K_{\sqrt{1+4i\kappa}}(\xi) K_1(\xi) d\xi \right.
$$

$$
\left. + K_{\sqrt{1+4i\kappa}}\left(2\sqrt{z}\right) \int_{0}^{2\sqrt{z}} \xi I_{\sqrt{1+4i\kappa}}(\xi) K_1(\xi) d\xi \right], \qquad \kappa = kL_c \tag{11.69}
$$

where K_α and I_α are the Bessel functions. The solutions of Eqs. (11.68, 11.69) satisfy the required boundary conditions: Both functions $\Psi(z)$ and $\Phi_k(z)$ decrease at infinity, function $\Psi(z)$ satisfies Eq. (11.67), and function Φ_k is integrable at small z.

By substituting Eqs. (11.68, 11.69) into Eq. (11.62) one can get the final result. It is more convenient to calculate the density–density correlator in the coordinate representation. The corresponding integral can be computed integrating first over k and then over z (for the chain model such a computation was done by Gogolin, Melnikov, and Rashba (1975) and Gogolin (1976)). The function Φ_k has the branching point $\kappa = i/4$ and therefore one should make the cut along the line $(i/4, i\infty)$. The contour of the integration over κ can be shifted such that integration is performed along the edges of the cut. The limit of low frequencies determines the density–density correlations at infinite time, $t \to \infty$. Because the function $Y^{00}(k, \omega)$ is proportional to i/ω at $t \to \infty$ the correlations remain finite. Then, in the real time and coordinate representation, the density–density correlation function Y^{00} in the limit $t \to \infty$ can be reduced to the following form:

$$Y^{00}(x, \varepsilon, t \to \infty) \equiv p_\infty(x) \tag{11.70}$$

$$= \frac{\pi^2}{16L_c} \int\limits_0^\infty \left(\frac{1+y^2}{1+\cosh \pi y} \right)^2 \exp\left(-\frac{1+y^2}{4L_c} |x| \right) y \sinh \pi y \, dy$$

The function $p_\infty(x)$ introduced in Eq. (3.33) characterizes the decay of wave functions. In the limit $|x| \gg L_c$ this function decays exponentially

$$p_\infty(x) \approx \frac{1}{4\sqrt{\pi} L_c} \left(\frac{\pi^2}{8} \right)^2 \left(\frac{4L_c}{|x|} \right)^{3/2} \exp\left(-\frac{|x|}{4L_c} \right) \tag{11.71}$$

The exponential form of $p_\infty(x)$ proves the localization of the wave functions; length L_c is the localization length. It is important to emphasize that localization length L_c is the only parameter characterizing the behavior of the density–density correlator Y^{00}, Eqs. (11.70, 11.71). Of course, the density–density correlation function Y^{00} is not the only universal function and one can construct many of them. Universality is very important because it yields a good possibility of making experimental and numerical checks of the ideas proposed in the present section for different systems. Some examples will be considered in the next section.

For weak disorder the dependence of the physical quantities on a single parameter confirms the idea of the one-parameter scaling discussed in Chapter 3. The doubling of the localization length L_c when changing from Model I to Model IIa and from Model IIa to Models IIb and III, Eq. (11.66), first discovered by Efetov and Larkin (1983), looks at first glance as if L_c were proportional to the index $\beta = 1, 2, 4$ characterizing the ensemble, $L_c(\beta) = \beta L_c(1)$. Such a proportionality was suggested by Pichard et al. (1990) and Stone et al. (1991) using the transfer-matrix approach based on maximal entropy ansatz, and was even claimed to be valid in a higher dimensionality.

However, Eqs. (11.66) do not quite agree with the claim. For example, the localization length is the same in systems with magnetic and spin-orbit impurities (Models IIb and III). One can have crossover from the symplectic to the unitary ensemble by applying a magnetic field. But this corresponds to changing from Model III to Model IIb, which does not lead to changing the localization length. So, the hypothesis of Pichard et al. (1990) and Stone et al. (1991) cannot be correct, and doubling of the localization length, Eqs. (11.66), is occasional. We have seen in Section 7.2 that an analogous curious doubling occurs in the quantum correction to the electric susceptibility of small metal particles, Eq. (7.59). A discussion of some other examples of systems where the relation $L_c(\beta) = \beta L_c(1)$ does not apply can be found in Lerner and Imry (1995). At the same time, applying the magnetic field to a system belonging to the orthogonal ensemble leads to doubling of localization length and this has been reliably established for systems with one-dimensional geometry including quantum-chaotic systems (Thaha, Blümel, and Smilansky (1993)).

So, in some cases doubling of the localization length does take place, but one should not think of it as of a universal property. This is especially clear in strongly disordered systems, where there is no doubling at all. As in the weak localization regime, the magnetic field tends to delocalize the system and this can be observed experimentally.

Using Eqs. (11.62, 11.68, 11.69) one can find the function $A(k)$ in the range of small momenta $k \ll L_c^{-1}$. The result is

$$A(k) = 1 - 4\zeta(3) k^2 L_c^2 \tag{11.72}$$

where $\zeta(y)$ is the Riemann ζ-function. The fact that $A(k) = 1$ at $k = 0$ means that $\int p_\infty(r, r')\, dr'$ with $p_\infty(r, r')$ from Eq. (3.33) equals the average density of states. But the quantity $p_\infty(r, r')$ contains a contribution of localized states only (Lifshitz, Gredescul, and Pastur (1988)) and one has to conclude that all states in systems with one-dimensional geometry are localized.

The function $A(k)$ at small momenta determines the static dielectric permeability ϵ

$$\epsilon = -4\pi e^2 v \frac{d^2 A(k)}{dk^2}\bigg|_{k=0} = 32\zeta(3) e^2 v L_c^2 \tag{11.73}$$

The permeability ϵ, Eq. (11.73), is proportional to the square of the number of transversal channels and therefore can be very large.

In the limit of high frequencies that can also be written as $\omega \gg D_0/L_c^2$, the classic diffusion formula, Eq. (11.59), holds and conductivity σ, according to the Einstein relation, is equal to

$$\sigma = 2ve^2 D_0 \tag{11.74}$$

Decreasing the frequency leads to a smooth crossover from Eq. (11.74) to Eq. (11.73). The dependence on the frequency is different for different models. In particular, the dependence is not monotonous in Model III, where the first correction to the conductivity at high frequencies is positive, Eq. (5.23). The dependence of localization length L_c on magnetic and spin-orbit interactions results in corresponding dependence of dielectric permeability. One can estimate the order of magnitude of the interactions corresponding to the crossover between Models I, IIa, IIb, and III by evaluating, as in Section 6.2, the first correction to the diffusion coefficient. The external magnetic field leads to the crossover from Model I, which is invariant under time reversal and spin rotations, to Model IIa. In the latter model the dielectric permeability is four times as large as that in Model I. This crossover occurs at fields of the order of $H \sim \phi_0(\sqrt{S}L_c)^{-1}$. If a sufficiently high concentration of magnetic or spin-orbit impurities is added (Models IIb and III)

$$\tau_s^{-1}, \ \tau_{so}^{-1} \gg \frac{D_0}{L_c^2} \tag{11.75}$$

the dielectric permeability is, according to Eqs. (11.66, 11.73), sixteen times as large as the one in a wire with no magnetic or spin-orbit impurities.

11.4 Inverse participation ratio in a finite sample

In the preceding section, it was shown how to calculate correlation functions in the limit $L \rightarrow \infty$, ω finite. However, the opposite limit, $\omega \rightarrow 0$, L finite, is also of considerable interest. In numerical simulations, one has to perform computations for a finite system. In experiments it is also easier to measure direct current of conductance of a finite wire and to study the dependence of the conductance on the length of the wire. The localization manifests itself by an exponential decay of the conductance with increase in the length of the system. I want to repeat that as soon as one considers the limit $\omega \rightarrow 0$, at finite sample size, physical quantities are no longer self-averaging and for a complete characterization of the system one has to know not only the averages of the physical quantities but their higher moments or the entire distribution functions. In this and the next section two problems that can be reduced to the one-dimensional σ-model for a sample with a finite length are considered.

The first problem considered in the present section concerns computation of the inverse participation ratio for an isolated system; studying this quantity is motivated by numerical simulations for the quantum kicked-rotator model and the models of the random band matrices. It is definitely interesting to obtain some quantities analytically by using the σ-model and to compare them with what is known for models of quantum chaos where traditionally all results are extracted from numerical works. Especially important is the study of universal relations that are not dependent on the microscopic parameters of the model because it usually provides the most reliable check.

The second problem considered in the next section is the calculation of the conductance of a finite disordered wire attached to ideal leads. This question is important because of its direct relation to experiments on localization in disordered wires.

In the preceding section it was mentioned that the kicked-rotator model used to study quantum chaos as well as some random band matrix models corresponded to a 1D disorder model, and the latter was mapped onto the 1D supersymmetric σ-model.

Intensive numerical investigations (Casati, Molinary, and Izrailev (1990), Evangelou and Economou (1990), Feingold, Leitner, and Wilkinson (1991)) revealed some nice scaling relations that hold for both the kicked-rotator and random band models. A set of generalized localization lengths ξ_m, $m = 1, 2, \ldots$, was introduced according to the formula

$$\ln \xi_m = \frac{1}{1-m} \ln \left[\overline{v^{-1}(\varepsilon) S^{-1} \sum_i |\varphi_i(r)|^{2m} \delta(\varepsilon - \varepsilon_i)} \right] \qquad (11.76)$$

Here ε_i is an energy level of the "Hamiltonian" (band matrix); $\varphi_i(r)$ represents the component of the corresponding eigenfunction at given site $r = 1, 2, \ldots, L$, where L is the sample length (matrix size); and the overbar denotes averaging over disorder and over all sites in the sample. The function $v(\varepsilon)$ is the density of states. The quantities ξ_m^{-1} for $m \geq 2$ are proportional to the coefficients of the inverse participation ratio discussed in Section 7.3.

It was found that in a wide region of parameters numerical results are well described by the following empirical scaling law:

$$\frac{\beta_m}{1 - \beta_m} = Cg, \qquad g = \frac{L_c}{L} \qquad (11.77)$$

In Eq. (11.77), $\beta_m = \xi_m / \xi_m^{\mathrm{ref}}$, L_c is proportional to the square of the bandwidth b^2 and C is some model-dependent constant. It has been mentioned in Section 11.1 that the constant D_0 entering the σ-model is proportional to b^2, and, hence, the parameter L_c in Eq. (11.77) is just the localization length. The parameter ξ_m^{ref} is a generalized localization length for some reference ensemble taken to be a Gaussian orthogonal ensemble. In the language of the σ-model, ξ_m corresponds to a wire with length L, whereas ξ_m^{ref} is the length calculated in the zero-dimensional limit, $L \ll L_c$. A relation equivalent to Eq. (11.77) has been claimed to be true also for genuine 1D Anderson and Lloyd tight-binding models (Casati et al. (1992)).

The relation Eq. (11.77) was confirmed by Fyodorov and Mirlin (1992) for $m = 2$ using the supermatrix σ-model, and the derivation presented later follows their work. A method of calculation of the coefficients of the inverse participation ratio was considered in Section 7.3, and $m = 2$ is the simplest case (although the calculation is not much more difficult for arbitrary m). The quantity ξ_2^{-1} can be written as

$$\xi_2^{-1} = (\nu(\varepsilon) L)^{-1} \int \left\langle \sum_i |\varphi_i(r)|^4 \, \delta(\varepsilon - \varepsilon_i) \right\rangle dr \tag{11.78}$$

$$= \lim_{\omega \to 0} \frac{-i\omega}{2\pi \nu(\varepsilon) L} \int \left\langle G_{\varepsilon+\omega}^R(r, r) \left(G_\varepsilon^A(r, r) - G_\varepsilon^A(r, r) \right) \right\rangle dr$$

With Eq. (11.78), by using the standard procedure one comes to the following integral over the supermatrices Q:

$$\xi_2^{-1} = (2L)^{-1} \int B(r) \, dr, \tag{11.79}$$

$$B(r) = -i\omega\pi\nu(\varepsilon) \int S(Q(r)) \exp(-F[Q]) \, dQ \tag{11.80}$$

where

$$S(Q(r)) = Q_{33}^{11}(r) Q_{33}^{22}(r) + 2Q_{33}^{12}(r) Q_{33}^{21}(r)$$

Now we have to write a proper recurrence equation using the same procedure as that used in Section 11.2. This can be done easily and one obtains for the discrete version of the σ-model

$$B(r) = -i\omega\pi\nu(\varepsilon) \int \Psi(Q, r) S(Q(r)) \Psi(Q, L - r) Z_0(Q) \, dQ \tag{11.81}$$

where the functions $\Psi(Q, L - r)$ and $\Psi(Q, r)$ are the partition functions of the segments located to the left and to the right of the point r. In contrast to Eq. (11.14), in Eq. (11.81) the central part Γ is absent. The function $\Psi(Q, r)$ depends on the coordinate r, and the corresponding equation generalizing Eq. (11.16) can be written as

$$\Psi(Q, r + 1) = \int N(Q, Q') Z_0(Q') \Psi(Q', r) \, dQ' \tag{11.82}$$

Eq. (11.82) should be supplemented by the boundary condition

$$\Psi(Q, 0) = 1 \tag{11.83}$$

We see that the computation of the function $B(r)$ is somewhat simpler than that of the function Y^{00}, Eq. (11.14), because now one has to find only one function, $\Psi(Q, r)$. Further manipulations are similar to those used for calculation of Y^{00}. In the continuous limit, one comes to a partial differential equation that dramatically simplifies in the limit $\omega \to 0$. Using the variable z, Eq. (11.61), one reduces Eqs. (11.79–11.82) to the following form:

$$\xi_2^{-1}(u) = \frac{3}{L} \int\limits_0^\infty dz \int\limits_0^u dv \Psi(z, v) \Psi(z, u - v) \equiv \left(\xi_2^{\text{ref}}\right)^{-1} \beta_2^{-1}(u), \qquad (11.84)$$

$$-\frac{\partial \Psi(z, v)}{\partial v} = \hat{R}\Psi(z, v), \qquad \hat{R} = z - z^2 \frac{\partial^2}{\partial z^2}, \qquad \Psi(z, v = 0) = 1 \qquad (11.85)$$

In Eq. (11.84) $u = L/(4L_c)$ is the dimensionless coordinate. Notice that far from the ends of the wire the function $\Psi(z, v)$ does not depend on v and coincides with the solution $\Psi(z)$ of Eq. (11.63). When deriving Eqs. (11.84, 11.85), it was used that in the limit $\omega \to 0$ the noncompact variables $\lambda_{1,2}$ are large, thus leading to the equality

$$Q_{33}^{11} Q_{33}^{22} \approx Q_{33}^{12} Q_{33}^{21} \qquad (11.86)$$

The length ξ_2^{ref} corresponding to the zero-dimensional limit can easily be obtained, for example, for the orthogonal ensemble from Eq. (6.87)

$$\xi_2^{\text{ref}} = \frac{L}{3} \qquad (11.87)$$

The most natural way to solve an equation like Eq. (11.85) is to seek its solution in the form of a generalized Fourier expansion in eigenfunctions of the operator $\hat{R}$. We know that the lowest eigenvalue is zero and the corresponding eigenfunction is given by Eq. (11.68). All other eigenfunctions $g_\rho(z)$ that decay at $z \to \infty$ are equal to

$$g_\rho(z) = 2z^{1/2} K_{i\rho}\left(2z^{1/2}\right) \qquad (11.88)$$

where $K_{i\rho}(z)$ are the MacDonald functions with imaginary index $i\rho$, $0 \leq \rho < \infty$. The corresponding eigenvalues ϵ_ρ are

$$\epsilon_\rho = \frac{1}{4}\left(\rho^2 + 1\right) \qquad (11.89)$$

The functions $g_\rho(z)$ form the complete orthonormal set and can be used as a suitable basis for the expansion. Such an expansion is known as the *Lebedev–Kontorovich transformation* (Marichev (1983)). This transformation was also used by Altshuler and Prigodin (1989) in solving analogous equations for one-dimensional chains. The solution of Eq. (11.85) with boundary condition $\Psi(z, v = 0) = 1$ can be written as

$$\Psi(z, v) = 2\sqrt{z}\left[K_1\left(2\sqrt{z}\right) + \int\limits_0^\infty d\rho b(\rho) K_{i\rho}\left(2\sqrt{z}\right) \exp\left\{-\frac{1 + \rho^2}{4}v\right\}\right], \qquad (11.90)$$

$$b(\rho) = \frac{2}{\pi^2}\rho \sinh \pi\rho \int\limits_0^\infty K_{i\rho}(y)\left[y^{-1} - K_1(y)\right]\frac{dy}{y}$$

$$= \frac{2}{\pi}\frac{\rho}{1 + \rho^2} \sinh\left(\frac{\pi\rho}{2}\right)$$

When calculating the coefficients $b(\rho)$ the identity

$$y^{-1} - K_1(y) = \int_0^\infty dx \exp(-x) \sin(y \sinh x) \tag{11.91}$$

was used.

To proceed further it is more convenient to use the Laplace transform $\beta_L(p)$ of the function $\beta_2^{-1}(u)$, Eq. (11.84),

$$\beta_L(p) = \int_0^\infty \exp(-pu) \beta_2^{-1}(u) \, du \tag{11.92}$$

With the notations $t = 2\sqrt{z}$, $\mu^2 = 4p + 1$ one obtains

$$\beta_L(p) = \frac{1}{2} \int_0^\infty t Y^2(t, p) \, dt \tag{11.93}$$

where

$$Y(t, p) = \frac{1}{p} \int_0^\infty dx \, x \frac{1 + \mu + x^2(\mu - 1)}{\left(x^2 + 1\right)^2} J_{\mu-1}(tx) \tag{11.94}$$

and $J_\mu(a)$ stands for the Bessel function. Substituting the function $Y(t, p)$, Eq. (11.94), into Eq. (11.93); changing the order of integration; and using the virtual orthogonality of the Bessel functions

$$\int_0^\infty t J_\mu(tx) J_\mu(tx') \, dt = \frac{1}{x} \delta(x - x') \tag{11.95}$$

we easily find the function $\beta_L(p)$, which enables us to obtain the function $\beta_2^{-1}(u)$

$$\beta_L(p) = \frac{1}{p} + \frac{1}{3p^2} \Rightarrow \beta_2^{-1}(u) = 1 + \frac{u}{3} \tag{11.96}$$

Recalling that $u = L/(4L_c)$ we see that Eq. (11.96) is equivalent to Eq. (11.77) provided $C = 12$ and, thus, the empirical law described by Eq. (11.77) is proved. The latter equation can be rewritten in the universal model-independent form proposed by Casati et al. (1992)

$$\frac{1}{\xi_2(L, L_c)} = \frac{1}{\xi_2(\infty, L_c)} + \frac{1}{\xi_2(L, \infty)} \tag{11.97}$$

where $\xi_2(\infty, L_c) = 4L_c$ is the generalized localization length defined by Eq. (11.76) but calculated for an infinitely long sample. The parameter $\xi_2(L, \infty) = \xi_2^{\text{ref}}$ is the length calculated in the limit $L \ll L_c$.

Although these results were derived for the orthogonal ensemble, the same can be done for the unitary one. The only difference is that the constant C for the unitary ensemble is twice as large as that for the orthogonal one. This fact is directly related to the effect of doubling the localization length in a magnetic field, which was discussed in the preceding section.

On the basis of numerical simulations the relation described by Eq. (11.77) was conjectured to be true also for $m > 2$. This assertion was checked by Mirlin and Fyodorov

(1993), using, again, the supersymmetric σ-model. It turns out that Eq. (11.77) is not valid for $m > 2$ but numerically the exact expression is extremely close to this equation. This can explain why the numerical results agree with Eq. (11.77) so well.

It is relevant to mention here that in one dimension one can calculate not only the first coefficients of the inverse participation ratio but entire distribution functions. For example, for the distribution function of the local density of states one can use Eq. (7.91). The symbol $\langle \ldots \rangle_Q$ for the one-dimensional case represents integration with the function $\Psi^2(Q)$. Such a computation of the distribution function was done by Mirlin and Fyodorov (1994).

11.5 Conductance of a finite sample

Now, let us consider the second problem mentioned at the beginning of the preceding section, namely, the problem of computing the conductance of a wire of finite length L with ideal metallic leads attached to the ends. This model has a relation to experiments on measuring direct current conductance of real disordered wires. For simplicity, the frequency ω is assumed to be zero. Numerous problems of this kind have been discussed in Chapter 9. In most cases for the description of relevant physical phenomena one could use the 0D σ-model. The only difference arising when describing the wires is that now one has to use the one-dimensional σ-model.

By now the average conductance and conductance variance have been calculated for an arbitrary length L of the wire (Zirnbauer (1992), Mirlin, Müller-Groeling, and Zirnbauer (1994)). These works contain a nice analysis of eigenfunctions and eigenvalues of the effective Schrödinger equation arising after reduction of the 1D functional integral to a 0D "quantum problem." However, following the corresponding mathematical details demands some additional mathematical training. Therefore, I present here only the main ideas and some results in limiting cases.

The σ-model in the presence of the leads has been discussed in detail in Chapter 9. The average conductance $\langle g \rangle$ was written in the form of, for example, Eqs. (9.56, 9.57) and in different versions in Section 9.6. For simplicity, both leads can be assumed to be broad so that the coupling is good. Then the average dimensionless conductance $\langle g \rangle$ (measured in units $2e^2/h$) can be written as

$$\langle g \rangle = -\pi^2 \nu \tilde{\nu} \int Q_{33}^{12}(0) \, Q_{33}^{21}(L) \exp\left(-F_0[Q]\right) \bar{\Psi}(Q(0)) \, \bar{\Psi}(Q(L)) \, DQ, \quad (11.98)$$

$$\bar{\Psi}(Q(a)) = \gamma \exp\left(-\frac{\gamma \tilde{\nu}}{4} \operatorname{str}(\Lambda Q(a))\right), \qquad a = 0, \, L \quad (11.99)$$

In Eq. (11.99) the free energy functional $F_0[Q]$ is the zero frequency limit of the functional $F[Q]$, Eq. (11.1). The function $\bar{\Psi}(Q(a))$ describes the coupling to the leads and contains the supermatrices Q at the end points; the parameter γ stands for the strength of the coupling. Ideal contacts correspond to the limit $\gamma \to \infty$.

Following the procedure of the transfer-matrix method one has to rewrite Eq. (11.98) in the form

$$\langle g \rangle = -\pi^2 \nu \tilde{\nu} \int Q_{33}^{12} \bar{\Psi}(Q) \, \bar{P}(0, Q) \, dQ \quad (11.100)$$

where the function $\bar{P}(r, Q)$ is introduced as

$$\bar{P}(r, Q) = \int \Gamma_0(r, L; Q, Q') (Q')^{21} \bar{\Psi}(Q') \, dQ' \quad (11.101)$$

with Γ_0 the zero-frequency limit of the kernel Γ, the latter satisfying Eq. (11.18). The function $\bar{P}$ can be found from the equation analogous to Eq. (11.23)

$$\bar{P}(r, Q) - \int N(Q, Q') \, \bar{P}(r+1, Q') \, dQ' = Q^{21} \bar{\Psi}(Q) \, \delta_{rL} \qquad (11.102)$$

Although Eqs. (11.100–11.102) are very similar to the corresponding equations written in Section 11.2 and the reduction to partial differential equations analogous to Eqs. (11.46–11.50) is also possible in the continuous limit, solving the latter equations is more difficult as a result of the absence of the translation invariance. The possibility of finding the solution of Eqs. (11.46–11.50) in a closed form in the limit $\omega \to 0$ is due to the simplification of the equations that occurs because the main contribution is from the region $\lambda_{1,2} \gg 1$. We know that the presence of leads results in a smearing of the levels, but the effective level width in the model considered depends on the distance from the leads. At distances r larger than the localization length L_c the effective width is really small and there the main contribution really comes from $\lambda_{1,2} \gg 1$. However, at small distances all $\lambda_{1,2}$ contribute and the equations have to be used in their exact form. Of course, one can try to solve the equations in the different regions and then match the solutions, but this is not simple.

Alternatively, one can calculate the average conductivity or its higher moments expanding the function $\bar{P}(r, Q)$ in a Fourier series in eigenfunctions of the operator $N(Q, Q')$, Eq. (11.17) (Zirnbauer (1992)). In the continuous version this operator is a generalized Laplacian in the space of the elements of supermatrix Q. Although the eigenvalues depend on the value of α in Eq. (11.17) the eigenfunctions are determined entirely by the symmetry of supermatrix Q and are the same in both the continuous and discrete limits. By denoting the Laplacian in the space of the matrix elements as Δ_Q one can formulate the problem of computing conductance as the mathematical problem of solving the equation

$$\left(\frac{\partial}{\partial u} - \Delta_Q\right) \Gamma_0(Q, Q'; \tau) = \delta(Q, Q') \, \delta(u) \qquad (11.103)$$

where $u = r/(4L_c)$, or, equivalently, as finding eigenfunctions and eigenvalues of the Laplacian Δ_Q. The theory of the eigenfunction expansions is well known for Riemannian symmetric spaces of the conventional type (Helgason (1984)). Zirnbauer (1991) generalized the classical analysis to treat superspace.

One important feature of the expansions is that the spectrum of the eigenvalues is a mixture of continuous and discrete fractions. The discrete part corresponds to the compact part of the symmetry of supermatrix Q. The corresponding eigenfunctions are analogous to the simple spherical functions that are eigenfunctions of the Laplace operator on a sphere. The continuous part of the spectrum consists of eigenfunctions that are analogous to solutions of the Laplace equation on a hyperboloid. As usual, a Fourier expansion is a series when the function is defined in a finite region and an integral when the region is infinite. So, the Fourier expansion of the function Γ, Eq. (11.103), implies summation over the discrete part of the spectrum and integration over the continuous part. Of course, because of the rotational symmetries of supermatrices Q only a certain subset of eigenfunctions contributes to the integral in Eq. (11.100) and is analogous to the usual selection rules of quantum mechanics.

A simplified additional discussion related to properties of the eigenfunctions of the Laplacian Δ_Q will be given in the next section; here I restrict myself to presenting the final results of the work of Zirnbauer (1992). The main series of eigenfunctions appearing in the Fourier expansion of Γ_0, Eq. (11.103), for Models I, II, and III have the eigenvalues $\mathcal{E}$

$$\mathcal{E}\,(l,\,\epsilon_1,\,\epsilon_2) = l^2 + \epsilon_1^2 + \epsilon_2^2 + 1, \qquad l = 2n+1, \quad \epsilon_1 > 0, \quad \epsilon_2 > 0, \quad \text{(I)}$$

$$\mathcal{E}\,(l,\,\epsilon) \;=\; l^2 + \epsilon^2 \qquad\qquad\qquad l = 2n-1, \quad \epsilon > 0, \quad \text{(II)}$$

$$\mathcal{E}\,(l_1,\,l_2,\,\epsilon) = l_1^2 + l_2^2 + \epsilon^2 - 1, \qquad l_1 = 2n_1 - 1, \quad l_2 = 2n_2 - 1, \quad \epsilon > 0, \quad \text{(III)}$$

$$(11.104)$$

where $n, n_1, n_2 = 1, 2, \ldots$, and $\epsilon_{1,2}$ are arbitrary positive numbers. We see explicitly that the eigenenergies are characterized by both discrete $l_{1,2}$ and continuous $\epsilon_{1,2}$ variables. For Models I and III there appear, in addition, the subsidiary series

$$\mathcal{E}\,(1,\,\epsilon,\,\epsilon) = 2\left(\epsilon^2 + 1\right), \qquad \epsilon > 0, \quad \text{(I)} \tag{11.105}$$

$$\mathcal{E}\,(i,\,l,\,l-2) = 2\,(l-1)^2\,, \qquad l = 2n+1 \quad \text{and} \quad \mathcal{E}\,(i,\,1,\,1) = 0, \quad \text{(III)} \tag{11.106}$$

Finally, the eigenvalue zero corresponding to the constant eigenfunction appears in all cases. This is due to the zero free energy for the supermatrix model and is a consequence of the cancellation of the fermion and boson contributions. The Fourier expansion of Γ_0 has the general form

$$W\left(Q,\,Q';\,\tau\right) = 1 + \int \exp\left(-\tau\mathcal{E}\,(\rho)\right) \sum_m \phi_\rho^m\,(Q)\,\bar\phi_\rho^m\,(Q')\,d\mu\,(\rho) \tag{11.107}$$

where the integral symbol stands for both integrations over $\epsilon_{1,2}$ and summations over $l_{1,2}$ and $\mu\,(\rho)$ is the proper measure. The sum over m takes into account all functions of degenerate states, where the degeneracies are due to the rotational invariance of the Laplacian in the space of supermatrices Q. (In the language of the conventional spherical functions the sum over m is the sum over different directions of the angular momentum.)

In the limit $\gamma \to \infty$ the average conductance $\langle g \rangle$ can be reduced to the following general form:

$$\langle g \rangle = \frac{1}{2q} \int \mathcal{E}\,(\rho)\,T\,(\rho)\,d\mu\,(\rho)\,, \qquad T\,(\rho) = \exp\left(-\frac{L}{2q\xi}\mathcal{E}\,(\rho)\right) \tag{11.108}$$

where $\xi = 4\pi\nu S D_0$ is four times the localization length for the orthogonal ensemble (cf. Eq. (11.66)) and $q = 1$ for Models I and IIa and $q = 2$ for Models IIb and III. Calculating the measure $d\mu\,(\rho)$ explicitly, Zirnbauer (1992) came to the final exact expressions valid for an arbitrary length L of the wire

$$\langle g \rangle = \frac{\pi}{2} \int\limits_0^\infty d\epsilon\,\tanh^2\left(\frac{\pi\epsilon}{2}\right) T\,(1,\epsilon,\epsilon)$$

$$+ 2^3 \sum_{l=2n+1} \int\limits_0^\infty d\epsilon_1 \int\limits_0^\infty d\epsilon_2 l\left(l^2 - 1\right) \epsilon_1\epsilon_2 \tanh\left(\frac{\pi\epsilon_1}{2}\right) \tanh\left(\frac{\pi\epsilon_2}{2}\right)$$

$$\times \left(l^2 + \epsilon_1^2 + \epsilon_2^2 + 1\right) \prod_{\sigma,\sigma_1,\sigma_2 = \pm 1} (-1 + \sigma l + i\sigma_1\epsilon_1 + i\sigma_2\epsilon_2)^{-1}\, T\,(l,\epsilon_1,\epsilon_2)\,, \quad \text{(I)}$$

$$(11.109)$$

$$\langle g \rangle = 2q^{-1} \sum_{l=2n-1} \int\limits_0^\infty d\epsilon\epsilon\,\tanh\left(\frac{\pi\epsilon}{2}\right) l\left(\epsilon^2 + l^2\right)^{-1}\, T\,(\epsilon,l)\,, \quad \text{(II)} \tag{11.110}$$

$$\langle g \rangle = \frac{1}{4} T(i, 1, 1) + \frac{1}{4} \sum_{l=2n+1} [T(i, l, l-2) + T(i, l-2, l)]$$

$$+ 2^3 \sum_{l_1=2n_1+1} \sum_{l_2=2n_2+1} \int_0^\infty d\epsilon\epsilon \left(\epsilon^2 + 1\right) \tanh\left(\frac{\pi\epsilon}{2}\right) l_1 l_2 \left(\epsilon^2 + l_1^2 + l_2^2 - 1\right)$$

$$\times \prod_{\sigma,\sigma_1,\sigma_2=\pm 1} (-1 + i\sigma\epsilon + \sigma_1 l_1 + \sigma_2 l_2)^{-1} T(\epsilon, l_1, l_2), \quad \text{(III)} \qquad (11.111)$$

In all the models, n, n_1, n_2 in Eqs. (11.109–11.111) are natural numbers. It is clear from Eq. (11.108) that the conductance $\langle g \rangle$ depends on the ratio L/L_c only. This confirms the hypothesis of one-parameter scaling expressed by Eq. (3.4). However, I have to repeat that Eq. (11.108) was obtained in the weak disorder limit. Wave functions in this limit vary at distances of the order of localization length L_c, which is much larger than the interatomic distances, and so, any dependence on short distances is negligible. One can expect that in the limit of strong disorder both the localization length and atomic distances are important and the universal relation Eq. (3.4) is no longer valid. This will be demonstrated in the next section.

The expressions for the average conductance described by Eqs. (11.109–11.111) cannot be further simplified at an arbitrary length L and one has to evaluate them numerically. Corresponding curves can be found in Zirnbauer (1992). Simple formulae are obtained in the limiting cases $L \ll L_c$ and $L \gg L_c$. Introducing the variable $u = L/\xi$ one can write the results in the limit $u \ll 1$ in the form

$$\langle g \rangle = \begin{cases} \frac{1}{2u} - \frac{1}{3} + \frac{2u}{45}, & \text{Model I} \\ \frac{1}{2u} - \frac{2u}{45}, & \text{Model IIa} \\ \frac{1}{2u} - \frac{u}{90}, & \text{Model IIb} \\ \frac{1}{2u} + \frac{1}{6} + \frac{u}{90}, & \text{Model III} \end{cases} \qquad (11.112)$$

Eq. (11.112) describes the average conductance in the weak localization limit and can also be obtained by the conventional diagrammatic technique. The first terms in the expansion give the classical conductance, whereas the second terms for Models I and III are the well-known cooperon contributions. Such a contribution is suppressed in Models II.

In the opposite limit $u \gg 1$, the corresponding formulae are written as

$$\langle g \rangle = \begin{cases} 2^{-5}\pi^{7/2}u^{-3/2}\exp(-u), & \text{Model I} \\ 2^{-1/2}\pi^{3/2}u^{-3/2}\exp\left(-\frac{u}{2}\right), & \text{Model IIa} \\ \pi^{3/2}u^{-3/2}\exp\left(-\frac{u}{4}\right), & \text{Model IIb} \\ \frac{1}{4} + 2^4 3^{-2}\pi^{3/2}u^{-3/2}\exp\left(-\frac{u}{4}\right), & \text{Model III} \end{cases} \qquad (11.113)$$

We see that exponents in Eq. (11.113) agree with Eq. (11.66), demonstrating again the doubling of the localization length when changing from Model I to Model IIa and then to Models IIb and III. Models I and IIa, b surely exhibit localization with localization length L_c, Eq. (11.66). Increasing the length of the sample leads in these models to an exponential decrease of conductance in contrast to that in short samples, where the dependence of the conductance on the sample length is close to the Ohm law, Eq. (11.112).

Surprisingly, the conductance in Model III in the limit $L \to \infty$ remains constant and in this limit equals $1/4$. From the formal point of view this constant is the contribution of the state with the eigenvalue $\mathcal{E}(i, 1, 1)$, which according to Eq. (11.106) is zero. At the same time, such a contribution does not show up in the density–density correlation function Y^{00}, Eqs. (11.70, 11.71), which was calculated for finite frequency ω but infinite sample length. Zirnbauer (1992) shows that at any finite frequency the state $\{i, 1, 1\}$ must be really discarded, and so the density–density correlation function Y^{00} must decay exponentially whereas the direct current conductance of a finite sample remains constant with increased length of the sample. Thus, the mathematical origin of the difference is clear but what is the physical explanation? If some states are not localized, why are they not seen in the function Y^{00} for the infinite sample? Zirnbauer attributes this discrepancy to a basic difference between isolated samples and samples coupled to leads. At the same time, a clear physical picture is not given.

Of course, as we know from Chapter 9, even a single extended ballistic state can give a conductance of the order of unity. But this can occur in the resonance regime only, and averaging over impurities would result in a vanishing contribution. At the same time, a finite fraction of the extended states would manifest itself in the function Y^{00}. Because of the absence of a clear physical picture I suspect that the existence of the state with eigenenergy $\mathcal{E}(i, 1, 1)$ is unstable against different perturbations. Possibly, taking into account fluctuations of the eigenvalues of the supermatrices Q (fluctuations of the density of states) would have the same effect as including the finite frequency, also resulting in discarding of the eigenstate.

11.6 Strong disorder

11.6.1 Fourier series for the density–density correlation function

In this section, I want to present a calculation of the density–density correlation function Y^{00}, Eq. (11.14), for an arbitrary disorder. As mentioned, the limit of the strong disorder can be modeled by the granulated σ-model, Eq. (11.6). This model is convenient because the average density of states, which is determined by $\langle Q \rangle$, is always a constant. It is common knowledge that, generally, the average density of states is a smooth positive function of disorder and is not related to localization properties. So, excluding fluctuations of the density of states does seem to result in neglecting interesting effects. In fact, the agreement of the results in the low-frequency limit for thick wires obtained in the preceding sections with the corresponding results for 1D chains demonstrates explicitly that the localization properties are not influenced by fluctuations of the density of states. At the same time, by changing the coupling parameter J one can change, according to Eq. (11.7), the classical macroscopic diffusion coefficient D_0, and the corresponding classical macroscopic mean free path can range from granule size to infinity, thus covering all interesting regions of disorder.

Of course, increasing disorder cannot lead in one dimension to new physical effects because all states are already localized in the limit of weak disorder and increasing the disorder can only strengthen the localization. The main purpose of this section is to develop formal schemes of calculation of correlation functions for arbitrary disorder. These schemes will be used not only in one dimension but also for the analysis in the next chapter of the Anderson metal–insulator transition that occurs necessarily at strong disorder. As concerns the 1D model, knowledge of the correlation functions at arbitrary disorder helps one to understand the limits of applicability of different universal expressions obtained in the preceding sections.

The basic equations have been obtained in Section 11.2 and the starting point of this section is Eqs. (11.16, 11.17, 11.23, 11.24). Using the symmetry of supermatrix Q one could reduce these equations to integral equations containing the variables $\lambda_{1,2}$, λ in the same way as when deriving the differential equations, Eqs. (11.46–11.50). However, such integral equations for function P_k are very complicated and cannot be solved analytically at an arbitrary length or frequency. In the limiting case of $L \to \infty$, one can again use Fourier expansion over eigenfunctions of a proper integral operator, but it is not the most economic way to do the calculations. It is more convenient to make the Fourier expansion in the beginning of the calculations in Eqs. (11.23, 11.24). We know from the preceding section that the general theory of such functions exists in the limit $\omega = 0$. Now, we have to deal with finite ω and some modifications are necessary. To avoid a complicated mathematical procedure an earlier simplified scheme following Efetov (1987b) will be suggested now. Although mathematically rigorous proofs are not given, I hope that the scheme looks natural and its validity can be accepted by physicists.

Making the substitution

$$P_k(Q) = Z_0^{-1/2}(Q)\,\tilde{P}_k(Q) \tag{11.114}$$

one brings Eq. (11.23) to the form

$$\tilde{P}_k(Q) - \exp(-ik)\,\hat{M}\tilde{P}_k(Q) = Q^{21} Z_0^{1/2}(Q) \tag{11.115}$$

where the operator $\hat{M}$ acts on an arbitrary function $\phi(Q)$ as

$$\hat{M}\phi(Q) = \int N(Q,\,Q')\left[Z_0(Q)\,Z_0(Q')\right]^{1/2}\phi(Q')\,dQ' \tag{11.116}$$

In Eqs. (11.114–11.116), the functions $N(Q,\,Q')$ and $Z_0(Q)$ are determined by Eq. (11.17). The eigenfunctions $\phi_\mathcal{E}(Q)$ of operator $\hat{M}$ satisfy the equation

$$\hat{M}\phi_\mathcal{E}(Q) = \mathcal{E}\phi_\mathcal{E}(Q) \tag{11.117}$$

In order to solve Eq. (11.115) we need eigenfunctions $\phi_\mathcal{E}(Q)$ in the class of 4×4 supermatrices with the same structure as Q^{21}. This means that the functions $\phi_\mathcal{E}(Q)$ should satisfy the condition

$$\phi^+(Q) = \bar{\phi}(Q)\,k \tag{11.118}$$

(All notations for the operations $^+$ and $^-$ operations of the "Hermitian" and "charge" conjugation as well as for the matrix k are the same as in Chapter 4 and in the rest of the book.) In particular, supermatrix iQ^{21} satisfies this condition.

To expand the solution $\tilde{P}$ in the eigenfunctions $\phi_\mathcal{E}$ it is necessary to introduce the scalar product in the space of the functions $\phi(Q)$ satisfying Eq. (11.118). Let us assume, by definition, that the scalar product of two matrices $\phi_1(Q)$ and $\phi_2(Q)$ is equal to

$$(\phi_1,\,\phi_2) = \int \text{str}\left[k\phi_1^+(Q)\,\phi_2(Q)\right]dQ \tag{11.119}$$

Using the definition, Eq. (11.119), and the property, Eq. (11.118), one can easily prove that

$$(\phi_1,\,\phi_2) = (\phi_2,\,\phi_1) \tag{11.120}$$

In addition, the scalar product, Eq. (11.119), is real

$$(\phi_1, \phi_2) = (\phi_1, \phi_2)^* \tag{11.121}$$

It is clear that the operator $\hat{M}$, Eq. (11.116), is self-adjoint

$$\left(\phi_1, \hat{M}\phi_2\right) = \left(\hat{M}\phi_1, \phi_2\right) \tag{11.122}$$

For real β corresponding to imaginary physical frequencies, Eq. (11.17), the operator $\hat{M}$ is real. In the usual way one can prove that eigenfunctions $\phi_\mathcal{E}$ of the operator $\hat{M}$ that correspond to different eigenvalues $\mathcal{E}$ are orthogonal to each other. The eigenvalues $\mathcal{E}$ for real β are real. The normalization condition for the eigenfunctions is written in the form

$$(\phi_{\mathcal{E}'}, \phi_\mathcal{E}) = \delta\left(\mathcal{E} - \mathcal{E}'\right) \tag{11.123}$$

The completeness of the set of functions $\phi_\mathcal{E}(Q)$ is written as

$$\sum_\mathcal{E} [\phi_\mathcal{E}(Q)]_{\alpha\beta} \left[\phi_\mathcal{E}^+(Q')\right]_{\gamma\delta} = \delta_{\alpha\delta}\delta_{\beta\gamma} \tag{11.124}$$

where the subscripts $\alpha, \beta, \gamma, \delta$ stand for matrix elements.

So, the functions $\phi_\mathcal{E}(Q)$ satisfy all the necessary conditions for writing the solution $\tilde{P}(Q)$ of Eq. (11.115) as

$$\tilde{P}_k(Q) = \sum_\mathcal{E} A_{k\mathcal{E}}\phi_\mathcal{E}(Q) \tag{11.125}$$

Substituting this expansion into Eq. (11.115) and using Eqs. (11.24, 11.114) one can represent the density–density correlator Y^{00} in the form

$$Y_k^{00} = 4\pi^2 v\tilde{v} \sum_\mathcal{E} B_\mathcal{E} \frac{1 - \mathcal{E}\cos k}{1 - 2\mathcal{E}\cos k + \mathcal{E}^2}, \tag{11.126}$$

$$B_\mathcal{E} = -\int \mathrm{str}\left[k\phi_\mathcal{E}^+(Q)Q^{21}\right]\Psi(Q)Z_0^{1/2}(Q)\,dQ \tag{11.127}$$

$$\times \int Q_{33}^{12} [\phi_\mathcal{E}(Q)]_{33}\Psi(Q)Z_0^{1/2}(Q)\,dQ$$

To calculate the density–density correlator $Y^{00}(r, r)$ at coinciding space points one should integrate the fraction in Eq. (11.126) over k, which gives unity, and use the completeness properties of the eigenfunctions $\phi_\mathcal{E}(Q)$. Performing the summation over $\mathcal{E}$ one obtains

$$Y^{00}(0) = -2\pi^2 v\tilde{v} \int Q_{33}^{12}Q_{33}^{21}\Psi^2(Q)Z_0(Q)\,dQ \tag{11.128}$$

which could also be obtained immediately from Eq. (11.14). At arbitrary distance the function Y^{00} can be written as

$$Y^{00}(r) = 2\pi^2 v\tilde{v} \sum_\mathcal{E} B_\mathcal{E}\mathcal{E}^{|r|} \tag{11.129}$$

which requires the inequality $\mathcal{E} < 1$ to be fulfilled.

Later, only eigenfunctions $\phi_\mathcal{E}$ of the form

$$\phi_\mathcal{E}(Q) = vR\left(\hat{\theta}\right)\bar{u}, \qquad R\left(\hat{\theta}\right) = \begin{pmatrix} f\left(\hat{\theta}\right) & 0 \\ 0 & if_1\left(\hat{\theta}\right) \end{pmatrix} \tag{11.130}$$

will be considered. Such a form is consistent with Eq. (11.116) because the kernel of the operator $\hat{M}$ is invariant under the simultaneous replacements

$$Q \to U_0 Q \bar{U}_0, \qquad Q' \to U_0 Q' \bar{U}_0 \qquad (11.131)$$

where U_0 has the same structure as U in Eq. (6.47). At the same time, it is clear that the set of these functions is sufficient to provide the expansion of the solution $\tilde{P}_k(Q)$ of Eq. (11.115). As concerns the solution Ψ of Eq. (11.16), because of the same invariance it depends on the variables $\hat{\theta}$ only.

All the formulae that have been written in the present section are of a general character and are correct for all three types of symmetry. The subsequent calculations are to be performed separately for each ensemble. The study of Eqs. (11.16, 11.116) for arbitrary frequencies is very difficult, but fortunately the equations can be solved in the most interesting, low-frequency limit.

11.6.2 Unitary ensemble: Integral equations in an explicit form

As usual, let us start with the unitary ensemble. As a result of the invariance, Eq. (11.131), in Eq. (11.16) one can write the central part Q_0 of the matrix Q (cf. Eq. (6.47)). For functions $\Psi(Q')$ depending on the central part Q_0' only one can integrate over the supermatrices u' and v'. It is more convenient first to subtract from Eq. (11.16) the identity

$$1 = \int N(Q, Q') \, dQ' \qquad (11.132)$$

which follows from the invariance of the function $N(Q, Q')$ under simultaneous rotation of supermatrices Q and Q'. This enables us to treat the anomalous contribution from the point $Q = \Lambda$ properly. The integration is straightforward and one reduces Eq. (11.16) to the form

$$\frac{\Psi(\lambda) - 1}{\lambda_1 - \lambda} = \int N(\mathbf{n}\mathbf{n}', \mathbf{n}_1\mathbf{n}_1') \frac{\Psi(\lambda') Z_0(\lambda') - 1}{\lambda_1' - \lambda'} \frac{d\mathbf{n}' d\mathbf{n}_1'}{(2\pi)^2} \qquad (11.133)$$

where

$$N(\mathbf{n}\mathbf{n}', \mathbf{n}_1\mathbf{n}_1') = \frac{\alpha^2}{2} [\exp(-\alpha\mathbf{n}_1\mathbf{n}_1') \frac{d}{d\alpha} \exp(\alpha\mathbf{n}\mathbf{n}')$$
$$- \exp(\alpha\mathbf{n}\mathbf{n}') \frac{d}{d\alpha} \exp(-\alpha\mathbf{n}_1\mathbf{n}_1')] \qquad (11.134)$$

and $\lambda = (\lambda, \lambda_1)$, $\lambda = \cos\theta$, $\lambda_1 = \cosh\theta_1$. The vectors $\mathbf{n}$ and $\mathbf{n}'$ are vectors on the unit sphere, whereas $\mathbf{n}_1$ and $\mathbf{n}_1'$ are vectors on the hyperboloid. Respectively, $\mathbf{n}\mathbf{n}'$ is the conventional scalar product, whereas $\mathbf{n}_1\mathbf{n}_1' = n_{1z}n_{1z}' - n_{1x}n_{1x}' - n_{1y}n_{1y}'$ is a scalar product on the hyperboloid. The components of the vectors $\mathbf{n}$ and $\mathbf{n}'$ equal

$$\mathbf{n} = (\sin\theta\cos\varphi, \ \sin\theta\sin\varphi, \ \cos\theta)$$
$$\mathbf{n}_1 = (\sinh\theta_1\cos\chi, \ \sinh\theta_1\sin\chi, \ \cosh\theta_1) \qquad (11.135)$$

The same expressions determine the vectors $\mathbf{n}$ and $\mathbf{n}_1'$ provided the substitution $\theta \to \theta'$, $\theta_1 \to \theta_1'$, $\varphi \to \varphi'$, $\chi \to \chi'$ is done.

The function $Z_0(\lambda)$ has the following form in the variables λ:

$$Z_0(\lambda) = \exp(\beta(\lambda - \lambda_1)) \qquad (11.136)$$

In Eq. (11.135) one can perform integration over φ' and χ', which reduces this equation to the equation that depends on two variables λ and λ_1 only

$$\Psi(\lambda) = \int\limits_{-1}^{1}\int\limits_{1}^{\infty} \mathcal{L}_0(\lambda, \lambda') \frac{\lambda_1 - \lambda}{\lambda_1' - \lambda'} Z_0(\lambda') \Psi(\lambda') d\lambda' d\lambda_1' + \exp[\alpha(\lambda - \lambda_1)] \quad (11.137)$$

where

$$\mathcal{L}_0(\lambda, \lambda') = \frac{\alpha^2}{2} \exp[\alpha(x - x_1)] [I_0(\alpha y_1) I_0(\alpha y)(x + x_1)$$
$$- I_1(\alpha y_1) I_0(\alpha y) y_1 + I_1(\alpha y) I_0(\alpha y_1) y] \quad (11.138)$$

I_0 and I_1 are the Bessel functions and

$$x_1 = \lambda_1 \lambda_1', \qquad x = \lambda \lambda',$$
$$y_1 = [(\lambda_1^2 - 1)(\lambda_1'^2 - 1)]^{1/2}, \qquad y = [(1 - \lambda^2)(1 - \lambda'^2)]^{1/2} \quad (11.139)$$

The kernel $\mathcal{L}_0(\lambda, \lambda')$ has the following important property

$$\int\limits_{-1}^{1}\int\limits_{1}^{\infty} \mathcal{L}_0(\lambda, \lambda') \frac{\lambda_1 - \lambda}{\lambda_1' - \lambda'} d\lambda' d\lambda_1' = 1 - \exp[\alpha(\lambda_1 - \lambda)] \quad (11.140)$$

Until now no approximations have been done and we see that Eqs. (11.137, 11.138) are very complicated. As concerns the orthogonal and symplectic ensembles the corresponding reduction is possible only with the help of computer programs like the one written by Hartmann and Davis (1989). However, even if the final integral equation is written, one has to solve it. Definitely, for an arbitrary frequency, it can be done by a complicated numerical computation only.

11.6.3 Low-frequency limit

So, if we want to do something we should simplify Eqs. (11.137, 11.138) and it can really be done, as in the continuous σ-model, in the most interesting limit of low frequencies ω corresponding to small β, Eq. (11.17). As in the continuous case, the main contribution is from large $\lambda_1 \sim 1/\beta \gg 1$ and for such λ_1 Eqs. (11.137, 11.138) become much simpler. Changing the variables λ_1 to $z = \beta\lambda_1 = \exp t$ one obtains instead of these equations

$$\Psi(t) = \int\limits_{-\infty}^{\infty} \bar{\mathcal{L}}(t - \tau) \exp\left(\frac{t - \tau}{2}\right) Z_0(\tau) \Psi(\tau) d\tau, \quad (11.141)$$

$$Z_0(\tau) = \exp(-e^\tau/2),$$

$$\bar{\mathcal{L}}(t) = \frac{\alpha^{3/2}}{(2\pi)^{1/2}} \exp(-\alpha \cosh t) \left[\frac{\sinh \alpha}{\alpha} \cosh t + \frac{\cosh \alpha}{\alpha} - \frac{\sinh \alpha}{2\alpha^2}\right] \quad (11.142)$$

Eqs. (11.141, 11.142) are applicable for arbitrary α. In the limit $\alpha \gg 1$ they reduce to the form of Eqs. (11.37, 11.38). One can check directly that

$$\int\limits_{-\infty}^{\infty} \bar{\mathcal{L}}(t) \exp(-t/2) dt = 1 \quad (11.143)$$

Using this property one can see that

$$\Psi(-\infty) = 1 \tag{11.144}$$

corresponding to Eq. (11.67). Asymptotics of Ψ at infinity can also be easily obtained from Eqs. (11.141, 11.142)

$$\Psi(\infty) = 0 \tag{11.145}$$

So, the function $\Psi(t)$ decays from 1 to 0 as the variable t varies from $-\infty$ to ∞ and this decay can be shown to be monotonous. Further information about this function can be obtained only numerically but, in fact, it is not very important for calculation of the function $Y^{00}(r)$, Eq. (11.129).

To compute the eigenvalues $\mathcal{E}$ of the operator $\hat{M}$, Eq. (11.116), and the coefficients $B_{\mathcal{E}}$, Eq. (11.127), one has to simplify these equations in the same limit $\lambda_1 \sim 1/\beta$. By using the form of the functions given by Eq. (11.130) and substituting it into Eq. (11.116) one can see that in the limit involved only the block $f_1(\lambda)$ with the function f_1 depending only on the variable λ_1 survives. This is in complete agreement with the corresponding result for the continuous model considered in Section 11.3. Integrating over the matrices v and u one obtains the following equation instead of Eqs. (11.116, 11.127):

$$B_{\mathcal{E}} = \beta^{-1} \left[\int_{-\infty}^{\infty} \Psi(t) \left[Z_0(t) \right]^{1/2} \exp\left(\frac{t}{2}\right) \chi_{\mathcal{E}}(t) \, dt \right]^2 \tag{11.146}$$

The functions $\chi_{\mathcal{E}}(t)$ are eigenfunctions of the operator $\bar{M}$

$$\int_{-\infty}^{\infty} \bar{M}(t, \tau) \chi_{\mathcal{E}}(\tau) \, d\tau = \mathcal{E} \chi_{\mathcal{E}}(t) \tag{11.147}$$

where the symmetric real kernel $\bar{M}$ is equal to

$$\bar{M}(t, \tau) = \bar{\mathcal{L}}(t - \tau) \left[Z_0(t) Z_0(\tau) \right]^{1/2} \tag{11.148}$$

The eigenvalues $\mathcal{E}$ can be computed by studying the asymptotics of the functions $\chi_{\mathcal{E}}(t)$ at $t \to -\infty$. In this limit, $Z_0(t \to -\infty) = 1$ and the kernel $\bar{M}(t, \tau)$ coincides with $\bar{\mathcal{L}}(t - \tau)$. Therefore, for determining the asymptotics of the function χ at $t \to -\infty$ instead of Eqs. (11.147, 11.148) we have

$$\int_{-\infty}^{\infty} \bar{\mathcal{L}}(t - \tau) \chi_{\mathcal{E}}(\tau) \, d\tau = \mathcal{E} \chi_{\mathcal{E}}(t) \tag{11.149}$$

The solution of Eq. (11.149)

$$\chi_{\mathcal{E}}(t) = \left(\frac{2}{\pi}\right)^{1/2} \sin\left[-\epsilon t + \delta(\epsilon)\right] \tag{11.150}$$

where the parameter ϵ is some function of $\mathcal{E}$ and $\delta(\epsilon)$ is the phase, which is a function of ϵ, can be found easily. At $t \to \infty$ the functions $\chi_{\mathcal{E}}(t)$ must decay at any $\mathcal{E}$ because of the fast-decaying function $\left[Z_0(t) Z_0(\tau) \right]^{1/2}$ in the expression for $\bar{M}(t, \tau)$, Eq. (11.148).

The spectrum of eigenvalues of the operator $\bar{M}(t, \tau)$ is continuous and the coefficient in Eq. (11.150) is chosen to satisfy the normalization

$$\int\limits_{-\infty}^{\infty} \chi_{\mathcal{E}}(t)\,\chi_{\mathcal{E}'}(t)\,dt = \delta\left(\epsilon - \epsilon'\right) \tag{11.151}$$

In the language of elementary quantum mechanics the wave functions $\chi_{\mathcal{E}}$, Eq. (11.150), correspond to free one-dimensional motion of a particle. The fact that the functions must decay at $t \to \infty$ means that there is a potential that does not allow the particle to go to $+\infty$. The explicit form of the potential determines the phase $\delta(\epsilon)$. The dependence of the eigenenergy $\mathcal{E}$ on ϵ can be found easily by writing the Schrödinger equation for the particle in the asymptotical region far to the left of the forbidden region. Of course, now we are studying properties of the solution of the integral equation, Eq. (11.147, 11.148), and not of the Schrödinger equation but the difference is not essential. I use the analogy with quantum mechanical motion because it helps to illustrate the general properties of the solutions immediately and shows the way to find the eigenvalues.

Substituting Eq. (11.150) into Eq. (11.149) one obtains

$$\mathcal{E}(\epsilon) = L_{\epsilon}(\alpha) \tag{11.152}$$

The function $L_{\epsilon}(\alpha)$ has the form

$$L_{\epsilon}(\alpha) = \alpha\left[K_{i\epsilon}(\alpha)\,\frac{d}{d\alpha}\,I_{1/2}(\alpha) - I_{1/2}(\alpha)\,\frac{d}{d\alpha}\,K_{i\epsilon}(\alpha)\right] \tag{11.153}$$

where $K_{\nu}(\alpha)$ and $I_{\nu}(\alpha)$ are the Bessel functions.

The function $L_{\epsilon}(\alpha)$ satisfies the inequality $L_{\epsilon}(\alpha) < 1$ and for $\epsilon = 0$ approaches 1 as $\alpha \to \infty$. Recalling Eq. (11.129) we see that in the limit $\alpha \to \infty$ the decay of the correlation function $Y^{00}(r)$ is slow, corresponding to a large localization length. In this limit one comes to Eq. (11.66) for localization length L_c. For $\alpha \sim 1$, the localization length is of the order of unity.

Now, let us obtain the explicit form of the function $Y^{00}(r)$. Usually, in the transfer-matrix technique, it is sufficient to find the largest eigenvalue $\mathcal{E}_0$, which determines the decay of the correlation function at large distances. This is true if the spectrum of the eigenvalues is discrete. However, Eqs. (11.152, 11.153) show that the spectrum of the eigenvalues is continuous, and so, instead of being restricted by the largest eigenvalue $\mathcal{E}_0$ in Eq. (11.129) one has to integrate over all ϵ. Nevertheless, the limit of large r simplifies the evaluation because the main contribution is from small ϵ. To calculate the integral in Eq. (11.129) one has to know not only the eigenvalues but also the eigenfunctions that determine the coefficients $B_{\mathcal{E}}$, Eq. (11.146). It is not sufficient to know only the asymptotics at $t \to -\infty$, Eq. (11.150), because the main contribution to the integral over t in Eq. (11.146) is from $t \sim 1$.

Although it is not possible to solve Eqs. (11.147, 11.148) exactly, the necessary information about the solution in the limit $\epsilon \to 0$ can be obtained rather easily. At $t \gg 1$ all the functions $\chi_{\mathcal{E}}(t)$ fall off fast and the behavior of the tails is not sensitive to ϵ provided this parameter is small. So, the functions $\chi_{\mathcal{E}}$ are proportional at $t \gg 1$ to a function $u(t)$. However, Eq. (11.147) is linear and so the asymptotics takes the form

$$\chi_E(t) \approx c(\epsilon)\,u(t), \qquad t \gg 1 \tag{11.154}$$

where $c(\epsilon)$ is a coefficient that cannot be determined by solving Eqs. (11.147, 11.148) in the region $t \gg 1$ only. It can be determined by matching the asymptotics at $t \gg 1$, Eq. (11.154), with the asymptotics at $t \to -\infty$, Eq. (11.150). The matching of the functions $\chi_{\mathcal{E}}$ and their derivatives with respect to t at $t \sim 1$ yields $c(\epsilon) \sim \epsilon$, $\delta \sim \epsilon t_0$, $t_0 \sim 1$. This reasoning enables us to write the function $\chi_{\mathcal{E}}(t)$ at $\epsilon \to 0$ and arbitrary t in the form

$$\chi_{\mathcal{E}}(t) = \epsilon v(t) \tag{11.155}$$

with a function $v(t)$ that does not depend on ϵ.

These semiqualitative arguments are sufficient for finding the long-distance asymptotics of function $Y^{00}(r)$. Expanding the function $L_{\epsilon}(\alpha)$, Eq. (11.153), in ϵ and substituting Eq. (11.155) into Eqs. (11.146, 11.129) one obtains

$$Y^{00}(r) = \frac{2\pi^2 v\tilde{v}}{\beta} a^2 \exp\left(-\frac{r}{4L_c}\right) \int_0^\infty \epsilon^2 \exp\left(-rb\epsilon^2\right) d\epsilon \tag{11.156}$$

where

$$a = \int_{-\infty}^{\infty} v(t) \Psi(t) [Z_0(t)]^{1/2} \exp\left(\frac{t}{2}\right) dt,$$

$$L_c^{-1} = -4\ln \mathcal{E}(0) = -4\ln L_0(\alpha),$$

$$b = \frac{1}{2}\left[\frac{\partial^2 \mathcal{E}(0)}{\partial \epsilon^2}\right] = -\frac{1}{2}\left[\frac{\partial^2 L_\epsilon(\alpha)}{\partial \epsilon^2}\right]_{\epsilon=0}$$

The parameters a and b as well as localization length L_c can be computed for arbitrary α only numerically, but they are certainly finite for finite α. Calculating the integral over ϵ in Eq. (11.156) and using Eq. (11.17) for the function $p_\infty(r)$, Eq. (3.33), characterizing the decay of the wave functions one obtains

$$Y^{00}(r, t \to \infty) = p_\infty(r) = \frac{\pi^{3/2} v}{2V} a^2 (br)^{-3/2} \exp\left(-\frac{r}{4L_c}\right) \tag{11.157}$$

V is the volume of the grains.

The asymptotic behavior, Eq. (11.157), is correct so long as $r \gg L_c$. We see that function $p_\infty(r)$ for arbitrary disorder, Eq. (11.157), is described at $r \gg L_c$ by the same law as the corresponding quantity calculated in the limit of weak disorder, Eq. (11.71), and the form of $p_\infty(r)$ proves the localization of all states. Only the parameters a, b, and L_c are new. Now we understand the formal origin of the preexponential $r^{-3/2}$. It is clear from Eq. (11.156) that the presence of this function is a consequence of the continuous spectrum of eigenvalues $\mathcal{E}$ of Eqs. (11.117, 11.147), which, in its turn, is the consequence of the noncompactness of the symmetry group of supermatrices Q.

Thus, it has been shown how to compute correlation functions in the limit of a strong disorder, and that is definitely a mathematical achievement. But does the study of the strong disorder limit change anything in our understanding of localization in systems with one-dimensional geometry? In fact, yes, because the function $p_\infty(r)$, Eq. (11.157), reveals an absence of universality that shows up in the weak disorder limit. This universality has been emphasized several times: The quantity $x p_\infty(x)$ from Eq. (11.71) is a function of the variable x/L_c only. Measuring the distance in units of L_c one obtains a completely universal relation. The same occurs with the dimensionless conductance $\langle g \rangle$ of a finite sample that changes with the sample length L according to Eqs. (11.109–11.111), which contain only the variable L/L_c. This is the universality that confirms the main relation, Eq. (3.4).

However, looking at Eqs. (11.156) no simple relation between the parameters L_c, a, and b is apparent and certainly the function $rp_\infty(r)$ cannot be expressed as a function of r/L_c only. Although the average conductance $\langle g \rangle$ of a finite sample was not calculated here in the limit of strong disorder, it is difficult to expect that this quantity can become universal. This means that the one-parameter scaling theory proposed by Abrahams et al. (1979) cannot operate with average conductance (it is fair to say that these authors did not claim that the scaling quantity had to be the average conductance, but the idea was implied in many other works). From the structure of the solution it seems to be unlikely that one can find a function of the conductance such that its average can be the scaling parameter. My feeling is that one has to operate with a nontrivial distribution of conductances and scale this quantity. However, this is only a guess.

Eqs. (11.156, 11.157) are obtained for the unitary case, but the same procedure can be used for both orthogonal and symplectic ensembles. One has to begin again with Eq. (11.16) and assume that the solution Ψ depends on matrix $\hat{\theta}$ only. Then, one should integrate over the matrices u and v. However, the proper manipulations are extremely lengthy, so that equations corresponding to Eqs. (11.133, 11.134) can hardly be obtained without computer programs. Fortunately, in the limit of low frequency $\beta \to 0$ we are interested in, everything is much simpler and it is better to go to this limit from the very beginning. We know already from Section 11.3 that in the symplectic ensemble the main contribution is from $\lambda \sim 1/\beta$, whereas in the orthogonal ensemble $\lambda_1 \lambda_2 \sim 1/\beta$. So, one can select the most important terms, thus making the computation feasible.

11.6.4 Symplectic and orthogonal ensembles

Let us first present the results for the symplectic ensemble. Integrating over the matrices u and v, collecting the main terms, and using the proper Jacobian, Eq. (6.68), one can reduce Eq. (11.16) to the form of Eq. (11.141). The only difference is that now function $\bar{\mathcal{L}}(t)$ has the form (Efetov 1987b)

$$\bar{\mathcal{L}}(t) = \left(\frac{\alpha}{2\pi}\right)^{1/2}\left[\frac{\alpha S(\alpha)}{2}\sinh^2 t + \left(\sinh\alpha - \frac{S(\alpha)}{\alpha}\right)\cosh t\right.$$

$$\left. + \cosh\alpha - \frac{\sinh\alpha}{2\alpha} - \frac{S(\alpha)}{8\alpha}\right]\exp(-\alpha\cosh t), \qquad S(\alpha) = \int_0^\alpha \frac{\sinh x}{x}dx \qquad (11.158)$$

By direct integration we can verify that with this $\bar{\mathcal{L}}(t)$ Eq. (11.143) is also fulfilled. Then one can repeat the procedure used for the unitary case and arrive at Eqs. (11.156, 11.157). In the symplectic case the function $L_\epsilon(\alpha)$ that determines the localization length L_c and coefficients a and b has the form

$$L_\epsilon(\alpha) = \left(\frac{\alpha}{2\pi}\right)^{1/2}\left\{\frac{i\epsilon S(\alpha)}{2}[K_{1+i\epsilon}(\alpha) - K_{1-i\epsilon}(\alpha)]\right. \qquad (11.159)$$

$$\left. + \sinh\alpha[K_{1+i\epsilon}(\alpha) + K_{1-i\epsilon}(\alpha)] + \left[2\cosh\alpha - \frac{\sinh\alpha}{\alpha} - \frac{S(\alpha)}{4\alpha}\right]K_{i\epsilon}(\alpha)\right\}$$

where $K_\nu(\alpha)$ is the MacDonald function.

Calculations for the orthogonal ensemble are most difficult. In the limit of small β, as in the preceding models, the large values of θ_1 and θ_2 are important. In this case $\theta_1' = \theta_1 + \theta_2 \sim \ln(1/\beta)$. At the same time, the important values of $\theta_2' = \theta_1 - \theta_2$ are of the order of unity. Despite the great simplifications that arise for small β, the necessary calculations are still very lengthy. The function Ψ in the limit of small β depends on θ_1' only. This follows from the fact that in the limit $\beta \to 0$ the variables θ and θ_2' do not enter the second term in Eq. (11.6) that violates the global rotational invariance. Therefore, by assuming that the function Ψ does not depend on θ and θ_2' and making a proper rotation of the supermatrices Q' one can show that the result of the integration in Eq. (11.16) contains only the variable θ_1', thus confirming the correctness of the assumption.

Changing the variables $t = \theta_1' + \ln(\beta/2)$ one then arrives at Eq. (11.141). The function $\bar{\mathcal{L}}(t)$ for the orthogonal ensemble takes the form (Efetov (1987b), Verbaarschot (1988))

$$\bar{\mathcal{L}}(t) = \frac{1}{16} \left(\frac{\alpha}{\pi}\right)^{3/2} \int\limits_1^\infty d\lambda_2 \int\limits_{-1}^1 d\lambda \frac{(1 - \lambda^2)}{(\lambda_2^2 - 1)^{1/2}(\lambda_2 - \lambda)^2}$$

$$\times \exp\left(-\frac{\alpha}{4}(\cosh t + \lambda_2 - 4\lambda)\right)\left[(2\lambda - \lambda_2 + \cosh t)\, K_0\left(\frac{\alpha}{4}(\cosh t + \lambda_2)\right)\right.$$

$$\left. + (\cosh t + \lambda_2)\, K_1\left(\frac{\alpha}{4}(\cosh t + \lambda_2)\right)\right] \qquad (11.160)$$

The gross properties of kernel $\bar{\mathcal{L}}(t)$, Eq. (11.160), are the same as those of the corresponding kernels for the unitary and symplectic ensembles, Eqs. (11.142, 11.158), and one again comes to Eqs. (11.156, 11.157) with function $L_\epsilon(\alpha)$ expressed through $\bar{\mathcal{L}}(t)$, Eq. (11.160), as

$$L_\epsilon(\alpha) = \int\limits_{-\infty}^\infty \bar{\mathcal{L}}(t)\cos(\epsilon t)\, dt \qquad (11.161)$$

In the limit $\alpha \ll 1$ Eq. (11.160) can be simplified because one can perform the integration explicitly. As a result one obtains

$$\bar{\mathcal{L}}_{\alpha \ll 1}(t) = \frac{1}{4}\left(\frac{\alpha}{\pi}\right)^{3/2}\cosh t \exp\left(-\frac{\alpha}{4}\cosh t\right) \qquad (11.162)$$

$$\times \left[K_0\left(\frac{\alpha}{4}\cosh t\right) + K_1\left(\frac{\alpha}{4}\cosh t\right)\right]$$

In the limit $\alpha \gg 1$ Eq. (11.141) for all three ensembles reduces to Eq. (11.63) and one can recover all formulae of Section 11.3. Using Eqs. (11.153, 11.156–11.161) one can find localization length L_c characterizing the asymptotic behavior of the density–density correlation function Y^{00} and see that the relation between the localization lengths L_c for different ensembles given by Eq. (11.66) is correct in the limit $\alpha \gg 1$ only.

Thus, the problems of disorder and quantum chaos corresponding to the 1D σ-model can be solved exactly at arbitrary disorder. For any weak disorder all wave functions are localized. Although the density–density correlation function Y^{00} can exhibit complicated behavior at short distances its decay at distances larger than localization length L_c is always described by Eq. (11.157). The localization leads to a dielectric response to an electric field applied along the wire. At the same time, the localization can lead to a large diamagnetic response to a magnetic field applied perpendicular to the wire. This curious phenomenon is considered in the next section.

11.7 Diamagnetism due to localization

11.7.1 Level crossing and difference between dynamic and thermodynamic quantities: Basic equations

In Chapter 8, in a discussion of properties of persistent currents in mesoscopic rings, it was demonstrated that the currents could exist in a finite volume. Increasing the volume results in decreasing the amplitude of the current, and in the limit of infinite size the density of the persistent current is zero. For a given configuration of impurities the persistent current corresponds to the minimum of free energy and this is why it does not decay in time. Averaging over impurities is not trivial because the possibility of crossing the chemical potential μ by energy levels when changing the impurity configuration is not allowed. However, by calculating the dynamic response one can estimate the amplitude of the thermodynamic persistent current, and the corresponding procedure was presented in Section 8.4.

Let us emphasize again that the dynamic approach was used to calculate thermodynamic quantities. For each particular impurity configuration the dynamic response coincides with the derivative of the thermodynamic current with respect to the magnetic field, and this coincidence is due to the fact that the energy levels as functions of the magnetic field do not cross each other. Only after averaging over the random potential does the difference between the results obtained by averaging the dynamic response and the thermodynamic quantities at fixed chemical potential arise. So, although averaged dynamic and thermodynamic responses for mesoscopic objects can differ after averaging, their nonaveraged values must coincide for any given impurity configuration.

However, a completely different situation is possible in wires, chains of coupled mesoscopic rings, or other objects with one-dimensional geometry (Kettemann and Efetov (1995)). Although level intersections are not possible in an isolated ring, they can occur in the 1D systems in the limit $L_x \to \infty$, where L_x is the total length. This is due to the fact that all wave functions in a disordered 1D chain (say, a chain of mesoscopic rings) are localized, and parts of the chain located far from each other do not interfere. This leads to an exponential suppression of the level repulsion, and, in the infinite chain, the levels can cross. This is similar to the situation in a disordered 1D chain (Gorkov, Dorokhov, and Prigara (1983), Sivan and Imry (1987)). The degeneracy that results from the level intersections already causes the dynamic response to be completely different from the corresponding thermodynamic response before averaging over impurities. In fact, in an infinite system one cannot expect any finite thermodynamic magnetization density except Landau diamagnetism, which is small compared to the effect considered.

In contrast to the thermodynamic result, the dynamic response of the chain of disordered rings does not vanish in the limit $L_x \to \infty$ and the total magnetization caused by the applied magnetic field is proportional to length L_x. Depending on the tunneling amplitude between the rings the magnetization can be measurably large and decay in time as a result of inelastic processes only. The effect is intimately related to the phenomenon of localization in 1D systems.

To understand the nature of the dynamic response that will be studied in the diffusive regime better let us first consider the corresponding clean limit. The chain of N_r rings is

Fig. 11.2. Chain of coupled mesoscopic rings.

assumed to be in the ballistic regime $L < l$, where L is the circumference of the ring and l is the mean free path. It is the ideal diamagnet and its magnetization M_{dyn} due to an applied flux ϕ through the rings is given by

$$M_{\text{dyn}} = \frac{1}{4\pi c^2} N_r V K_d \phi_\omega, \qquad K_d = -\frac{ne^2}{m} \qquad (11.163)$$

where ϕ_ω is a small time-dependent part of the flux through the ring, n is the electron density, and V is the volume of a single ring. This magnetization does not correspond to the free energy minimum and, because of inelastic processes, decays to the smaller thermodynamic value.

What happens with the dynamic magnetization M_{dyn}, Eq. (11.163), in the presence of random potential of impurities? According to classical physics the magnetization must vanish with the decay time τ, which is just the mean free time for scattering on the impurities. But it turns out that the classical picture breaks down at low temperature, when the quantum interference becomes important. Nevertheless, it is important to emphasize that the effect considered later is not a weak localization effect, and although the decay time for the magnetization remains finite it is determined by inelastic processes and can much exceed the elastic mean free time τ.

Let us specify the model under consideration. A chain of disordered mesoscopic rings with the circumference L, represented in Fig. 11.2, is assumed to be placed into an external magnetic field $\tilde{H}$ of the form

$$\tilde{H} = H + H_\omega \cos(\omega t) \qquad (11.164)$$

directed perpendicular to the ring. In Eq. (11.164), H is the static part. Using the Kubo formula the frequency-dependent magnetization M_ω is then obtained from the response function $K(\omega)$ as

$$M_\omega = \frac{1}{4\pi c^2} N_r V \, \text{Re} \left(K(\omega) \phi_\omega \right) \qquad (11.165)$$

It will be shown that in the absence of inelastic processes the dissipative part of the magnetization vanishes in the zero frequency limit but the real part of the response can remain finite. In the limit of small frequency the diagrammatic technique leads to divergencies and one should use the supersymmetry method.

The responses to the vector potential both along the chain of rings with cross section $S = V/L$ and in the azimuthal direction of the rings can be reduced to computation of functional integrals over supermatrices Q. Here, the response in the azimuthal direction of an ensemble of chains of rings will be calculated with the assumption that the inequality $L \ll L_c \ll L_x$ is fulfilled. For the problem of persistent currents in disconnected mesoscopic rings the dynamic response was calculated in Chapter 8 using the 0D version of the σ-model. In the general case of an arbitrary hopping amplitude between the rings in the free energy functional one has to add a term describing the coupling between the rings. This coupling is analogous to the coupling between the granules in the model of granulated metal considered in the preceding section. Changing the probability of tunneling from ring to ring one can describe the crossover from the case of isolated rings to the homogeneously weakly disordered chain of rings. Free energy $F[Q]$ for the model involved is analogous to that in Eq. (11.6). The only difference is that now one should explicitly include the vector potential **A**, which leads to the following form:

$$F[Q] = -\sum_{i,j} J_{ij} \,\text{str}\, Q_i Q_j + \sum_i F_r[Q_i] \qquad (11.166)$$

In Eq. (11.166), the first term describes the coupling between the rings, whereas the second is the free energy of isolated rings. The subscripts i and j enumerate the rings in the chain and $F_r[Q_i]$ is the free energy of a single ring (see also Eqs. (8.42–8.44))

$$F_r[Q_i] = \frac{\pi \nu V}{8} \,\text{str} \left[-D_0 \left(\frac{e}{c} \mathbf{A} [Q_i, \tau_3] \right)^2 + 2i\omega \Lambda Q_i \right] \qquad (11.167)$$

where D_0 is the classical diffusion coefficient inside the rings. Only nearest neighbors are assumed to be coupled, so that $J_{ij} = J$ for neighbors and $J_{ij} = 0$ otherwise. The limit $J = 0$ corresponds to the chain of disconnected rings. In the limit $J \gg 1$, the model on the lattice, Eq. (11.166), becomes the continuous 1D σ-model. This can be a model for a metallic strip with holes in it. Very often such holes are called *antidots* and the antidot arrays can be fabricated experimentally. In such a situation the coupling constant J should be expressed in terms of D_0 as

$$J = \frac{\pi \nu V D_0}{8 L^2} \qquad (11.168)$$

Because of the one-dimensionality of the model one can again use the transfer matrix technique in the same way as in preceding sections. In terms of integrals over the supermatrices Q the resulting expression is a combination of formulae of the present chapter and Chapter 8. The response $K(\omega)$ can be written in the form of Eqs. (8.78, 8.79, 8.85–8.90). Using the transfer-matrix method one can replace integration over supermatrices Q_i of all rings by integration with a function $\Psi(Q)$. So, one can imply by the symbol $\langle \ldots \rangle_Q$ in Eqs. (8.88–8.90) the following integral:

$$\langle \ldots \rangle_Q = \int (\ldots) \Psi^2(Q) \exp\left(-F_r[Q]\right) dQ \qquad (11.169)$$

The function $\Psi(Q)$ in Eq. (11.169) is the solution of the equation

$$\Psi(Q) = \int \exp\left(2J \,\text{str}\, QQ'\right) \exp\left(-F_r[Q']\right) \Psi(Q') dQ' \qquad (11.170)$$

Although the preceding equations are very similar to those written in Section 8.4, I want to emphasize again that now the dynamic response is being discussed and this quantity, as a result of the possibility of level crossings, is completely different from the corresponding thermodynamic derivative. Therefore, relations like Eq. (8.84) that were derived under the assumption of the absence of level crossing are not applicable here. At the same time, working with a finite frequency one may take the limit $L_x \to \infty$ and replace the physical response by its average over the impurity configurations.

The solution of Eq. (11.170) depends on the vector potential **A** in $F_r[Q]$, Eq. (11.167). In principle, it can be solved, at least numerically, for an arbitrary magnetic field using the magnetic field parametrization discussed in Chapter 8, but this has not yet been done and later only results in limits of zero and high enough magnetic fields are presented. These limits correspond to the orthogonal and unitary ensembles. In both cases one can omit the first term in Eq. (11.167) because in the unitary case supermatrix Q commutes with matrix τ_3. As in the preceding sections the function Ψ depends only on the variables λ, λ_1, λ_2, and integration over the matrices u and v can be performed at the very beginning.

11.7.2 Diamagnetism from the one-dimensional σ-model

The computation of the response is, as usual, most simple for the unitary ensemble because there is only one noncompact variable λ_1. The corresponding manipulations are not very different from those performed in Chapter 8 and one obtains

$$K^{\text{unit}}(\omega) = -i\omega\sigma_0\left\{1 - \frac{1}{2}\int_{-1}^{1}d\lambda\int_{1}^{\infty}d\lambda_1\exp\left[\left(\frac{i\pi\omega}{\Delta} - \delta\right)(\lambda_1 - \lambda)\right]\Psi^2(\lambda, \lambda_1)\right\}$$

(11.171)

where $K^{\text{unit}}(\omega)$ is the average response along the circumference of the rings in the unitary ensemble and σ_0 is the classical conductivity, Eq. (3.1). There can be a finite response in the limit $\omega \to 0$ if at low frequencies ω the second term in the brackets in Eq. (11.171) is proportional to $1/i\omega$. We already know from preceding sections that, in the limit $\omega \to 0$, solving Eq. (11.170) becomes simpler because the main contribution to the integral over λ_1 is from $\lambda_1 \sim 1/\omega$. Then, the function $\Psi(Q)$ depends only on the variable $\omega\lambda_1$ and one obtains as a response

$$K^{\text{unit}}(\omega \to 0) = -\frac{\sigma_0\Delta_{\text{eff}}(J)}{\pi}$$

(11.172)

where the effective mean level spacing Δ_{eff} is a nontrivial function of the coupling between the rings

$$P(J) \equiv \frac{\Delta_{\text{eff}}}{\Delta} = \int_{0}^{\infty}\exp(-u)\Psi_J^2(u)\,du$$

(11.173)

This function describes the probability that an electron does not leave a ring forever and is known numerically for arbitrary J (Zirnbauer (1986)). At high frequencies the second term in Eq. (11.171) is small and one obtains the classical response function $K(\omega) = -i\omega\sigma_0$. In this limit the conductance is classical in all directions. We see from Eqs. (11.171, 11.172) that the characteristic frequency of the crossover from the quantum to the classical regime is of the order of Δ_{eff}.

In the limit $J = 0$ corresponding to disconnected rings, $\Psi = 1$ and $\Delta_{\text{eff}}(0) = \Delta$. The magnetization of the chain of disconnected rings is then, using Eq. (11.165),

$$M_{\omega}^{\text{unit}}(J = 0) = \frac{1}{4\pi^2 c^2}N_r V K_d\tau\Delta\phi_{\omega}$$

(11.174)

Of course, one can argue that for disconnected rings one should use the results of Chapter 8 obtained in the absence of level crossing. Then, what does Eq. (11.174) mean? The answer is that the result given by Eq. (11.174) can be used in the limit $J \to 0$ but J has to remain finite. This suppresses fluctuations of the chemical potential and averaging over impurities can be done by keeping the chemical potential fixed, as was assumed in deriving Eq. (11.174). Only if the rings are completely isolated is this assumption no longer correct. Taking into account Coulomb interaction between electrons can lead to the existence of a finite critical J_c such that for $J < J_c$ the rings are effectively isolated. However, consideration of this effect is not simple and a proper theory does not exist.

In the opposite limit $J \gg 1$, the function $\Psi(u)$ is the solution of the differential equation, Eq. (11.63), which should be written in the variable u in the form

$$u \frac{d^2 \Psi_J}{du^2} - 16 J \Psi_J = 0 \tag{11.175}$$

Solving Eq. (11.175) and calculating the integral, (11.173), we find that there is still a finite probability

$$P(J) = (96J)^{-1} \tag{11.176}$$

that the electron stays in one ring. This is a direct consequence of the localization of the electrons along the chain. The localization length L_c is given by Eqs. (11.66, 11.168) and can also be written for the model under consideration as

$$L_c = 32JL \tag{11.177}$$

In Eq. (11.177) an additional factor of 2 is included since the effective cross section of the chain of rings is $2S$. Calculating the integral in Eq. (11.173) that is proportional to the inverse participation ratio for the effective level spacing one obtains

$$\Delta_{\text{eff}}(J \gg 1) = (3 v S L_c)^{-1} \tag{11.178}$$

We see from Eq. (11.178) that in the limit of large J the response $K^{\text{unit}}(0)$, Eq. (11.172), looks as if the chain of N_r rings effectively consists of $L_x / (3 L_c(J))$ rings. In this limit, with Eq. (11.165) one obtains that magnetization does not depend on disorder

$$M_\omega^{\text{unit}}(J \gg 1) = \frac{K_d}{4\pi c^2} \frac{N_r V}{(8\pi M_T)^2} \phi_\omega \tag{11.179}$$

where $M_T = p_0^2 S / (4\pi^2)$ is the number of transverse channels in a single ring.

Decreasing the coupling constant J makes the localization length shorter, thereby leading to a larger response. The response $K(\omega)$, Eq. (11.172), describes the dynamic magnetization per unit length along the chain, and therefore the total dynamic magnetization is proportional to the length of the chain. The corresponding thermodynamic response is small, as is the difference between the canonical and grand canonical ensembles. To make an explicit estimate one can use the expansion in terms of fluctuations of the chemical potential. Using the approaches developed in Chapter 8 one can show that the total thermodynamic magnetization remains constant in the limit $L_x \to \infty$ and so, the thermodynamic magnetization per unit length vanishes. Of course, this result concerns only the contribution due to the difference between the canonical and grand canonical ensembles. The standard Landau diamagnetism contributes in the usual way.

11.7.3 Possibility of experimental observation

The finite dynamic magnetization calculated previously does not correspond to the minimum of energy and must decay in time if inelastic processes are included. Dynamic magnetization is a consequence of the electron localization, and as soon as localization is lifted and the electrons can travel through the chain the magnetization should vanish. At temperatures

$T > \Delta_{\text{eff}}$ and at frequencies ω smaller that $t^{-1}(J)$, where $t(J)$ is a time due to inelastic processes, the dynamic magnetization decays because the direct current conductivity is finite (and proportional to $t^{-1}(J)$). At higher frequencies the inelastic processes have no influence on magnetization.

The simplest way to include the time $t(J)$ in formulae obtained previously is to substitute $\omega \to \omega + it^{-1}(J)$ in all expressions for $K(\omega)/\omega$, where $K(\omega)$ can stand for the responses in both the longitudinal and azimuthal directions. ($K(\omega)/\omega$ is proportional to the product of two Green functions, and it is this quantity that is calculated in the linear response theory.) With this substitution one obtains finite longitudinal direct current conductivity proportional to $t^{-1}(J)$ and magnetization decaying in time with time $t(J)$. This substitution is a hypothesis and can be used for rough estimates only. The finite inelastic time can be due to an interaction with phonons that leads to hopping between the localization centers (Mott and Davis (1971)). Explicit calculations for disordered chains (Gogolin, Melnikov, and Rashba (1975)) at temperatures that are not too low corresponding to $T > \Delta_{\text{eff}}$ lead to the result $t(J) = \tau_{ph}$, where τ_{ph} is the electron–phonon scattering time. To the best of my knowledge nobody has performed analogous calculations for thick wires that are equivalent to the chain of the rings, but possibly this estimate can give a reasonable estimate for the case considered here. Then, the inverse decay time is proportional to T^3. Thus, at temperatures much below the Debye temperature ω_D but above the effective level spacing $\Delta_{\text{eff}} < T \ll \omega_D$, one can have $1/t(J) < \Delta_{\text{eff}}$ and it should be possible to observe a large diamagnetic response.

At lower temperatures $T \ll \Delta_{\text{eff}}$ the exponential Mott law

$$t(J) \sim (\omega_D)^{-1} \exp\left(4\left(\frac{3\Delta_{\text{eff}}}{T}\right)^{1/2}\right) \tag{11.180}$$

might serve as a lower limit for the decay time (the exponent $1/2$ is due to one-dimensionality).

Eq. (11.171) was obtained for the unitary ensemble that corresponds to the limit of a strong magnetic field. Analogous calculations can be performed also for the orthogonal ensemble corresponding to the zero static component H of the magnetic field in Eq. (11.164). In this limit, as in Chapter 8, one obtains

$$M_{\omega \to 0}^{\text{orth}}(0) = 0 \tag{11.181}$$

By changing the static component H of the magnetic field we can have a crossover between Eqs. (11.181) and (11.172). The characteristic flux of this crossover is of the order of $\phi_0(\Delta_{\text{eff}}/\mathcal{E}_c)^{1/2}$, where $\mathcal{E}_c = \pi^2 D_0/L^2$ is the Thouless energy. The dependence of the response on flux ϕ is periodic with the period $\phi_0/2$. The flux average is finite and one can expect a large diamagnetic response not only in the chain of rings but also in a wire.

Adding magnetic or spin-orbit impurities changes Eqs. (11.181, 11.172). The system with magnetic impurities corresponds to Model IIb. One can use, as before, Eq. (11.181) provided $\Delta_{\text{eff}}(J)$ is replaced by $(1/2)\Delta_{\text{eff}}(2J)$. If the magnetic impurities are absent, spin-orbit impurities lead to a different function $\bar{\Delta}_{\text{eff}}(J)$. However, this difference is only numerical and does not change the sign of the response, which is in all cases diamagnetic.

Although the dynamic magnetism suggested in this section has not been observed yet, such a possibility apparently exists. One can get a quite measurable magnetization provided the coupling constant J is not very large and localization length L_c is of the same order as the size of the rings. The dephasing rate should not exceed the effective level spacing Δ_{eff} which gives for dephasing length $l_\phi = \sqrt{D_0\,(J)\,\tau_\phi}$ the condition $l_\phi > \sqrt{L_c\,(J)\,L}$. For $J < 1$ this is the usual mesoscopic regime met in Lévy et al. (1990). For $J \gg 1$, when the localization length L_c much exceeds the size of the rings one may need to have phase coherence over several rings. Measuring the magnetization may yield a new contactless method of probing the localization. The preceding presentation follows the work of Kettemann and Efetov (1995).

The results presented in this chapter demonstrate that many completely different problems reduce to different versions of the 1D σ-model. The most interesting information concerning low-frequency behavior is obtained from exact solutions. So, as soon as a physical problem of disorder or chaos is effectively one-dimensional, one can hope that it can be solved. Higher-dimensional problems are usually much more difficult. In the next chapter I present an attempt to attack a very difficult and interesting problem of the Anderson metal–insulator transition.

12

Anderson metal–insulator transition

12.1 Description of phase transitions

12.1.1 Phenomenological approach

Some basic information concerning theory of the Anderson metal–insulator transition has been given in Chapters 3 and 5. The agreement of the one-parameter scaling hypothesis of Abrahams et al. (1979) with the results of the renormalization group treatment of the nonlinear replica and supermatrix σ-models in $2 + \epsilon$ dimensions was considered by many researchers final proof that the transition was a conventional second-order transition. The only thing that remained to be done was to compute critical exponents, and that could be done by making an expansion in ϵ and putting $\epsilon = 1$ at the end. Other approximate schemes (Götze (1981, 1985), Vollhardt and Wölfle (1980, 1992)) lead to similar results. Although agreement between the exponents computed analytically and those extracted from numerical simulations or experiments was not always good, the validity of the one-parameter scaling description was not usually questioned.

Of course, on the basis of what is known about mesoscopic systems one cannot speak of the average conductance of a finite system, and, possibly, the entire distribution function of the conductances should be scaled. However, renormalization group treatment of the σ-model in $2 + \epsilon$ dimensions does not lead to such a scenario. If one accepts that this approach is appropriate for studying the Anderson transition, the conclusion that the transition is a conventional second-order phase transition is inevitable.

But can one expect anything else from the renormalization group in $2 + \epsilon$ dimensions? In fact, this scheme always predicts a second-order phase transition for any system (or the absence of any transition). It was invented to study universal properties in field theories. From this point of view the second-order phase transitions are much more interesting because the behavior near, for example, a first-order phase transition is not universal. So, for field theorists the disadvantage that only second-order transitions are usually obtained is not essential and this did not attract much attention.

On the contrary, in statistical physics the first question that should be asked concerns the type of phase transition. Generally speaking, the transition does not need to be just a conventional first- or second-order phase transition. For example, the Kosterlitz–Thouless phase transition (Kosterlitz and Thouless (1973)) cannot be classified in these terms. If it is clear that the transition is not of the conventional second-order type, it is meaningless to try to use the $2 + \epsilon$ scheme to study it. But are we sure that the Anderson transition belongs to the class of conventional second-order phase transitions? Definitely not, because most of the information came from the $2 + \epsilon$ expansion, the one-parameter scaling hypothesis, and approximate schemes that were designed assuming an analogy between the Anderson transition and conventional second-order phase transitions. Therefore, to understand the nature of the Anderson transition other schemes of studying phase transitions are highly desirable. In this section, I want to review briefly some approaches commonly used in the theory of critical phenomena. I hope it can help in understanding the logic of the study of the Anderson transition presented in the next sections.

When investigating a new transition first one usually tries to construct a mean field theory. This approximation is called the *Curie–Weiss molecular field method* in the theory of ferromagnetism, the *Bragg–Williams method* in the theory of binary alloys, and the *Bardeen–Cooper–Schrieffer–Bogoliubov method* in the theory of superconductivity. This method was also used, for example, by Kurchatov (1933) in ferroelectricity theory and by Edwards and Anderson (1975) in theory of spin glasses. All theories of these types are unified in the Landau theory of phase transitions (Landau and Lifshitz (1968)).

The central concept of this theory is that of an order parameter M. This quantity is zero in the disordered phase and takes nonzero values in the ordered phase. For example, in the theory of ferromagnetism one should take magnetization as the order parameter, in the theory of superconductivity the superconducting gap plays this role, and so on. Depending on the system undergoing the phase transition the order parameter can be a real scalar (binary alloys), vector (ferromagnets), complex scalar (superconductivity), or matrix (liquid crystals, phases of ^{3}He). In the Landau theory one has to write a free energy function $F(T, M)$, where T is, for example, temperature. The temperature is an independent variable but the order parameter M has to be found from the condition of the minimum of the function $F(T, M)$ with respect to this variable. Near the phase transition the order parameter M is small and it is assumed that the function $F(T, M)$ can be expanded in M

$$F(T, M) = AM^2 + BM^4 + DM^3 \tag{12.1}$$

The minimum can exist provided $B > 0$. For some systems such as magnets or superconductors the third term in Eq. (12.1) is forbidden by symmetry (the energy must be invariant under the substitution $\mathbf{M} \to -\mathbf{M}$ in ferromagnets or under rotation of the phase of the order parameter in superconductors). For models with, for example, a matrix order parameter the cubic term is allowed and can be important. In a situation when $D = 0$, the function $F(T, M)$ is shown in Fig. 12.1a. For $A > 0$, it has the minimum at $M = 0$ and this is the disordered phase. At $A < 0$ the minimum corresponds to $M \neq 0$. The transition occurs at the point T_c where $A(T_c) = 0$. If A is linear in temperature near T_c

$$A = \alpha(T - T_c)$$

the dependence of the order parameter obeys the law

$$M = \begin{cases} 0, & T > T_c \\ \left(\frac{\alpha}{2B}\right)^\beta (T_c - T)^\beta, & T < T_c \end{cases} \tag{12.2}$$

with the critical exponent

$$\beta = 1/2 \tag{12.3}$$

This is typical behavior of second-order phase transitions. Nonzero values of D change the critical behavior drastically.

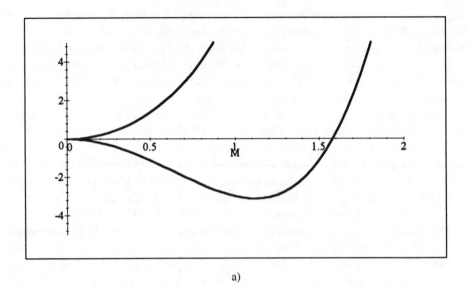

a)

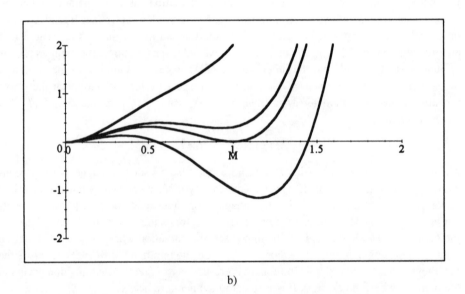

b)

Fig. 12.1. Free energy F as function of the order parameter M (arbitrary units): a) for $D = 0$ (the upper curve corresponds to $A > 0$, the lower to $A < 0$); b) for several values of $A > 0$ and $D < 0$.

The case with $D < 0$ is depicted in Fig. 12.1b. Changing the parameter A one can have a region where the energy $F(T, M)$ has two minima, one of them the absolute minimum. The transition occurs when both minima have the same energy. Then the order parameter corresponding to the state with the lowest energy has a jump at the transition. At the critical point the two different states coexist. This is a first-order phase transition.

In a situation when $D > 0$ but M can take only positive values, the dependence of F on M is similar to that represented in Fig. 12.1a. The energy can have only one minimum and the transition occurs when $A(T)$ changes sign. Near the transition the cubic term is more

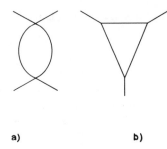

a)   b)

Fig. 12.2. The lowest-order corrections to the coupling constants: a) correction to constant B, b) correction to D.

important than the quartic one and one obtains

$$M = \begin{cases} 0, & T > T_c \\ \left(\frac{2\alpha}{3D}\right)^{\gamma} (T_c - T)^{\gamma}, & T < T_c \end{cases} \tag{12.4}$$

where

$$\gamma = 1 \tag{12.5}$$

We see from Eqs. (12.1–12.5) that the cubic term in the expansion Eq. (12.1) is really very important for the critical behavior. Although the mean field theory helps to explain the nature of the transition qualitatively, it is usually not correct quantitatively because one has to take fluctuations into account. This can be done by computing the partition function Z expressed in terms of the functional integral

$$Z = \int \exp\left(-\frac{F[M]}{T}\right) DM \tag{12.6}$$

where $F[M]$ is the free energy functional

$$F[M] = \int \left[C (\nabla M (\mathbf{r}))^2 + AM^2 (\mathbf{r}) + DM^3 (\mathbf{r}) + BM^4 (\mathbf{r}) \right] d\mathbf{r} \tag{12.7}$$

As soon as the partition function Z is known, one can obtain all thermodynamic quantities. The functional integral cannot be computed exactly and one should use different approximate schemes. The Landau theory of phase transitions is obtained by calculating the integral in the saddle-point approximation. Then, one has to make an expansion near the saddle point and calculate corrections. This procedure is described in many books and reviews (see, e.g., Wilson and Kogut (1974), Ma (1976), Amit (1978), Patashinskii and Pokrovskii (1979)).

Consideration above the transition point is somewhat simpler because the saddle point corresponds to $M = 0$. By representing the functional $F[M]$ in the form

$$F[M] = F_0[M] + F_1[M] \tag{12.8}$$

where

$$F_0[M] = \int \left[C (\nabla M)^2 + AM^2 \right] d\mathbf{r} \tag{12.9}$$

and $F_1[M]$ describes the cubic and quartic interactions one can expand the exponential in Eq. (12.6) in $F_1[M]$ and obtain a series. Each term of this perturbation theory can be represented by a diagram containing lines and vertices. The lines correspond to the bare Green functions $g(\mathbf{k})$,

$$g(\mathbf{k}) = \frac{1}{C\mathbf{k}^2 + A} \tag{12.10}$$

and the vertices to the anharmonic terms in $F_1[M]$. Typical graphs for the vertex corrections are represented in Fig. 12.2. The graph in Fig. 12.2a corresponds to a correction to the coupling constant B, and the graph in Fig. 12.2b contributes to the constant D. One can write integrals corresponding to the graphs in Fig. 12.2. The correction to the constant B is proportional to

$$B^2 \int \frac{d^d\mathbf{k}}{\left(C\mathbf{k}^2 + A\right)^2} \tag{12.11}$$

As concerns the cubic term one obtains

$$D^3 \int \frac{d^d\mathbf{k}}{\left(C\mathbf{k}^2 + A\right)^3} \tag{12.12}$$

When approaching the transition the parameter A goes to zero and both corrections, Eqs. (12.11 and 12.12), diverge in three dimensions. The integral, Eq. (12.11), converges if the dimensionality $d > 4$. This means that in the absence of cubic interaction in Eq. (12.7) the mean field approximation can be good only if the dimensionality exceeds 4. In four-dimensional space the corrections to the mean field approximation are logarithmically divergent. This dimensionality is called the *upper critical dimensionality*. In the presence of cubic terms the upper critical dimensionality is 6. It is easy to see that terms of higher order in M in Eq. (12.7) can really be neglected because they have less divergent contributions.

In the upper critical dimensionality, one can use parquet equations (Larkin and Khmelnitskii (1969)) or the renormalization group scheme (Wilson and Kogut (1974)), which makes it possible to describe the critical behavior completely. As concerns the three-dimensional case, one can use $4 - \epsilon$ or $6 - \epsilon$ expansion for small ϵ, substituting $\epsilon = 1$ or $\epsilon = 3$ at the end. This leads to power law behavior of physical quantities but with exponents that are different from their mean field values. Below the transition, Eqs. (12.2, 12.4) remain valid, but the critical exponents no longer obey Eqs. (12.3, 12.5). What is important is that the results are universal, depending only on the symmetry of the variable M. In some cases when the mean field approximation gives a second-order phase transition the fluctuations can convert the transition to a first-order one.

12.1.2 Phase transitions in spin models

It is clear that the scheme of the study of phase transitions starting with the mean field approximation and then using the expansion near the upper critical dimensionality makes it possible to describe a larger variety of phase transitions than the $2 + \epsilon$ expansion that always gives second-order ones. However, Eqs. (12.6–12.7) are written phenomenologically. In many cases, it is not difficult to understand what the symmetry of the variable M is and whether the cubic term is allowed or not. As a result of universality this information is sufficient to obtain the proper critical exponents. But what should one do if the symmetry of the order parameter is not clear in advance? This question is very important, for example, for the problem of the Anderson metal–insulator transition we are interested in because it is not even clear what the order parameter is.

A proper procedure for performing the calculations starting from a microscopic Hamiltonian is well known in the theory of phase transitions and can be demonstrated with the example of a classical spin model of a ferromagnet. The Hamiltonian $\mathcal{H}$ of the system is taken in the form

$$\mathcal{H} = -\frac{1}{2} \sum_{i,j} J_{ij} \mathbf{S}_i \mathbf{S}_j, \qquad \mathbf{S}_i^2 = 1 \tag{12.13}$$

where $\mathbf{S}_i$ is a vector on the n-dimensional unit sphere and the sum is calculated over different sites i and j (not necessarily nearest neighbors). The free energy $\mathcal{F}$ of the system is expressed through the Hamiltonian $\mathcal{H}$, Eq. (12.13), in the usual way

$$\mathcal{F} = -T \ln \int \exp\left(-\frac{\mathcal{H}}{T}\right) D\mathbf{S} \tag{12.14}$$

with the integral over the spins of all sites. The first step is to try a mean field approximation. For the spin model under consideration the relevant order parameter is a mean magnetic field $\mathbf{M}$ created by the spins or equivalently the magnetization, which is equal to the average spin in each site. In the mean field approximation the interaction between the spins in Eq. (12.13) is replaced by interaction of the spins with the magnetic field

$$\mathcal{H} \to \mathcal{H}_0 = -\mathbf{M} \sum_i \mathbf{S}_i + \frac{N}{2J(0)} \mathbf{M}^2, \qquad J(0) = \sum_j J_{ij} \tag{12.15}$$

where N is the total number of sites. The mean magnetic field $\mathbf{M}$ has to be found in a self-consistent way, which leads to the following equation:

$$\mathbf{M} = J(0) \langle \mathbf{S} \rangle, \qquad \langle \mathbf{S} \rangle = \frac{\int \mathbf{S} \exp(\mathbf{MS}/T) \, d\mathbf{S}}{\int \exp(\mathbf{MS}/T) \, d\mathbf{S}} \tag{12.16}$$

Computation of the integral in Eq. (12.16) is simple and can be found in textbooks. Near the transition the mean field is small and one should expand $\langle \mathbf{S} \rangle$ in $\mathbf{M}$ up to cubic terms. It is not difficult to see that one obtains an equation that is just the condition for an extremum of the function F, Eq. (12.7), with

$$D = 0, \qquad A = 1 - \frac{J(0)}{nT}, \qquad B = \frac{1}{12}\left(\frac{3}{n^2} - \left\langle (S^\alpha)^4 \right\rangle_0 \right) \tag{12.17}$$

Eq. (12.17) gives the macroscopic parameters of the Landau expansion. The absence of the cubic term in the expansion is a consequence of the rotational invariance of the Hamiltonian $\mathcal{H}$, Eq. (12.13), which leads to the invariance of the Landau expansion in $\mathbf{M}$. The latter would be inevitably violated by any cubic term. The mean field approximation gives a good description if the function J_{ij} decays slowly with $|i - j| \to \infty$ (long-range interaction). However, fluctuations are inevitably large very close to the transition.

The fluctuations near the mean field can be treated by using the Hubbard–Stratonovich transformation

$$Z = \int \exp\left(\frac{1}{2T} \sum_{ij} J_{ij} \mathbf{S}_i \mathbf{S}_j \right) D\mathbf{S} \tag{12.18}$$

$$= Z_0^{-1} \int \exp\left(\frac{1}{T} \sum_i \mathbf{M}_i \mathbf{S}_i - \frac{1}{2T} \sum_{i,j} \left(J^{-1}\right)_{ij} \mathbf{M}_i \mathbf{M}_j \right) D\mathbf{M} D\mathbf{S},$$

$$Z_0 = \exp\left(-\frac{1}{2T}\sum_{i,j}\left(J^{-1}\right)_{ij}\mathbf{M}_i\mathbf{M}_j\right)D\mathbf{M}$$

By integrating over $\mathbf{S}_i$ in each site one obtains

$$Z = Z_0^{-1}\exp\left(\sum_i f\left(\mathbf{M}_i\right) - \frac{1}{2T}\sum_{i,j}\left(J^{-1}\right)_{ij}\mathbf{M}_i\mathbf{M}_j\right)D\mathbf{M}, \qquad (12.19)$$

$$f\left(\mathbf{M}\right) = \ln\left(\int\exp\left(\frac{\mathbf{MS}}{T}\right)d\mathbf{S}\right)$$

The mean field approximation corresponds to the condition of the maximum of the exponent in Eq. (12.19), which gives Eq. (12.16). Near the transition only small values of $|\mathbf{M}|$ are essential. Making an expansion in this quantity one can find that the free energy F, Eq. (12.14), is described by Eq. (12.7) with $D = 0$.

So, in the preceding scheme, one should first find a mean field, and, in principle, the further procedure is quite standard. It is possible to use from the beginning not the mean field approximation when only one site is singled out but more sophisticated schemes. For example, when considering the spin model, Eq. (12.13), instead of Eq. (12.15), which contains only one spin, one can write an equation with two interacting spins and an effective medium. The interaction can be expressed as

$$\mathcal{H}_{\text{int}}\left(\mathbf{S}_1, \mathbf{S}_2\right) = -J_{12}\mathbf{S}_1\mathbf{S}_2 - (K - 1)\left(Y\left(\mathbf{S}_1\right) + Y\left(\mathbf{S}_2\right)\right) \qquad (12.20)$$

where K is the coordination number that depends on the symmetry and dimensionality of the lattice. Eq. (12.20) means that interaction of, for example, spin $\mathbf{S}_2$ with all other spins is replaced by interaction with spin $\mathbf{S}_1$ and an effective medium Y depending on $\mathbf{S}_2$, and each bond contributes equally (therefore, the coefficient $K - 1$). The function $Y\left(\mathbf{S}\right)$ must change sign under the substitution $\mathbf{S} \to -\mathbf{S}$. The condition of self-consistency can be expressed as

$$\exp\left(\frac{KY\left(\mathbf{S}_1\right)}{T}\right) = Z_1^{-1}\int\exp\left(-\frac{\mathcal{H}_{\text{int}}\left(\mathbf{S}_1, \mathbf{S}_2\right)}{T}\right)d\mathbf{S}_2 \qquad (12.21)$$

$$Z_1 = \int\exp\left(\frac{J_{12}\mathbf{S}_1\mathbf{S}_2}{T}\right)d\mathbf{S}_2$$

This means that in integrating over the spin at site 2 one has to obtain the interaction of the spin at site 1 with the same effective medium. Eq. (12.21) can also be rewritten as

$$\Psi\left(\mathbf{S}_1\right) = Z_1^{-1}\int\exp\left(\frac{J_{12}\mathbf{S}_1\mathbf{S}_2}{T}\right)\Psi^m\left(\mathbf{S}_2\right)d\mathbf{S}_2 \qquad (12.22)$$

where

$$\Psi\left(\mathbf{S}\right) \equiv \exp\left(\frac{Y\left(\mathbf{S}\right)}{T}\right), \qquad m = K - 1 \qquad (12.23)$$

The transition point can be found by assuming that the interaction with the effective medium in the vicinity of this point is weak. In the main approximation one can seek the solution $Y\left(\mathbf{S}\right)$ representing it in the form

$$Y\left(\mathbf{S}\right) = \mathbf{MS} + \left(\mathbf{MS}\right)^3 \qquad (12.24)$$

where $\mathbf{M}$ is a vector with a small modulus. Substituting Eq. (12.24) into Eqs. (12.22, 12.23) one should make an expansion in $\mathbf{M}$ and calculate corresponding integrals over $\mathbf{S}_2$. Keeping the linear terms, one obtains an equation for the transition temperature that can be written as

$$\mathbf{n} = m Z_1^{-1} \int \exp\left(\frac{J_{12}\mathbf{nS}}{T}\right) \mathbf{S} d\mathbf{S} \tag{12.25}$$

where $\mathbf{n}$ is a unit vector. Eq. (12.25) gives a transition temperature different from $T_c = J(0)/n$ obtained in the mean field approximation, Eq. (12.17). However, both methods give the same result in a high-dimensional space when the number m is large. Then, J_{12}/T_c is small and one may expand the exponential in Eq. (12.25), thus obtaining the mean field value of T_c. Making expansion an Eq. (12.22) up to cubic terms in $|\mathbf{M}|$ one arrives at the critical behavior determined by the Landau theory, Eq. (12.2).

The approximation scheme described by Eq. (12.20) is very similar to the Bethe–Peierls approximation (Huang 1987) in statistical physics or the effective medium approximation (EMA) in percolation theory (Kirkpatrick (1973)). We see that although this approximation gives a different critical temperature, it is not essentially different from the mean field theory and can be considered a modification of the latter. The scheme of this approximation is presented here because this method will be used for studying the Anderson transition in the rest of the chapter. Here I want to emphasize only that EMA is a very natural scheme and that it leads to the Landau theory of phase transitions for conventional models of statistical physics.

The schemes considered in the present section are definitely superior to the $2 + \epsilon$ expansion. Let me present one more example demonstrating this claim. Consider a model with the following Hamiltonian:

$$\tilde{\mathcal{H}} = -\frac{1}{2}\sum_{ij} J_{ij}\left(\mathbf{S}_i\mathbf{S}_j\right)^2, \qquad \mathbf{S}^2 = 1 \tag{12.26}$$

assuming that J_{ij} decay slowly such that the mean field approximation can provide a good description. The models given by Eqs. (12.13) and (12.26) coincide in the long-wavelength limit, where they take the form of a simple σ-model

$$\mathcal{H} = \int D_s\left(\nabla\mathbf{S}\right)^2 d\mathbf{r} \tag{12.27}$$

where D_s is a coefficient that can be expressed through J_{ij}. Therefore, for both models the $2 + \epsilon$ treatment leads to the same result predicting a second-order phase transition. Although the result is correct for the model given by Eq. (12.13), it is wrong for the model Eq. (12.26). The basic difference between the models is that Eq. (12.26) is invariant with respect to the substitution $\mathbf{S}_i \rightarrow -\mathbf{S}_i$ for any i, whereas Eq. (12.13) does not possess this invariance. The model Eq. (12.26) describes an interaction of rods that do not have direction and can be relevant to the description of some liquid crystals. So, the models Eq. (12.13) and (12.26) have the same perturbation theory but are topologically different.

Let us now construct the mean field theory for the model Eq. (12.26). The long-range interaction helps to identify the order parameter and in the mean field approximation the Hamiltonian $\tilde{\mathcal{H}}$ takes the form

$$\tilde{\mathcal{H}} = -\sum_i \left(\mathbf{S}_i Q \mathbf{S}_i\right) + \frac{N}{2J(0)}\,\mathrm{tr}\,Q^2 - NJ(0) \tag{12.28}$$

where Q is a symmetric $n \times n$ matrix with the constraint tr $Q = 0$, and N is the number of sites. The self-consistency equation reads

$$Q^{\alpha\beta} = J(0) Z^{-1} \int \left(S^\alpha S^\beta - \frac{\delta_{\alpha\beta}}{n} \right) \exp\left(\frac{(SQS)}{T} \right) dS \qquad (12.29)$$

where Z is the weight denominator. In the usual way we can expand in the integrand in Q and get a transition point in the linear approximation. Expanding up to cubic terms in Q one comes to an equation corresponding to the condition for the minimum of the free energy F, Eq. (12.1). But now the coefficient D is not zero! In contrast to the model of a ferromagnet, Eq. (12.13), the integrand in Eq. (12.29) is invariant under the substitution $S \to -S$ and quadratic in Q terms on the right-hand side are not forbidden by symmetry. But we know that the presence of a cubic term can lead to a first-order transition and so, the critical behavior of the two models is different in spite of the same $2 + \epsilon$ expansion. The model with the Hamiltonian $\tilde{\mathcal{H}}$, Eq. (12.26), has been studied in more detail by Kunz and Zumbach (1989).

Phase transitions can also be studied by using high-temperature expansions. A high-temperature series can be obtained by expanding the exponential in Eq. (12.14) in $\mathcal{H}/T$. Each term can be calculated explicitly and one obtains a sufficiently large number of terms to allow one to obtain information about the transition with good accuracy.

The schemes described in the present section can be classified as regular schemes for studying phase transitions. In some cases phase transitions cannot be described by the Landau theory even in a rough approximation and special methods should be applied. Famous examples are the two-dimensional Ising and XY models. Their solutions can be found in Landau and Lifschitz (1968) (2D Ising model) and in Berezinsky (1970), Kosterlitz and Thouless (1973) and Kosterlitz (1974) (XY model). However, such special methods cannot be used for studying the metal–insulator transition.

To complete the discussion of critical behavior near phase transitions it is relevant to mention phase transitions in quantum systems. The quantum effects can be discussed conveniently in the model of a granulated superconductor. It is assumed that the temperature is below and not close to the BCS transition temperature T_{c0}. Then in each granule the superconducting order parameter Δ is well defined but can fluctuate, taking different values in different grains. Macroscopic superconductivity is possible if the average $\langle \Delta \rangle$ is not equal to zero. Fluctuations of the phase ϕ of the order parameter are most important because their spectrum is gapless. The fluctuations can be suppressed by Josephson coupling of the phases in different grains.

In expressing this coupling one comes to a classical XY model on the lattice of the granules (Deutscher, Imry, and Gunter (1974)). However, adding one Cooper pair to a granule changes the energy of the granule by $2e^2/C$, where C is the capacitance. Therefore, there is a tendency for the number of electrons to be a constant in each granule. But the number of electrons and the phase are conjugate variables, so phase fluctuations are extremely enhanced. The free energy functional of the model, taking into account the charging effects, was first derived by Efetov (1980) and this allowed the study of superconductor–nonsuperconductor transition. This functional F has the form

$$F[\phi] = \sum_{i,j} \int_0^{1/T} \left[\frac{1}{2} C_{ij} \dot{\phi}_i(\tau) \dot{\phi}_j(\tau) + J_{ij} \left(1 - \cos\left(\phi_i(\tau) - \phi_j(\tau) \right) \right) \right] d\tau \qquad (12.30)$$

where C_{ij} is the capacitance matrix, τ is the imaginary time, and T is the temperature. Because of the time dependence of phase ϕ the problem is now essentially quantum. How does it influence the phase transition?

The model with the free energy functional, Eq. (12.30), is a quantum version of the XY model and can be studied again by using the Hubbard–Stratonovich scheme to decouple the Josephson interaction. One obtains as usual a functional $F[M]$ of an auxiliary complex field M. The minimum of this functional determines the transition point, and the transition can occur even at zero temperature. Near the transition $|M|$ is small and one can make an expansion in M and its derivatives with respect to coordinates and time. The result can be written as

$$F[M] = \int_0^{1/T} \int \left[K |\dot{M}|^2 + C |\nabla M|^2 + A |M|^2 + B |M|^4 \right] d\tau d\mathbf{r} \qquad (12.31)$$

with A going linearly to zero when approaching the transition, and the other coefficients being noncritical at the transition. In the zero-order approximation, by neglecting the last term, for the Green function $g(\mathbf{k}, \omega)$ corresponding to the functional $F[M]$ one obtains

$$g(\mathbf{k}, \omega) = \frac{1}{K\omega_n^2 + C\mathbf{k}^2 + A} \qquad (12.32)$$

where $\omega_n = 2\pi n T$ is the Matsubara frequency.

Writing the next terms of the expansion in B one obtains graphs that are usual in the theory of phase transitions. The only difference is that now one has to sum in each graph over the Matsubara frequencies. The summation is simpler in the limits $KT^2 \ll A$ and $KT^2 \gg A$. In the former limit the sums over the frequencies can be replaced by integrals and the result is the same as for a classical problem with one additional dimension. In the latter limit one may keep $\omega_n = 0$ only and then the behavior is just classical. This means that only at $T = 0$ is the critical behavior essentially different from the classical. At any finite temperature the critical behavior is always classical provided one is close enough to the transition point.

We can learn from this example that as soon as the sample length in one of the dimensions is finite (in the preceding example $1/T$ plays the role of the length) this dimension can be excluded from consideration of the critical behavior. Another important fact is that the critical behavior of the granulated superconductor at finite temperature is the same as that of a conventional superconductor. This is because when neglecting the time dependence the functional $F[M]$ reduces the standard form of the Ginzburg–Landau functional.

The preceding discussion of the methods of phase transition theory can help to illuminate the logic of the study presented in the next sections because, in fact, no other ideas will be used. The unusual critical behavior obtained later for the Anderson metal–insulator transition is argued to be a consequence of some special features of the supermatrix σ-models and not of the approximations used.

12.2 Effective medium approximation

12.2.1 Perturbation theory: Need for partial summation

Now we can try to use one of the schemes discussed in the previous section to study the Anderson metal–insulator transition. All the schemes are applicable to spinlike models and are suitable for investigation of the properties of the supermatrix σ-model on a lattice, which corresponds to a granulated metal. It was argued in Section 12.1 that the critical behavior should not depend on the structure of the model at short distances, and therefore, the model of granulated metal is no worse for a description of the metal–insulator transition than, say, the Anderson tight binding model. Intuitively, it is clear that the question whether the system is macroscopically a metal or an insulator depends on two parameters: the mean level spacing Δ in a single grain and the intergranule tunneling amplitude T_{ij} (cf. Eq. (11.5)). In the limit $T_{ij} \ll \Delta$ the system must be an insulator, whereas in the opposite limit the levels in granules are totally smeared and the system conducts. So, one can expect the metal–insulator transition at $T_{ij} \sim \Delta$.

Assuming for simplicity that all granules are macroscopically equal and the disorder is due to impurities inside the granules or a random surface one can rewrite the σ-model, Eq. (11.6), as

$$F = -q \sum_{i,j} J_{ij} \, \text{str} \, Q_i Q_j - \frac{\beta}{4} \sum_i \text{str} \, \Lambda Q_i, \qquad \beta = -i \, (\omega + i\delta) \, / \Delta \qquad (12.33)$$

where

$$J_{ij} = \begin{cases} J_{12}, & i, j \text{ -nearest neigbors} \\ 0, & \text{otherwise} \end{cases}$$

The condition $T_{ij} \sim \Delta$ corresponds to $J_{12} \sim 1$. In contrast to the conventional spin models discussed in the preceding section in the supersymmetric σ-model the free energy cannot indicate anything about the transition because it is identically zero. Therefore, one has to calculate correlation functions. The difference between the metal and the insulator is most explicitly seen from the correlation function $Y^{00} \, (\mathbf{r}, \omega)$, Eqs. (3.24, 3.49, 4.57, 11.13). In the metallic regime this function has the diffusion form, Eq. (3.52), whereas in the insulating regime its form is determined by Eq. (3.32). The function $Y^{00} \, (\mathbf{r}, \omega)$ has been written in terms of an integral over Q supermatrices, but to make the present chapter more self-contained it is convenient to present it here once more in slightly different notations

$$Y^{00} \, (\mathbf{r}, \omega) = -2\pi^2 \nu^2 q \int Q_{33}^{12} \, (0) \, Q_{33}^{21} \, (\mathbf{r}) \exp \left(-F \, [Q] \right) \prod_i dQ_i \qquad (12.34)$$

with $F \, [Q]$ from Eq. (12.33). The model described by Eqs. (12.33, 12.34) is valid for any type of lattice.

From what we have learned in the previous section we understand that one should try, first, the mean field approximation making a decoupling analogous to that in Eq. (12.15), writing

$$\sum_{i,j} J_{ij} \, \text{str} \, Q_i Q_j \to 2 \, \text{str} \sum_j J_{ij} Q_j \, \langle Q \rangle \qquad (12.35)$$

However, we immediately fail in this attempt because

$$\langle Q \rangle \equiv \Lambda \qquad (12.36)$$

That means that the mean field constructed in this way does not depend on the parameters of the Hamiltonian and cannot describe the transition. This fact has a natural physical meaning. As we know from Eq. (4.60) the quantity $\langle Q \rangle$ determines the average density of states, which is not a critical quantity. One of the advantages of the granule model is that the average density of states is a constant that does not depend on the parameters of the model, and Eq. (12.36) confirms this property.

As another attempt one can try to make a "high temperature" expansion, which in the present case is the expansion of $\exp(-F[Q])$ in J_{ij}. This would correspond to taking deviations from the insulating state into account. As soon as the deviations become large and the series divergent one expects the insulator–metal transition. But again, the procedure is not simple because of the noncompactness of the symmetry group of the supermatrices Q. Performing the expansion one inevitably obtains integrals of the type

$$J_{12}^n \int (\cosh\theta_1)^n \exp(-\beta\cosh\theta_1)\sinh\theta_1 d\theta_1 \qquad (12.37)$$

In the limit of small frequencies $\beta \to 0$, this integral grows and the high-temperature expansion is not valid because the series is always divergent.

Let us then try a "low-temperature" expansion, which is an expansion in deviations from a good metal. In the model under consideration the latter series is just the expansion in the diffusion modes performed in Chapter 5, which was based on Eq. (5.1) and subsequent expansion in the supermatrices B, $\bar{B}$. This procedure, however, leads to errors that are due to replacing all the limits of the integration over the elements of the supermatrix B in Eq. (5.1) by infinity. We understand that the noncompactness can be important but the symmetry group is not perceived in the expansions of Chapter 5. This disadvantage can be prevented by a procedure suggested by Vaks, Larkin, and Pikin (1966) for the investigation of spin systems. Following this method one should expand near the mean value of the spin; for the present case that means that one should expand near Λ. Rewriting the function $F[Q]$, Eq. (12.33), in the form

$$F[Q] = -2q\tilde{J}\sum_i \text{str}\,\Lambda Q_i + F_1, \qquad (12.38)$$

$$F_1[Q] = -q\sum_{i,j} J_{ij}\,\text{str}\,(Q_i - \Lambda)(Q_j - \Lambda),$$

$$\tilde{J} = J + \beta/8, \qquad J = \sum_j J_{ij}$$

one has to make an expansion of $\exp(-F[Q])$ in $F_1[Q]$ calculating integrals over Q in each term of the series. At first glance, at large $\tilde{J}$ one should obtain a well convergent series because in this limit deviations from the average are small. Notwithstanding the decrease of the first terms of the perturbation theory with increase of the order, starting with a certain number that depends on the value of J_{ij}, the perturbation theory series terms begin to increase. The reason for this increase is again the noncompactness of the symmetry group of the supermatrices Q, and no such increases occur in compact σ-models. The growth of the terms of the perturbation theory series is observed in the calculation of any of the mean values of a product of several matrices Q. To estimate the magnitude of the terms of the perturbation theory series one can choose any two neighboring sites 1 and 2 and calculate the integral

$$I = \int \text{str}\,[(Q_1 - \Lambda)\,k\Lambda]\,\text{str}\,[(Q_2 - \Lambda)\,k\Lambda]\exp(-F[Q])\prod_i dQ_i \qquad (12.39)$$

Using Eq. (12.38) and expanding the exponential in Eq. (12.39) in $F_1[Q]$ one can write all the terms of the series. It is convenient to represent each J_{ij} in the graphs by a thin line

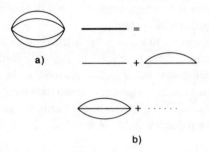

a)

b)

Fig. 12.3. Perturbation theory in deviations $(Q_i - \Lambda)$.

connecting the sites i and j. Some typical examples are represented in Fig. 12.3. The graph in Fig. 12.3a consists of n lines joining sites i and j and the contribution of the graph is given by the integral

$$I_n = \frac{(2J_{12})^n}{n!} \int \text{str}\left[(Q_1 - \Lambda)\, k\Lambda\right] \text{str}\left[(Q_2 - \Lambda)\, k\Lambda\right] \tag{12.40}$$

$$\times \left\{\text{str}\left[(Q_1 - \Lambda)(Q_2 - \Lambda)\right]\right\}^n \exp\left[2\tilde{J}\, \text{str}\, \Lambda\,(Q_1 + Q_2)\right] dQ_1 dQ_2$$

The evaluation of the integral in Eq. (12.40) must be separately performed for the orthogonal, unitary, and symplectic ensembles. The growth mechanism of the perturbation theory series terms is the same for all three ensembles, and it is sufficient to present explicit results for the unitary ensemble only. For large $n \gg \tilde{J}$, the region of large $\theta_1 \gg 1$ provides the main contribution, which can be written as

$$I_n \approx \frac{(4J_{12})^n\,(-1)^n \cosh^2 8\tilde{J}}{(n-1)!} \left[\int_0^\infty \exp\left(n\theta_1 - 8\tilde{J}\cosh\theta_1\right) d\theta_1\right]^2 \tag{12.41}$$

The integral in Eq. (12.41) can be calculated by the saddle-point method. The saddle-point value θ_{1s} is equal to

$$\theta_{1s} \approx \ln\left(n/4\tilde{J}\right) \tag{12.42}$$

Integrating in Eq. (12.41) near the saddle point, Eq. (12.42), one gets

$$I_n \approx (-1)^n\,(n-1)!\left(J_{12}/\tilde{J}\right)^n \left(4\tilde{J}\right)^{-n} \cosh^2 8\tilde{J} \tag{12.43}$$

The value of $|I_n|$ increases for any large $\tilde{J}$ with increase of n provided n is large enough. An essential condition of this growth is the noncompactness of the model, as a result of which the nontrivial saddle-point θ_{1s}, Eq. (12.42), exists at large n. Calculation by expanding in the supermatrices P, Eq. (5.1), corresponds to calculation near $\hat{\theta} = 0$, which misses the saddle point θ_{1s}, Eq. (12.42).

A similar computation can also be carried out for models that are invariant to time reversal. The perturbation theory series terms also increase in calculations of other correlators, different from I, Eq. (12.39). Of course, besides the diagram in Fig. 12.3a, there are other graphs of the same order in $\tilde{J}^{-1}$. It will be shown that at high dimensionality d of space these graphs have additional smallness in d^{-1}. It is difficult to assume, however, that they can cancel out the increasing term obtained at $d \sim 1$, since their contribution should depend

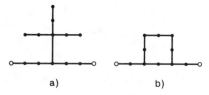

Fig. 12.4. a) Typical graph without loops; b) graph with one loop. Confluence of lines is not allowed.

on the lattice geometry, whereas the contribution of the graph in Fig. 12.3a does not depend on it.

The difficulty with the divergence of the perturbation theory series can be circumvented by a partial summation over repeated lines and by conversion to graphs that contain only effective lines. Denoting the effective line by a thick bar we obtain for it the equality illustrated by the graph in Fig. 12.3b. Carrying out such a summation in all the perturbation theory graphs we can change to a new expansion whose graphs contain only effective lines. An analogous partial summation is also the starting point of the calculation of the partition function in the Ising model (Landau and Lifschitz (1968)), where it was done for convenience. In the localization problem, however, this partial summation is necessary.

"High-temperature" expansion also necessitates summing of the graph sequence in Fig. 12.3b. In this case the sequence of diagrams singled out is most divergent as $\omega \to 0$ since the nth order contains $n!\omega^{-2n}$.

12.2.2 Principal approximation

Of course, partial summation does not solve the problem exactly, and approximations are necessary. In carrying out partial summation one can represent the function $-Y^{00}(\mathbf{r}, \omega) / 2\pi^2 \nu^2$ in the form of a sum of graphs containing only thick lines. Typical graphs are shown in Fig. 12.4. To calculate the contribution of any particular graph it is necessary to set each line with ends at the sites i and j in correspondence with the expression $\exp\left(2J_{ij}\,\text{str}\,Q_i Q_j\right)$. Each left blank circle differs from the dark circle by an additional factor $\left(Q_{33}^{12}\right)_i$ and the right circle by the factor $-\left(Q_{33}^{21}\right)_j$. After writing down the corresponding expression we must integrate over Q in each site.

The graphs in Fig. 12.4 correspond to the case of a simple cubic d-dimensional lattice with nearest-neighbor interaction, and this is the case studied. The length of each segment on any graph is equal to the edge length of the elementary cube, and any two sites can be connected by no more than one such segment.

The graphs obtained in this manner can be classified according to the number of closed loops. For example, the graph in Fig. 12.4a contains no loops, and the graph in Fig. 12.4b contains one loop. Denoting the sum of all the n-loop graphs by $K_n(0, \mathbf{r})$ one can represent the exact density–density correlator $Y^{00}(\mathbf{r}, \omega)$ in the form

$$Y^{00}(\mathbf{r}, \omega) = \sum_{n=0}^{\infty} K_n(\mathbf{r}) \tag{12.44}$$

It is impossible to calculate $K_n(\mathbf{r})$ for arbitrary n. The simplest approximation takes into account only the zero-loop graphs $K_0(\mathbf{r})$ in Eq. (12.44). In a real three-dimensional space

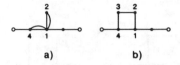

Fig. 12.5. Simplest graphs with thin lines.

this approximation cannot be justified since the diagrams with loops are not small compared with those without loops. The approximation becomes exact, however, in the limit of high dimensionality of space.

To check this statement it would be necessary to estimate the contribution of the graphs with loops and compare it with that of graphs without loops. The explicit calculations are quite difficult. It is simpler to recall the initial perturbation theory in $F_1 [Q]$, to which graphs with thin lines correspond, and see when merging several lines into one is more expedient than formation of a loop. By way of example we can compare the graph in Fig. 12.5a with the graph in 12.5b, the former obtained from the latter by merging points 3 and 1. The cause of the difference is that the graph in Fig. 12.5b contains one more integration with respect to Q. Changing over to the variable $\hat{\theta}$ gives rise to the additional Jacobian and the elementary volume for the unitary model is proportional to

$$\frac{d\lambda \, d\lambda_1}{(\lambda_1 - \lambda)^2} \tag{12.45}$$

If J in Eq. (12.38) is small, the main contribution in the limit $\omega \to 0$ comes from large $\lambda_1 \sim J^{-1}$. Each additional integration with respect to Q introduces, according to Eq. (12.45), a small factor J. The same occurs for the orthogonal and symplectic ensembles. It can be verified, by comparing more complicated graphs, that for small J it is more profitable to merge the lines than to make loops. One can conclude from these arguments that the zero-loop approximation is good in the limit

$$J \ll 1 \tag{12.46}$$

As will be shown later, provided the dimensionality d is high ($d \gg 1$), the metal–insulator transition occurs in the region where this inequality is fulfilled. Let me emphasize, however, that although the dimensionality d should be high, this by no means implies that the limit $d \to \infty$ should be taken. In the limit of infinite dimension, a transition is absent and even rescaling of J in a power of d cannot help in obtaining a finite transition point.

So, it remains now to sum the contribution of the zero-loop graphs that have the form of trees that grow out of sites of the broken lines connecting the points 0 and $\mathbf{r}$ in Fig. 12.4a. In the approximation considered one should assume that the trees and branches are not linked, so that all the trees are independent. From each site on the broken line $m - 1$ trees can grow, where m is connected with the dimensionality of the space by the relation $m = 2d - 1$. Each branch of the tree can have m branches of its own.

The structure of the trees obtained enables us immediately to write an integral equation (Efetov (1988, 1990)) that determines their contribution $\Psi (Q)$, where Q pertains to the base of the trees

$$\Psi (Q) = \int N (Q, \, Q') \, Z (Q') \, \Psi (Q') \, dQ' \tag{12.47}$$

where the function $N(Q, Q')$ has been introduced in Eq. (11.17), and the function $Z(Q)$ has the form

$$Z(Q) = \Psi^{m-1}(Q) \exp\left(\frac{\beta}{4} \operatorname{str} \Lambda Q\right), \qquad m = 2d - 1 \qquad (12.48)$$

Eq. (12.47, 12.48) is remarkably similar to Eq. (11.16) written for one-dimensional geometry and coincides with the latter at $m = 1$. This means that by summing the loopless graphs we make a transfer matrix approximation in high dimensionality. However, there is a dramatic difference between the nonlinear Eq. (12.47) at $m \neq 1$ and Eq. (11.16), which is linear. It is the nonlinearity that yields the possibility of phase transition. In mathematical language the metal–insulator transition corresponds to a bifurcation point.

So, Eqs. (12.47, 12.48) are absolutely natural and exact at $m = 1$. Is their form unusual at $m \neq 1$? In fact, no! Comparing them with Eqs. (12.22, 12.23) we see that they have exactly the same form (the partition function Z_1 is now unity because of the supersymmetry). The procedure of summing the most divergent at small J diagrams leads to the effective medium approximation! This approximation is very natural because the σ-model originates from the study of transport. Any self-consistent scheme should operate with a probability of electron transfer from site to site. But the effective medium approximation takes into account a bond explicitly replacing the remaining interactions by an effective medium, so it is really a natural self-consistent scheme for a quantity like conductivity. Although the domain described by the inequality (12.46) can contain the metal–insulator transition only in a high dimensionality, the effective medium approximation is good whenever this inequality is fulfilled.

With the function $\Psi(Q)$, Eq. (12.47), one can derive a recurrence equation for the function $Y^{00}(\mathbf{r}, \omega)$, Eq. (12.34). In the zero-loop approximation one rather easily obtains

$$Y^{00}(\mathbf{r}, \omega) = -2\pi^2 v^2 q \int Q_{33}^{12} P_{33}(\mathbf{r}, Q) Z(Q) \Psi(Q) dQ \qquad (12.49)$$

where the function $P(\mathbf{r}, Q)$ satisfies the following equation:

$$P(\mathbf{r}, Q) - \sum_{\mathbf{r}'} W(\mathbf{r} - \mathbf{r}') \int N(Q, Q') P(\mathbf{r}', Q') Z(Q') dQ' \qquad (12.50)$$

$$+ m \int N_2(Q, Q') P(\mathbf{r}, Q') Z(Q') dQ' = \delta(\mathbf{r}) Q^{21} \Psi(Q)$$

In Eq. (12.50) the function $N_2(Q, Q')$ is equal to

$$N_2(Q, Q') = \int N(Q, Q'') N(Q'', Q') Z(Q'') dQ''$$

and

$$W(\mathbf{r} - \mathbf{r}') = \begin{cases} 1, & |\mathbf{r} - \mathbf{r}'| = 1 \\ 0, & |\mathbf{r} - \mathbf{r}'| \neq 1 \end{cases}$$

The third term on the left-hand side of Eq. (12.50) takes into account the fact that two segments of a broken line cannot coincide. Eqs. (12.49, 12.50) are very similar to Eqs. (11.23, 11.24) for the 1D case, although an assumption of high dimensionality was used for their derivation.

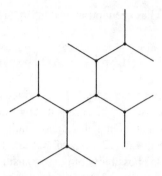

Fig. 12.6. Bethe lattice with the coordination number $K = 3$.

Eqs. (12.47–12.50) constitute a closed system of integral equations, and as soon as their solution is found one can completely describe the metal–insulator transition that exists for any $m > 1$. For a 3D lattice these equations are approximate, although as discussed the approximation is natural. However, there is a special lattice where Eqs. (12.47, 12.48) become exact, namely, the Bethe lattice (Efetov (1984)). This lattice has the structure represented in Fig. 12.6. The parameter m entering these equations is related to the coordination number K, which is equal to the number of nearest neighbors as $m = K - 1$. Eq. (12.49) can also be used and one can write an exact equation for the function $P(r, Q)$ that differs, however, from Eq. (12.50). Although the structure of the Bethe lattice is rather unusual, it may, possibly, model some highly conducting polymers. This question will be discussed in Section 12.6, but now let us concentrate on solving Eqs. (12.47–12.50).

12.3 Metal–insulator transition and functional order parameter

12.3.1 General properties of the main equation

This section is devoted to solving Eq. (12.47), which is expected to describe the Anderson metal–insulator transition. This equation appears to be very similar to Eq. (12.22), and the first thing that comes to mind is to seek the solution in the form of Eq. (12.24), introducing an order parameter M, which now has to be a matrix. Then, Eq. (12.25) would be easily obtained and one would have a Landau-type expansion in the order parameter. This procedure would be really correct if the symmetry group of the supermatrices Q were compact. But we know and it has been discussed several times in the text that it is not compact and noncompactness is an intrinsic property of the model without which one would not be able to obtain, for example, the correct form of the density–density correlation function in the localized regime. As a result of the noncompactness the expansion suggested in Eq. (12.24) makes no sense because all higher-order terms in M become important. This means that the solution has to be more sophisticated, and one should not expect to get any Landau expansion.

As in the 1D case one can assume that the solution $\Psi(Q)$ of Eq. (12.47) depends on the variables $\hat{\theta}$ only, as a consequence of the symmetry of the kernel. For the unitary ensemble, by integrating over all other variables one comes to the following equation:

$$\Psi(\lambda) = \int\limits_{-1}^{1} \int\limits_{1}^{\infty} \mathcal{L}_0(\lambda, \lambda') \frac{\lambda_1 - \lambda}{\lambda'_1 - \lambda'} Z_0(\lambda') \Psi^m(\lambda') d\lambda' d\lambda'_1 + \exp\left[\alpha(\lambda - \lambda_1)\right] \quad (12.51)$$

where the functions $\mathcal{L}_0 \left(\lambda, \lambda' \right)$ and $Z_0 \left(\lambda \right)$ of the vectors $\lambda = (\lambda, \lambda_1)$ and $\lambda' = \left(\lambda', \lambda'_1 \right)$ are introduced in Eqs. (11.133–11.136, 11.138, 11.139). At $m = 1$ Eq. (12.51) coincides with Eq. (11.137).

Using the identity Eq. (11.140) one can see that in the limit $\beta \to 0$ there exists the solution

$$\Psi_{\beta \to 0} = 1 \tag{12.52}$$

independently of the disorder strength. This is the only solution in the localized regime. The transition to the metallic state occurs at a point where a nontrivial solution appears. It will be seen that near the transition the solution is a function of only the variable λ_1, whereas far from the transition, dependence on the second variable λ becomes important. So, one can speak of a functional order parameter $\Psi \left(Q \right)$ (or $\ln \Psi \left(Q \right)$) that is equal to unity (zero) above the transition and is a function below the transition.

It is clear that at very large λ_1 the function Ψ must decay to zero. At the same time, near the transition this function must be close to unity. These two properties can be compatible if the function $\Psi \left(\lambda_1 \right)$ near the transition is very close to unity for $\lambda_1 \leq A^{-1}$, where A is a parameter characterizing the closeness to the transition, and decays to zero when $\lambda_1 \gg A^{-1}$. So, to analyze the solution it is convenient to represent $\Psi \left(\lambda \right)$ as

$$\Psi \left(\lambda \right) = 1 - \left(\lambda_1 - \lambda \right) U \left(\lambda \right) \tag{12.53}$$

and furthermore define

$$\Phi \left[U \right] = \frac{\left[1 - \left(\lambda_1 - \lambda \right) U \right]^m + m \left(\lambda_1 - \lambda \right) U - 1}{\lambda_1 - \lambda} \tag{12.54}$$

Then for $\beta = 0$, Eq. (12.51) reduces to the more compact form

$$\left[\hat{\mathcal{L}}_0^{-1} \left(\alpha \right) - m \right] U = -\Phi \left[U \right] \tag{12.55}$$

where $\hat{\mathcal{L}}_0 \left(\alpha \right)$ is the integral operator defined by its kernel, Eq. (11.138). The latter equation was obtained from Eq. (11.134), which is explicitly invariant under rotations on a sphere for the compact sector of the variables and on a hyperboloid for the noncompact sector. For any invariant integral operator on the sphere the Legendre polynomials $P_n \left(\lambda \right)$ form a complete set of eigenfunctions. For operators on a hyperboloid the polynomials P_n cannot be eigenfunctions because they do not decrease as $\lambda_1 \to \infty$. However, for hyperbolic symmetry the eigenfunctions are also known and are called *cone functions*. These functions are Legendre functions P_ν with the index $\nu = -1/2 + i\epsilon$, $0 < \epsilon < \infty$. Because of the noncompactness of the symmetry group the spectrum of eigenvalues is continuous and they are determined by the parameter ϵ. The cone functions $P_{-1/2+i\epsilon} \left(z \right)$ satisfy the conditions of orthogonality and completeness (Vilenkin (1986), Gradshteyn and Ryzhik (1966))

$$\int_1^\infty P_{-1/2+i\epsilon} \left(z \right) P_{-1/2+i\epsilon'} \left(z \right) dz = \frac{\delta \left(\epsilon - \epsilon' \right)}{\epsilon \tanh \pi \epsilon} \tag{12.56}$$

$$\int_0^\infty P_{-1/2+i\epsilon} \left(z \right) P_{-1/2+i\epsilon} \left(z' \right) \epsilon \tanh \pi \epsilon d\epsilon = \delta \left(\epsilon - \epsilon' \right) \tag{12.57}$$

and are real at real ϵ.

Using the Legendre polynomials and the cone functions one can write the function $\mathcal{L}_0 \left(\lambda, \lambda' \right)$ in the form

$$\mathcal{L}_0 \left(\lambda, \lambda' \right) = \sum_{n=0}^{\infty} \int L_{n\epsilon} \left(\alpha \right) P_n \left(\lambda \right) P_n \left(\lambda' \right)$$

$$\times P_{-1/2+i\epsilon} \left(\lambda_1 \right) P_{-1/2+i\epsilon'} \left(\lambda_1' \right) \epsilon \tanh \pi \epsilon \, d\epsilon \tag{12.58}$$

In order to obtain the expansion coefficients

$$L_{n\epsilon} \left(\alpha \right) = \int_{-1}^{1} \int_{1}^{\infty} \mathcal{L}_0 \left(\lambda, 1 \right) P_n \left(\lambda \right) P_{-1/2+i\epsilon} \left(\lambda_1 \right) d\lambda d\lambda_1 \tag{12.59}$$

one can use the following integral relations between the Legendre polynomials (cone functions) and Bessel functions (Gradshteyn and Ryzhik (1966))

$$\int_{1}^{\infty} \exp \left(-\alpha x \right) P_\nu \left(x \right) dx = \sqrt{2/\pi\alpha} \, K_{\nu+1/2} \left(\alpha \right) \tag{12.60}$$

$$\int_{-1}^{1} \exp \left(\alpha x \right) P_n \left(x \right) dx = \sqrt{2\pi/\alpha} \, I_{n+1/2} \left(\alpha \right) \tag{12.61}$$

Substituting Eq. (11.138) into Eq. (12.59) and calculating the integrals one obtains

$$L_{n\epsilon} \left(\alpha \right) = \alpha \left[K_{i\epsilon} \left(\alpha \right) \frac{d}{d\alpha} I_{n+1/2} \left(\alpha \right) - I_{n+1/2} \left(\alpha \right) \frac{d}{d\alpha} K_{i\epsilon} \left(\alpha \right) \right] \tag{12.62}$$

Analogously for the Fourier transforms of the functions U and $\Phi \left[U \right]$ one obtains

$$U_{n\epsilon} = \int_{-1}^{1} \int_{1}^{\infty} U \left(\lambda \right) P_n \left(\lambda \right) P_{-1/2+i\epsilon} \left(\lambda_1 \right) d\lambda d\lambda_1 \tag{12.63}$$

$$\Phi_{n\epsilon} = \int_{-1}^{1} \int_{1}^{\infty} \Phi \left[U \right] P_n \left(\lambda \right) P_{-1/2+i\epsilon} \left(\lambda_1 \right) d\lambda d\lambda_1 \tag{12.64}$$

so that eventually one ends up with the equation

$$\left[L_{n\epsilon}^{-1} \left(\alpha \right) - m \right] U_{n\epsilon} = -\Phi_{n\epsilon} \left[U \right] \tag{12.65}$$

As mentioned, $\Psi \equiv 1$ is the trivial solution of Eq. (12.55) and both sides of this equation vanish identically. Eq. (12.65) is satisfied because for this solution $U_{n\epsilon} = \Phi_{n\epsilon} \equiv 0$.

12.3.2 Transition point and solutions

The condition for a nontrivial solution to appear is that the bracket on the left-hand side of Eq. (12.65) must vanish at the point where such a solution appears for the first time. At this point the function Φ can be neglected because it contains only higher-order terms in

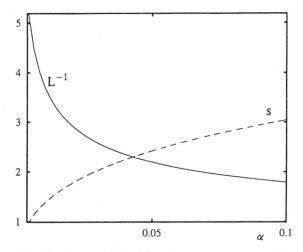

Fig. 12.7. The solutions of $L_{00}^{-1}(\alpha) = m$ determine the critical hopping amplitude α_c. From the dashed curve the proportionality constants occurring in Eq. (12.77), $s\,(m = 2,\,3,\,5) = 2.68,\ 1.72,\ 0.99$, can be extracted. (From Efetov and Viehweger (1992).)

U. Since $L_{n\epsilon}^{-1}(\alpha)$ grows monotonically with ϵ and n, one has to seek the first zero with $\epsilon = n = 0$, which gives

$$L_{00}^{-1}(\alpha_c) = m \tag{12.66}$$

Fig. 12.7 shows a plot of $L_{00}^{-1}(\alpha)$ and one can immediately extract the values of the critical disorder depending on dimensionality. In the limit $m \gg 1$, the transition occurs at $\alpha_c \ll 1$, thereby simplifying Eqs. (12.62, 12.66). In this limit the transition point α_c is determined by

$$m\left(\frac{\alpha_c}{2\pi}\right)^{1/2} \ln\frac{2}{\alpha_c} = 1 \tag{12.67}$$

On a 3D cubic lattice $m = 5$ and one can use Eq. (12.67) instead of Eq. (12.66) with good accuracy. As a result of the monotonous decay of the function $L_{n\epsilon}^{-1}(\alpha)$ with growing α, only the trivial solution $\Psi \equiv 1$ exists for $\alpha < \alpha_c$.

At $\alpha > \alpha_c$ a nontrivial solution becomes possible. To make the discussion easier it is convenient to show first what the solution looks like. Numerical study of Eq. (12.51) (Zirnbauer (1986), Efetov and Viehweger (1992)) gives a solitonlike form for the solution represented in Fig. 12.8. The position τ_c of the front of the soliton moves to infinity as $\alpha \to \alpha_c$, but the shape of this "domain wall" is not sensitive to small changes of α. The solution has the asymptotics $\Psi(1,\,1) = 1$, $\Psi(\lambda_1 \to \infty) = 0$. The value $\Psi(1,\,1)$ follows immediately from Eqs. (12.47, 12.48) and the assumption that the solution depends on the variables θ, θ_1 only. The shape of the front is quite sharp. In the limit $\lambda_1 \to \infty$ its shape is described by the asymptotics (Efetov (1984))

$$\Psi(\lambda_1) = \Psi\left(e^{\theta_1}/2\right) \sim \exp\left(-b\left(\theta_1^{1/2} + \gamma\right)\exp\left(c\theta_1^{1/2}\right)\right) \tag{12.68}$$

where b and c are numbers depending on m and α and γ is related to the position of the kink. Using the fact that the form of the kink is not sensitive to α near the transition and

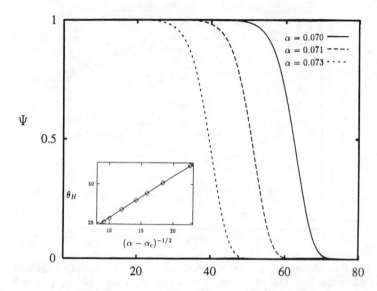

Fig. 12.8. Numerical solution $\Psi\,(1, \lambda_1)$ of Eq. (12.51) for $m = 2$ and different hopping amplitudes in the critical metallic regime. The functional order parameter is characterized by a front at a coordinate τ_c that moves to infinity as $\tau_c = s\,(\alpha - \alpha_c)^{-1/2}$ without altering its shape. The inset shows data points (diamonds) for θ_H defined by $\Psi\,(\lambda_H) = 0.5$, from which the slope $s \simeq 2.7$ can be obtained via linear interpolation. (From Efetov and Viehweger (1992).)

only its position τ_c is a critical quantity, one can extract analytically from Eq. (12.51) how τ_c depends on $\alpha - \alpha_c$ and this determines the critical behavior of all physical quantities.

Near the transition τ_c is large and, in variables λ_1, the position of the kink is located at large $\lambda_1 \sim A^{-1} \gg 1$, $A = \exp\,(-\tau_c)$. Introducing the variable $z = A\lambda_1$ one can reduce Eq. (12.51) for $z \sim 1$ to the form

$$\Psi\,(z) = \int\limits_0^\infty \bar{\mathcal{L}}\left(\frac{z}{z'}\right) \Psi^m\,(z')\,\frac{dz'}{z'} \tag{12.69}$$

where $\bar{\mathcal{L}}\,(t)$ is given by Eq. (11.142). Eq. (12.69) determines the shape of the domain wall $\Psi_s\,(z)$ but cannot determine its position because $\Psi_s\,(Cz)$ is the solution of Eq. (12.69) for arbitrary C. This means that one should make a matching with the region $\lambda_1 \sim 1$. The proper solution can be written at $\lambda_1 \gg 1$ as

$$\Psi\,(\lambda_1) = \Psi_s\,(A\lambda_1) \tag{12.70}$$

Using the function U, Eq. (12.53), one can represent the shape of the soliton $\Psi_s\,(z)$ by

$$\Psi_s\,(z) = 1 - zU_s\,(z)\,A^{-1} \tag{12.71}$$

which gives in the limit $z \gg 1$ the following asymptotics of the solution U_s:

$$U_s\,(z) = \frac{A}{z}, \qquad z \gg 1 \tag{12.72}$$

which is not sensitive to $\alpha - \alpha_c$.

The function U, Eq. (12.53), must satisfy Eq. (12.65). In the region $A \ll z \ll 1$, the function U is small and one may neglect the right-hand side of Eq. (12.65), which contains higher powers of U. Using in Eq. (12.59) the kernel $\bar{\mathcal{L}}(t)$, Eq. (11.142), instead of the exact kernel $\mathcal{L}_0(\lambda, \lambda')$, for the asymptotics of the domain wall in the region $A \ll z \ll 1$ one obtains

$$U_s(z) = a(\epsilon) \epsilon^{-1} z^{-1/2} \sin\left[-\epsilon \ln z + \delta(\epsilon)\right] \qquad (12.73)$$

where $a(\epsilon)$ and $\delta(\epsilon)$ are unknown parameters that have to be determined by matching with the asymptotics at $z \gg 1$ given by Eqs. (12.68, 12.71), in which one should set $\gamma = 0$ and $\theta_1 = \ln(2z)$. In the limit $\epsilon \to 0$ this asymptotics has no singularities. Comparing its value at $z \sim 1$ with $U_s(1)$ from Eq. (12.73) one obtains the limiting expressions

$$a(\epsilon) \to a, \qquad \delta(\epsilon)/\epsilon \to \delta_0 \quad \text{at} \quad \epsilon \to 0 \qquad (12.74)$$

The parameter ϵ is determined by the equation

$$L_{0\epsilon}^{-1}(\alpha) = m \qquad (12.75)$$

In the critical region $0 < \alpha - \alpha_c \ll 1$, one can expand $L_{0\epsilon}^{-1}(\alpha)$ around $\alpha = \alpha_c$ and $\epsilon = 0$ to the lowest nonvanishing order, which, together with Eq. (12.66), gives

$$L_{0\epsilon}^{-1}(\alpha) = m - m^2(\alpha - \alpha_c) \frac{\partial L_{00}(\alpha_c)}{\partial \alpha} - \frac{m^2 \epsilon^2}{2} \frac{\partial^2 L_{0\epsilon}(\alpha_c)}{\partial \epsilon^2}\Big|_{\epsilon=0} \qquad (12.76)$$

Consequently, the condition $L_{0\epsilon}^{-1}(\alpha) = m$ is fulfilled provided

$$\epsilon = \frac{\pi}{s}(\alpha - \alpha_c)^{1/2}, \qquad s = \pi\left(\frac{b(\alpha_c)}{2\partial L_{00}(\alpha_c)/\partial \alpha}\right)^{1/2} \qquad (12.77)$$

where $b(\alpha_c)$ is determined by Eq. (11.156) ($L_\epsilon(\alpha) \equiv L_{0\epsilon}(\alpha)$). So, we know how the form of the domain wall looks provided it is located far from the region $\lambda_1 \sim 1$ (in fact, we know only the asymptotics, but this is sufficient for further calculations). What remains is to determine the location by matching with the solution at $\lambda_1 \sim 1$.

In the region $\lambda_1 \ll A^{-1}$ the order parameter Ψ deviates only slightly from unity and $\Phi[U] \simeq m(m-1)\lambda_1 U^2/2 \ll 1$. Thus, $\Phi[U]$ in Eq. (12.55) can be neglected again and one obtains

$$U(\lambda, \lambda_1) \simeq a_0 P_{-1/2+i\epsilon}(\lambda_1) \qquad (12.78)$$

with ϵ given by Eq. (12.77). In Eq. (12.78), the zero harmonics in λ is only taken because we have to match with a function that does not depend on this variable. Using the asymptotics of the cone function $P_{-1/2+i\epsilon}(\lambda_1)$ at $\lambda_1 \gg 1$ (Gradshteyn and Ryzhik (1966))

$$P_{-1/2+i\epsilon}(\lambda_1) \approx (2/\pi\epsilon)\lambda_1^{-1/2}\sin\left[\epsilon \ln(2\lambda_1)\right] \qquad (12.79)$$

and comparing Eq. (12.79) with Eqs. (12.70, 12.73, 12.74) in the limit $\epsilon \ll 1$ one obtains

$$2a_0/\pi = A^{1/2}, \qquad \epsilon \ln(1/A) = \pi \qquad (12.80)$$

Using Eqs. (12.53, 12.77–12.80) one can write the solution of Eq. (12.51) for $\lambda_1 \gg 1$ in the simple form

$$\Psi(\lambda_1) = 1 - (A\lambda_1)^{1/2}\ln(2\lambda_1), \qquad A = p_0 \exp(-\tau_c) \qquad (12.81)$$

where p_0 is a numerical coefficient that only can be found numerically. The position τ_c of the kink depends on the closeness to the transition $\alpha - \alpha_c$ as follows:

$$\tau_c = s \left(\alpha - \alpha_c\right)^{-1/2} \tag{12.82}$$

An explicit calculation of the parameter s from Eqs. (12.77, 12.62) yields the values of s given in the caption of Fig. 12.7. The independent numerical solution presented in Fig. 12.8 agrees with this result, as can be seen from the inset, where $\theta_H \simeq \tau_c$ is plotted against $(\alpha - \alpha_c)^{-1/2}$. The asymptotics given by Eq. (12.81) is sufficient for the calculation of physical quantities characterizing the metallic regime and this will be done in the next section.

The orthogonal and symplectic ensembles can be considered in the same way as the unitary ensemble (Efetov (1987b), Verbaarschot (1988)). The solution Ψ of Eq. (12.47) depends again on the elements of the matrix $\hat{\theta}$ only. Although, because of the large number of variables of supermatrices Q, it is extremely difficult to obtain analogs of Eq. (12.51), this general form is not necessary to study critical behavior. In all cases the function Ψ has a solitonlike shape and near the transition the front moves to infinity. This enables us to carry out the calculation by assuming from the beginning that the noncompact variable λ for the symplectic case or $\lambda_1 \lambda_2$ for the orthogonal case is large.

The transition point α_c is determined as for the unitary ensemble by Eq. (12.66). The function $L_{0\epsilon}(\alpha)$ is determined by Eq. (11.161) with $\bar{\mathcal{L}}(t)$ given by Eqs. (11.158) and (11.160) for the symplectic and orthogonal ensembles, respectively. For the symplectic ensemble, by substituting Eq. (11.158) into Eq. (11.161) one obtains $L_{0\epsilon}(\alpha)$, Eq. (11.159), which should be substituted into Eq. (12.66). In the limit $m \gg 1$ one has

$$\frac{3}{4}m \left(\frac{\alpha_c}{2\pi}\right)^{1/2} \ln \frac{2}{\alpha_c} = 1 \tag{12.83}$$

The left-hand side of Eq. (12.83) differs by a factor of 3/4 from that of Eq. (12.67); that means that $\alpha_{cu} < \alpha_{cs}$, where α_{cu} and α_{cs} are the critical points for the unitary and symplectic ensembles. Comparison of the exact functions $L_{0\epsilon}(\alpha)$ for the unitary and symplectic ensembles makes it possible to conclude that for an arbitrary parameter m (including the formal limit $m \to 1$) the inequality

$$9/16 < \alpha_{cu}/\alpha_{cs} < 1 \tag{12.84}$$

is valid. This means that the metallic region is broader in the unitary case than in the symplectic one. One should distinguish between unitary models with a magnetic field (Model IIa) and those with magnetic impurities (Model IIb). The parameters α in these models, for the same density of states and the same hopping amplitudes, differ by a factor of 2 (the presence of q in Eqs. (11.6, 12.33)). Model IIb can correspond to a system with spin-orbit impurities and a magnetic field. We see from Eq. (12.84) that applying a magnetic field to a system with spin-orbit impurities makes the metallic region broader. At the same time, the metallic region in a system with a magnetic field but without spin-orbit impurities is narrower than that in a time-reversal-invariant system with the spin-orbit impurities.

To compute the critical value α_c for the orthogonal ensemble in the limit $m \gg 1$ one should substitute Eq. (11.162) into Eq. (11.161). Using Eq. (12.66) one obtains

$$\frac{2^{3/2}m}{\pi} \left(\frac{\alpha_c}{2\pi}\right)^{1/2} \ln \frac{\gamma}{\alpha_c} = 1 \tag{12.85}$$

where γ is a number of the order of unity.

Comparing Eq. (12.85) with Eq. (12.67) we see that the inequality

$$\alpha_{co} > \alpha_{cu} \tag{12.86}$$

is fulfilled (α_{co} is the critical point for the orthogonal ensemble). This means that applying a magnetic field to a time-reversal-invariant system without spin-orbit impurities leads to a broadening of the metallic region. Comparing Eq. (12.86) with Eq. (12.83) one can also conclude that adding spin-orbit impurities to a time-reversal-invariant system broadens the metallic region.

From the discussion of the present section, we see that the transition is characterized by the function Ψ. By analogy with the Landau theory of phase transitions we can still keep the concept of order parameter, but now the order parameter is the entire function Ψ and one can speak of a functional order parameter. This idea was proposed in Efetov (1988, 1990a), where it was guessed that a Laplace transform of the function Ψ could be related to a conductance distribution. Mirlin and Fyodorov (1991) considered the Anderson model on the Bethe lattice and also came to the concept of the functional order parameter that could be related to a joint distribution of the real and imaginary parts of Green functions. In the granulated model the function Ψ is related to the distribution function of density of states (Mirlin and Fyodorov (1994)).

The description of the metal–insulator transition in terms of an order parameter function bears some formal resemblance to the theory of Ising spin-glasses in a model with an infinite range interaction between spins that was proposed by Parisi (1979) (see also Binder and Young (1986)). Low-temperature equilibrium phases in the latter theory are described by the Parisi order parameter function $q(x)$, $0 \le x \le 1$, where x is an additional variable. It is possible that quite a deep relation between these systems exists but this question is not clear yet.

12.4 Correlation functions

12.4.1 General formulae

To calculate the density–density correlation function $Y^{00}(\mathbf{r}, \omega)$ in the effective medium approximation one should start from Eqs. (12.49–12.50). These equations can be studied by using an expansion in eigenfunctions ϕ_ε of an integral operator. An appropriate procedure has been developed in Section 11.6 for calculations in one dimension and can be applied here with a slight modification. Making Fourier transformation

$$P_{\mathbf{k}}(Q) = V \sum_{\mathbf{r}} P(\mathbf{r}, Q) \exp(i\mathbf{k}\mathbf{r})$$

one can reduce Eq. (12.50) to the form

$$\tilde{P}_{\mathbf{k}}(Q) - W(\mathbf{k}) \hat{M} \tilde{P}_{\mathbf{k}}(Q) + m \hat{M}^2 \tilde{P}_{\mathbf{k}}(Q) = i Q^{21} \Psi(Q) [Z(Q)]^{1/2},$$

$$\tilde{P}_{\mathbf{k}}(Q) = i [Z(Q)]^{1/2} P_{\mathbf{k}}(Q) \tag{12.87}$$

where the function $Z(Q)$ is defined in Eq. (12.48). The operator $\hat{M}$ acts on an arbitrary function $\phi(Q)$ as follows:

$$\hat{M}(Q)\phi(Q) = \int N(Q, Q') [Z(Q) Z(Q')]^{1/2} \phi(Q') dQ' \tag{12.88}$$

Using 4×4 supermatrix eigenfunctions $\phi_{\mathcal{E}}(Q)$ and eigenvalues $\mathcal{E}$ introduced in Eqs. (11.117–11.124) with $\hat{M}$ from Eq. (12.88) one can represent the correlation function $Y^{00}(\mathbf{k}, \omega)$ in the form

$$Y^{00}(\mathbf{k}, \omega) = 2\pi^2 v^2 q V \sum_{\mathcal{E}} B_{\mathcal{E}} \left[1 - \mathcal{E}W(\mathbf{k}) + m\mathcal{E}^2\right]^{-1}, \qquad (12.89)$$

$$B_{\mathcal{E}} = -\int \mathrm{str}\left[k\phi_{\mathcal{E}}^+(Q) Q^{21}\right] \Psi(Q)[Z(Q)]^{1/2} dQ$$

$$\times \int Q_{33}^{12} [\phi_{\mathcal{E}}(Q)]_{33} \Psi(Q)[Z(Q)]^{1/2} dQ \qquad (12.90)$$

where

$$W(\mathbf{k}) = 2\sum_{i=1}^{d} \cos k_i$$

and k_i are components of the vector $\mathbf{k}$.

The functions $\phi_{\mathcal{E}}$ satisfy the equation

$$\hat{M}\phi_{\mathcal{E}} = \mathcal{E}\phi_{\mathcal{E}} \qquad (12.91)$$

with the operator $\hat{M}$ determined by Eq. (12.88).

Eqs. (12.89–12.90) look very similar to Eqs. (11.126, 11.127). The expression in the square brackets in Eq. (12.89) must be positive for any momentum $\mathbf{k}$; that implies that the eigenvalues $\mathcal{E}$ satisfy the inequality $\mathcal{E} > 1/m$. As will be shown, in the limit $\omega \to 0$ the quantity $\mathcal{E} - 1/m$ is positive in the insulating regime and turns to zero in the metallic region. At the same time, the coefficients $B_{\mathcal{E}}$ are proportional to $1/\omega$ in the insulating region and remain constants in the metallic one. According to the general properties of the function $Y^{00}(\mathbf{k}, \omega)$ discussed in Chapter 3 it must have $1/\omega$ singularity in the insulating regime (cf. Eqs. (3.32, 3.33)). The coordinate dependence is determined by the function $p_\infty(\mathbf{r}, \mathbf{r}', \varepsilon)$ from Eq. (3.33) that characterizes the spatial decay of wave functions. The $1/\omega$ singularity is the most general property of any system with a discrete spectrum.

In the metallic regime, in the limit of a weak disorder the function $Y^{00}(\mathbf{k}, \omega)$ has the form of the diffusion propagator, Eq. (3.52). This form is expected to be general for diffusion motion, and the correlation function $Y^{00}(\mathbf{k}, \omega)$ can generally be written as

$$Y^{00}(\mathbf{k}, \omega) = \frac{4\pi v}{D\mathbf{k}^2 - i\omega} \qquad (12.92)$$

with the diffusion coefficient D dependent on disorder. In the limit of weak disorder it must coincide with D_0, whereas in the limit of strong disorder it turns to zero and the system no longer conducts.

Although very useful for proving diffusion motion and computing the diffusion coefficient, the function $Y^{00}(\mathbf{k}, \omega)$ does not give much information about the characteristic form of wave functions. A more complete characterization of those functions is provided by the correlation function of the density of states $S(y, y', \varepsilon, \omega)$ defined as

$$S(y, y', \varepsilon, \omega) = \sum_{\sigma,\sigma'} \mathrm{Im}\, G_{\varepsilon}^R(y, y)\, \mathrm{Im}\, G_{\varepsilon-\omega}^R(y', y') \qquad (12.93)$$

$$= \sum_{\sigma,\sigma'} |\phi_k(y)|^2 \left|\phi_{k'}(y')\right|^2 \delta(\varepsilon - \varepsilon_k)\,\delta(\varepsilon - \varepsilon_{k'} - \omega)$$

which is rather similar to the function $Y^{00}(y, y', \varepsilon, \omega)$, Eq. (3.24), but now the Green functions should be taken at coinciding arguments $y = (\mathbf{r}, \sigma)$. In the insulating regime the main contribution in the limit $\omega \to 0$ is that of coinciding levels and one obtains for a system without spin interactions

$$\lim_{\omega \to 0} S(\mathbf{r}, \omega) = \lim_{\omega \to 0} Y^{00}(\mathbf{r}, \omega) \tag{12.94}$$

where

$$S(\mathbf{r} - \mathbf{r}', \omega) = \left\langle \sum_{\sigma, \sigma'} S(y, y', \varepsilon, \omega) \right\rangle - \left\langle \sum_{\sigma} \operatorname{Im} G_\varepsilon^R(y, y) \right\rangle^2 \tag{12.95}$$

We see that in the insulating regime the function $S(y, y', \varepsilon, \omega)$ does not give new information. At the same time, this function can help in a discussion of the characteristic behavior of the wave functions of conducting states.

The function $S(\mathbf{r}, \omega)$ can be written in terms of an integral over the supermatrices Q as

$$S(\mathbf{r}, \omega) = -2\pi^2 \nu^2 q^2 \int \left[Q_{33}^{11}(0) Q_{33}^{22}(\mathbf{r}) + 1 \right] \exp(-F[Q]) \prod_i dQ_i \tag{12.96}$$

which is very similar to the integral for the function $Y^{00}(\mathbf{r}, \omega)$, Eq. (12.34). This allows one to write equations analogous to Eqs. (12.49, 12.50). One can also write an expression in terms of a sum over eigenstates $\phi_\varepsilon^{(1)}$ of the operator $\hat{M}$, Eq. (12.88). The supermatrix functions $\phi_\varepsilon^{(1)}$ have to have the same structure as the blocks Q^{11} and Q^{22}. As a result, one obtains expressions that are practically the same as Eqs. (12.89, 12.90). The only difference is that the blocks Q^{12} and Q^{21} should be replaced by $Q^{11} - 1$ and $Q^{22} + 1$, and one should use the eigenfunctions $\phi_\varepsilon^{(1)}$ with the proper symmetry, which yield eigenvalues $\mathcal{E}^{(1)}$ different from eigenvalues $\mathcal{E}$ contributing to the function $Y^{00}(\mathbf{r}, \omega)$.

So, the main problem is to solve Eqs. (12.88–12.90) and calculate the coefficients $B_\mathcal{E}$. This should be done separately for the insulating and metallic regimes and for the functions $Y^{00}(\mathbf{r}, \omega)$ and $S(\mathbf{r}, \omega)$.

12.4.2 Insulator

The correlation functions in the insulating regime in the limit of low frequencies can be calculated in practically the same way as used in Section 11.6 for 1D systems. One can derive equations analogous to Eqs. (11.141–11.151). The only modification is that one should replace the function $Z_0(Q)$ entering these equations by the function $Z(Q)$, Eq. (12.48). For example, the coefficients $B_\mathcal{E}$ entering Eqs. (12.90) can be written as

$$B_\mathcal{E} = \beta^{-1} \int \Psi(t) [Z(t)]^{1/2} \exp\left(\frac{t}{2}\right) \chi_\mathcal{E}(t) dt \tag{12.97}$$

and the equation for the function $\Psi(t)$ takes the form

$$\Psi(t) = \int_{-\infty}^{\infty} \bar{\mathcal{L}}(t - \tau) \exp\left(\frac{t - \tau}{2}\right) Z(\tau) \Psi(\tau) d\tau \tag{12.98}$$

with

$$Z(\tau) = \Psi^{m-1}(\tau) \exp\left(-\frac{e^\tau}{2}\right) \tag{12.99}$$

Using the asymptotics of the function $\Psi(t)$, which are the same as in Eq. (11.144), we can conclude that $Z(t \to -\infty) = 1$ and therefore one comes to Eqs. (11.149–11.155). As a result, the correlation function $Y^{00}(\mathbf{k}, \omega)$ can be reduced to the form

$$Y^{00}(\mathbf{k}, \omega) = \frac{2\pi \nu}{-i\omega} \int_0^\infty \frac{a^2(\epsilon)\,\epsilon^2 d\epsilon}{1 - W(\mathbf{k})\,\mathcal{E}(\epsilon) + m\mathcal{E}^2(\epsilon)} \tag{12.100}$$

where $\mathcal{E}(\epsilon) = L_\epsilon(\alpha)$. The function $L_\epsilon(\alpha)$ is determined by Eqs. (11.153, 11.159) and (11.160ths–11.162) for the unitary, symplectic, and orthogonal ensembles, respectively. The spatial dependence of the density–density correlation function Y^{00} can be found by Fourier transformation of Eq. (12.100). At large distances $r \gg 1$ the main contribution is from small k and ϵ and the correlation function can be written in a rather simple form

$$-i\omega Y^{00}(\mathbf{r}, \omega) = p_\infty(r) \tag{12.101}$$

$$= 2\pi \nu c \int_0^\infty \left(\epsilon^2 + (4L_c)^{-2}\right)^{(d-2)/2} G\left(r\left(\epsilon^2 + (4L_c)^{-2}\right)^{1/2}\right) \epsilon^2 d\epsilon$$

In Eq. (12.101) the function $G(u)$ satisfies the equation

$$-\Delta_d G(u) + G(u) = \delta(u) \tag{12.102}$$

where Δ_d is the d-dimensional Laplacian. The length L_c entering Eq. (12.101) is the localization length, and its dependence on the parameter α is given by the following expression:

$$(4L_c)^{-2} = \frac{(1 - mL_0(\alpha))(1 - L_0(\alpha))}{L_0(\alpha)} \tag{12.103}$$

The coefficient c in Eq. (12.101) equals

$$c = L_0^{-d/2}(\alpha)\, b^{-3/2}\,(1 + m - 2mL_0(\alpha))^{3/2} \tag{12.104}$$

The function $L_0(\alpha)$ grows monotonously with α, being restricted from above by 1. Localization length L_c is small at small α and grows with α. This length diverges when function $L_0(\alpha)$ reaches the value $1/m$ and this gives the transition point α_c (cf. Eq. (12.66)). The critical behavior of the localization length near the transition point is determined by the following relation

$$L_c = \text{const}\,(\alpha_c - \alpha)^{-1/2} \tag{12.105}$$

In the limits $r \ll L_c$ and $r \gg L_c$, the asymptotics of the function $p_\infty(r)$ can be obtained explicitly. For $r \gg L_c$ the quantity $r\left(\epsilon^2 + (4L_c)^{-2}\right)^{1/2}$ is large and the main contribution to the integral, Eq. (12.101), is from small values of $\epsilon \sim (L_c r)^{-1} \ll L_c^{-2}$. Making an expansion of the expression $\left(\epsilon^2 + L_c^{-2}\right)^{1/2}$ in ϵ and using the asymptotics of the function $G(u)$ at $u \gg 1$:

$$G(u) \sim u^{(1-d)/2} \exp(-u)$$

one obtains

$$p_\infty(r) = \text{const}\, r^{-(d+2)/2} L_c^{-d/2} \exp\left(-\frac{r}{4L_c}\right) \tag{12.106}$$

Eq. (12.106) demonstrates explicitly that the wave functions are localized. Although the effective medium approximation has to be correct in a high dimensionality, Eq. (12.106) works very well even in one dimension. Substituting $d = 1$ we see that even the preexponential behavior $r^{-3/2}$ (cf. Eq. (11.157)) is reproduced.

The exponent $1/2$ in the power law behavior of the localization length L_c, Eq. (12.105), is somewhat surprising. Such a value is characteristic of the localization length of the wave function of a particle in a quantum well. If the particle is in a state with eigenenergy $\mathcal{E}$ and the continuous spectrum starts from $\mathcal{E}_c$, localization length L_c for this state obeys the simple law

$$L_c \sim (\mathcal{E}_c - \mathcal{E})^{-1/2} \tag{12.107}$$

In one of the early works on metal–insulator transition Mott and Davis (1971) and Mott (1972) suggested that Eq. (12.107) might describe the critical behavior in the vicinity of the Anderson transition. Although later many other values of the exponent of the localization length were proposed (see, e.g., Kramer and MacKinnon (1993)), the agreement of Eqs. (12.105) and (12.107) may indicate that the description of the localization near the transition within the effective medium approximation is fairly close to the description in terms of an effective quantum well.

In the region $1 \ll r \ll L_c$ existing at $|\alpha_c - \alpha| \ll \alpha_c$, the quantity L_c^{-2} in the integrand in Eq. (12.101) can be neglected. In this limit one obtains

$$p_\infty (r) = A r^{-d-1}, \qquad A = 2\pi vc \int\limits_0^\infty u^d G(u)\, du \tag{12.108}$$

Eq. (12.108) shows that at distances smaller than the localization length the wave functions decay according to power law. At the transition point the localization length diverges and this results in power law behavior at arbitrary distances $r \gg 1$. The function $p_\infty (r) r^{d-1} S_d$, where S_d is the volume of the d-dimensional unit sphere, is proportional to the probability that a particle located at $t = 0$ at some point will be found at $t \to \infty$ at distance r from this point. The integral of $p_\infty (r)$ over the volume is convergent for all $\alpha \le \alpha_c$ and remains finite in the limit $\alpha \to \alpha_c$, indicating that the wave functions at the transition point decay rather fast. At the same time, all moments of this quantity diverge in this limit. The second moment determines the electric susceptibility

$$\kappa \delta_{\alpha\beta} = e^2 \int r_\alpha r_\beta\, p_\infty (r)\, dV \tag{12.109}$$

where e is the electron charge.

The integral in Eq. (12.109) can be calculated by writing the function p_∞ in the momentum space. Using Eq. (12.100) one can see that in the limit $\alpha_c - \alpha \ll \alpha_c$ the main contribution comes from small ϵ and one obtains

$$\kappa = 4\pi^2 vc L_c \tag{12.110}$$

This equation shows that the susceptibility in the critical region is proportional to the localization length. In this region the dielectric permeability coincides with the susceptibility. This result contradicts the picture of one-parameter scaling discussed in Chapter 3. Within that approach everything has to be determined by one length and therefore the susceptibility near the transition would have to be proportional to L_c^2. From the formal point of view the unusual dependence of κ on L_c in Eq. (12.110) arises from the anomalous exponent $((d + 2)/2$ and not $(d/2))$ in the power law behavior of the preexponent in Eq. (12.106).

Eq. (12.110) shows that near the transition not only localization length but intersite distance (which was chosen to be unity) determines the electric susceptibility.

As discussed earlier, in the insulating regime the correlation function of the density of states $S(\mathbf{k}, \omega)$ must coincide in the limit $\omega \to 0$ with the function $Y^{00}(\mathbf{k}, \omega)$, Eq. (12.94). This fact can be proved with Eqs. (12.34, 12.96). In the limit $\omega \to 0$ the main contribution comes from $\cosh \theta_1 \sim 1/\omega$. In this region one has approximately

$$Q_{33}^{11} Q_{33}^{22} \approx Q_{33}^{12} Q_{33}^{21} \tag{12.111}$$

which proves the correspondence of the correlation functions.

Interesting information comes from the inverse participation ratio that can be obtained from the correlation function of the density of states at coinciding points. In the effective medium approximation, Eq. (12.96) can be reduced to the form

$$S(0, \omega) = -2\pi^2 \nu^2 q^2 \int \left(Q_{33}^{11} Q_{33}^{22} + 1 \right) \Psi^{m+1}(Q) \exp\left(\frac{\beta}{4} \Lambda Q \right) dQ \tag{12.112}$$

As usual, in the limit $\omega \to 0$ integration over Q becomes simpler and for the unitary ensemble one obtains

$$-i\omega V S(0, \omega) = p_\infty(0) = 2\pi \nu \int_{-\infty}^{\infty} \Psi^{m+1}(\tau) \exp\left(\frac{-e^\tau}{2} \right) \exp(\tau)\, d\tau \tag{12.113}$$

Analogous formulae can be obtained for the orthogonal and symplectic ensembles. The quantity $p_\infty(0)$ describes the probability that the particle located at a space point at time $t = 0$ will be found at this point after an infinite time. Of course, this quantity is equal to zero in the metallic phase. In the insulating regime $p_\infty(0)$ decreases with increase of the parameter α but remains constant in the limit $\alpha \to \alpha_c$. This means that $p_\infty(0)$ as a function of α has a jump at the transition point α_c. This statement can be correct in the thermodynamic limit only. Apparently, in a sample with a finite volume $p_\infty(0)$ is a smooth function of α. Unfortunately, it is not clear how one can extend the effective medium approximation to a finite volume, and this leads to difficulties when trying to make a comparison with results of a numerical study.

12.4.3 Density–density correlation function in the metallic region

In the metallic region one expects the density–density correlation function $Y^{00}(\mathbf{k}, \omega)$ to be described by the general formula, Eq. (12.92). This form must be valid everywhere in the metallic regime; the diffusion coefficient depends on disorder or other parameters like α in the model under consideration. Therefore, the diffusion form of the function $Y^{00}(\mathbf{k}, \omega)$ has to be a consequence of general symmetries of the σ-model rather than of some specific details of the system.

In fact, the diffusion form of Eq. (12.92) can be derived (Efetov (1987b)) using only the invariance of the free energy functional $F[Q]$, Eq. (12.33), at $\omega = 0$ under the substitution

$$Q_i \to V Q_i \bar{V} \tag{12.114}$$

where V is an arbitrary supermatrix satisfying the condition

$$\bar{V} V = 1 \tag{12.115}$$

and having a structure imposed by Eq. (4.49). The invariance described by Eq. (12.114) makes it possible to confirm Eq. (12.92) and obtain an explicit expression for the diffusion coefficient D in terms of the solution $\Psi(Q)$ of Eq. (12.47) taken at $\beta = 0$.

To derive Eq. (12.92) let us note that if $\Psi_0(Q)$ is a certain solution of Eq. (12.47) for $\beta = 0$, then $\Psi_0(VQ\bar{V})$ is also a solution of this equation for any V satisfying the conditions, Eqs. (12.115, 4.49). This fact is a direct consequence of Eq. (12.114).

The solution $\Psi_0(Q)$ depends, in fact, only on Q_0 in the parametrization, Eq. (6.47). This implies that $\Psi_0(Q)$ can be represented in the form

$$\Psi_0(Q) = K\left(\text{str } f_1(\Lambda Q), \text{ str } f_2(\Lambda Q), \dots, \text{ str } f_{n_0}(\Lambda Q)\right) \tag{12.116}$$

where K is a function of n_0 variables, and f_k are functions of supermatrices. The number n_0 in Eq. (12.116) depends on the dimension of the matrix Q. The supermatrix V in Eqs. (12.114, 12.115) can be represented in the form

$$V = (1+iH)(1-iH)^{-1}, \qquad H = \begin{pmatrix} 0 & h \\ \bar{h} & 0 \end{pmatrix} \tag{12.117}$$

Now, let us use the fact that $\Psi_0(VQ\bar{V})$ is a solution of Eq. (12.47) with $\beta = 0$ for any H including small H. Expanding $\Psi_0(VQ\bar{V})$ in Eq. (12.47) in H, comparing the terms linear in H, and using the representation, Eqs. (6.42–6.47), one obtains

$$\sum_{k=1}^{n_0} v \left[\exp i \begin{pmatrix} 0 & \hat{\theta} \\ \hat{\theta} & 0 \end{pmatrix} \cdot f_k' \left(\exp i \begin{pmatrix} 0 & \hat{\theta} \\ \hat{\theta} & 0 \end{pmatrix} \right) \right]^{21} \bar{u} K_k'$$

$$= m \int \exp \left(\frac{\alpha}{4} \text{ str } QQ' \right) \Psi_0^{m-1}(Q') \tag{12.118}$$

$$\times \sum_{k=1}^{n_0} v' \left[\exp i \begin{pmatrix} 0 & \hat{\theta}' \\ \hat{\theta}' & 0 \end{pmatrix} f_k' \left(\exp i \begin{pmatrix} 0 & \hat{\theta}' \\ \hat{\theta}' & 0 \end{pmatrix} \right) \right]^{21} \bar{u}' K_k' dQ'$$

where K_k' denotes the derivative with respect to the kth argument. On the left-hand side of Eq. (12.118), K_k' depends on the elements of the matrix $\hat{\theta}$, whereas on the right-hand side F_k' depends on the elements of matrix $\hat{\theta}'$.

As the next step we introduce the function $\Phi_\beta(Q)$

$$\Phi_\beta(Q) = v \frac{\partial \Psi}{\partial \hat{\theta}} \bar{u} Z^{1/2}(Q) = v \begin{pmatrix} \partial\Psi/\partial\theta_{11} & 0 \\ 0 & \partial\Psi/\partial\theta_{22} \end{pmatrix} \bar{u} Z^{1/2}(Q) \tag{12.119}$$

The functions $\Psi(Q)$ and $Z(Q)$, Eq. (12.48), are written in Eq. (12.119) for arbitrary β. In the unitary model,

$$\partial\Psi/\partial\theta_{11} = \partial\Psi/\partial\theta, \qquad \partial\Psi/\partial\theta_{22} = \partial\Psi/\partial\theta_1, \tag{12.120}$$

in the orthogonal model,

$$\partial\Psi/\partial\theta_{11} = \partial\Psi/\partial\theta, \qquad \partial\Psi/\partial\theta_{22} = \partial\Psi/\partial\theta_1 + \tau_1 \partial\Psi/\partial\theta_2, \tag{12.121}$$

and in the symplectic model,

$$\frac{\partial\Psi}{\partial\theta_{11}} = \frac{\partial\Psi}{\partial\theta_1} + \frac{\partial\Psi}{\partial\theta_2}\tau_1, \qquad \frac{\partial\Psi}{\partial\theta_{22}} = \frac{\partial\Psi}{\partial\theta}, \qquad \tau_1 = \begin{pmatrix} 0 & 1 \\ 1 & 0 \end{pmatrix} \tag{12.122}$$

Calculating the derivatives with respect to the elements of the matrix $\hat{\theta}$ in Eq. (12.119) and comparing this equation with Eq. (12.118) one can rewrite Eq. (12.118) in the form

$$\Phi_0(Q) = m \int \exp\left(\frac{\alpha}{4} \text{ str } QQ'\right) \left[Z_0(Q) Z_0(Q')\right]^{1/2} \Phi_0(Q') dQ' \qquad (12.123)$$

where

$$\Phi_0(Q) = \left[\Phi_\beta(Q)\right]_{\beta=0}$$

We turn now to the calculation of the correlator $Y^{00}(\mathbf{k}, \omega)$ using Eqs. (12.89, 12.90). In the metallic region the eigenvalue spectrum of the operator $\hat{M}$, Eq. (12.88), is discrete, and at least the difference between the zeroth and the first value does not tend to zero at $\beta = 0$. This is a consequence of the fact that the solution Ψ of Eqs. (12.47, 12.48) decreases to zero as $\theta \to \infty$ in the symplectic model and as $t \to \infty$ in the orthogonal model. In the unitary model, Ψ decreases to zero as $\theta_1 \to \infty$. This property has been discussed in Section 12.3. The decrease of the solution Ψ leads to the result that the kernel that arises after the integration of $\hat{M}$ over U differs substantially from zero in a finite range of variation of the elements of the matrix $\hat{\theta}$. From this follows the discreteness of the eigenvalues of the operator $\hat{M}$, Eq. (12.88).

At large distances the main contribution in the sum in Eq. (12.89) is made by just the state with the largest eigenvalue $\mathcal{E}_0(\beta)$. For $\beta = 0$ the largest eigenvalue and the eigenfunction $\phi_0(Q)$ corresponding to it can be found exactly. Comparing Eqs. (12.91, 12.88) with the identity Eq. (12.123) one obtains

$$\phi_0(Q) = N_\phi^{-1} \Phi_0(Q), \qquad \mathcal{E}_0(0) = \frac{1}{m} \qquad (12.124)$$

where N_ϕ is a normalization factor, equal to

$$N_\phi^2 = (\Phi_0, \Phi_0) \equiv \int \text{str}\left[k\Phi_0^+(Q) \Phi_0(Q)\right] dQ \qquad (12.125)$$

However, it is not yet quite sufficient to know the solution at zero frequency, and one has to determine the leading order of the frequency corrections to $\mathcal{E}_0$. To calculate the largest eigenvalue $\mathcal{E}_0(\beta)$ for finite values of β let us take the scalar product of both sides of Eq. (12.91) with the function $\Phi_\beta(Q)$ from Eq. (12.119). By the scalar product the operation defined by Eq. (11.119) is implied. The result of the scalar multiplication is written in the form

$$\left(\Phi_\beta, \hat{M}\phi_\varepsilon\right) = \mathcal{E}(\beta)\left(\Phi_\beta, \phi_\varepsilon\right) \qquad (12.126)$$

The function Φ_β in Eq. (12.126) contains derivatives with respect to $\hat{\theta}$, Eq. (12.119). Integrating over $\hat{\theta}$ by parts in the left-hand side of Eq. (12.126) one can bring this equation to the form

$$\mathcal{E}(\beta)\left(\Phi_\beta, \phi_\varepsilon\right) = -\frac{i}{m}\beta \int \text{str}\left[kQ^{12}k\phi_\varepsilon(Q')\right] Z^{1/2}(Q') \Psi^m(Q)$$

$$\times \exp\left[\frac{\beta}{4}\text{ str } \Lambda Q + \frac{\alpha}{4}\text{ str } QQ'\right] dQ'dQ - \frac{1}{m}\int \exp\left(\frac{\beta}{4}\text{ str } \Lambda Q\right)\Psi^m(Q)$$

$$\times Z^{1/2}(Q') \text{ str}\left[ku\frac{\partial}{\partial\theta}\bar{u}k\phi_\varepsilon(Q')\right]\exp\left(\frac{\alpha}{4}\text{ str } QQ'\right) dQ'dQ \qquad (12.127)$$

The second term on the right-hand side of Eq. (12.127) contains the matrix $\partial/\partial\hat{\theta}$, which is in fact defined in Eqs. (12.120–12.122). It is assumed that the operator of derivatives with respect to the elements of the matrix $\hat{\theta}$ acts on all the factors to the right of $\partial/\partial\hat{\theta}$ (including the Jacobian that arises in the integration over $\hat{\theta}$, u, and v). Comparing Eq. (12.123) with the derivative with respect to $\hat{\theta}$ of both sides of Eq. (12.47) taken for $\beta = 0$ one finds

$$\int v \frac{\partial}{\partial\hat{\theta}} \bar{u} \exp\left(\frac{\alpha}{4} \operatorname{str} QQ'\right) \Psi_0^m (Q') \, dQ'$$

$$= -\int \Psi_0^m (Q') \, v' \frac{\partial}{\partial\hat{\theta}'} \bar{u}' \exp\left(\frac{\alpha}{4} \operatorname{str} QQ'\right) dQ' \qquad (12.128)$$

However, Eq. (12.128) is not only valid for the solution $\Psi_0 (Q)$ of Eq. (12.47). In fact, it is fulfilled for all functions Ψ_s that depend only on the central part Q_0 of supermatrix Q. This statement can be verified by substituting an arbitrary function Ψ_s in place of Ψ_0 on the left-hand side of Eq. (12.47) taken for $\beta = 0$. Repeating the transformations, Eqs. (12.116–12.123), one can derive Eq. (12.128), but now Ψ_0 in Eq. (12.128) is not necessarily a solution of Eq. (12.47).

Varying with respect to Ψ_0 in Eq. (12.128), substituting the result into Eq. (12.127), and using Eq. (12.47, 12.120–12.122) for $\beta = 0$, one obtains

$$\left[\mathcal{E}(\beta) - \frac{1}{m}\right] (\Phi_\beta, \phi_\mathcal{E}) = -i\frac{\beta}{m} \left(Z^{1/2}(Q) \Psi(Q) Q^{12}, \hat{M}\phi_\mathcal{E}(Q)\right)$$

$$= \frac{i\beta\mathcal{E}(\beta)}{m} \left(Z^{1/2}(Q) \Psi(Q) Q^{12}, \phi_\mathcal{E}(Q)\right) \qquad (12.129)$$

Eq. (12.129), which is valid for all β, is greatly simplified for values of β that are small in comparison with the difference between the eigenvalues. Retaining only terms linear in β and using Eqs. (12.124, 12.125), one comes to the result

$$m\mathcal{E}(\beta) = 1 - \frac{i N_\phi^{-2}\beta}{m} \int \operatorname{str}\left(kQ^{12}kv \frac{\partial\Psi_0}{\partial\hat{\theta}}\bar{u}\right) \Psi_0^m (Q) \, dQ \qquad (12.130)$$

The subsequent calculations are rather simple. First it is necessary to integrate over u and v in Eq. (12.130). The resulting expressions must be integrated over the elements of matrix $\hat{\theta}$. The integrand then turns out to be an exact divergence in the space of the elements of the matrix $\hat{\theta}$ and the integral is transformed to a surface integral over an infinitesimal surface about the coordinate origin $\theta = \theta_1 = \theta_2 = 0$. At this point $\Psi_0 = 1$, and this makes it possible to calculate the integral in explicit form. For the unitary model more detailed calculations are given in Efetov (1987a) and Efetov and Viehweger (1992). As a result one obtains

$$\mathcal{E}_0 (\beta) = \frac{1}{m}\left[1 - \frac{2\beta}{m (m + 1) N_\phi^2}\right] \qquad (12.131)$$

The integrals in Eq. (12.90) are the same as the one in Eq. (12.130). All nontrivial contributions come from the normalization factor N_ϕ^2. Substituting the result of the manipulations into Eq. (12.89) one comes to Eq. (12.92). The diffusion coefficient D in Eq. (12.92) can be written in a general form as

$$D = \frac{m\Delta}{2\pi} \int \left[\left(\frac{\partial\Psi_0}{\partial\theta}\right)^2 + \sum_i \left(\frac{\partial\Psi_0}{\partial\theta_i}\right)^2\right] \Psi_0^{m-1} \bar{J}\left(\hat{\theta}\right) d\theta \prod_i d\theta_i \qquad (12.132)$$

where $i = 1$, 2 for the orthogonal and symplectic models and $i = 1$ for the unitary model, and Δ is the mean level spacing in a single granule. In Eq. (12.132) Ψ_0 is the solution of Eq. (12.47) taken at $\omega = 0$. The Jacobians $\bar{J}$ entering Eq. (12.132) should be taken for each model separately, and they are proportional to the Jacobians, Eqs. (6.66–6.68). For the orthogonal ensemble $\bar{J}\left(\hat{\theta}\right)$ equals

$$\bar{J}\left(\hat{\theta}\right) = [\cosh(\theta_1 - \theta_2) - \cos\theta]^{-2} [\cos(\theta_1 + \theta_2) - \cos\theta]^{-2} \tag{12.133}$$
$$\times \sin^3\theta \sinh\theta_1 \sinh\theta_2$$

The Jacobian for the symplectic ensemble can be obtained from Eq. (12.133) by the substitution $\sinh \leftrightarrow \sin$, $\cosh \leftrightarrow \cos$. In the unitary ensemble

$$\bar{J}\left(\hat{\theta}\right) = (\cosh\theta_1 - \cos\theta)^{-2} \sin\theta \sinh\theta_1 \tag{12.134}$$

We see that Eq. (12.92) is correct everywhere in the metallic region. The diffusion coefficient is finite because the solution Ψ_0 is not trivial. In the insulating regime $\Psi_0 = 1$ and the diffusion coefficient D vanishes. Eq. (12.132) enables us to calculate the diffusion coefficient D at arbitrary α. At large $\alpha \gg 1/m^2$ the solution Ψ of Eq. (12.47) is rather simple

$$\Psi(Q) = \exp\left(\frac{1}{4}\alpha \operatorname{str}\Lambda Q\right) \tag{12.135}$$

Substituting Eq. (12.135) into Eq. (12.132) and performing integration one obtains

$$D_0 = \frac{\alpha\Delta}{\pi} = 8J_{12}(\pi\nu V)^{-1} \tag{12.136}$$

The quantity D_0 is the classical diffusion coefficient for the system of granules under consideration. When deriving Eq. (12.47) the condition Eq. (12.46) was essentially used. This condition corresponds to the inequality $\alpha \ll 1/m$. Of course, Eq. (12.136) is valid in the region $\alpha \geq 1/m$, too. However, in this region the effective medium approximation does not properly reproduce the effects of weak localization. These effects can be taken into account in the region $\alpha \ll 1/m$. In the latter limit quantum corrections to the classical diffusion coefficient D_0, Eq. (12.136), can be found by solving Eq. (12.47) more precisely.

Using the solution for the order parameter function Ψ near the metal–insulator transition found in Section 12.3 one can describe the critical behavior of diffusion coefficient D. Near the transition this function depends essentially on the variable λ_1 only and has the form of a kink, where the position of the front of the kink moves to infinity when approaching transition point α_c. Although the explicit form of the solution can be found only numerically the main contribution to the integral in Eq. (12.132) is from the region $1 \ll \lambda_1 \ll A^{-1}$, where Ψ is close to unity. This is correct provided the front of the kink is sharp; the property is confirmed both numerically, Fig. 12.8, and analytically, Eq. (12.68). In the unitary case, one can use Eq. (12.81) for the solution Ψ. Substituting Eq. (12.81) into Eq. (12.132) one obtains the following integral

$$D = \frac{m\Delta}{4\pi} \int_1^{\cosh\tau_c} \left[\left(\frac{A}{\lambda_1}\right)^{1/2} \ln 2\lambda_1\right]^2 d\lambda_1 = \frac{m\Delta}{12\pi} A\ln^3\left(\frac{1}{A}\right) \tag{12.137}$$

(To prevent confusion I emphasize that the distance between the grains is assumed to be 1, and m is given by Eq. (12.48).) Although Eq. (12.137) is obtained for the unitary ensemble,

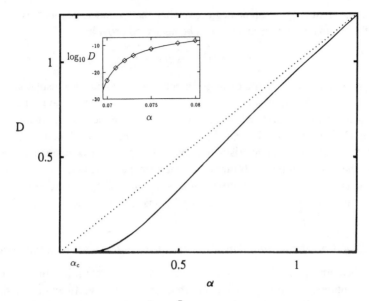

Fig. 12.9. Dimensionless diffusion coefficient $\bar{D}$ as a function of the hopping amplitude for $m = 2$. For $\alpha > 1$ the graph approaches asymptotically $\bar{D} = \alpha$. The inset shows $(\alpha, \bar{D})$ data (diamonds) together with a fit to $\bar{D} = p\tau_c^3 \exp(-\tau_c)$, $p = 0.01$. (From Efetov and Viehweger (1992).)

similar formulae can be written for the orthogonal and symplectic ensembles, too (Efetov (1987b)). Using Eq. (12.82) one can write the final formula for diffusion coefficient D valid for all three ensembles

$$D = \frac{\Delta}{\pi}\bar{D}, \qquad \bar{D} = \frac{p\exp\left[-s\,(\alpha - \alpha_c)^{-1/2}\right]}{(\alpha - \alpha_c)^{3/2}} \qquad (12.138)$$

In Eq. (12.138), p and s are numerical coefficients that should be calculated for each model separately. In the unitary ensemble the coefficient s is determined by Eq. (12.77). To check the analytical solution for the diffusion coefficient one can numerically find the function Ψ (Zirnbauer (1986)); substitute the solution into the integral, Eq. (12.132); and compute the integral (Efetov and Viehweger (1992)). This procedure does not require any approximations for solving Eq. (12.51) and calculating the integral. The result of the computation for the unitary ensemble is represented in Fig. 12.9 for $m = 2$ (this value of m does not correspond to physical dimensionality of a real space, but the critical behavior must be the same for all $m > 1$; Eqs. (12.47, 12.51) can also be obtained on the Bethe lattice (see Section 12.6) with the coordination number $K = m + 1 \geq 3$).

The critical behavior of the diffusion coefficient, Eq. (12.138), is very unusual and does not agree with the power law behavior predicted from one-parameter scaling, Section 5.2. I want to emphasize that both results were obtained within the same nonlinear supermatrix σ-model although different calculational schemes were used. The discrepancy will be discussed in Section 12.5, where it will be demonstrated that from the mathematical point of view the noncompactness of the symmetry group of supermatrix Q plays a crucial role.

For the consideration presented it was extremely important that the frequency ω was assumed to be small. How small should it be? The characteristic scale of the decay of Ψ as a function of λ_1 is equal to A, Eq. (12.81). The other characteristic scale for the variable

λ_1 is $1/\beta$. The frequency ω can be neglected when the second scale much exceeds the first one. In other words, a finite frequency can be neglected provided

$$\omega \ll \Gamma = \Delta \exp(-\tau_c) \qquad (12.139)$$

In the opposite limit, one cannot distinguish between the metal and insulator. Apparently, the quantity Γ plays the role of an effective width of a conduction band near the transition.

In the scaling theory of localization, Section 5.2, conductivity σ (and the diffusion coefficient D) is proportional to ξ^{2-d}, where ξ is a correlation length. Does the result, Eq. (12.138), imply that characteristic lengths in the system grow exponentially when approaching the transition point? It turns out that near the transition not one but two characteristic lengths exist and both diverge in a power law. They can be extracted from the correlation function of the density of states, which is calculated next.

12.4.4 Correlation function of the density of states in the metallic region

Although the correlation function S, Eq. (12.93), does not determine directly measurable physical quantities, it gives important information about wave functions of the system. This helps one to understand the physical picture. Besides, wave functions are simpler for numerical study than kinetic coefficients.

Calculation of the density of states correlator S, Eqs. (12.93, 12.96), is not very different from the corresponding procedure for the density–density correlation function Y^{00}. One can use Eq. (12.87–12.91), in which one should replace the matrix elements of Q^{12}, Q^{21} by those of $Q^{11} - 1$, $Q^{22} + 1$. The proper matrix eigenfunctions $\phi_{\mathcal{E}}^{(1)}$ should now have the structure of the matrices Q^{11} and Q^{22} and this leads to a new set of eigenvalues $\mathcal{E}^{(1)}$ in Eq. (12.91). In contrast to the function Y^{00}, keeping the frequency ω finite is not important when calculating the function S and we can take the limit $\omega \to 0$ just at the beginning. Using an expansion similar to Eqs. (12.89, 12.90) it is very important to know the maximum eigenvalue $\mathcal{E}_0^{(1)}$. The maximum eigenvalue $\mathcal{E}_0$ arising in calculation of the density–density correlation function at $\omega = 0$ is exactly equal to $1/m$. This value is a consequence of the rotational symmetry of the σ-model and leads to the gapless form of the diffusion propagator, Eq. (12.92). For the density-of-states correlation function all eigenvalues $\mathcal{E}^{(1)}$ turn out to be smaller than $1/m$, and the corresponding propagator that can be derived from an equation corresponding to Eq. (12.89) has a gap. That is why one can set $\omega = 0$ when calculating this quantity.

What is the characteristic distance between the eigenvalues $\mathcal{E}$ in Eq. (12.91)? To answer this question one should know the scale of the decay of the kernel in the integral operator in Eq. (12.88), which, in its turn, is determined by the decay of the order parameter function Ψ. We have found that this scale is just τ_c, Eq. (12.82), and the next step is to estimate the eigenenergies. This was done by Zirnbauer (1986). The main idea is that as a result of the decay of the kernel in the integral equation the eigenenergies must be quantized; this is analogous to quantization in a confining potential in the Schrödinger equation. At large λ_1 all eigenfunctions can be approximated by the cone functions $P_{-1/2+\epsilon}(\lambda_1)$. The parameter ϵ is analogous to momentum in standard quantum mechanics. Using the Bohr–Sommerfeld quantization rule one can write allowed values ϵ_n of the "momentum" ϵ as

$$\epsilon_n \tau_c = (n+1)\pi \qquad (12.140)$$

At large τ_c the distance between the eigenvalues is small and one can make an expansion of the kernel in ϵ. The first correction is proportional to ϵ^2 (the "energy" is proportional to the square of the "momentum") and one comes to the estimate

$$\mathcal{E}_n - \mathcal{E}_{n+1} \propto \alpha - \alpha_c \tag{12.141}$$

This estimate is correct for calculation of both function Y^{00} and S. This means that the eigenvalue next to $1/m$ for the function Y^{00} is separated by $\alpha - \alpha_c$. The distance between the eigenvalues, Eq. (12.141), determines a length scale. For example, the diffusion form of this function, Eq. (12.92), is valid only for small momenta $k \ll \zeta^{-1}$, where ζ is a length that can be estimated as

$$\zeta \propto (\alpha - \alpha_c)^{-1/2} \tag{12.142}$$

The maximum eigenvalue $\mathcal{E}_0^{(1)}$, in contrast to $\mathcal{E}_0$, differs from $1/m$. At first glance, the difference must be of the order of $\alpha - \alpha_c$. Surprisingly, numerical study of an equation for the correlation function S analogous to Eq. (12.87) showed (Zirnbauer (1986)) that the difference had to be proportional to

$$\Delta\mathcal{E}_0 = \frac{1}{m} - \mathcal{E}_0^{(1)} \propto (\alpha - \alpha_c)^{3/2} \tag{12.143}$$

which is much smaller than $\mathcal{E}_n - \mathcal{E}_{n+1}$. The result, Eq. (12.143), was later obtained analytically by Efetov and Viehweger (1992), who also discussed physical consequences.

To derive Eq. (12.143) one can assume from the beginning that $\Delta\mathcal{E}_0$ is small compared to the characteristic eigenvalue spacing $\alpha - \alpha_c$. This makes it possible to use a perturbation theory expanding near the state with the eigenfunction ϕ_0 and eigenvalue $\mathcal{E}_0$, Eq. (12.124). The eigenfunction $\phi_0(Q)$ can be written in the form Eq. (11.130), which allows one to rewrite Eq. (12.91) as

$$\hat{T} R_0\left(\hat{\theta}\right) = \mathcal{E}_0 R_0\left(\hat{\theta}\right) \tag{12.144}$$

where $R_0\left(\hat{\theta}\right)$ is the central part of the function $\phi_0\left(\hat{\theta}\right)$ in the representation, Eq. (11.130) (see also Eqs. (12.119, 12.124)) and the operator $\hat{T}$ acts on $R_0\left(\hat{\theta}\right)$ as follows:

$$\hat{T} R_0\left(\hat{\theta}\right) \equiv \int N\left(Q_0, Q'\right)\left(Z\left(Q_0\right) Z\left(Q_0'\right)\right)^{1/2} v' R_0\left(\hat{\theta}'\right) \bar{u}' dQ' \tag{12.145}$$

In Eq. (12.145), Q_0 is the central part of the supermatrix Q (cf. Eq. (6.47)). Analogous equations can be written for the functions $\phi^{(1)}$. Representing them in the form

$$\phi^{(1)}(Q) = u R^{(1)}\left(\hat{\theta}\right) \bar{u} \tag{12.146}$$

(one more set of eigenfunctions can be obtained by replacing the supermatrices u in Eq. (12.146) by v) one can write the equation for the maximum eigenvalue $\mathcal{E}_0^{(1)}$ as

$$\hat{T}^{(1)} R_0^{(1)}\left(\hat{\theta}\right) = \mathcal{E}_0^{(1)} R_0^{(1)}\left(\hat{\theta}\right) \tag{12.147}$$

where the operator $\hat{T}^{(1)}$ acts as

$$\hat{T}^{(1)} R^{(1)}\left(\hat{\theta}\right) = \int N\left(Q_0, Q'\right)\left(Z\left(Q_0\right) Z\left(Q_0'\right)\right)^{1/2} u' R^{(1)}\left(\hat{\theta}'\right) \bar{u}' dQ' \tag{12.148}$$

Assuming that the difference between the operators $\hat{T}$ and $\hat{T}^{(1)}$ is small, one can just write the first order of the standard quantum mechanical perturbation theory

$$\Delta \mathcal{E}_0 = \int R_0 \left(\hat{\theta}\right) \left(\hat{T} - \hat{T}^{(1)}\right) R_0 \left(\hat{\theta}\right) \bar{J} \left(\hat{\theta}\right) d\hat{\theta} \qquad (12.149)$$

where $\bar{J} \left(\hat{\theta}\right)$ is the Jacobian, determined, depending on the ensemble, by Eqs. (12.133, 12.134).

In order to provide an estimate for Eq. (12.149) it is very important to note that the kernel of $\hat{T} - \hat{T}^{(1)}$ decays very rapidly for large values of noncompact variables $\theta_{1,2}$. Of course, this property can be checked directly by performing integrations over u and v in Eqs. (12.145, 12.148), but this would require lengthy calculations. A simpler proof comes from the fact that in the insulating regime the correlation functions Y^{00} and S coincide, being determined by $\lambda_{1,2}, \lambda'_{1,2} \sim 1/\beta$. The kernels $\hat{T}$ and $\hat{T}^{(1)}$ differ in the region $\lambda_{1,2}$, $\lambda'_{1,2} \sim 1$. In the metallic region near the transition the kernels still have this property and the main contribution to the integral in Eq. (12.149) is that from the region $\lambda_{1,2}, \lambda'_{1,2} \sim 1$. Using the explicit form of function $\phi_0 (Q)$, Eqs. (12.119, 12.124), and Eq. (12.81) for the order parameter function $\Psi (Q)$, and calculating the integral in Eq. (12.149) one obtains the result

$$\Delta \mathcal{E}_0 \propto N_\phi^{-2} A \propto \ln^{-3} \left(\frac{1}{A}\right) \qquad (12.150)$$

Using Eqs. (12.81, 12.82) for α one finally obtains

$$\mathcal{E}_0^{(1)} - \frac{1}{m} \propto (\alpha - \alpha_c)^{3/2} \qquad (12.151)$$

which is small compared to the spacing between higher levels described by Eq. (12.141). Inserting these results into the expansion in the eigenstates analogous to Eqs. (12.89, 12.90) and considering the long-range limit we find that the decay of the correlation function of the density of states is characterized by two length scales. The shorter length ζ is characterized by Eq. (12.142), whereas the longer one ξ obeys the law

$$\xi \propto (\alpha - \alpha_c)^{-3/4} \qquad (12.152)$$

The decay of the function $S (\mathbf{r}, 0)$ at large distances r can be written as

$$S (\mathbf{r}, \omega) \propto r^{(1-d)/2} \left(a \exp \left(-\frac{r}{\zeta}\right) + b \exp \left(-\frac{r}{\xi}\right)\right) \qquad (12.153)$$

where a and b are some coefficients. We see from Eqs. (12.138, 12.153) that although the diffusion coefficient D decays exponentially, near the transition the characteristic lengths characterizing the system obey power laws. This is again in striking contrast to the behavior near conventional second-order phase transitions and contradicts the predictions of the one-parameter scaling theory discussed in Section 5.2. Comparing Eqs. (12.138, 12.142, 12.152) it is not difficult to see that diffusion coefficient D is related to the lengths ζ and ξ in a curious way

$$D \propto \Delta \xi^2 \exp (-\zeta) \qquad (12.154)$$

Eq. (12.154) shows that the quantity Γ, Eq. (12.139), acting as the width of a conduction band, can also be considered as the Thouless energy of a sample with the size ξ.

Calculation of the density-of-states correlation function for coinciding points is comparatively easy. After integrating over the anticommuting variables one obtains from Eq. (12.112)
(unitary ensemble)

$$S(0, 0) = \pi^2 \nu^2 \int\limits_{-1}^{\infty} d\lambda_1 \int\limits_{-1}^{1} d\lambda \Psi^{m+1}(\lambda) \propto \frac{\pi \nu}{\Gamma V} \tag{12.155}$$

with Γ from Eq. (12.139). The same result can be derived for the other ensembles and
for function Y^{00}. To confirm the interpretation of Γ as the width of a band or as a level
broadening one can consider a single granule connected with a surrounding medium. As a
result of the connection each level in the granule is assumed to acquire a small but finite
width Γ. Using the results of Chapter 6 one can write the corresponding correlation function
$S(0, 0)$ in the form

$$S(0, 0) = \pi^2 \nu^2 \int\limits_{1}^{\infty} d\lambda_1 \int\limits_{-1}^{1} d\lambda \exp\left(\frac{\pi \Gamma}{\Delta}(\lambda - \lambda_1)\right) \propto \frac{\pi \nu}{\Gamma V} \tag{12.156}$$

Comparing Eq. (12.155) with Eq. (12.156) we see that the quantity Γ, Eq. (12.139), can really
be interpreted as the width of a band. Note again that all results of the present section and,
in particular, Eq. (12.155) are valid provided frequency ω does not exceed the broadening
Γ. At larger frequencies any difference between the metal and insulator disappears.

12.5 Interpretation of the results

12.5.1 Noncompactness: Formal reason for the unconventional behavior

The critical behavior near the Anderson metal–insulator transition obtained in the preceding
sections appears to be very unusual. It definitely disagrees with the conventional picture
of a second-order phase transition discussed in Section 12.1. But what are the arguments
supporting the idea that the critical behavior near the metal–insulator transition should be
described by one-parameter scaling and all relevant physical quantities obey a power law?

The description of the metal–insulator transition in terms of one-parameter scaling was
suggested by Wegner (1976) using summation of diagrams. Abrahams et al. (1979) proposed
nice physical arguments leading to the same results. The invention of the nonlinear σ-
models (Wegner (1979), Efetov, Larkin, and Khmelnitskii (1980), Schäfer and Wegner
(1980), Efetov (1982a, 1983)) make it possible to use the renormalization group scheme.
Using this scheme in $2 + \epsilon$ dimensions with small ϵ and putting at the end $\epsilon = 1$ one can
confirm the existence of the metal–insulator transition (at least for the orthogonal and unitary
ensembles), prove the power law behavior near the transition, and calculate exponents (cf.
Section 5.2). For a long time this has been a generally accepted scenario for the Anderson
transition.

Can the renormalization group treatment in $2 + \epsilon$ dimensions at small ϵ give a good
description of a transition in a three-dimensional system? In principle, it can, and the classical $O(3)$ model (classical Heisenberg model) can be an example. At the same time, a
slightly different model, Eq. (12.26), with the same renormalization group equations gives
completely different critical behavior. The only difference between these models is that
the latter is invariant with respect to changing $\mathbf{S}_i \to -\mathbf{S}_i$. So, consideration of perturbation

theory only cannot give complete information about the phase transition and it is clear that
the topology of the symmetry group can be very important.

The symmetry group of the supermatrix nonlinear σ-model used throughout this book
is not compact. This is a very unusual property, and, from the beginning, one could expect
something new near the transition. At the same time, the renormalization group treatment
is not sensitive to noncompactness and therefore the results obtained in this scheme cannot
be taken for granted. As discussed in Section 12.1, the effective medium approximation
used in the preceding sections is quite a natural scheme and is, in fact, a modified version
of the mean field theory. It becomes exact on the Bethe lattice and in high dimensionality.
The effective medium approximation corresponds to the saddle point of a functional in an
extended space (cf. Appendix 4), which may provide a way of treating fluctuations. For
compact spin models, the use of this scheme leads to the conventional second-order phase
transition. As concerns the results obtained for the supermatrix σ-model, they are rather
exotic and it is very important to understand the origin of the unconventional behavior.

In principle, one can imagine three different formal reasons for the exotic results (the
exponential behavior of the diffusion coefficient, Eq. (12.138), is most unusual): 1) the
approximation scheme is not good, 2) the unusual results are obtained because the model
contains a mixture of commuting and anticommuting variables, 3) the model is not compact.
In fact, it is not difficult to demonstrate that it is the noncompactness that leads to the exotic
behavior. This can be done by making some changes in the initial model.

Let us assume that the new model is described as before by Eq. (12.33) with supermatrices
Q obeying the constraint $Q^2 = 1$. However, this constraint does not specify the symmetry
group unambiguously and one can introduce the new model by making a formal substitution
in the boson–boson block Q_{022} of the central part Q_0 of supermatrix Q, Eq. (6.47), for the
unitary ensemble

$$Q_{022} = \begin{pmatrix} \cosh\theta_1 & -\sinh\theta_1 \\ \sinh\theta_1 & -\cosh\theta_1 \end{pmatrix} \rightarrow \begin{pmatrix} f(z)z & -x \\ x & -f(z)z \end{pmatrix}, \qquad x > 0 \quad (12.157)$$

where the variables z and x are related to each other by the equation

$$f^2(z)z^2 - x^2 = 1 \tag{12.158}$$

If $f(z) \equiv 1$ we return to the initial noncompact model, since then $x = \sinh\theta_1$, $z = \cosh\theta_1$.
To construct a compact model one has to choose the function $f(z)$ in such a way that
Eq. (12.158) describes a closed curve. As $f(z)$ one can take, for example, the function

$$f(z) = \frac{1}{z}\left\{1 + \left(z^2 - 1\right)\left[1 - 2\exp\left(-\frac{R}{z^2 - 1}\right)\right]\right\}^{1/2} \tag{12.159}$$

where R is a number. The curve described by Eq. (12.158) with $f(z)$ from Eq. (12.159) is
represented in Fig. 12.10. In the limit $R \to \infty$ one obtains the initial model corresponding
to the hyperbola in Fig. 12.10. It should be emphasized, however, that the new model has no
physical meaning and is designed formally to demonstrate the role of the noncompactness.

Having changed the original model in this way one can obtain, as before, Eqs. (12.47,
12.51). The representation for N described by Eqs. (11.133, 11.134) can be used again, but
now $\mathbf{n}_1$ is a vector not on a hyperboloid but on the closed surface obtained from the curve,
Eq. (12.159), by rotation about the z-axis. Correspondingly, as the scalar product of the

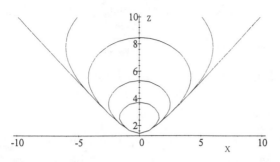

Fig. 12.10. Graphical representation of Eqs. (12.158, 12.159). The curves are drawn for $R = 10, 20, 50, 100$, and 1000. For large R the curves are very close to a hyperbola.

vectors $\mathbf{n}_1$ and $\mathbf{n}_1'$ one should use the expression

$$\mathbf{n}_1 \mathbf{n}_1' = f(z) f(z') zz' - \mathbf{xx}', \qquad \mathbf{n}_1 = (z, \mathbf{x}) \tag{12.160}$$

With Eq. (12.158), the vectors $\mathbf{n}_1$ and $\mathbf{n}_1'$ are now unit vectors in the sense of the scalar product, Eq. (12.160), and $\lambda_1 = f(z) z$. In the weakly disordered regime $\alpha \gg 1$, the main contribution in the integral in Eq. (12.51) is given by values of λ and λ_1 close to unity, and this gives the solution, Eq. (12.135). Moreover, the entire perturbation theory in the limit $\alpha \gg 1$ in the compact model being considered with $f(z)$ from Eq. (12.159) coincides with the perturbation theory for the correct noncompact model that was studied in the preceding sections, since for values of z close to unity the difference between $f(z)$ and unity is exponentially small. However, despite the coincidence of the results obtained within perturbation theory, the critical behavior is entirely different for these two models.

To elucidate the critical behavior it is necessary to find the eigenfunctions of the kernel $N\left(\mathbf{nn}', \mathbf{n}_1 \mathbf{n}_1'\right)$, Eq. (11.134), for the new model. In the previous case the eigenfunctions could be written in the form of the product $P_n(\lambda) P_{-1/2+\epsilon}(\lambda_1)$. As a result of noncompactness the set of eigenvalues was continuous. In the general case, it is also not difficult to construct the eigenfunctions of the changed operator $\tilde{N}$. Naturally, the variables are again separable and the eigenfunctions are written in the form of products $P_n(\lambda) \varphi_l(z)$. The function $\varphi_l(z)$ is a generalized spherical harmonics, corresponding to transformations g that transform the set of unit (in the sense of the scalar product, Eq. (12.160)) vectors into itself, so that $\mathbf{n}_1' = g \mathbf{n}_1$. The procedure of constructing such spherical harmonics by means of representations $T^l(g)$ of the group of g is described by Vilenkin (1986). Any function on a compact set of g can be expanded in a Fourier series in these spherical harmonics (on a noncompact set, a Fourier integral is necessary).

Solutions of Eq. (12.51) can be sought in a form analogous to Eq. (12.135)

$$\Psi = \exp\left((\lambda - \lambda_1) c(\lambda, z)\right) \tag{12.161}$$

with c a function of λ and z. The function c can be expanded in a Fourier series

$$c(\lambda, z) = \sum_{n,l} c_{nl} P_n(\lambda) \varphi_l(z) \tag{12.162}$$

Substituting Eqs. (12.161, 12.162) into Eq. (12.51) one can reduce it to the form of Eq. (12.65), in which one should replace $P_{-1/2+i\epsilon}$ by φ_l. In the critical region, as usual, only the zero harmonics is important (cf. discussion at the end of Section 12.1). The corresponding coefficient c_{00} in Eq. (12.162) is small. Expanding Eq. (12.65) in c_{00} and discarding the other harmonics one obtains

$$a_2 c_{00}^2 + (\alpha_c - \alpha) a_1 c_{00} = a_0 \beta \qquad (12.163)$$

where a_0, a_1, and a_2 are numerical factors of order unity.

Eq. (12.163) coincides in form with the equation for the mean field in standard theories of phase transitions (in spin models, the term c_{00}^2 is absent and it is necessary to write c_{00}^3). We see that the compactification leads immediately to power law behavior near the transition. Eq. (12.163) holds for an arbitrarily large R. In the limit $R \to \infty$ the distance $\Delta\epsilon$ between the eigenvalues goes to zero, but for any finite value of this quantity one can be close enough to the transition (such that the distance to the transition is smaller than $\Delta\epsilon$), and, in this region, one again obtains Eq. (12.163). Only noncompactness can invalidate this equation. At the same time, the toy model considered in the present section was chosen in such a way that the perturbation theory in the limit $\alpha \gg 1$ for this model exactly coincides with the theory for the exact noncompact model (the difference is exponentially small).

So, we have to conclude that the unusual results obtained in the preceding sections are due to noncompactness. Any compactification would change the results. However, the noncompactness of the model is an unavoidable property. In my opinion, the significance of the results obtained in the effective medium approximation is analogous to that of the results obtained within the mean field approximation in the theory of more conventional phase transitions. Of course, it is difficult to guarantee that the results obtained in the preceding sections can be used in three dimensions and one has to consider fluctuations. For that purpose it is very important to determine the upper critical dimensionality. Then one could hope to construct a renormalization group scheme near this dimensionality analogous to the Wilson approach (cf. Wilson and Kogut (1974)). The reformulation of the σ-model in terms of a functional integral in an extended space presented in Appendix 4 can be the starting point of such a theory because, as in the Wilson scheme, in this approach the effective medium approximation is the saddle point of a functional containing the order parameter variable. The only difference is that now the order parameter is a function. This leads to technical difficulties and the problem has not been solved yet.

Returning to the compactified toy model considered previously one may ask about what this model would give for the localization problem. Strictly speaking, the model is not able to describe localization because the most general form of, for example, the density–density correlation Y^{00}, Eq. (3.32), containing the frequency ω in the denominator can be obtained from the noncompact model only. However, one may try to do extrapolations from the region of weak disorder.

In fact, this is the way results are obtained in the renormalization group scheme in $2 + \epsilon$ dimensions or in self-consistent approaches (Götze (1981), Belitz and Götze (1983), Vollhardt and Wölfle (1980, 1992)). Comparing Eqs. (12.135, 12.136, 12.161, 12.162) we see that in the limit of large coupling constants α (weak disorder) the quantity $c(1, 1)$ and, hence, c_{00} is proportional to the diffusion coefficient D. The next step is to assume that this relation holds for an arbitrary disorder; this step is equivalent to replacing $\alpha \, \mathrm{str} \, (\Lambda Q)$ in Eq. (12.135) by $\tilde{\alpha} \, \mathrm{str} \, P^2$, where P is determined by Eq. (5.1) and $\tilde{\alpha}$ is a new parameter (proportional to the diffusion coefficient). This is in the spirit of summing the diagrammatic expansions when one assumes that the result of the summation for a correlation function can be expressed in the form of the bare correlation function but with renormalized parameters. Then, Eq. (12.163) can be interpreted as the equation for the diffusion coefficient. An equation analogous to Eq. (12.163) was obtained by using a self-consistent scheme (Götze (1981), Belitz and Götze (1983)).

At $\alpha > \alpha_c$ and $\beta \to 0$ one obtains from Eq. (12.163)

$$c_{00} \sim \alpha - \alpha_c \qquad (12.164)$$

From Eq. (12.164), one has to conclude that the diffusion coefficient decreases linearly when approaching the transition. The same linear behavior was obtained by Vollhardt and Wölfle (1980) and by Wegner (1985) from the renormalization group in $2 + \epsilon$ dimensions.

For $\alpha < \alpha_c$, one obtains

$$c_{00} \sim \beta \, (\alpha - \alpha_c)^{-1} \qquad (12.165)$$

Eq. (12.165) shows that in this region the diffusion coefficient is proportional to frequency; this dependence corresponds to a finite electric susceptibility. This susceptibility diverges when approaching the transition with the exponent equal to unity.

So, it is not difficult to obtain power law behavior from the effective medium approximation provided artificial compactification is done. This distortion of the original model can be done without changing perturbation theory in the limit of weak disorder. This means that the approximate schemes mentioned earlier might give wrong critical behavior, or, at least, the results obtained with these schemes cannot exclude the applicability of the results of the previous sections (possibly modified by fluctuations) to the description of the Anderson transition in three dimensions.

12.5.2 Physical picture: Quasi-localized states

It is clear from the preceding discussion that noncompactness takes into account some features of the physical system that are neglected in compactified versions of the same σ-model and in perturbational schemes. What are these features?

Although very efficient in explicit computations the supersymmetry technique does not allow the possibility of visualizing the physical picture directly because it provides results only for averaged quantities. In this situation the only way to understand the underlying physics is to hypothesize a picture and then compare results that might be derived qualitatively from it with the corresponding quantitative results obtained with the σ-model.

From the analysis of the preceding sections we know that the characteristic values of the noncompact variable λ_1 in the insulating regime are of the order of the inverse frequency ω^{-1}. Therefore, one may consider λ_1 as the characteristic time the particle spends at a point. In the limit $\omega \to 0$ one obtains $\lambda_1 \to \infty$; that means that the particle is localized at this point. In the metallic region, the characteristic λ_1 are finite but exponentially large near the transition. This can mean that the particle is located at a point surrounded by barriers. The probability of tunneling through the barriers is exponentially small, and this leads to the exponentially large time the particle remains at the point. So, one arrives at the notion of quasi-localized states. It is clear that all the perturbational schemes neglect such states.

Now, I want to suggest a simple intuitive picture that helps one to understand all the results obtained with the effective medium approximation. Let us start the discussion from the localized regime. The simplest and most commonly accepted picture for localized states is that the particle is located in an effective quantum well, which is represented in Fig. 12.11a. If the particle stays in a state with eigenenergy $\mathcal{E}$ and the edge $\mathcal{E}_c$ is identified with the mobility edge, the wave function of this state decays on the length L_c related to the energy

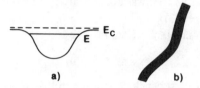

Fig. 12.11. Insulating regime: a) effective potential well; the particle has the energy $\mathcal{E} < \mathcal{E}_c$. b) A schematic representation of the space dependence of the amplitude of the wave function; the black area denotes the region where the amplitude is high.

$\mathcal{E}$ by Eq. (12.107). This elementary quantum mechanical result agrees perfectly with the result for the localization length, Eq. (12.105), and the agreement is certainly not accidental.

If the energy $\mathcal{E}$ exceeds $\mathcal{E}_c$, one gets into the continuous spectrum of extended states. This is again the standard picture. However, now I want to make a new assumption. I assume that the particle becomes delocalized in *one direction* only. This means that there are barriers at the edges of the well in all directions except this one. The well has this form already for $\mathcal{E} < \mathcal{E}_c$ because the wave function has to decay slowly in one direction only. Denoting high values of the amplitude of the wave function by the color black one may draw a typical configuration of the density represented in Fig. 12.11b. The black area has the form of a sausage, and the length of the sausage grows when approaching the mobility edge. Localization length L_c must diverge at this point according to Eq. (12.107), but this divergence is along one direction only. So, at the transition one gets an infinitely long tube. In principle, at the mobility edge, the amplitude may decay in a power law along this direction.

How can one confirm this hypothesis? Of course, the form of the function $p_\infty (r)$, Eq. (12.106), alone is not sufficient to extract this picture because, even if the wave function extends in one direction, it should be averaged over all directions to get $p_\infty (r)$. So, I just assume the one-dimensional shape, hoping that in the future the hypothesis will be confirmed by a more sophisticated analytical or numerical study.

What happens when one crosses the transition? The next assumption is that many tubes with finite density appear. These tubes bend in the space in a complicated way, but what is important is that they do not touch each other. It is natural to think that the transition should be in many respects continuous, and so, the density has to vanish when approaching the transition from the metallic side. This gives correlation length ζ, which is a characteristic distance between the tubes. This length has to diverge at the transition, but I do not know how to determine its dependence on $\mathcal{E}$ near $\mathcal{E}_c$. It may obey, for example, Eq. (12.142). The wave functions are extended along the tubes, but the density of states is very inhomogeneous because of the decay of the wave functions in directions perpendicular to the tubes. This density is represented in Fig. 12.12. The strong fluctuations of the density of states can partially average only at distances $r > \zeta$. This agrees with the first term in Eq. (12.153) for the correlation function of the density of states.

Although the wave functions are extended along the tubes, that does not trivially lead to macroscopic conductivity because the tubes are assumed not to touch each other. At the same time, we are considering a quantum problem and tunneling between the tubes is always possible. Using the fact that the tubes are separated on average by correlation

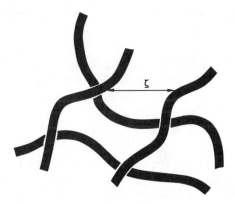

Fig. 12.12. Probability amplitude of an eigenstate in a metallic regime.

length ζ we may conclude that the overlap of the wave functions of different tubes yields a conduction band with the width

$$\Gamma \propto \exp(-\zeta) \qquad (12.166)$$

Eq. (12.166) agrees with Eqs. (12.139, 12.142, 12.82). It is clear that if the frequency exceeds the bandwidth, the difference between the band and a discrete level is negligible. At such a frequency one does not distinguish between metal and insulator, and this property is in accordance with the analogous property for the solution obtained in the preceding sections. So, in the preceding picture, the exponential behavior of the characteristic energy Γ is due to the tunneling through barriers, and the width ζ of the barriers obeys a power law. Of course, the diffusion coefficient D must also be exponentially small, and this corresponds to Eq. (12.154). However, this equation contains another length $\xi > \zeta$, Eq. (12.152), which also enters the second term in Eq. (12.153). What is the meaning of this length?

In fact, this is a phase coherence length. Its physical meaning can be clarified from the following arguments: Let us consider two points at a distance r from each other. Suppose a particle moves diffusively from the first point to the second along a path of length x. The phase difference φ between these two points can be estimated as

$$\varphi = k(\mathcal{E})\, x = \mathcal{E}t$$

where $k(\mathcal{E})$ and $\mathcal{E}$ are the wave vector and the energy $\mathcal{E}$, respectively, and t is the time the particle travels between the points. Eq. (12.167) is written under the assumption that the particle is scattered elastically by impurities. Particles with different energies acquire different phases. The maximal phase difference $\Delta\varphi$ for the particles in the band of width Γ is described then by the following relation:

$$\Delta\varphi = \Gamma t \qquad (12.167)$$

Since the particle moves diffusively, the time t is related to the distance r as $t \simeq r^2/D$. Substituting this relation into Eq. (12.167) we obtain

$$\Delta\varphi \simeq \Gamma r^2/D \qquad (12.168)$$

From Eq. (12.168) one can easily obtain that the characteristic scale ξ on which the phase difference $\Delta\varphi$ becomes of the order of unity is determined by the following relation

$$\xi \sim (D/\Gamma)^{1/2} \tag{12.169}$$

It is clear from Eq. (12.169) that the coherence length ξ is just the well-known Thouless length (Thouless (1977)). The correlation function of the density of states $S(\mathbf{r}, 0)$ defined by Eqs. (12.93, 12.95) vanishes as soon as the coherence is lost. Therefore, length ξ, Eq. (12.169), should be identified with the second length in Eq. (12.153). We see also that Eq. (12.169) is consistent with Eqs. (12.154, 12.166); Eq. (12.152) cannot be obtained from such simple arguments.

Thus, the results obtained in the effective medium approximation can be explained with the preceding physical picture. I want to emphasize once more that the main ingredient of the qualitative description is the existence of quasi-localized states in the metallic regime. The noncompactness of the σ-model is very important because it apparently takes these states into account. Using a perturbative scheme one neglects the noncompactness and, as a result, the quasi-localized states. Although such schemes can work in the limit of weak disorder they fail to describe the metal–insulator transition.

It is worth mentioning that some "confluence states" that are a mixture of localized and extended states have been introduced by Srivastava (1990) using completely different arguments. Possibly, they are related to the quasi-localized states discussed. In fact, the quasi-localized states exist at arbitrarily weak disorder and can be studied by using the σ-model (Altshuler, Kravtsov, and Lerner (1991), Muzykantskii and Khmelnitskii (1995), Falko and Efetov (1995)). This question will be discussed in the next chapter.

The scaling picture of Abrahams et al. (1979) might be modified by taking into account the idea of strong fluctuations. We understand that as a result of the fluctuations the notion of conductance is not well defined for a random system. Although in the limit of weak disorder the variance is much smaller than the conductance itself, this inequality is no longer valid near the metal–insulator transition. Apparently, in the critical region one has to scale not a conductance but an entire distribution function. This possibility was discussed by Altshuler, Kravtsov, and Lerner (1986, 1989) and by Shapiro (1986).

12.5.3 Effective medium approximation and reality

As has been mentioned, the effective medium approximation is definitely correct in high dimensionality. How do the results look in real three-dimensional space? Unfortunately, the solution of the problem does not yet exist and one can only try to guess the correct formulae using the qualitative picture suggested. The density–density correlation function Y^{00} in the insulating regime, Eqs. (12.101, 12.106), must remain exponential although the exponent of the preexponential power law function may change. The exponent of localization length L_c in Eq. (12.105) may also change.

The same can happen with lengths ζ and ξ in the metallic region. They can be characterized by some exponents ν and μ different from $1/2$ and $3/4$. At the same time, the existence of two lengths seems to be quite probable. Most difficult is the question about the critical behavior of the diffusion coefficient D. Its exponential decay, Eq. (12.138), can occur if the tubes do not intersect each other. It is not difficult to imagine this possibility in a high dimensionality, but what about real three-dimensional space?

Two scenarios are possible. If these tubes do not intersect or, at least, do not form connected infinite clusters, the conductivity is possible only as a result of tunneling and one can obtain Eq. (12.154). However, if they can contact each other and electrons can pass from one tube to another through the junctions, diffusion coefficient D should be proportional to a combination of lengths ζ and ξ. This would lead to a power law decay of D. Nevertheless, even in this case the results obtained by the conventional renormalization group technique seem unlikely to be applicable because that would imply, at least, the confluence of the lengths ζ and ξ.

In the renormalization group scheme the metallic region is characterized near the transition by only one correlation length ξ (cf. Chapter 5) that diverges near the transition in a power law, and the exponent ν is the same as the exponent for localization length L_c. Conductivity σ (and diffusion coefficient D) in this theory are related to length ξ by the simple relation $\sigma \propto \xi^{2-d}$. From this relation one would obtain

$$\sigma \propto (\mathcal{E} - \mathcal{E}_c)^{\nu(d-2)} \tag{12.170}$$

Then, according to folklore, the exponential behavior of the diffusion coefficient D, Eq. (12.138), corresponds to the limit $d \to \infty$ of Eq. (12.170) at a finite ν. In the effective medium approximation $\nu = 1/2$, and so, the exponential behavior is "explained" and it is announced that the "upper critical dimensionality" for the Anderson transition is infinity. I hope that the reader who has read the present chapter can consider this type of argument a joke. It is an attempt to explain a really new type of phase transition in terms of conventional second-order phase transitions. Of course, by making some parameters diverging one can in this way "explain" everything.

At the same time, the exponential decay of D, Eq. (12.138), has been observed neither in numerical simulations nor in real experiments on semiconductors, and this is used as an argument against the scenario that follows from the effective medium approximation. I emphasize again that I do not claim that all the results obtained in this approximation can be directly applied to 3D systems. However, even if they were valid in 3D, would one be able to observe the exponentially decaying diffusion coefficient? I am very pessimistic about that.

In the commonly used scaling method of MacKinnon and Kramer (1981) one studies the scaling of a length. As concerns the localized regime, a localization length that diverges near the transition in a power law is found (for a review see Kramer and MacKinnon (1993)). Although the exponent is not equal to $1/2$, this result does not contradict the effective medium approximation. In the metallic regime, one again studies numerically a *length*, which is found to obey a power law. The diffusion coefficient is written then by using the conventional scaling relations, but in the effective medium approximation, one does obtain the length ζ that obeys the power law dependence, Eq. (12.142). So, there is no contradiction. As concerns the second length ξ, Eq. (12.152), it can hardly be seen in the rather small systems that can be studied with the present computer power.

The intuitive picture developed in the present section can work near the transition in large samples only. In a finite sample, it is always possible that a finite fraction of all tubes just passes through the sample. Then, the main contribution to the conductance may be due to transport along the tubes. Only in very large samples should the tunneling mechanism prevail. This would make direct observation of the exponential behavior through computers very difficult. However, the existence of the tubes can be extracted from numerical study of wave functions. If the wave functions are concentrated in the tubes, this means that the states are not space filling and one can observe a fractal dimension.

In fact, a fractal dimension of eigenstates has been found in many numerical simulations for disordered systems near metal–insulator transitions. The idea of a fractal wave function with a filamentary structure like a net over the whole sample at the mobility edge was suggested originally by Aoki (1983). This idea was numerically exploited by many authors, determining a fractal dimension from the density–density correlation function (Soukoulis and Economou (1984)), the participation number (Schreiber (1985, 1990), Kramer, Ono, and Ohtsuki (1988), Ono, Ohtsuki, and Kramer (1989)), or the amplitude of the wave function (Aoki (1986)). Further investigation (Schreiber and Grussbach (1991), Grussbach and Schreiber (1992)) showed that the wave functions are even more complicated entities, and a more general concept of multifractality (Halsey et al. (1986), Hentschel and Procaccia (1983), Feder (1988)) has to be employed, yielding a set of generalized dimensions. An introductory review on the multifractality of eigenstates can be found in Janßen et al. (1994).

We return to multifractality in the next chapter. Now, I want to emphasize only that the picture of the strongly fluctuating amplitudes of the wave function in the vicinity of the metal–insulator transition with a formation of tubes that was suggested for an interpretation of the results of the present chapter may correlate well with what is known from the numerical computations. Another numerical finding that contradicts the analysis in $2 + \epsilon$ dimensions but correlates with the effective medium approximation is that the critical behavior is not sensitive to breaking of the time-reversal symmetry (Ohtsuki, Kramer, and Ono (1993), Henneke, Kramer, and Ohtsuki (1994)).

As concerns experiments on semiconductors the situation seems to be rather intricate. First, Coulomb interaction can be important, and in, principle, it can completely change the critical behavior (for a review, see Belitz and Kirkpatrick (1994)). There is a large amount of work devoted to the study of metal–insulator transitions (see, e.g., Thomas (1986), Katsumoto (1991)), but I do not want to give a review here. Concerning some recent work it is worth mentioning that in some cases the critical behavior seems to be determined by electron correlations rather than by disorder (Bogdanovich et al. (1995)). In other cases the electron–electron interaction is believed to be less important (Hornung et al. (1994)). However, in my opinion, the most difficult problem is that the measurements are done at temperatures that are not very low. The zero temperature results are extracted, then, by interpolation. Usually, the temperature-dependent conductivity $\sigma(T)$ can be well described by the formula

$$\sigma(T) = \sigma(0) + m\sqrt{T} \tag{12.171}$$

where m is a coefficient. The square root dependence on temperature is attributed to the electron–electron interaction because in the weak disorder limit the interactions give a correction to the classical conductivity that obeys this law (see, e.g., Altshuler et al. (1980), Lee and Ramakrishnan (1985)). However, nobody knows how the conductivity should depend on temperature in the critical region near the transition. One cannot exclude the possibility that the temperature dependence given by Eq. (12.171) would be governed by another law if one were able to come closer to the transition. It is clear from the data by Hornung et al. (1994) that the best fit on the theoretical curve, Eq. (12.171), is found very close to the transition. But it may be true that approaching the transition one needs lower and lower temperatures to study the genuine critical regime.

Suppose that the scenario derived from effective medium approximation is applicable to experimental systems. Then, it follows that in order to get into the critical region, one must have a temperature lower than Γ, Eq. (12.166). So, close to the transition the temperature must be exponentially low! This is hardly possible in existing experiments. If temperature is not low enough, it may turn out that the quantum problem of the Anderson metal–insulator transition is replaced by a classical percolation problem. The exponent for the conductivity extracted by Hornung et al. (1994) from the measurements is approximately equal to 1.3, which is rather close to the classical value 1.5 (Kirkpatrick (1973)). This may signal in favor of the classical picture. The quantum limit can then be achieved only at much lower temperatures.

Thus, although no evidence in favor of the exponential decay of the diffusion coefficient has been seen in either numerical computations or real experiments, in my opinion, it could not be seen. So, I believe that the scenario of the effective medium approximation can be a good candidate for description of real 3D systems. Of course, it may give only a rough picture that has to be somehow modified, but one should keep the picture in mind when thinking of the Anderson metal–insulator transition.

12.6 Bethe lattice, highly conducting polymers, sparse matrices

12.6.1 Models on the Bethe lattice

Although it is not clear to what extent the effective medium approximation can be used to describe the Anderson metal–insulator transition in a 3D disordered system, this approximation becomes exact on the Bethe lattice. The treelike structure of this lattice is represented in Fig. 12.6. Denoting by $\Psi(Q)$ the partition function of a branch of the tree with a fixed supermatrix Q at the base of the branch and comparing this partition function with the one calculated at the neighboring site one immediately obtains the recurrence equation, Eqs. (12.47, 12.48). By its construction, the effective medium approximation includes the contribution of loopless graphs only. The Bethe lattice itself does not have any loops, and that is why Eqs. (12.47, 12.48) are exact in such a model.

At the same time, the correlation functions Y^{00} and S have a form that is somewhat different from that given by Eqs. (12.92, 12.106). The exponents of the characteristic lengths are two times larger than those given by Eqs. (12.105, 12.142, 12.152). Corresponding formulae can be found in Efetov (1987a,b). This difference has a purely geometrical origin. For example, the average distance the electron can diffuse in a time t on the Bethe lattice is proportional to t and not to $t^{1/2}$ as on usual lattices.

The σ-model on the Bethe lattice is rather artificial because it corresponds to a system of granules arranged in a treelike structure. Within the σ-model approach the average density of states is excluded from consideration, as manifested by the constraint $Q^2 = 1$. Is this constraint important for the critical behavior? In Chapter 11 it was shown that the correlation functions for a disordered wire in the limit of low frequency have the same behavior as the corresponding quantities for one-channel chains.

It turns out that the same equivalence holds for the Bethe lattice. One may start, for example, from the Anderson tight-binding model described by the Hamiltonian

$$\hat{\mathcal{H}} = \sum_i \epsilon_i c_i^+ c_i + \sum_{ij} t_{ij} c_i^+ c_j, \qquad t_{ij} = t_{ij}^* \tag{12.172}$$

where the site energies ϵ_i or the hopping matrix elements t_{ij} are assumed to be random variables. The σ-model can be derived from the model given by Eq. (12.172) if one attributes to each site a large number N of orbitals (Schäfer and Wegner (1980)), whereas in the original Anderson model $N = 1$.

The Bethe lattice version of the Anderson model was studied for the first time by Abou-Chacra, Anderson, and Thouless (1973), who proved the existence of the metal–insulator transition and found the position of the mobility edge. These analytical results were confirmed by numerical Monte Carlo calculations (Abou-Chacra and Thouless (1974), Girvin and Johnson (1980)). Moreover, Kunz and Suillard (1980, 1983) showed that some correlation length (whose physical meaning is not sufficiently clear) displayed a power law critical behavior within the region of localized states.

Recently, using the supersymmetry approach, Mirlin and Fyodorov (1991) managed to solve the Anderson model on the Bethe lattice and determine the critical behavior. They started with the effective Lagrangian $\mathcal{L}$, Eq. (4.14), with the Hamiltonian $\mathcal{H}_0$ standing for nearest-neighbor hopping. In principle, the distribution of the site energies does not need to be Gaussian and one may use an arbitrary function $\gamma(u)$. A recurrence equation analogous to Eq. (12.47) can be written almost immediately. Denoting the partition function of a branch of the tree by $G(\psi)$ ($G(\psi)$ is the analogue of the function Ψ in Eq. (12.47)) and putting $t_{ij} = 1$ one obtains

$$G(\phi) = \int D\psi \int du\, \gamma(u) \exp\left[-i\bar{\phi}\psi + \bar{\psi}\left(\tilde{\varepsilon} - \frac{1}{2}\Lambda\omega - u\right)\psi\right] G^m(\psi) \quad (12.173)$$

where the supervector ϕ has the same structure as ψ and $\tilde{\varepsilon} = \varepsilon - \omega/2$.

Again, the symmetry of the exponent in Eq. (12.173) with respect to rotations of the supervectors ψ and ϕ drastically simplifies the solution of the integral equation. Because of this symmetry one can seek its solution in the form

$$G(\psi) = g\left(\bar{\psi}\psi,\ \bar{\psi}\Lambda\psi\right) \quad (12.174)$$

By substituting Eq. (12.174) into Eq. (12.173) and integrating over the anticommuting components of the supervector ψ one can obtain an integral equation containing conventional integrations only. The procedure of solving the equation derived in this way is analogous to that used in the preceding sections. One can determine the transition point and find correlation functions in both the metallic and localized regimes. The solution $g(x,\ y)$ plays, as before, the role of a functional order parameter. For the model under consideration Mirlin and Fyodorov (1991) identify this function with a joint probability distribution function of the real and imaginary parts of the on-site Green function. The critical behavior near the transition point proves to be exactly the same as that obtained for the σ-model. In particular, in the metallic region the diffusion coefficient decays exponentially as in Eq. (12.138). The agreement of the results shows that the critical behavior is universal although the location of the transition point is model-dependent. This is a common phenomenon in the theory of phase transitions.

So, there are theoretical models in which the critical behavior near the metal–insulator transition is precisely known. At first glance, the models do not seem to be sufficiently realistic and one may argue that the results obtained on the Bethe lattice are purely academic. Fortunately, it is not so, and some interesting physical objects may be described well by models on lattices with loopless structure.

12.6.2 Highly conducting polymers

Among the examples of systems with loopless structure are highly conducting polymers. Recently, such polymers as polyacetylene, polyaniline, and polypyrrole have attracted

Fig. 12.13. Schematic view of the fibril structure of polymer. The rings indicate the interfibril cross-links.

considerable interest in both applied and fundamental research (Heeger et al. (1988)). Their common feature is that the conductivity can be increased by a few orders of magnitude upon doping. In the heavily doped Tsukamoto polyacetylene the room temperature conductivity (σ_{RT}) has already reached that of copper. However, in spite of the large σ_{RT}, the transport properties of these conducting polymers are still far from being traditionally metallic.

First of all, in contrast to that of metals, the conductivity of the polymers decreases with lowering of the temperature. Depending on σ_{RT} (i.e., on the level of doping and the degree of disorder) this decay varies from activation-type behavior to weak logarithmic behavior. For the most highly conducting samples the conductivity even approaches the residual value at low temperatures (Javadi et al. (1991), Nogami et al. (1991)). At the same time, their thermoelectric power and Pauli susceptibility suggest a metallic density of states at the Fermi level in the whole temperature interval. It is noteworthy also that at low temperatures there is a significant magnetoresistance and its sign correlates with the preceding temperature dependence of conductivity, being negative for highly conducting samples.

On the basis of these observations, Javadi et al. (1991), Nogami et al. (1991), and Ishiguro et al. (1992) suggested that the highly conducting polymers were close to a metal–insulator transition driven by disorder. Since their σ_{RT} greatly exceeds those of all known systems near the metal–insulator boundary, one can conclude that the highly conducting polymers exhibit an unusual type of transition.

For an explanation of the unusual transport properties of the polymers their chain nature seems to be very important. Electrons move primarily along polymer chains over large distances without scattering, hopping from time to time from chain to chain. At the same time, the chains do not form a regular array. The most well-known polymer, polyacetylene (Heeger et al. (1988)), has a so-called fibril structure: Single chains are coupled into fibrils that occupy a part of the entire volume. The fibrils bend in space in a very complicated way. They come in contact with each other, forming a random cross-linked network. The fibril structure is schematically illustrated in Fig. 12.13.

At first glance, the "spaghetti" structure makes the evaluation of the transport properties for polymers extremely difficult. However, a model can be proposed (Prigodin and Efetov (1993)) that not only takes into account the main features of the system under consideration but also can be solved exactly. To define the model one assumes that each fibril is a weakly disordered metallic wire and that the cross-links between the fibrils can be described by interwire junctions. As a result, one has a network of randomly coupled metallic wires. In the absence of the junctions all the electronic states of the wires are localized by any weak disorder (cf. Chapter 11). The states become extended over the whole network only if the interwire coupling is strong enough. Within the present model one can determine the position of the metal–insulator transition and the critical behavior as functions of the intrinsic disorder and interwire coupling.

The existence of the delocalized phase in the network is a nontrivial phenomenon because, as a result of its irregularity, the random network might, in principle, remain always insulating. Whether the electronic states are localized or extended can be seen in conductivity at low temperature only. At high temperature, because of inelastic scattering, electrons are delocalized and the network always exhibits metallic conductivity even in the absence of cross-links. This property of the model can reconcile the observable discrepancy between the high- and low-temperature conductivity of polymers. It is important also to note that thermodynamic characteristics of the network always correspond to a metallic density of states at the Fermi energy in both the metallic and insulating phases.

Let us specify the model under consideration. A network of weakly disordered metallic wires like that represented in Fig. 12.13 is studied. Two apparent spatial parameters can be introduced to characterize the network: the concentration of the junctions per unit length along the wire n_l and the concentration of wires per unit area n_2. The ratio n_2/n_l^2 plays the role of the mean number of "neighbors" in the network. Indeed, a segment of a wire of length $1/n_l$ is surrounded by approximately n_2/n_l^2 other wires, one of which is touched by the wire over this length. In the limit

$$\frac{n_2}{n_l^2} \gg 1 \tag{12.175}$$

the probability that any two wires contact each other more than once is small. In other words, because of this inequality the statistical weight of closed paths can be negligible. What we get in this limit is reminiscent of the Bethe lattice and one can try to write recurrence equations.

The fibrils of the network are well specified by the density of states at the Fermi level per unit of wire length $\tilde{\nu}$, Eq. (11.2), and the classical diffusion coefficient along the wire D_0. In terms of these parameters the localization length L_c in the wire is determined by Eq. (11.65). The junctions between the wires can be described by pointlike contacts with the amplitude of electron transfer integral T_J. The dimensionless parameter characterizing the intensity of the interwire transitions at the contact is the Born cross section of the "capture" by the contact

$$\alpha_J = (\pi T_J \tilde{\nu})^2 \tag{12.176}$$

To complete the definition of the model let us assume that the contacts are randomly distributed over the wires with a low linear concentration n_l such that the following inequality holds:

$$l_J \gg l, \qquad l_J \simeq \max\left(n_l^{-1}, (\alpha_J n_l)^{-1}\right) \tag{12.177}$$

where l_J is the characteristic length of scattering by the junctions. In this limit the localization is mainly caused by the intrawire scattering and the contribution of the disorder due to the random cross-links to the localization is negligible. The principle effect of the presence of the cross-links is the extension of the localized wave functions over an increasingly large number of wires. So, the delocalization transition is expected at some critical concentration n_{lc}. Provided the number of chains in each fibril is large the transition occurs in the region determined by the inequalities (12.177).

The two types of disorder in the system should be essentially distinguished. As concerns irregularities inside the wires, one can average over them by using the supersymmetry method. An isolated wire j is thus described by the free energy functional $F[Q_j]$, Eq. (11.1). The electron hops between wires i and j lead to Josephson-type terms F_J in the total functional

$$F_J[Q_i, Q_j] = -\frac{\alpha_J}{4} \, \mathrm{str} \, Q_i(x_i) \, Q_j(x_j) \tag{12.178}$$

where x_i and x_j are the cross-link coordinates along the ith and jth wires. The density–density correlation function $Y_j^{00}(x, \omega)$ for the jth wire of the network involved reads

$$Y_j^{00}(x, \omega) = -2(\pi \tilde{\nu})^2 \int (Q_j(0))_{33}^{12} (Q_j(x))_{33}^{21} \exp(-F[Q]) \prod_i DQ_i \tag{12.179}$$

(The coupling between the wires is assumed to be weak and so we can really speak in terms of coordinates along single wires.) The observable one-wire correlation function can be obtained from Eq. (12.179) after averaging $Y_j^{00}(x, \omega)$ over all realizations of the random network

$$Y^{00}(x, \omega) = \left\langle Y_j^{00}(x, \omega) \right\rangle_J \tag{12.180}$$

The absence of closed paths in the network drastically simplifies the computation of the functional integral in Eq. (12.179) because one can use the transfer-matrix technique. In the limit of weak interwire contacts generalization of Eq. (11.37) is straightforward and after averaging over positions of the junctions one obtains

$$\mathcal{H}_0 \Psi(Q) \tag{12.181}$$

$$= n_l \Psi(Q) \int \left[\exp\left(\frac{\alpha_J}{4} \, \mathrm{str} \, (QQ' + QQ'')\right) - 1 \right] \Psi(Q') \, \Psi(Q'') \, dQ' dQ''$$

where the "Hamiltonian" $\mathcal{H}_0$ is determined by Eq. (11.38). Although appearing somewhat more complicated, Eq. (12.181) is analogous to Eq. (12.47) obtained in the effective medium approximation or on the Bethe lattice. I want to emphasize that no uncontrolled approximations were used when deriving Eq. (12.181) for the random network of fibrils, and this is possible because of the absence of closed paths. I would like to draw attention to an interesting similarity between the network represented in Fig. 12.13 and the configuration of the probability density, Fig. 12.12, conjectured to describe the vicinity of the metal–insulator transition. It may indicate once more that the physical picture suggested in Section 12.5 is close to reality.

Corresponding transfer-matrix equations for function $Y^{00}(x, \omega)$ can be derived in the same way as Eq. (12.181). All these equations can be solved by using the scheme developed in the present chapter. At low concentration n_l of the cross-links the right-hand side of Eq. (12.181) can be neglected and one obtains Eqs. (11.70, 11.71). In this limit one has a system of decoupled fibrils and this is the insulating phase. Generally, the density–density correlation function $Y^{00}(x, \omega)$ in the insulating regime is written for an arbitrary α_J as

$$Y^{00}(x, \omega \to 0) \propto \frac{1}{-i\omega} x^{-3/2} \exp\left[-\frac{x}{4L_c(\alpha_J)} \right] \tag{12.182}$$

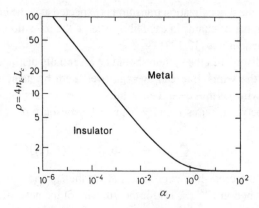

Fig. 12.14. Phase diagram of the random network of fibrils. (From Prigodin and Efetov (1993).)

where $L_c\,(\alpha_J)$ is an effective localization length, which is dependent on the parameter α_J. If the concentration of the links is large enough, one enters the conducting phase. The critical concentration $n_{lc}\,(\alpha_J)$ at which the metal–insulator transition occurs is given by the equation (Prigodin and Efetov (1993))

$$[4\alpha_J n_{lc}\,(\alpha_J)\,L_c]^{-1} = K_0\,(\alpha_J)\left(\overleftarrow{\partial}_{\alpha_J} - \overrightarrow{\partial}_{\alpha_J}\right) I_{1/2}\,(\alpha_J) \qquad (12.183)$$

where K_0 and $I_{1/2}$ are the Bessel functions (cf. Eqs. (12.62, 12.66)). Eq. (12.183) is written for the unitary ensemble corresponding to the system in a strong magnetic field. The metal–insulator boundary obtained from Eq. (12.183) is shown in Fig. 12.14.

The dependence of $n_{lc}\,(\alpha_J)$ on α_J at values of this parameter is rather weak. At $\alpha_J \gg 1$ the critical concentration $n_{lc}\,(\alpha_J)$ approaches the value

$$n_{lc} = (4L_c)^{-1} \qquad (12.184)$$

This result can be interpreted in the following way: The strong interwire coupling leads to the appearance of mixed states located on two interacting wires. The energy of these states is randomly spread around the Fermi energy within the interval $\Delta\mathcal{E}$

$$\Delta\mathcal{E} = \omega_0 = \frac{4D_0}{L_c^2} \qquad (12.185)$$

Being near the links these states are also randomly distributed over the network. The typical overlap between them equals

$$\delta\mathcal{E} = \omega_0 \exp\left(-\frac{2}{n_l L_c}\right) \qquad (12.186)$$

because the states are exponentially localized within the wire. According to the arguments of Thouless (1977) the metal–insulator transition occurs at $\delta\mathcal{E} \approx \Delta\mathcal{E}$, i.e., at $n_l L_c \approx 1$, in agreement with Eq. (12.184). In the limit of weak coupling, $\alpha_J \ll 1$, Eq. (12.183) yields

$$n_{lc} = (4L_c)^{-1}\left(\frac{2\pi}{\alpha_J}\right)^{1/2}\left[2 - C + \ln\left(\frac{2}{\alpha_J}\right)\right]^{-1} \qquad (12.187)$$

where $C = 0.577$ is Euler's constant. Eq. (12.187) can be approximately rewritten as $n_{lc}L_c \approx 1/\sqrt{\alpha_J}$. The product $n_{lc}L_c$ gives the number $m \gg 1$ of states with which a localized state interacts. On the other hand, the dimensionless parameter α_J characterizing the interwire coupling can also be written as $\alpha_J \sim (\tilde{T}_J / \Delta)^2 \ll 1$, where $\tilde{T}_J$ is the overlap integral and Δ is the energy spacing between localized states. Thus, the critical concentration of links n_{lc} again corresponds to the case when the energy separation between the interacting states Δ / m becomes comparable with their overlap $\tilde{T}_J$. The parameter dependence of the transition point in this limit is similar to Eq. (12.67) derived in the effective medium approximation in the limit $m \gg 1$.

The behavior described by Eqs. (12.182) is valid in the entire region of the insulating phase. The localization length $L_c (\alpha_J)$ has a weak dependence on α_J and does not diverge at the critical point. This property is specific for the model and quantity under consideration (the function Y^{00} determines the correlation along one wire, which does not directly give a macroscopic dielectric permeability).

In the metallic region near the transition, the function $Y^{00} (x, \omega)$ takes the form

$$Y^{00} (x, \omega) \propto \exp \left[\frac{2\pi}{\delta} - \frac{x}{4 L_c} \right], \qquad \delta^2 = \left[\frac{n_l}{n_{lc}} - 1 \right] \ll 1 \qquad (12.188)$$

The dynamics of electron propagation over the system is more explicitly described by the function

$$P (k, \omega) = \int Y^{00} (x, \omega) \exp (ikx) \, dx$$

At zero momentum k along the wire this function takes the form

$$P (0, \omega) = [-i\omega + W]^{-1} \qquad (12.189)$$

Here W is the frequency of interwire hopping and, correspondingly, W^{-1} is the time the electron spends in one wire. Eqs. (12.188, 12.189) show that near the transition the frequency W decreases exponentially

$$W = W_0 \exp \left[-\frac{2\pi}{\delta} \right] \qquad (12.190)$$

To understand the meaning of the parameter W_0 one should consider the deeply metallic region, $n_l \gg n_{lc} (\alpha_J)$, where the function $P (0, \omega)$ can be written as

$$P (k, \omega) = \left[-i\omega + D_0 k^2 + W_0 \right]^{-1} \qquad (12.191)$$

Eq. (12.191) is valid as long as ω or W_0 is larger than ω_0, Eq. (12.185). At $\alpha_J \gg 1$, W_0 is the inverse mean time of the classical diffusion between the neighboring links such that

$$W_0 = n_l^2 D_0$$

The condition $W_0 \gg \omega_0$ is equivalent to the requirement $n_l \gg n_{lc} \sim 1 / L_c$. In the limit of weak coupling, $\alpha_J \ll 1$, one can obtain the expression $W_0 = \pi \tilde{\nu} T_J^2 n_l$, which holds at least at $n_l L_c \gg 1 / \alpha_J$.

Although Eq. (12.183) and subsequent formulae are written for the unitary ensemble, all calculations can be repeated for the orthogonal ensemble, which is applicable in the absence of a magnetic field. We know that the localization length of a wire without a magnetic field is two times smaller than in a strong magnetic field (cf. Eq. (11.66)). The right-hand side of Eq. (12.183) also changes, and one can show that applying a magnetic field results in a shift of the transition to a lower concentration of cross-links. Because of a random orientation of wires with respect to the field direction the region of crossover magnetic field H_c is very broad

$$H_c = \phi_0 S^{-1/2} \left(L_c^{-1} \div S^{-1/2} \right) \qquad (12.192)$$

where ϕ_0 is the flux quantum and $S \ll L_c^2$ is the cross section of the wire.

A metal–insulator transition in the intrafibril disorder produced by aging is observed in heavily doped polyacetylene and polypyrrole (Ishiguro et al. (1992)). In the model considered in the present section such disorder is incorporated into L_c. Being inversely proportional to the concentration of the internal defects, L_c decreases with disorder and at the critical disorder determined by Eq. (12.183) the system becomes macroscopically dielectric. According to the phase diagram in Fig. 12.14 the conducting region is more sensitive to the value $n_l L_c$ than to the strength of the interwire coupling α_J and can be broadened by an increase of $n_l L_c$ more readily. Experimentally, this tendency is seen in an appreciable enhancement of the low-temperature conductivity without signs of saturation under application of the magnetic field (Javadi et al. (1991), Nogami et al. (1991)), pressure (Andersson et al. (1992)), and stretching (Ahmed et al. (1992)). The fact that the conductivity is controlled by the cross-linking was explicitly demonstrated by Mac-Diarmid et al. (1992).

Thus, the model considered in the present section can describe the metal–insulator transition in highly conducting polymers quite well. Some additional work is necessary to calculate the macroscopic conductivity and dielectric susceptibility.

12.6.3 Sparse random matrices

One more interesting application of the theory of metal–insulator transition developed in the present chapter was found by Mirlin and Fyodorov (1991a) and Fyodorov and Mirlin (1991a), who studied a transition in ensembles of sparse matrices. We have discussed properties of completely random matrices in Chapter 6. It was shown that the ensembles of the random matrices can be well described by the 0D supermatrix σ-model. Ensembles of banded random matrices were introduced to describe such physical objects as the quantum kicked rotator. The latter systems are equivalent to the 1D σ-model; this mapping was established in Chapter 11. Now we encounter sparse random matrices. Where do these mathematical objects come from?

Interest in studying the ensembles of sparse matrices originates from the theory of spin glasses, in which models with a long-range exchange interaction J_{ij} are very often used. The coupling constants J_{ij} can be considered as random elements of a symmetric matrix. The system can be diluted; this is modeled by making some part of all J_{ij} zero and one thus obtains sparse matrices. Spin models of this type were considered by Viana and Bray (1985), Kanter and Sompolinsky (1987), and Mezard and Parisi (1987). Sparse matrices are also relevant to some combinatorial optimization problems (Mezard and Parisi (1985), Fu and Anderson (1986), Wong and Sherrington (1987)).

The applications of random sparse matrices motivated Rodgers and Bray (1988) to start studying the properties of these mathematical objects. Using the replica trick they managed to compute the density of states. Although the trick helps to obtain the density of states it does not work for more complicated correlation functions. Fortunately, by using super-symmetry technique one can compute the correlation functions of interest and demonstrate that depending on the "average connectivity" p, that is, on the mean number of nonzero elements per row, one can have a sharp transition from the Wigner–Dyson distribution of eigenvalues at large p to the uncorrelated Poisson distribution at small p.

The distribution function $D(\mathcal{H}_{mn})$ for the orthogonal ensemble of the sparse matrices $\mathcal{H}$ ($\mathcal{H}_{mn} = \mathcal{H}_{nm}$) is introduced as

$$D(\mathcal{H}_{mn}) = \left(1 - \frac{p}{N}\right)\delta(\mathcal{H}_{mn}) + \frac{p}{N}h(\mathcal{H}_{mn}) \tag{12.193}$$

where $h(u) = h(-u)$ is any even distribution function nonsingular at $u = 0$ and normalized to unity, and N is the size of the matrices. It is assumed that fluctuations of different matrix elements $\mathcal{H}_{mn}$ are not correlated. The ensembles of completely random matrices studied in Chapter 6 are obtained from Eq. (12.193) by setting $p = N$.

In order to discuss the properties of eigenvalues of sparse random matrices it is convenient to introduce Green functions. That can be done in the same way as in Chapter 6, using Eq. (6.92). All other correlation functions are expressed in terms of the Green functions as previously. Calculating a product of two Green functions and using the supervectors ψ_m, $m = 1, 2, \ldots, N$ one comes to the Lagrangian $\mathcal{L}$, Eq. (6.96). The next step is averaging over symmetric matrices with the distribution D, Eq. (12.193). To avoid complicated intermediate expressions let us assume that p remains finite in the limit $N \to \infty$ and keep the main orders in $1/N$. Then, instead of Eq. (6.98), we obtain the following expression for the effective Lagrangian $\mathcal{L}$

$$\mathcal{L} = i\sum_{m=1}^{N}\bar{\psi}_m\left(\tilde{\varepsilon} - \frac{\omega + i\delta}{2}\Lambda\right)\psi_m + \frac{p}{2N}\sum_{m,n=1}^{N}\left[\tilde{h}(2\bar{\psi}_m\psi_n) - 1\right] \tag{12.194}$$

where $\tilde{h}(z)$ is the Fourier transform of the function $h(x)$

$$\tilde{h}(z) = \int h(x)\exp(-ixz)\,dx \tag{12.195}$$

(when writing Eq. (12.194) it was used that, for the orthogonal ensemble, $\bar{\psi}_m\psi_n = \bar{\psi}_n\psi_m$). Because of the large parameter N it is natural to decouple the second term in Eq. (12.194), integrating over auxiliary variables as in Section 6.5. This can be carried out by using a generalized Hubbard–Stratonovich transformation (Mirlin and Fyodorov (1991a)), which is written as

$$\int Dg\exp\left\{-\frac{Np}{2}\int d\psi\,d\psi'\,g(\psi)\,C(\psi,\psi')\,g(\psi) + p\int d\psi\,g(\psi)\,v(\psi)\right\}$$

$$= \exp\left\{\frac{p}{2N}\int d\psi\,d\psi'\,v(\psi)\,C^{-1}(\psi,\psi')\,v(\psi')\right\} \tag{12.196}$$

where $C^{-1}(\psi,\psi')$ stands for the kernel of an integral operator inverse to $\hat{C}$ with the kernel $C(\psi,\psi')$. The identity Eq. (12.196) can be proved by expanding the functions $g(\psi)$ and $v(\psi)$ in a series in eigenfunctions of the operator $\hat{C}$ and calculating Gaussian integrals over the coefficients of the expansion.

Choosing

$$C^{-1}(\psi,\psi') = \tilde{h}(\bar{\psi}\psi') - 1 \tag{12.197}$$

one comes to a theory with Lagrangian $\tilde{\mathcal{L}}$ in an extended space

$$\tilde{\mathcal{L}} = \frac{Np}{2} \int d\psi d\psi' g(\psi) C(\psi, \psi') g(\psi') \tag{12.198}$$

$$- N \ln \int d\psi \exp\left[i\bar{\psi}\left(\tilde{\varepsilon} - \frac{\omega + i\delta}{2}\Lambda\right)\psi + pg(\psi)\right]$$

In order to calculate physical quantities one has to evaluate functional integrals over the function $g(\psi)$. This formulation is very similar to the approach presented in Appendix 4. In the limit $N \to \infty$, the computation can be carried out by using the saddle-point approximation. Minimizing the functional $\tilde{\mathcal{L}}$, Eq. (12.198), one comes to the equation for the function $g(\psi)$

$$g(\psi) = \frac{\int d\phi \left[\tilde{h}(2\bar{\phi}\psi) - 1\right] \exp\left[i\bar{\phi}(\tilde{\varepsilon} - (\omega + i\delta)\Lambda/2)\phi + pg(\phi)\right]}{\int d\phi \exp\left[i\bar{\phi}(\tilde{\varepsilon} - (\omega + i\delta)\Lambda/2)\phi + pg(\phi)\right]} \tag{12.199}$$

This is a nonlinear integral equation similar to those given by Eqs. (12.47, 12.173) obtained in the effective medium approximation, on the Bethe lattice or as a result of minimizing the effective Lagrangian $\mathcal{L}$ in Appendix 4. As usual, one can seek its solution in the form

$$g(\psi) = g_0(\bar{\psi}\psi, \bar{\psi}\Lambda\psi) \tag{12.200}$$

where g_0 is a function of two variables. The form Eq. (12.200) follows from the symmetry of the integrand in Eq. (12.199) with respect to rotations in the space of supervectors. Using Eq. (2.94) for integration over invariant expressions we conclude that the weight denominator in Eq. (12.199) is equal to unity, which further simplifies the equation

$$g(\psi) = \int d\phi \left[\tilde{h}(2\bar{\phi}\psi) - 1\right] \exp\left[i\bar{\phi}\left(\tilde{\varepsilon} - \frac{(\omega + i\delta)\Lambda}{2}\right)\phi + pg(\phi)\right] \tag{12.201}$$

Investigating Eq. (12.201) Fyodorov and Mirlin (1991a) proved the existence of a transition at a critical mean number of nonzero elements per row $p_c > 1$. Up to this critical value eigenvalues of sparse matrices remain uncorrelated on the energy scale $\delta\mathcal{E} \propto 1/N$ despite the existence of a connected block of size proportional to N. In the language of the Anderson localization, this corresponds to a situation when the eigenstates are localized but the average density of states remains finite. At $p > p_c$ all eigenvectors of the infinite block become delocalized and one arrives at the Wigner–Dyson level–level correlation function, Eq. (6.13), with the full density of states in the mean level spacing Δ replaced by the infinite cluster contribution to it.

As mentioned, Eq. (12.201) is reminiscent of Eq. (12.173) written for the Bethe lattice with on-site disorder. Can one modify the model on the Bethe lattice such that the exact mapping on the model of random sparse matrices is obtained? This can really be done for a model on a Bethe lattice with random bonds and a random connectivity number m.

Suppose that the t_{ij} in Eq. (12.172) are random and their distribution is described by the function $h(t)$, Eq. (12.193). Averaging over this disorder one arrives at the following equation, instead of Eq. (12.173):

$$G(\psi) = \int d\phi \tilde{h}(2\bar{\phi}\psi) \exp\left(i\bar{\phi}\left(\tilde{\varepsilon} - \frac{(\omega + i\delta)\Lambda}{2}\right)\phi\right) G^m(\phi) \tag{12.202}$$

with $\tilde{h}$ from Eq. (12.195). Now, let us assume that the coordination number m at any site is a random variable independently distributed according to the Poisson law with the mean value equal to p. Performing this averaging in Eq. (12.202) and denoting $g(\phi) = G(\phi) - 1$ we come immediately to Eq. (12.201). This shows that the random sparse matrix model is intimately related to the models on the Bethe lattice.

13

Disorder in two dimensions

13.1 Electron in a strong magnetic field

13.1.1 General remarks

Some aspects of electron motion in a two-dimensional (2D) disordered metal have been considered in Chapter 5. These were effects that could be studied by using the perturbation theory in diffusion modes. The renormalization group scheme is a way to sum up a certain class of graphs and is a straightforward extension of the simple perturbation theory. Although the first quantum correction, Eq. (5.23), contains many interesting effects that can be and have been confirmed by numerous experiments, some other interesting effects cannot be obtained in this simple manner.

One of the most interesting phenomena occurring in two dimensions is the quantum Hall effect (von Klitzing, Dorda, and Pepper (1980)). Since its discovery a lot of theoretical and experimental activity has been devoted to studying this effect and related properties of 2D electron gases. It makes no sense to review this direction of research here because many interesting and comprehensive reviews and books already exist (see, e.g., Prange and Girvin (1990), Büttiker (1992), Stone (1992), Aoki (1986), Janßen et al. (1994)). The aim of this section is to demonstrate only how the supersymmetry technique can help in studying electron motion in a 2D disordered metal.

In fact, in discussing theoretical aspects of the quantum Hall effect one should distinguish between the integer quantum Hall effect (occurring at integer filling factor ν_f) and the fractional effect discovered by Tsui, Stormer, and Gossard (1982). Observation of the fractional quantum Hall effect requires high-mobility samples. It is generally accepted that this effect is due to strong electron correlations, and disorder does not seem to be important for its explanation. Therefore, the supersymmetry technique is not very useful in this case although there can be one exception.

According to a popular concept (Jain (1989, 1990)), real interacting electrons can be replaced by some composite fermions in a fictitious gauge field. From this point of view, the values of $\nu_f = p/(2p \pm 1)$ (p is integer) at which the fractional quantum Hall states appear can be interpreted as ordinary $\nu_f = p$ Shubnikov–de Haas oscillations for the composite fermions in the gauge field. The case of half-filling of the Landau level, $\nu_f = 1/2$, corresponds to the zero field and one can expect metallic behavior (Halperin, Lee, and Read (1993)). This interesting picture seems to be supported by findings of a number of experiments (Willet et al. (1990, 1993), Kang et al. (1993), Du et al. (1993, 1994)). If there are inhomogeneities in the system, the filling factor may fluctuate in the space and this can be described by space fluctuations of the gauge field. So, one arrives at a picture of fermions noninteracting with each other but interacting with the gauge field. To study transport in this system one can use the supersymmetry technique. As a result one comes to the conventional supermatrix σ-model with unitary symmetry (Aronov, Mirlin, and Wölfle (1994)). This is the case when the supersymmetry method can be useful in the regime of the fractional quantum Hall effect.

The integer quantum Hall effect can be explained within a model of noninteracting particles moving in a magnetic field and in a potential of impurities. Again, one can distinguish between the limit when the potential of the impurities is smooth and the opposite limit of short-range impurities. In the former case one may use a classical percolation picture (Iordansky (1982), Ono (1982), Lurui and Kazarinov (1983), Trugman (1983), Apenko and Lozovik (1985)). In this approach one can prove the existence of the quantization of the Hall conductivity.

However, in many experimental situations short-range impurities can be important and this leads to the necessity of considering models analogous to those studied in the preceding chapters. For this class of systems one can derive an appropriate σ-model. However, from the discussion of properties of the two-dimensional σ-model presented in Chapter 5 one can learn only that a magnetic field changes the universality class of the Q-matrices from the orthogonal to the unitary. Nothing like a quantization could be seen from the results of that chapter. The reason is that the consideration was restricted to the case of a not very strong magnetic field. In the opposite limit the results should be modified and later it is shown how the modification can be done.

13.1.2 Density of states

First, let us understand what happens with the average density of states in a strong magnetic field. As in Chapter 4, calculation of averaged quantities can be reduced to a functional integral over supermatrices Q. The free energy functional $F[Q]$, Eq. (4.45), may still be used without changes. By minimizing this functional one obtains Eq. (4.46) with the function $g(\mathbf{r}, \mathbf{r}')$ obeying Eq. (4.42). When calculating the average density of states one may put $\omega = 0$. As usual, one can represent the supermatrix Q in the form

$$Q(\mathbf{r}) = V(\mathbf{r}) S(\mathbf{r}) \Lambda \bar{V}(\mathbf{r}), \qquad V(\mathbf{r}) \bar{V}(\mathbf{r}) = 1 \tag{13.1}$$

where $S(\mathbf{r})$ has the structure

$$S(\mathbf{r}) = \begin{pmatrix} S^{11}(\mathbf{r}) & 0 \\ 0 & S^{22}(\mathbf{r}) \end{pmatrix} \tag{13.2}$$

The unitary supermatrix V is the same as in Eq. (4.25) and has the structure determined by Eq. (4.49). For the calculation of the density of states one may forget about the supermatrices $V(\mathbf{r})$ and understand Eqs. (4.42, 4.46) as if they were written for the supermatrices $S^{11}(\mathbf{r})$ and $S^{22}(\mathbf{r})$ only. Seeking a homogeneous solution of these equations in the form

$$S^{11}(\mathbf{r}) = S^{22}(\mathbf{r}) = \begin{pmatrix} s_0 \cdot \mathbf{1} & 0 \\ 0 & s_0 \cdot \mathbf{1} \end{pmatrix} \tag{13.3}$$

one obtains the following equation for s_0:

$$s_0 = \frac{i}{\pi} \sum_{n=0}^{\infty} \frac{\omega_c}{\varepsilon - \varepsilon_n + i s_0 / 2\tau}, \qquad \varepsilon_n = \omega_c \left(n + \frac{1}{2} \right) \tag{13.4}$$

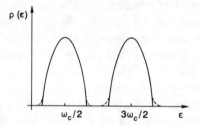

Fig. 13.1. Density of states $\rho(\varepsilon)$ as a function of energy. The solid line corresponds to the semicircle law, Eq. (13.8). The dashed line describes tails.

In Eq. (13.4), $\omega_c = eH/mc$ is the cyclotron frequency, and the sum is taken over all Landau levels. Eq. (13.4) is written for a two-dimensional system and coincides with the well-known self-consistent Born approximation (SCBA) equation (Ando and Uemura (1974)). In the limit of a weak magnetic field $\omega_c\tau \ll 1$, the sum over n can be replaced by an integral and one obtains $s_0 = 1$, thus returning to Eq. (4.25).

In the opposite limit $\omega_c\tau \gg 1$, overlapping between the Landau levels is small and one may keep in the sum in Eq. (13.4) only one term corresponding to such an n when ε_n is closest to ε. All the calculations presented are carried out in this limit. Then, we obtain easily

$$\frac{s_0}{2\tau} = \frac{\varepsilon - \varepsilon_n}{2i} + \frac{1}{2}\left(\Gamma^2 - (\varepsilon - \varepsilon_n)^2\right)^{1/2} \tag{13.5}$$

where Γ equals

$$\Gamma = \omega_c \left(\frac{2}{\pi\omega_c\tau}\right)^{1/2} \ll \omega_c \tag{13.6}$$

The imaginary part of s_0 renormalizes the energy ε and is not important. The real part of s_0 leads to the broadening of the Landau levels. It is not zero only in the region $|\varepsilon - \varepsilon_n| < \Gamma$. The density of states $\rho(\varepsilon)$ (per one spin) can be obtained by writing a formula analogous to Eq. (4.60)

$$\rho(\varepsilon) = \frac{\nu}{8}\operatorname{Re}\operatorname{str}k\left(S^{11}(\mathbf{r}) + S^{22}(\mathbf{r})\right) \tag{13.7}$$

which is valid as soon as Eq. (4.46) is fulfilled. Using Eqs. (13.2, 13.3, 13.5) for the density of states $\rho_n(\varepsilon)$ of the nth level one obtains

$$\rho_n(\varepsilon) = \frac{2\nu\omega_c}{\pi\Gamma}\left(1 - \frac{(\varepsilon - \varepsilon_n)^2}{\Gamma^2}\right)^{1/2} \tag{13.8}$$

Eq. (13.8) holds in the bands $|\varepsilon - \varepsilon_n| \leq \Gamma$. Outside these bands the density of states in the preceding approximation vanishes; that means that a more sophisticated scheme should be used there. The semicircular shape of the Landau levels is represented in Fig. 13.1. It is possible to check that the saddle-point approximation that is equivalent to SCBA can be used for high Landau levels $n \gg 1$. To study the lowest Landau levels one should apply other methods.

Using the supersymmetry technique Ziegler (1982, 1983) obtained an analogous semi-circle law for an N-orbital model without a magnetic field in the limit of large N.

In fact, the average density of states for the first Landau level can be found exactly provided the overlap between different levels is neglected (Wegner (1983)). It can also be done in a nice way by using the supersymmetric approach (Brezin, Gross, and Itzykson (1984)). The derivation has been presented in several books (see, e.g., Itzykson and Drouffe (1989), Janßen et al. (1994)) and therefore is not given here. The final result for the density of states of the lowest Landau level reads

$$\rho(\varepsilon) = \frac{2v\omega_c}{\pi^{3/2}\Gamma} \frac{\exp(|\tilde{\varepsilon}|/\Gamma)^2}{1 + \left(2\pi^{-1/2} \int_0^{|\tilde{\varepsilon}|/\Gamma} \exp t^2 dt\right)^2} \tag{13.9}$$

where $\tilde{\varepsilon} = \varepsilon - \omega_c/2$. We see from this exact formula that the density of states does not vanish identically in any region. Far from the center it has Gaussian tails.

Unfortunately, it is not clear how to extend the method of Wegner (1983) or Brezin, Gross, and Itzykson (1984) to calculation of transport quantities. Therefore, for this purpose it is natural to try to use the SCBA scheme. However, before applying the SCBA it is important to improve this approach such that a finite density of states beyond the bands $|\varepsilon - \varepsilon_n| < \Gamma$ can be obtained. With Eq. (13.9) in hand one can expect that this quantity has to be different from zero in these regions but exponentially small.

In fact, Eq. (4.46) for the saddle point with $g(\mathbf{r}, \mathbf{r}')$ from Eq. (4.42) is still sufficient for this purpose. A nonzero contribution can come from coordinate-dependent solutions of the saddle-point equations (Efetov and Marikhin (1989)) and so, one should use the general form, Eq. (13.1). Only solutions that correspond to saddle points that are on contours of integration over elements of the supermatrices Q contribute.

The nontrivial solution for the block S^{11} can be sought in the form

$$S^{11} = w \begin{pmatrix} s(\mathbf{r}) & 0 \\ 0 & is(\mathbf{r})_1 \end{pmatrix} \bar{w}, \qquad w\bar{w} = 1 \tag{13.10}$$

where w and $\bar{w}$ do not depend on coordinates.

As concerns $s(\mathbf{r})$, only the trivial saddle point

$$s = i\bar{s}_0, \qquad \frac{\bar{s}_0}{2\tau} = -\frac{\varepsilon - \varepsilon_n}{2} + \frac{1}{2}\left[(\varepsilon - \varepsilon_n)^2 - \Gamma^2\right]^{1/2} \tag{13.11}$$

gives a contribution beyond the bands. This leads to renormalization of the energy ε. To find a saddle point when integrating over $s_1(\mathbf{r})$ one can represent this variable in the form

$$s_1(\mathbf{r}) = \bar{s}_0 + q(\mathbf{r}) \tag{13.12}$$

and then Eqs. (4.42, 4.46) are reduced to the form

$$q(\mathbf{r}) = (\pi v)^{-1} \int \tilde{G}_0(\mathbf{r}, \mathbf{r}') q(\mathbf{r}') \tilde{G}(\mathbf{r}', \mathbf{r}) d\mathbf{r}' \tag{13.13}$$

where $\tilde{G}(\mathbf{r}, \mathbf{r}')$ satisfies the following equation

$$\left(\tilde{H} - s_1(\mathbf{r})\right) \tilde{G}(\mathbf{r}, \mathbf{r}') = \delta(\mathbf{r} - \mathbf{r}') \tag{13.14}$$

and $\tilde{G}_0(\mathbf{r}, \mathbf{r}')$ is obtained from $\tilde{G}(\mathbf{r}, \mathbf{r}')$ by setting $q(\mathbf{r}) = 0$. Besides the solution $q(\mathbf{r}) = 0$, which does not contribute to the density of states, Eq. (13.13) has a nontrivial solution. This solution corresponds to an extremum of the free energy functional of $F[Q]$, which has finite action beyond the bands. If $|\varepsilon - \varepsilon_n|$ beyond a band is close to Γ, the solution

$q(\mathbf{r})$ of Eq. (13.13) varies in space slowly and is small. In this limit the integral equation, Eq. (13.13), reduces to the differential equation

$$C \Delta q - Aq + Bq^2 = 0 \qquad (13.15)$$

where

$$C = (2n + 1) l_H^2, \qquad A = 2 \left(\frac{2 |\varepsilon - \Gamma|}{\Gamma} \right)^{1/2}, \qquad B = \frac{1}{\Gamma}$$

and Δ is the Laplacian. Using the notations

$$q(r) = q_0 y \left(\frac{r}{r_0} \right), \qquad q_0 = \Gamma A, \qquad r_0^2 = (2n + 1) l_H^2 A^{-1} \qquad (13.16)$$

one can rewrite Eq. (13.15) as

$$\Delta y - y + y^2 = 0 \qquad (13.17)$$

In Eq. (13.16) l_H is the magnetic length, $l_H^2 m = \omega_c^{-1}$. Nontrivial solutions of Eq. (13.17) have the "instanton form" (Langer (1967)). They are finite for all r and have the following asymptotics at $r \to \infty$:

$$y_0(r) = r^{-1/2} \exp(-r) \qquad (13.18)$$

As soon as the form of the "instanton" is found, using Eqs. (13.1, 13.11–13.13) one can write supermatrix Q_{cl} at the saddle point. Substituting this supermatrix into the free energy functional $F[Q]$, Eq. (4.45), one obtains the energy $F[Q_{\text{cl}}]$ of the "instanton"

$$F[Q_{\text{cl}}] = 4(2n + 1) \frac{|\varepsilon - \Gamma|}{\Gamma} f, \qquad f = \int y_0^3(\xi) \, d\xi \qquad (13.19)$$

Then, one should expand the free energy functional $F[Q]$ near $F[Q_{\text{cl}}]$ up to quadratic terms and calculate Gaussian integrals. Special care should be taken about zero modes, which exist as a result of translational symmetry and symmetry with respect to rotations in the space of the supermatrices Q of the functional $F[Q]$. As a result one obtains (Efetov and Marikhin (1989)) the density of states near the band edge Γ of the nth Landau level

$$\rho(\varepsilon) = a\nu\omega_c \frac{|\varepsilon - \Gamma|}{\Gamma^2} \exp \left(-4(2n + 1) \frac{|\varepsilon - \Gamma|}{\Gamma} f \right) \qquad (13.20)$$

where a is a number of order one. The "instanton" contribution to the density of states is represented in Fig. 13.1 by the dashed line. We see that the tails of the density of states for high Landau levels are also exponentially small. The saddle-point approximation is valid when the exponent in Eq. (13.20) is large; that gives the criterion of the applicability of Eq. (13.20)

$$|\varepsilon - \Gamma| \gg \Gamma n^{-1} \qquad (13.21)$$

The same inequality makes it possible to take into account only one "instanton." The expansion in gradients and in q when deriving Eq. (13.15) from Eq. (13.13) is valid provided

$$|\varepsilon - \Gamma| \ll \Gamma \qquad (13.22)$$

Both of the inequalities can be satisfied if n is large. The "instanton" approximation is applicable even if Eq. (13.22) is not fulfilled. In this case the expansion in gradients and in q is not valid and one should find another saddle-point solution.

The Gaussian tail for the density of states in the exact formula, Eq. (13.9), for $n = 1$ can also be reproduced by a saddle-point method (Affleck (1984), Viehweger and Efetov (1991)). This form is different from the form given by Eq. (13.20). However, Eq. (13.20) is written for the intermediate regime, Eqs. (13.21, 13.22). For the extreme tail region, the density of states was found to be Gaussian, too (Benedict (1987), Broderix, Heldt, and Leschke (1991)).

13.1.3 Integer quantum Hall effect

Now, let us discuss the integer quantum Hall effect. An introduction to the theory of this effect can be found in Janßen et al. (1994). Although qualitatively the effect is quite well understood the necessity of using sophisticated approaches for a detailed quantitative description seems to be inevitable.

The quantization of the Hall conductivity as a function of energy (or filling factor v_f) for a pure 2D gas can be demonstrated rather easily. For this purpose it is convenient to represent the Hall part of the general Kubo formula, Eqs. (3.26–3.30), in the form (Smrcka and Streda (1977))

$$\sigma_{xy} = \sigma_{xy}^I + \sigma_{xy}^{II}, \tag{13.23}$$

$$\sigma_{xy}^I = \frac{1}{2\pi V} \int d\mathbf{r} d\mathbf{r}' \Big\langle \big[\hat{\pi}_y G_\varepsilon^R (\mathbf{r}, \mathbf{r}') \hat{\pi}_{x'} G_\varepsilon^A (\mathbf{r}', \mathbf{r})$$

$$-\hat{\pi}_x G_\varepsilon^R (\mathbf{r}, \mathbf{r}') \hat{\pi}_{y'} G_\varepsilon^A (\mathbf{r}', \mathbf{r}) \big] \Big\rangle,$$

$$\sigma_{xy}^{II} = \frac{ie}{2\pi V} \lim_{\mathbf{r}_1 \to \mathbf{r}} \int d\mathbf{r} \, (x\hat{\pi}_y - y\hat{\pi}_x) \Big\langle \big[G_\varepsilon^R (\mathbf{r}, \mathbf{r}_1) - G_\varepsilon^A (\mathbf{r}, \mathbf{r}_1) \big] \Big\rangle$$

Eqs. (13.23) are written at zero temperature. They contain the single energy ε and are valid at arbitrary disorder, and the angular brackets indicate averaging over the disorder. It is assumed that spin interactions are absent and one can consider subsystems with different spin directions separately. Eq. (13.23) is written for one of the subsystems. Discussing Eqs. (13.23) it is convenient to use the spectral representation for the Green functions, Eq. (3.17).

In the clean limit the Landau levels are infinitely narrow and the result of computation of the integrals is strongly dependent on whether the energy ε is located between the levels or coincides with one of them. If the energy does not coincide with eigenenergies of states centered inside the the sample, these states do not contribute. However, the eigenenergy of states located near edges grows when the center of the states approaches the edges. Some states near the edges have eigenenergy that coincides with ε. Such states contribute to the integrals in Eq. (13.23).

If the system size is large σ_{xy}^I becomes very small because the contribution of the edge states to the integral is proportional to the perimeter of the sample. The integral in Eqs. (13.23) for the quantity σ_{xy}^{II} contains the additional factor x or y in the integrand and therefore this part of the Hall conductivity can be finite. In principle, it is not difficult to evaluate σ_{xy}^{II} directly from Eq. (13.23) but one can also write another representation

$$\sigma_{xy}^{II} = -ec \int_{-\infty}^{\varepsilon} \frac{\partial}{\partial H} \rho (\varepsilon') \, d\varepsilon' \tag{13.24}$$

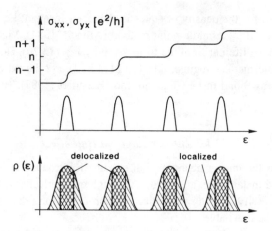

Fig. 13.2. Qualitative picture for the quantum Hall effect. In the upper part of the figure the longitudinal and Hall conductivities are depicted as a function of the Fermi energy. In the lower part of the figure the same is done for the density of states. The width of the region of the extended states is believed to turn zero in the thermodynamic limit.

containing the average density of states $\rho\,(\varepsilon)$. To calculate σ_{xy}^{II} using Eq. (13.24) one does not need any information about the edge states but it is necessary to integrate over all states below ε. The degeneracy per Landau level equals $eH/\,(2\pi c)$, which for the Hall conductivity gives

$$\sigma_{xy} = \frac{e^2 v_f}{h} \tag{13.25}$$

where v_f is the number of Landau levels below the Fermi energy.

So, the reason for the Hall quantization in a clean system is simple. However, there is always a disorder in real samples and one should explain the perfect quantization. Experimentally, the Hall quantization is also observed by changing the magnetic field, keeping the filling factor constant. In this situation Eq. (13.25) cannot serve as an explanation because the Hall conductivity σ_{xy} as a function of the field would not change at all.

By now, there is overwhelming evidence that quantization of Hall conductivity in disordered systems is due to localization of eigenstates. In fact, it is possible to show explicitly that localized states do not contribute to Hall conductivity (Ando and Aoki (1981)). The transitions between the plateaus correspond to localization–delocalization transitions (Laughlin (1981), Halperin (1982)). Thus, one may try to give a quantitative description of the quantum Hall effect starting from the standard models with disorder used everywhere in the present book. A qualitative picture explaining the integer quantum Hall effect is drawn in Fig. 13.2. As soon as an electron–electron interaction is not included, the supersymmetry technique can become very useful.

Using this method one can rather easily compute the conductivities in the tails of the density of states discussed in the preceding subsection that are depicted by the dashed lines in Fig. 13.1 (Efetov and Marikhin (1989)). With the commutation relations $v_\alpha = i\,[\mathcal{H}, r_\alpha]$, where $\mathcal{H}$ is the Hamiltonian, and the spectral expansion of the Green functions, Eq. (3.17), the formula for the conductivities $\sigma_{\alpha\beta}$, Eq. (3.26–3.30) (per one spin), can be rewritten as

$$\sigma_{\beta\alpha}\,(\omega) = \frac{\omega}{iV} \int r_\alpha \left(r_\beta - r_\beta' \right) \left[K_1^{00}\,(\mathbf{r} - \mathbf{r}',\, \omega) + K_2^{00}\,(\mathbf{r} - \mathbf{r}',\, \omega) \right] d\mathbf{r} d\mathbf{r}' \tag{13.26}$$

where the correlation functions $K_1 \left(\mathbf{r} - \mathbf{r}', \varepsilon, \omega \right)$ and $K_2 \left(\mathbf{r} - \mathbf{r}', \varepsilon, \omega \right)$ are defined in Eqs. (3.28–3.31). Calculation of these correlation functions can be performed by writing a proper integral over supermatrices Q. The only difference with respect to what was done in the preceding chapters is that now the eigenvalues of the supermatrix Q can depend on coordinates. In calculating the longitudinal conductivity the main contribution is from the function K_1^{00}, which contains the product of the retarded and advanced Green functions. The second term in Eq. (13.26) contains an additional factor $[(\varepsilon - \Gamma) / \Gamma]^{1/2}$, which is small provided the condition, Eq. (13.22), is fulfilled.

Using integration over the supermatrices Q one reduces the function R^{00}, Eq. (3.31), to the form

$$R^{00} \left(\mathbf{r}, \varepsilon, \omega \right) = - \int g_{33}^{12} (0, 0) \, g_{33}^{21} (\mathbf{r}, \mathbf{r}) \exp \left(-F \left[Q \right] \right) dQ \tag{13.27}$$

where $g \left(\mathbf{r}, \mathbf{r}' \right)$ and $F [Q]$ are determined by Eqs. (4.42) and (4.45), respectively. The other possible pairing that arises when reducing the function R^{00} to the integral over the supermatrices gives a function that decays faster than that in Eq. (13.27) and therefore gives a smaller contribution to $\sigma_{\beta\alpha} (\omega)$. Using the saddle-point method one can further simplify Eq. (13.27) with the help of Eq. (4.46).

As usual, in the limit $\omega \to 0$ the saddle-point solution Q_s for supermatrix Q is degenerate. Using the representation, Eq. (13.1), we see that supermatrix V can be arbitrary. This means that in the limit of low frequencies one should first take the saddle-point values of S^{11} and S^{22}, integrate in the Gaussian approximation over the elements of S^{11} and S^{22} near the saddle point, and then integrate over the supermatrices V. This procedure is analogous to the one used in Chapter 4. The only difference is that now S^{11} and S^{22} at the saddle point depend on coordinates.

In the limit described by Eq. (13.21), just as in the preceding subsection, the one-"instanton" approximation is applicable. Two possibilities exist: Either $S^{11} = i s_0$, $S^{22} = i s_0 + i q$ or $S^{11} = i s_0 + q$, $S^{22} = i s_0$, where q contains one "instanton." Integrating in Eq. (13.27) over q it is convenient first to fix a position $\mathbf{R}_0$ of the "instanton," thus reducing the integral in Eq. (13.27) to an integral over $\mathbf{R}_0$ and V. Finally, this integral can be reduced to the form

$$R^{00} \left(\mathbf{r}, \varepsilon, \omega \right) = \frac{\nu \rho \left(\varepsilon \right) \omega_c \phi \left(\mathbf{r} \right)}{2 b \Gamma^2} \int \left(V \Lambda \bar{V} \right)_{33}^{12} \left(V \Lambda \bar{V} \right)_{33}^{21}$$

$$\times \exp \left[\frac{2 i \nu b \omega_c}{\Gamma^2} \operatorname{str} \left(V \Lambda \bar{V} \Lambda \right) \right] dV \tag{13.28}$$

where

$$\phi \left(\mathbf{r} - \mathbf{r}' \right) = \int q \left(\mathbf{r} - \mathbf{R}_0 \right) q \left(\mathbf{r}' - \mathbf{R}_0 \right) d^2 \mathbf{R}_0, \qquad b = \int q \left(\mathbf{r} \right) d^2 \mathbf{r} \tag{13.29}$$

In Eq. (13.29) $q \left(\mathbf{r} \right)$ is the "instanton" solution determined in Eqs. (13.16–13.18). Calculation of the integral over the supermatrices V in Eq. (13.28) is simple because it contains combinations that are just the conventional supermatrix Q with the constraint $Q^2 = 1$ (cf. Eq. (4.25)). In fact, the integral in Eq. (13.28) is the same as integrals for the 0D σ-model. Of course, this similarity is not accidental. The "instanton" solution corresponds to states localized in some domain. These states are physically equivalent to states in a metallic granule or quantum dot, and so on.

Evaluating the integral in Eq. (13.28) one obtains

$$R^{00}(\mathbf{r}, \varepsilon, \omega) = \frac{\rho(\varepsilon)}{-i\omega b^2} \phi(\mathbf{r}) \qquad (13.30)$$

The correlation function R^{00}, Eq. (13.30), falls off exponentially at large distances. The form of the correlation function proves the localization of the wave functions. It is relevant to notice that if the condition, Eq. (13.22), is not fulfilled and the "instanton" is not determined by Eqs. (13.16–13.18), Eq. (13.30) is still valid provided one implies by q the exact solution of the saddle-point equation. It is clear from the preceding discussion that the localization in the regime of the tails of the density of states is due to the fast decay of the "instanton" solutions at large distances.

Using Eqs. (3.29–3.31) and Eq. (13.30) it is not difficult to find the longitudinal conductivity σ_{xx}

$$\sigma_{xx} = -\frac{i\omega e^2 \rho(\varepsilon_F)}{2b^2} \int \phi(r) r^3 dr = -i\omega e^2 \rho(\varepsilon) r_0^2 \bar{c} \qquad (13.31)$$

with r_0 from Eq. (13.16), where $\bar{c}$ is a numerical factor of order unity, and ε_F is the Fermi energy. The frequency dependence, Eq. (13.31), is typical for an insulator: $i\sigma_{xx}/\omega$ is the electric susceptibility. The length r_0 plays the role of a typical dipole moment. The density of states $\rho(\varepsilon)$ should be taken from Eq. (13.20), and this is what one could expect. So, the localization of states far from the centers of the Landau levels can be proved by the supersymmetry technique, and, in fact, the work of Efetov and Marikhin (1989) on which the present consideration is based was the first analytical calculation demonstrating the localization of the states in two dimensions in the limit of a not very strong disorder.

The presence of the localized states is sufficient to prove the existence of the plateaus with the quantized values $e^2 n/h$ of the Hall conductivity. The computation can be carried out in a similar way. Because of the problem with the edge states it is more convenient to consider the quantity

$$\Delta\sigma_{xy} = \sigma_{xy} - \sigma_{xy}^{(0)} \qquad (13.32)$$

where $\sigma_{xy}^{(0)}$ is the Hall conductivity of a clean system, which is determined by Eq. (13.25). Writing integrals for $\Delta\sigma_{xy}$ one can see that only configurations with a nonzero number of "instantons" contribute to this quantity. Then, all the integrals over coordinates converge and one comes, as for the longitudinal conductivity, to the conclusion that this quantity vanishes in the limit $\omega \to 0$ and Eq. (13.25) holds in the presence of impurities too.

So, we see that although in the clean limit the total density of particles and the Hall conductivity as a function of Fermi energy are proportional to each other, in a disordered system they are different. The electron density as a function of Fermi energy is a continuous function, and therefore Hall conductivity as a function of electron density has plateaus too.

Similar calculations have also been performed for the tails of the lowest Landau level (Viehweger and Efetov (1990, 1991); see also Janßen et al. (1994)). Although the form of the "instanton" in this case is somewhat different, it was possible to arrive at the conclusion about the localization and the quantization of the Hall conductivity.

13.1.4 Hall insulator

Everywhere in this section the longitudinal σ_{xx} and Hall σ_{xy} conductances have been considered. Experimentally somewhat different quantities, namely, resistivities ρ_{xx} and ρ_{xy}, are measured. The relation between the conductivities and resistivities is simple

$$\rho_{xx} = \frac{\sigma_{xx}}{\sigma_{xx}^2 + \sigma_{yx}^2}, \qquad \rho_{xy} = \frac{\sigma_{yx}}{\sigma_{xx}^2 + \sigma_{yx}^2} \qquad (13.33)$$

However, these simple relations lead to quite curious consequences. For example, if there is at least one Landau level below the Fermi surface, the Hall conductivity σ_{yx} is not equal to zero. At the same time, in the plateau regions $\sigma_{xx} = 0$. Then, Eq. (13.33) shows us immediately that in the plateau regions

$$\rho_{xx} = 0, \qquad \rho_{xy} = \sigma_{yx}^{-1} = h \left(ne^2 \right)^{-1} \qquad (13.34)$$

So, both longitudinal conductivity and resistivity turn to zero simultaneously.

Another interesting and very general result that can be derived from Eq. (13.33) concerns the Hall coefficient R. In classical Drude transport theory the conductivities have the form

$$\sigma_{xx} = \frac{\sigma_0}{1 + (\omega_c \tau)^2}, \qquad \sigma_{yx} = \frac{\omega_c \tau \sigma_0}{1 + (\omega_c \tau)^2} = \omega_c \tau \sigma_{xx} \qquad (13.35)$$

where σ_0 is the classical conductivity, Eq. (3.1). Substituting Eqs. (13.35) into Eqs. (13.33) and using Eq. (3.1) one obtains for the resistivities

$$\rho_{xx} = \sigma_0^{-1}, \qquad \rho_{xy} = RH, \qquad R = (n_e ec)^{-1} \qquad (13.36)$$

where n_e is the electron density. Putting more disorder into the system leads to increasing ρ_{xx} but does not change the Hall coefficient R, which is dependent on the electron density n_e only. The classical theory is applicable in the limit of weak disorder, and one may ask what happens if the disorder is so strong that the system is no longer conducting. We understand that in the insulating state both the longitudinal and Hall conductivity turn to zero. (In the regime of the quantum Hall effect this statement is correct if there are no Landau levels below the Fermi surface.) Does that mean that both resistivities diverge?

This question was addressed in works by Viehweger and Efetov (1990, 1991). By calculating the conductivities σ_{xx} and σ_{xy} in the region of the tail of the density of states at finite although low frequencies in the "instanton" approximation it was found that they behaved as

$$\sigma_{xx} \sim -i\omega\alpha, \qquad \sigma_{yx} \sim -\beta\omega^2 H \qquad (13.37)$$

where α and β are coefficients depending on disorder. The form of the longitudinal conductivity, Eq. (13.37), is usual for an insulator, and α represents the polarizability. The frequency dependence of σ_{yx} can also be simply understood from the Onsager relation $\sigma_{xy}(\omega, H) = \sigma_{yx}(-\omega, -H)$. This relation is also valid for nonaveraged quantities. If the rotational symmetry is restored by averaging, an additional relation $\sigma_{xy}(\omega, H) = -\sigma_{yx}(\omega, H)$ holds. Thus, one obtains

$$\sigma_{xy}(\omega, H) = -\sigma_{xy}(-\omega, -H) \qquad (13.38)$$

This means that the term linear in H must be even in ω. As soon as the Hall conductivity vanishes in the insulating state, one comes to Eq. (13.37). Substituting Eq. (13.37) into Eq. (13.33) we find

$$\rho_{xx} = (i\alpha\omega)^{-1}, \qquad \rho_{xy} = RH, \qquad R = \frac{\beta}{\alpha^2} \qquad (13.39)$$

Eq. (13.39) shows that in the insulating state the longitudinal resistivity diverges whereas the Hall coefficient R remains finite. In the tail region both coefficients α and β are exponentially small, so the Hall coefficient R is exponentially large. At the same time, the electron density n_e is exponentially small; that means that R is approximately inversely proportional to this quantity. Further discussion of the possibility of having a finite Hall coefficient in an insulator can be found in Zhang, Kivelson, and Lee (1992) and Imry (1993).

Experimentally, measurements are usually carried out at zero frequency but finite temperature T. Apparently, the question whether the Hall coefficient remains finite or diverges in the limit $T \to 0$ depends on inelastic processes and both types of behavior are possible. Still, this problem remains not very well understood. A finite Hall coefficient in an insulating state with a diverging longitudinal resistivity has already been observed in a number of experiments in both 3D (Hopkins et al. (1989)) and 2D (Goldman, Shayegan, and Tsui (1988), Dorozhkin et al. (1993), Kravchenko, Furneaux, and Pudalov (1994)) electron gases. Zhang, Kivelson, and Lee (1992) suggested calling the state with a diverging longitudinal but finite Hall resistivity "Hall insulator."

13.1.5 Transition between Hall plateaus

We see from the preceding analysis that a disordered 2D metal in a strong magnetic field can be described rather simply in the tail regions. Proceeding in this way one can prove the localization of wave functions and the quantization of Hall conductivity. However, one of the most interesting questions, namely, about the transitions between the Hall plateaus, cannot be answered by studying the tails. It is even difficult to answer the question whether the slope of the Hall conductivity between the plateaus is finite or infinite at zero temperature $T = 0$.

The first idea that comes to mind is to use the conventional σ-model, Eq. (4.55), written for the unitary ensemble for the description. It is clear from the analysis of the present section that in the limit $\omega_c \tau \gg 1$ one may try to use this model at least for high Landau levels, $n \gg 1$, where the fluctuations of the density of states can be neglected. In fact, as we have learned from Chapter 12, the fluctuations of the density of states are not important for the critical behavior of transport quantities, and so, one might try to use Eq. (4.55) for the lowest Landau level, too.

However, this model is also applicable in the limit $\omega_c \tau \ll 1$ where one does not expect Hall quantization. Besides, according to a general belief, the σ-model in the form of Eq. (4.55) must always lead in two dimensions to the localization of all states, and so, Hall quantization would be impossible. A very interesting possibility for resolving this paradox was suggested (for the compact replica version of the σ-model) by Levine, Libby, and Pruisken (1983, 1984) and Pruisken (1985).

These authors noticed that all important effects of a magnetic field were not taken into account in the conventional σ-model. For example, Eq. (4.55) is invariant under the substitution $(x, y) \to (-x, y)$. This invariance is also seen in the standard expansion in diffusion modes. At the same time, the presence of a magnetic field must break such an invariance and so, if something is missed when deriving the σ-model, this "something" must be nonperturbative.

Indeed, it turns out that a term of topological origin is neglected in Eq. (4.55). This term can be obtained by a more careful expansion of the logarithm in Eq. (4.45) in gradients of the Q-matrix. Such a procedure can be carried out in a more elegant way than that used when deriving Eq. (4.55). Assuming from the beginning that the eigenvalues of the supermatrix Q do not fluctuate and the magnetic field is strong enough that the unitary ensemble should be used, one can rewrite the free energy functional in the form

$$F[Q] = -\frac{1}{2} \text{str} \int d\mathbf{r} \ln \left[-\left(\varepsilon + \frac{\omega}{2} \right) + \frac{1}{2m} \bar{V} \hat{\mathbf{P}}^2 V + \frac{\omega \bar{V} \Lambda V}{2} + \frac{i\Lambda}{2\tau} \right] \qquad (13.40)$$

where $\hat{\mathbf{P}} = \hat{\mathbf{p}} - e\mathbf{A}/c$, and $\hat{\mathbf{p}}$ is the momentum operator. Using the identity

$$\bar{V} \hat{\mathbf{P}}^2 V = \hat{\mathbf{P}}^2 - 2i \delta V \hat{\mathbf{P}} - \bar{V} \Delta_{\mathbf{r}} V, \qquad \delta V \equiv \bar{V} \nabla V \qquad (13.41)$$

where $\Delta_{\mathbf{r}}$ is the Laplacian, one can perform direct expansion of the logarithm in δV and ω. The last term with $\Delta_{\mathbf{r}}$ in Eq. (13.41) can be neglected as soon as the limit of weak disorder is under consideration. The terms present in Eq. (4.57) are from the linear term in ω and quadratic terms in $\delta_x V$ and $\delta_y V$. After an elementary calculation of the contribution of the latter terms one obtains the combination $\text{str} [\Lambda, \delta V]^2$, which reduces to $\text{str} [\nabla Q]^2$. The coefficient in front of this term is proportional to the longitudinal conductivity $\bar{\sigma}_{xx}^{(0)}$ calculated in the self-consistent Born approximation ($\bar{\sigma} = \sigma h/e^2$). The term with ω is proportional as usual to $\text{str} \Lambda Q$.

However, the expansion in Eq. (13.40) also yields two additional terms: The first contains the product $\delta V_x \delta V_y$; the second contains a linear combination of δV_x and δV_y. These are the terms that generate the topological term in the σ-model. The final result for the free energy functional F of the effective σ-model can be written in the form

$$F = \frac{1}{16} \text{str} \int \left[\bar{\sigma}_{xx}^{(0)} (\nabla Q)^2 - \bar{\sigma}_{xy}^{(0)} Q [\nabla_x Q, \nabla_y Q] + 4\pi v \omega i \Lambda Q \right] d^2 r \qquad (13.42)$$

In Eq. (13.42), the parameters $\bar{\sigma}_{xx}^{(0)}$ and $\bar{\sigma}_{xy}^{(0)}$ are the classical longitudinal and Hall conductivities expressed in the units of e^2/h. If the magnetic field is weak, $\omega_c \tau \ll 1$, these conductivities can be found from Eq. (13.35). In the opposite limit, $\omega_c \tau \gg 1$, the Hall conductivity is given by Eq. (13.25) with

$$\bar{\sigma}_{xy}^{(0)} = v_f = \frac{N_0 e c}{H} \qquad (13.43)$$

where N_0 is the electron density. Because of the impurity smearing of the Landau levels the filling factor v_f is a smooth function of energy and the conductivity $\sigma_{xy}^{(0)}$ does not show sharp quantization. The longitudinal conductivity $\sigma_{xx}^{(0)}$ in the limit of strong fields takes the form

$$\bar{\sigma}_{xx}^{(0)} = \frac{2}{\pi} \left(n + \frac{1}{2} \right) \qquad (13.44)$$

The supermatrix Q in Eq. (13.42) obeys the usual constraint $Q^2 = 1$; this enables us to write the second term F_{top} in Eq. (13.42) in another form

$$F_{\text{top}} = -\frac{\bar{\sigma}_{xy}^{(0)}}{8} \text{str} \int Q \nabla_x Q \nabla_y Q d^2 r \qquad (13.45)$$

At arbitrary disorder the Hall conductivity $\bar{\sigma}_{xy}^{(0)}$ in Eqs. (13.42, 13.45) consists of two parts, $\bar{\sigma}_{xy}^{I}$ and $\bar{\sigma}_{xy}^{II}$, Eq. (13.23). When deriving Eq. (13.42), the contribution from the term $\delta V_x \delta V_y$ in the expansion of Eq. (13.40) is proportional to $\bar{\sigma}_{xy}^{I}$, whereas the terms linear in δV_x and δV_y have a contribution proportional to $\bar{\sigma}_{xy}^{II}$. Both terms yield similar combinations of supermatrix Q and its gradients that are identical to those entering Eq. (13.45). That is why the total Hall conductivity $\bar{\sigma}_{xy}^{(0)}$ finally enters Eq. (13.42). The dependence of the longitudinal conductivity $\bar{\sigma}_{xx}^{(0)}$ on energy in the preceding approximation is similar to the dependence of the density of states $\rho\,(\varepsilon)$, Eq. (13.8), and obeys a semicircle law. The σ-model, Eq. (13.42), is applicable in the regions close to the centers of the Landau levels where $\bar{\sigma}_{xx}^{(0)}$ is not small. The transitions between the Hall plateaus occur in these regions and so, the σ-model is suitable for their description.

The term F_{top} violates the symmetry with respect to changing $(x,\,y) \to (x,\,-y)$. The presence of this term cannot be observed in any order of perturbation theory in the cooperons and diffusons. The topological origin of the term F_{top} can be explicitly seen by using the representation $Q = V \Lambda \bar{V}$ with unitary supermatrices V. Then, one can easily derive the identity

$$\text{str}\left(Q\nabla_x Q\nabla_y Q\right) = 2\,\text{curl}\left[\text{str}\left(\Lambda\bar{V}\nabla V\right)\right] \tag{13.46}$$

Using Eq. (13.46) one can reduce the term F_{top} to an integral along the contour C confining the sample

$$F_{\text{top}} = -\frac{\bar{\sigma}_{xy}^{(0)}}{4}\int\limits_{C}\text{str}\left(\Lambda\bar{V}\nabla V\right)d\mathbf{l} \tag{13.47}$$

So, only nontrivial configurations that are not seen in perturbation theory can lead to non-vanishing contributions in F_{top}. Therefore, trying to apply a renormalization group scheme analogous to that developed in Chapter 5 one should find nontrivial minima of the free energy functional $F\,[Q]$, Eq. (13.42). Levine, Libby, and Pruisken (1983, 1984) found such configurations (instantons) for the compact version of the replica σ-model suggested by Efetov, Larkin, and Khmelnitskii (1980). The solution for an arbitrary number of replicas n is rather complicated. Besides, one should integrate near these solutions to take fluctuations into account and this is even more difficult.

The same procedure can be carried out in a somewhat simpler way (Weidenmüller (1987), Weidenmüller and Zirnbauer (1988)) by using the supermatrix σ-model with the topological term, Eq. (13.42). Nontrivial solutions for this σ-model exist in the compact fermion–fermion block of the supermatrices Q. It is convenient to use Eq. (6.56) written for the supermatrix $\tilde{Q}_0$ containing all commuting variables. The fermion–fermion block $\tilde{Q}_{0F}$ of this supermatrix in the unitary ensemble has the form

$$\tilde{Q}_{0F} = \begin{pmatrix} \cos\theta & i\sin\theta\exp\left(i\phi\tau_3\right) \\ -i\sin\theta\exp\left(-i\phi\tau_3\right) & -\cos\theta \end{pmatrix} \tag{13.48}$$

with $0 < \theta < \pi$, $0 < \phi < 2\pi$. Assuming that the variables θ and ϕ entering Eq. (13.48) depend on the coordinates but the other variables parametrizing the supermatrix Q do not, and substituting Eq. (13.48) into Eq. (13.42), in the limit $\omega \to 0$ one obtains the corresponding energy $F_{0F}\,[Q]$

$$F_{0F}\,[Q] = \frac{\bar{\sigma}_{xx}^{(0)}}{4}\int\left[(\nabla\theta)^2 + \sin^2\theta\,(\nabla\phi)^2\right]d^2r$$

$$\quad - \frac{i\bar{\sigma}_{xy}^{(0)}}{2}\int\left(\nabla_x\phi\nabla_y\theta - \nabla_x\theta\nabla_y\phi\right)\sin\theta\,d^2r \tag{13.49}$$

The first line of Eq. (13.49) is exactly the free energy functional of a 2D ferromagnet with three spin components n^α provided the vector $\mathbf{n}$ is on the unit sphere, $\mathbf{n}^2 = 1$. Such a model is often called the $O(3)$-*model*. The integral (divided by 2π) in the second line measures the degree q of mapping of the unit sphere onto the two-dimensional real space (Belavin and Polyakov (1975)). The number q must be an integer but can be either positive or negative. Using the polar coordinates r and α the degree of the mapping q can be rewritten as

$$q = -\frac{1}{4\pi} \int \frac{\partial \phi(r, \alpha)}{\partial \alpha} \frac{\partial \theta(r, \alpha)}{\partial r} \sin \theta(r, \alpha) \, dr \, d\alpha \tag{13.50}$$

Seeking minima of free energy $F_{0F}[Q]$, Eq. (13.49), one may consider the first term only because the second one can be reduced to an integral over the boundary. Then, we obtain equations determining the minima

$$\nabla \left(\nabla \phi \sin^2 \theta \right) = 0, \qquad -2\Delta\theta + (\nabla\phi)^2 \sin 2\theta = 0 \tag{13.51}$$

The simplest nontrivial solutions of Eqs. (13.51) with finite energy were found by Skyrme (1961). This solutions can be obtained by assuming that the variable ϕ depends only on α whereas θ depends only on r. Then, we easily obtain for the simplest solutions

$$\phi = \mp\alpha, \qquad -\frac{1}{r}\frac{d}{dr}r\frac{d(2\theta)}{dr} + \frac{1}{r^2}\sin 2\theta = 0 \tag{13.52}$$

The solution for θ of Eq. (13.52) with finite energy has the form

$$\theta = 2\arctan r \tag{13.53}$$

We see that this solution changes from 0 to π when changing the radial variable r from 0 to ∞. Substituting Eqs. (13.52, 13.53) into Eq. (13.50) one obtains

$$q = \pm 1 \tag{13.54}$$

The solution, Eqs. (13.52, 13.53), is often called a *skyrmion* or *instanton*. The free energy $F_{0F}[Q_{\text{sk}}]$ corresponding to the skyrmion equals

$$F_{0F}[Q_{\text{sk}}] = 2\pi |q| \bar{\sigma}_{xx}^{(0)} + 2\pi i q \bar{\sigma}_{xy}^{(0)} \tag{13.55}$$

where q is given by Eq. (13.54). In principle, solutions with an arbitrary integer q (Belavin and Polyakov (1975)) that correspond to multiinstanton solutions can be found. At a large separation of the instantons (these instantons are completely different from the "instantons" obtained previously for the density of states), Eq. (13.55) is still valid.

Why can the instantons be important to the quantum Hall effect? Substituting Eq. (13.43) into Eq. (13.55) we see that the second term changes by $2\pi i$ under the substitution $\nu_f \rightarrow \nu_f + 1$. The "partition function" does not change, and this is exactly what one expects for the quantum Hall effect. Using renormalization group arguments one should consider as the scaling parameter not only the longitudinal conductivity $\bar{\sigma}_{xx}$ but also the Hall conductivity $\bar{\sigma}_{xy}$. Thus, the one-parameter renormalization group considered in Chapter 5 should be replaced by a scheme with two parameters. Using the semiclassical instanton approximation Pruisken (1985, 1987) proposed a two-parameter renormalization group flow with unstable fixed points at $\bar{\sigma}_{xy}^{(0)} = 1/2 + m$, with integer m, and some $\bar{\sigma}_{xx}^{(0)} = \bar{\sigma}_{xx}^{*}$.

In principle, a renormalization group flow can be drawn assuming only that $\bar{\sigma}_{xx}$ and $\bar{\sigma}_{xy}$ are the only relevant scaling parameters and the picture must be periodic in $\bar{\sigma}_{xy}$ (Khmelnitskii (1983)). Making such an assumption one does not need to relate the scaling parameters to the coupling constants in the free energy, Eq. (13.42). The flow diagrams are represented schematically in Fig. 13.3. If the proposed picture is also applicable in the limit $\omega_c \tau \ll 1$

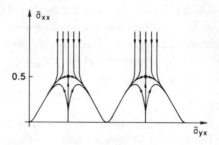

Fig. 13.3. Flow diagram in the $\bar{\sigma}_{xx}$–$\bar{\sigma}_{xy}$ plane. The solid dots indicate unstable fixed points.

one should use Eq. (13.35) for the bare conductivities. Then, the dependence of $\bar{\sigma}_{xy}^{(0)}$ on the magnetic field is more complicated and this led Khmelnitskii (1983) to the conclusion that additional oscillations could occur when varying the magnetic field.

Although the preceding discussion helps a lot in understanding how the quantum Hall effect can be obtained from the supermatrix σ-model, it does not help to clarify the critical behavior near the transitions between the plateaus without making assumptions. It is difficult even to say whether the slope of the Hall conductivity is finite or infinite at the transition. This is equivalent to the question whether there is a band or just a single extended state between the plateaus. In the replica formulation of the σ-model one can use some methods and results from field theory and then try to extrapolate them to $n = 0$. This is the main line of study followed by Pruisken (1985, 1987). However, as was argued recently by Zirnbauer (1994), the dependence of the critical behavior on n of the model with the topological term is not necessarily smooth, and the transition may be of the first order for $n \geq 2$. If this is true, the extrapolation to $n = 0$ would make no sense.

Instead, Zirnbauer (1994) made an attempt to solve the problem of the transition between the plateaus by using the supermatrix σ-model, Eq. (13.42). With a version of the transfer-matrix technique he reduced calculation of the functional integral with free energy $F[Q]$, Eq. (13.42), to a study of the corresponding 1D quantum problem. The reduction of dimensionality by transforming a classical problem to the corresponding quantum problem is by now a quite standard trick that often helps to solve problems. In fact, it has already been used in Chapter 11 for studying wires. The topological term F_{top}, Eq. (13.45), is transformed into an alternating sum of Wess–Zumino terms.

As a result, for values of Hall conductivity $\bar{\sigma}_{xy}$ close to $m + 1/2$, where m is an integer, where delocalization transition occurs, Zirnbauer obtained the Hamiltonian of a quantum superspin chain. The latter Hamiltonian proves to be closely related to an anisotropic version of a model proposed by Chalker and Coddington (1988). Starting from a quantum Hall system in the semiclassical high-field limit, where electrons drift slowly along the equipotential lines of a smooth random potential while exhibiting rapid cyclotron motion, these authors formulated a network model that mimicked the relevant effects of quantum tunneling near saddle points. Recalling that the σ-model was derived from a short-range potential, the agreement of both approaches that was discovered is quite encouraging because it serves as confirmation of the hypothesis of low-energy universality at the critical point.

Although the model of the superspin chains obtained by Zirnbauer does not seem to be integrable, the critical behavior is expected to be universal. This raises the hope of finding another, simpler model with the same symmetries and, therefore, with the same critical behavior. If this new model is integrable, it can lead to the solution of the quantum Hall problem, but this remains to be done. At the same time, the supersymmetry technique seems to be the most suitable analytical tool for studying the problem.

Numerical simulations for noninteracting systems (Huckestein and Kramer (1989, 1990), Miek (1990), Huckestein (1990, 1992)) show some second-order phase transition with a single extended state in the middle of each Landau level. The localization length L_c diverges in a power law, $L_c \sim |\varepsilon - \varepsilon_c|^{-\nu}$, with exponent $\nu = 2.3 \pm 0.1$. Comparison of the data with those obtained from different models (Huo and Bhatt (1992)) including the quantum mechanically treated quasi-classical percolation limit (Chalker and Coddington (1988), Milnikov and Sokolov (1988)) indicates that independent of the microscopic details of the potential the result is truly universal. Thus, the numerical studies also demonstrate that the critical properties are highly universal.

The discussion presented in this section shows that electron motion in 2D disordered systems in a strong magnetic field is a very interesting problem that can be adequately studied by the supersymmetry technique. Although the existence of Hall plateaus is in no doubt, either experimentally or theoretically, the transition between the plateaus deserves further analytical study. This seems to demand an application of techniques combining the supersymmetry method with methods of conformal field theory.

13.2 Multifractality of eigenstates in 2D disordered metals

13.2.1 What is multifractality?

As one can see from the discussion of the preceding section analytical description of the transitions between the Hall plateaus is difficult. Therefore, a lot of effort has been put into getting information about the transitions through numerical study. In the plateau regions, the eigenstates should be localized and numerical study confirms this property. The localization length diverges near the transitions and can exceed the sample size; that means that the wave functions spread all over the sample. Does it mean that they appear similar to plane waves?

Numerous numerical works show that the structure of the wave functions near the transitions is multifractal (Aoki (1983, 1986), Schreiber (1985), Schreiber and Grussbach (1991), Grussbach and Schreiber (1993, 1995), Kramer, Ono, and Ohtsuki (1988), Karawarabayashi and Ohtsuki (1995), Janßen et al. (1994), Pook and Janßen (1991), Fastenrath, Janßen, and Pook (1992)): that is, that it exhibits very complex behavior with splashes and slow decay. The fractality of some unusual objects of condensed matter physics has been intensively discussed during the last decade (see, e.g., Mandelbrot (1983), Benzi et al. (1984, 1985), Halsey et al. (1986), Paladin and Vulpiani (1987)). This notion has also been used in the analytical treatment of the Anderson metal–insulator transition. For example, using the coefficients of the inverse participation ratio (IPR) known from the renormalization group treatment in $2 + \epsilon$ dimensions (Wegner (1980)), Castellani and Peliti (1986) suggested that the structure of the wave functions at the transition is multifractal. It is the nontrivial character of the wave functions that probably manifests itself in the essentially non-Gaussian distribution of fluctuations of the density of states or conductances of weakly disordered metals predicted by Altshuler, Kravtsov and Lerner (1985, 1986, 1989, 1991) using a sophisticated renormalization group scheme for an extended σ-model.

I do not want to introduce here general definitions of multifractality, which can be found in, for example, Janßen et al. (1994). Instead, the discussion that follows is concentrated on properties of wave functions only. The concept of the multifractality of wave functions was introduced as a way to characterize their complexity in a prelocalized regime and originated in the numerically nontrivial dependence of the coefficients of IPR $t_n(V)$ on the volume of a system. By definition, the latter are the moments of the distribution function $f(t)$ of local amplitudes $t \equiv |\varphi_\alpha(\mathbf{r})|^2$ of wave functions at an arbitrary point $\mathbf{r}$ inside a sample

$$t_n = \int_0^\infty t^n f(t)\, dt \tag{13.56}$$

$$f(t) = (Vv)^{-1} \left\langle \sum_\alpha \delta\left(t - |\varphi_\alpha(\mathbf{r})|^2\right) \delta(\varepsilon - \varepsilon_\alpha) \right\rangle \tag{13.57}$$

Eq. (13.57) is similar to Eq. (9.182). The variables t and v are related to each other as $v = Vt$ and correspondingly, $f(t) = VW(tV)$. In the present section it is somewhat more convenient to use the variable t and the function $f(t)$. The distribution function $f(t)$ and the wave functions $\varphi_\alpha(\mathbf{r})$ are properly normalized such that

$$t_0 = 1, \qquad t_1 = V^{-1} \tag{13.58}$$

The coefficients t_n can be written explicitly as

$$t_n = (Vv)^{-1} \left\langle \sum_\alpha |\varphi_\alpha(\mathbf{r})|^{2n} \delta(\varepsilon - \varepsilon_\alpha) \right\rangle \tag{13.59}$$

These coefficients very sensitively indicate the degree of localization of states through their dependence $t_n(V)$ on the volume of the system. In a pure metal or a ballistic chaotic box where the wave functions extend over the whole system one has

$$t_n \propto V^{-n} \tag{13.60}$$

If disorder makes the localization length L_c much shorter than the sample size $L \sim V^{1/d}$, the coefficients t_n are insensitive to L. However, very interesting information about the development of localization can be gained through an analysis of $t_n(V)$ for small samples with $L < L_c$.

As mentioned, the latter situation can occur in the vicinity of a localization–delocalization transition, when the length $L_c(\varepsilon)$ can be larger than L. In this critical regime, the multifractality is a manifestation of a prelocalization in a piece of matter that still has predominantly metallic properties. As soon as localization length L_c exceeds the sample size, any length scale disappears and, in the language of the coefficients t_n, this is described as

$$V t_n(\varepsilon) \propto L^{-\tau(n)}, \qquad \tau(n) = (n-1)\, d^*(n) \tag{13.61}$$

where $d^*(n)$ may differ from the physical dimension d of the system and be a function of n. This function gives the values of the fractal dimensions $d^*(n)$ for each n. If the behavior of the wave functions is described by Eq. (13.60), the fractal dimension $d^*(n)$ coincides with the physical dimension d. What is very important, Porter–Thomas-type formulae, Eqs. (9.170, 9.171), which describe statistical properties of the local amplitudes in a regime of chaotic dynamics, also lead to Eq. (13.60). This means that the fractal dimension of a system obeying random matrix theory coincides with the physical dimension. In such a situation, although the amplitude fluctuations are possible, they are not very strong.

Once we assume that the envelope of a typical wave function at a length scale shorter than L_c obeys a power law $\varphi(r) \propto r^{-\mu}$ with a single fixed exponent $\mu < d/2$, the set of coefficients t_n reveals $d^* = d - 2\mu$ different from d but the same for all $n > d/(2\mu)$. This is when one speaks of fractal behavior with fractal dimension d^*.

If $d^*(n)$ is not a constant, that signals a more sophisticated structure of the wave functions. They can be imagined as splashes of multiply interfering waves at different scales and with various amplitudes, and possibly, with a self-similarity characterized by a relation between the amplitude t of the local splash of the wave function and the exponent $\mu(t)$ of the envelope of its extended power law tail.

To distinguish between the fractal and multifractal behavior in a numerical study, one has to compute and compare the values $d^*(n)$ of at least several of the lowest coefficients of the IPR. In the pioneering works by Aoki (1983, 1986), the assertion about fractality at criticality was made on the basis of studies of $d^*(2)$. The same quantity was studied by Schreiber (1985), who observed nontrivial dependence of the dimensionality $d^*(2)$ on disorder in 2D metals. Only later did it become possible to resolve different $d^*(n)$ for electron states in different situations (at the transition between the plateaus in the regime of the quantum Hall effect, at the point of the Anderson metal–insulator transition, and in 2D disordered conductors) and, thus, establish multifractality.

The goal of the present section is to demonstrate that the multifractality of the wave functions of 2D weakly disordered conductors is the most general property of these systems as soon as the sample size L does not exceed the localization length L_c (Falko and Efetov (1995a,b)). In 2D weakly disordered metals localization length L_c is, at least, exponentially large, and therefore the result can be applicable to quite large samples.

13.2.2 Reduced σ-model and its saddle points

The multifractality problem can be adequately formulated and studied by using the super-symmetry technique. First, one has to express the distribution function $f(t)$, Eq. (13.57), in terms of the Green functions, and this has been done in Eq. (9.184). The next step is to express the functions $f(t)$ in terms of an integral over supermatrix Q. For the 0D σ-model one could use Eq. (9.186); however, approximating by the 0D σ-model is good only if the parameter t is not very large. It will be seen that the distribution function $f(t)$ in the limit of large values of t should be described by the σ-model in higher dimensions, and one has to derive a proper formula for this case. Generalization of Eq. (9.186) is simple. The corresponding formula can be written for any ensemble and everywhere in the crossover regions between the ensembles (Falko and Efetov (1995a,b)). However, all results prove to be similar for all the ensembles and, to avoid cumbersome formulae, let us consider the unitary ensemble (Model IIa) only. Proceeding in the same way as in Section 9.7 one can obtain

$$f(t) = \lim_{\gamma \to 0} \int DQ \int \frac{d\mathbf{r}}{4V} \, \mathrm{str}\left(\pi_b^{(1)} Q(\mathbf{r})\right) \delta\left(t - \frac{\pi \nu \gamma}{4} \, \mathrm{str}\left(\pi_b^{(2)} Q(\mathbf{r}_o)\right)\right)$$
$$\times \exp\left(-F[Q]\right) \tag{13.62}$$

where the free energy functional $F[Q]$ has the form

$$F = \frac{\pi \nu}{8} \int \mathrm{str}\left[D_0 \left(\nabla Q(\mathbf{r})\right)^2 - \gamma \Lambda Q(\mathbf{r})\right] d\mathbf{r} \tag{13.63}$$

D_0 is the diffusion coefficient and $\mathbf{r}_o$ is the observation point. When the system can be described by the 0D σ-model the distribution function $f(t)$ does not depend on $\mathbf{r}_o$. However, beyond the 0D approximation, this function can also be a function of the coordinates.

The matrices $\pi_b^{(1,2)}$ in Eq. (13.62) select from the supermatrix Q its boson–boson sector and have the form

$$\pi_b^{(1)} = \begin{pmatrix} \pi_b & 0 \\ 0 & 0 \end{pmatrix}, \qquad \pi_b^{(2)} = \begin{pmatrix} 0 & 0 \\ 0 & \pi_b \end{pmatrix}, \qquad \pi_b = \begin{pmatrix} 0 & 0 \\ 0 & 1 \end{pmatrix} \qquad (13.64)$$

The limit $\gamma \to 0$ in Eq. (13.64) corresponds to a closed system with elastic scattering only. Because of the noncompactness of the σ-model one should, first, calculate the integral over $Q(\mathbf{r})$ and take the limit $\gamma \to 0$ at the end. However, it is not very convenient to keep an additional free parameter, and it is better to get rid of the parameter γ at an earlier stage. This can be done by integrating over the zero space harmonics of Q in the very beginning of the calculations.

If we restricted ourselves to integration over this harmonics only, we would arrive at the 0D result of Section 9.7, Eq. (9.170). If the parameter t is not large, this approximation works well, but for large t a less trivial procedure is necessary. With the scheme used later, integration over the zero space harmonics gives a reduced σ-model, which is studied in the saddle-point approximation.

To derive the reduced σ-model one should represent supermatrix $Q(\mathbf{r})$ in the form

$$Q(\mathbf{r}) = V(\mathbf{r}) \Lambda \bar{V}(\mathbf{r}), \qquad V(\mathbf{r}) \bar{V}(\mathbf{r}) = 1$$

and change the variables of integration $V(\mathbf{r})$ to $\tilde{V}(\mathbf{r})$ as $V(\mathbf{r}) = V(\mathbf{r}_o) \tilde{V}(\mathbf{r})$. This gives for supermatrices $\tilde{Q}$: $Q(\mathbf{r}) = V(\mathbf{r}_o) \tilde{Q}(\mathbf{r}) \bar{V}(\mathbf{r}_o)$. In terms of the new variables $\tilde{V}(\mathbf{r})$ and $\tilde{Q}(\mathbf{r})$ the gradient term in Eq. (13.63) preserves its form, but now the condition

$$V(\mathbf{r}_o) = 1, \qquad \tilde{Q}(\mathbf{r}_o) = \Lambda \qquad (13.65)$$

has to be fulfilled. Changing the variables of the integration for all points $\mathbf{r} \neq \mathbf{r}_o$ from $Q(\mathbf{r})$ to $\tilde{Q}(\mathbf{r})$ one obtains a new free energy functional that does not contain $V(\mathbf{r}_o)$ or $Q(\mathbf{r}_o)$. These variables enter the preexponential only, and, hence, the integral over $V(\mathbf{r}_o)$ can be computed without making approximations. The result of the integration contains only the variables $\tilde{Q}(\mathbf{r})$ with the boundary condition, Eq. (13.65). This means that the reduced σ-model obtained in this way operates only with relative variations of the field Q with respect to its value at the observation point.

The limit $\gamma \to 0$ simplifies the computation because the main contribution to the integral over the variable θ_{1o} entering the parametrization, Eqs. (6.42, 6.44, 6.47), is from $\cosh \theta_{1o} \sim 1/\gamma$. The integration over the other elements of supermatrix Q_o is a slight modification of the calculations carried out in Section 9.7. After standard manipulations one can express the distribution function $f(t)$ in the form

$$f(t) = \frac{1}{V} \frac{d^2 \Phi(t)}{dt^2}, \qquad \Phi(t) = \left\{ \int\limits_{\tilde{Q}(\mathbf{r}_o)=\Lambda} \exp\left(-\tilde{F}\left[\tilde{Q}, t\right]\right) D\tilde{Q}(\mathbf{r}) \right\} \qquad (13.66)$$

where the free energy $\tilde{F}\left[\tilde{Q}, t\right]$ has the following form:

$$\tilde{F}[Q, t] = \frac{1}{8} \int \mathrm{str} \left[\pi \nu D_0 \left(\nabla \tilde{Q}\right)^2 - 2t \Lambda n \tilde{Q} \right] d\mathbf{r} \qquad (13.67)$$

The matrix Π selects from $\tilde{Q}$ its noncompact boson–boson sector $Q_b = V_b \Lambda \bar{V}_b$

$$\Pi = \begin{pmatrix} \pi_b & \pi_b \\ \pi_b & \pi_b \end{pmatrix}, \qquad V_b = \exp \begin{pmatrix} 0 & u_2 \pi_b \theta_1/2 \\ u_2^+ \pi_b \theta_1/2 & 0 \end{pmatrix} \qquad (13.68)$$

and $u_2 = \exp(i\chi\tau_3)$, where $\theta_1 \geq 0$ and $0 \leq \chi < 2\pi$.

The 0D result, Eq. (9.170), can be obtained just by putting $\tilde{Q}(\mathbf{r}) = \Lambda$ for all $\mathbf{r}$. In this approximation, the inverse participation numbers t_n obey Eq. (13.61) and show neither multifractal nor fractal behavior. Nonetheless, this would be only an approximate procedure because the value $\tilde{Q}(\mathbf{r}) = \Lambda$ does not correspond to the minimum of the functional $\tilde{F}$ when $t \neq 0$. The second term in Eq. (13.67) acts on the supermatrix $\tilde{Q}$ as if an external field tended to "align" the supermatrix Q_b along such a direction that $\theta_1 \to \infty$. However, the condition $\tilde{Q}(\mathbf{r}_o) = \Lambda$ prevents that, and the minimum corresponds to a nonhomogeneous configuration of a finite $Q_b(\mathbf{r})$. An equation for the minimum can be found by substituting Q_b for $\tilde{Q}$ in Eq. (13.67) and varying the variable θ_1 under the conditions

$$\mathbf{n}\nabla\theta_1 = 0, \qquad \theta_1(\mathbf{r}_o) = 0 \qquad (13.69)$$

The first condition in Eq. (13.69) is written on the boundary of the sample, where $\mathbf{n}$ is the unit vector perpendicular to it. The origin of this condition will be explained later, whereas the second condition is obvious. Using Eqs. (13.69) one writes the equation for the extremal solution θ_t in the form

$$\Delta_{\mathbf{r}}\theta_t(\mathbf{r}) = -\frac{t}{\pi\nu D_0} \exp(-\theta_t(\mathbf{r})), \qquad \chi(\mathbf{r}) = \pi \qquad (13.70)$$

where $\Delta_{\mathbf{r}}$ is the Laplacian. The solution $\theta_t(\mathbf{r})$ of Eq. (13.70) has to be substituted into the free energy functional $\tilde{F}$ in Eq. (13.67), which takes the form

$$F_t = \frac{1}{2} \int \left[\pi\nu D_0 (\nabla\theta_t)^2 + 2t \exp(-\theta_t) \right] d\mathbf{r} \qquad (13.71)$$

Substituting Eq. (13.71) into Eq. (13.66) one can obtain the final result with exponential accuracy. Let us emphasize that nontrivial solutions are obtained in the noncompact boson–boson sector, whereas in the preceding section nontrivial solutions in the compact sector were considered.

Nontrivial saddle points in the noncompact sector were discovered by Muzykantskii and Khmelnitskii (1995) when studying a σ-model in a time representation obtained for calculating current relaxation in an open disordered conductor. These authors attributed the nontrivial saddle points to the existence of nearly localized states. Calculating the distribution function of wave functions $f(t)$ one can probe such states directly and this will be done later. Eq. (13.70) is somewhat different from the one obtained by Muzykantskii and Khmelnitskii (1995).

Let us note that Eq. (13.58) can be obtained from Eq. (13.66) exactly. As concerns the coefficient t_1 one easily obtains $t_1 = \Phi(0)$. At $t = 0$ free energy functional $\tilde{F}\left[\tilde{Q}, t\right]$ is completely invariant under rotations of supermatrix $\tilde{Q}$. This means that the value of the integral for $\Phi(0)$ can be obtained just by putting $\tilde{Q} = \Lambda$, which gives

$$\Phi(0) = 1 \qquad (13.72)$$

Calculating t_0 one obtains $d\Phi(0)/dt$. Then, one arrives at averaging of $\tilde{Q}$ with the functional $\tilde{F}\left[\tilde{Q}, 0\right]$ and this again gives unity.

In principle, fluctuations near the saddle-point solution could be important. Therefore, before solving Eq. (13.70) explicitly, it makes sense to discuss conditions under which the fluctuations can be neglected. It is useful also to clarify the procedure of integrating near the saddle point.

In fact, the calculations are straightforward. When computing the integral over $\tilde{Q}$ in Eq. (13.66) one can represent the supermatrix $\tilde{Q}$ in the form (cf. Eq. (5.1))

$$\tilde{Q}(\mathbf{r}) = V_t(\mathbf{r}) \Lambda (1 + iP(\mathbf{r})) (1 - iP(\mathbf{r}))^{-1} \bar{V}_t(\mathbf{r}) \qquad (13.73)$$

The supermatrix $V_t(\mathbf{r})$ is obtained from the supermatrix V_b, Eq. (13.68), substituting the solution $\theta_t(\mathbf{r})$ of Eq. (13.70) for θ_1. The supermatrix $P(\mathbf{r})$ has the structure determined by Eq. (5.1) and describes the fluctuations. To evaluate the contribution of the fluctuations one has to substitute $\tilde{Q}(\mathbf{r})$ from Eq. (13.73) into Eqs. (13.66, 13.67) and expand the latter equations in $P(\mathbf{r})$. In the zeroth order in P, Eq. (13.73) corresponds to the minimum solution that defines F_t, Eq. (13.71). The next term of the expansion $F^{(1)}$ is, generally speaking, linear in P. Using Eq. (5.1) one can write this term in the form

$$F^{(1)} = \frac{i}{2} \int \mathrm{str}\, \pi_b \left[-\pi \nu D_0 \nabla \theta_t \left(\nabla B + \nabla \bar{B} \right) + t \exp(-\theta_t) \left(B + \bar{B} \right) \right] d\mathbf{r} \qquad (13.74)$$

An application of the saddle-point method implies absence of linear terms and it is natural to choose θ_t in such a way that they vanish. Integrating the first term in Eq. (13.74) by parts we see easily that the terms that are linear in B vanish provided the first equalities in Eqs. (13.69, 13.70) are satisfied. The function $\Phi(t)$, Eq. (13.66), can be written as

$$\Phi(t) = J(t) \exp(-F_t) \qquad (13.75)$$

where the function $J(t)$ includes the fluctuations and has the form

$$J(t) = \int DB \exp\left[-F^{(2)} - F^{(3)} - F^{(4)} \ldots \right] \qquad (13.76)$$

At $t = 0$ one easily obtains the trivial solution $\theta_t = 0$. With Eq. (13.72) this gives $J(0) = 1$. To calculate the integral in Eq. (13.76) one may proceed by making an expansion of the exponent in all nonquadratic terms, thus reducing the computation to evaluation of Gaussian integrals. All the higher-order terms are small as soon as the sample size L is smaller than the localization length L_c. Calculation of these terms is analogous to calculation of the weak localization corrections considered in Chapter 5. Thus, in the limit $L \ll L_c$, the function $J(t)$ reduces to a superdeterminant of a Hamiltonian related to fluctuations around the saddle point. The evaluation of the superdeterminant is presented in Falko and Efetov (1995b). In the main approximation the superdeterminant can be replaced by a constant.

13.2.3 Distribution function and coefficients of the inverse participation ratio

So, to compute the distribution function $f(t)$ and the coefficients of the inverse participation ratio t_n one has to solve Eq. (13.70) with the boundary conditions, Eq. (13.69), and substitute the solution into Eqs. (13.71, 13.66). Although the basic results are not sensitive to the geometry of the sample and the position of the observation point, all formulae are somewhat simpler if the sample has the form of a disc and the observation point is located at the center. In this case Eq. (13.70) can be solved exactly and the solution is presented in Falko and Efetov (1995a,b). However, the solution is rather cumbersome. Alternatively, one can solve this equation by using a perturbative scheme that can also be applied in higher dimensions.

In this scheme, the zero order is obtained, neglecting the right-hand side of Eq. (13.70). The solution $\theta_t^{(0)}$ in this approximation is equal in 2D to

$$\theta_t^{(0)} = 2\mu \ln u \tag{13.77}$$

where μ is a constant and $u = r/l$ is the dimensionless distance from the center. The function $\theta_t^{(0)}$ turns to zero at $r = l$ and formally diverges as $r \to 0$. However, the σ-model was derived for lengths exceeding the mean free path l and therefore one should not use Eq. (13.77) for $r < l$. Instead, one should cut off the radii $r < l$ and replace the boundary condition at $r = 0$ by the same boundary condition at $r = r_0 \sim l$. Then, the second condition in Eq. (13.69) has to be written as

$$\theta_t (1) = 0 \tag{13.78}$$

and the zero-order approximation $\theta_t^{(0)}$, Eq. (13.77), is chosen to satisfy this condition.

The exact solution θ_t of Eq. (13.70) is sought in the form

$$\theta_t = \theta_t^{(0)} + \theta_t^{(1)} \tag{13.79}$$

Substituting θ_t in the right-hand side of Eq. (13.70) by $\theta_t^{(0)}$, Eq. (13.77), and using Eq. (13.78) we obtain for $\theta_t^{(1)}$ the following formula

$$\theta_t^{(1)} (u) = \frac{1}{4\rho^2 (1 - \mu)^2} \left(1 - u^{2(1-\mu)} \right) \tag{13.80}$$

where

$$\rho^2 = \frac{\pi \nu D_0}{t l^2}$$

To find the parameter μ one has to use the condition at the sample edge, Eq. (13.69). Differentiating $\theta_t^{(0)}$ and $\theta_t^{(1)}$ over r one arrives at the following equation for μ:

$$z \exp z = T \equiv \frac{t V \ln (L/l)}{2\pi^2 \nu D_0}; \qquad \mu = \frac{z (T)}{2 \ln (L/l)} \tag{13.81}$$

It is important to check that the perturbative scheme works for the value of μ determined by Eq. (13.81). Eq. (13.80) is obtained by neglecting $\theta_t^{(1)}$ in the exponential in Eq. (13.70); this is valid provided $\theta_t^{(1)} (u) \ll 1$ for all u. Using this condition one obtains

$$\left(\frac{L}{l} \right)^{1-\mu} \ll \rho (1 - \mu) \tag{13.82}$$

Combining this inequality with Eq. (13.81) we see easily that the approximations made are justified in the limit

$$\mu \ll 1 \tag{13.83}$$

This is the inequality (13.83) that gives the possibility of making the cutoff at short distances and using all formulae at an arbitrary position of the observation point and the shape of the sample. Let us note that this inequality does not prevent the parameter z from being much larger than unity. However, the parameter z is restricted from above and this leads to the inequality

$$t \ll (\lambda_F l)^{-1} \ln^{-2} \left(\frac{L}{l} \right) \tag{13.84}$$

where λ_F is the wavelength. The leading terms of the expansion of free energy F_t corresponding to the extremal solution can be written as

$$F_t \approx (2\pi)^2 \, \nu D_0 \left(\mu + \mu^2 \ln \frac{L}{l} \right) \tag{13.85}$$

Solving Eq. (13.81) and substituting the solution into Eqs. (13.85, 13.75, 13.66) one finds the distribution function $f(t)$. In the limiting cases of $T \ll 1$ and $T \gg 1$, with the exponential accuracy this function has the form

$$f(t) = AV \begin{cases} \exp\left(-Vt\left[1 - \frac{T}{2} + \cdots\right]\right), & T \ll 1 \\[2mm] \exp\left(-\frac{\pi^2 \nu D_0}{\ln(L/l)} \ln^2 T\right), & T \gg 1 \end{cases} \tag{13.86}$$

where A is a normalization constant.

Eq. (13.86) shows that the disorder makes high-amplitude splashes of wave functions much more probable than one would expect from the Porter–Thomas formula. Deviations from this distribution become essential for $Vt \geq \sqrt{2\pi\nu D_0}$. Only in the opposite limit can the 0D result serve as a good approximation. In the latter limit one can make an expansion of the distribution function in the parameter $Vt/\sqrt{\nu D_0}$ (Fyodorov and Mirlin (1994)).

At very large t the function $f(t)$ has log-normal asymptotics that is strikingly similar to the asymptotics of the distribution functions of the local density of states or conductances discovered by Altshuler, Kravtsov, and Lerner (1985, 1986, 1989, 1991). Even the numerical coefficients in the exponentials are the same, although, of course, the logarithms contain different variables. It appears that the log-normal form is really universal. Obviously, this behavior is due to localization effects. At the same time the tails of the states responsible for this asymptotics do not decay exponentially, as one would expect for a particle localized in a quantum well or in a wire at distances exceeding the localization length. This is seen from the second line of Eq. (13.86): Even in the regime $T \gg 1$ the size L of the system enters the distribution function. The splashes look as if they were formed by focusing the waves of some rare configurations of scatterers.

Information about the structure of these states can be extracted from the way their distribution is sensitive to the boundary or by calculation of more complicated cross-correlation functions. It turns out (Falko and Efetov (1995b)) that the envelope of the tail of a state associated with a large-amplitude splash $|\varphi(\mathbf{r}_o)|^2 = t$ follows the form of the minimal solution and, therefore, has a power-law tail $|\varphi(\mathbf{r}_o)|^2 \propto \exp(-\theta_t(\mathbf{r})) \approx (l/r)^{2\mu}$ with an individual exponent $\mu = \mu(t) < 1$ for each amplitude, approaching the r^{-2} dependence for the limiting values of $t \sim (l\lambda_F)^{-1}$. At the same time, a clear physical explanation of the log-normal tail is absent as yet.

Eq. (13.86) for the distribution function $f(t)$ enables us to demonstrate a scaling of the coefficients t_n with the size of the system and, ultimately, the multifractal nature. Alternatively, one can write these coefficients in the form

$$t_n = \frac{n(n-1)}{V^n} \int dt \, J(t) \exp\left(-F_t + (n-2)\ln(tV)\right) \tag{13.87}$$

and use it for calculation of the integral over t in the saddle-point method. The validity of the approximation is easily justified for large n. In the leading order the coefficients t_n take

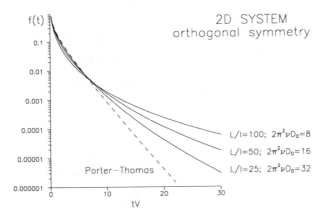

Fig. 13.4. The distribution function of the amplitudes of the wave functions for the orthogonal ensemble at several values of the mean free path l. The values of the function f in the tail region increase with an increase in the disorder. The dimensionless conductance g for the parameters given in the figure takes the values $g = 2\pi^2 \nu D_0 / \ln (L/l) = 1.7, 4.1, 9.9$, the largest value corresponding to the weakest disorder.

the form

$$t_n \approx \frac{\min \left[n!, \, (2\pi \nu D_0 / \ln (L/l))^n \right]}{l^{2\delta}} \left(\frac{1}{V}\right)^{n-\delta}, \qquad \delta \approx \frac{n^2}{8\pi^2 \nu D_0} \qquad (13.88)$$

Eq. (13.88) explicitly demonstrates the multifractality of the eigenstates. Comparing this equation with Eq. (13.61) we obtain the fractal dimension $d^*(n)$

$$d^*(n) = 2 - n \left(2\pi^2 \beta_0 \nu D_0\right)^{-1} \qquad (13.89)$$

where $\beta_0 = 2$. Because of the limitation on t, Eq. (13.84), the preceding formula works for $n \leq 2\pi D_0$, so that $n - \delta > 0$, which means that the fractal dimension $d^*(n)$ must be positive. At small $n \ll \sqrt{2\pi \nu D_0}$ the dependence of t_n on the sample size can be found by using a finite polynomial expansion of $f(t)$ in a series in T (Fyodorov and Mirlin (1994)). In the leading order in $(\nu D_0)^{-1} \ln (L/l)$ the result of the calculation agrees with Eqs. (13.61, 13.89).

Eq. (13.89) is also correct for orthogonal and symplectic ensembles provided a proper value of β_0 is used. For the orthogonal ensemble $\beta_0 = 1$, whereas for the symplectic ensemble $\beta_0 = 4$. For the unitary ensemble, in which time-reversal symmetry is broken by magnetic impurities, one should use the value $\beta_0 = 4$.

The entire distribution function $f(t)$ can be calculated numerically for all ensembles. For the orthogonal ensemble it is depicted for different values of parameters characterizing the disorder in Fig. 13.4. At comparatively small values of the amplitudes t, the function is close to the Porter–Thomas distribution, whereas at larger values the deviation from this law is considerable.

Studying statistical properties of wave functions is no doubt an interesting theoretical problem, but can one confirm the predictions experimentally? Usually it is not simple because in most situations wave functions cannot be probed directly. However, very recently an experiment in which the wave functions were really studied has been performed (Kudrolli, Kidambi, and Sridhar (1995)). It was done by using thin microwave cavities,

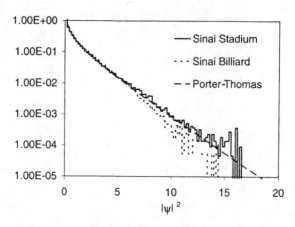

Fig. 13.5. Density distribution $f(|\psi|^2)$ for the Sinai stadium and Sinai billiard compared with the Porter–Thomas distribution for the orthogonal ensemble. While the Sinai stadium is in excellent agreement, the Sinai billiard data show slight deviations due to states influenced by bouncing-ball orbits. (From Kudrolli, Kidambi and Sridhar (1995).)

which exploits the correspondence between the Maxwell and Schrödinger equations. In these systems the amplitudes of the eigenfunctions can really be directly measured by using a cavity perturbation technique. By changing the shapes of the cavities one can have different analogs of quantum billiards, which allow the possibility of testing the predictions quantum chaos theory. At the same time, one can fabricate cavities in which tiles are placed to act as hard scatterers and study effects of localization.

The distribution function of the local amplitudes $f(|\psi|^2)$ for clean cavities used to study quantum chaos was found to obey the Porter–Thomas distribution very well for the orthogonal ensemble, Eq. (9.171). The results for two shapes of the cavity are shown in Fig. 13.5. The distribution function in the disordered cavities is more complicated. At weak disorder, the low-amplitude part of the distribution function is very close to the Porter–Thomas distribution. However, at higher densities the distribution function decays more slowly, signaling that high local densities corresponding to nearly localized states are quite probable. The decay is especially slow in highly disordered cavities. Such behavior is in agreement with Fig. 13.4, although a detailed comparison has not been performed.

The theory presented previously was not available when Kudrolli, Kidambi, and Sridhar (1995) made their experiment, and so, these authors tried to make a fit to the perturbative results of Fyodorov and Mirlin (1994). The corresponding experimental and theoretical curves are represented in Fig. 13.6. Although good agreement can be achieved, the use of the perturbative results in the region where the deviations from the Porter–Thomas distribution are considerable is not legitimate and one should try to make a fit to the nonperturbative results of Fig. 13.4.

Thus, we see that properties of weakly disordered 2D conductors are highly nontrivial even for sample sizes L much smaller than localization length L_c. As concerns the opposite limit $L \geq L_c$, the problem becomes even more difficult because one must take into account fluctuations near the extremal solution. Apparently, one should try to use a renormalization group scheme analogous to the one developed in Chapter 5. However, after each step of the renormalization one may obtain a new extremal solution. So, it is quite probable that

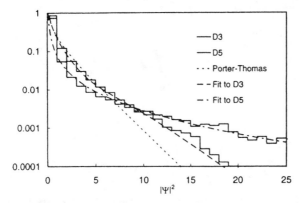

Fig. 13.6. The distribution function for two realizations of the disorder (D3 and D5), showing deviations from Porter–Thomas distribution that are due to localization. (From Kudrolli, Kidambi, and Sridhar (1995).)

instead of renormalizing the conductance of the sample one should renormalize a function that can describe, for example, a distribution of the conductances. This would mean that one has to consider a functional renormalization group. However, this does not seem to be an easy task.

14

Afterword

It is clear from the material presented here that the supersymmetry technique, in particular, the supermatrix nonlinear σ-model, is an extremely efficient way of studying various problems. A natural question may be asked: Why does all this work? What is the physical meaning of the invariance of the σ-model under rotations of supermatrices Q?

These questions are not easy to answer. I cannot explain why the supermatrix σ-model enables us to get some nontrivial results that cannot be obtained by other methods currently available. The supersymmetry formalism was derived from the Schrödinger equation with a random potential. All symmetries of the σ-model appeared in the process of the derivation, but the initial Schrödinger equation does not contain them. My attitude to these questions is that all nice symmetry features of the σ-model are formal and it is difficult to attribute to them a clear physical sense.

I want to emphasize that the Grassmann anticommuting variables χ_i were introduced in a completely formal way. The initial Schrödinger equation did not contain them and they were introduced with the hope that they would help in the calculations. I cannot explain why the variables χ_i that are completely formal mathematical objects helped to get the results. However, it is not unusual for abstract mathematical objects to be useful for explicit computations.

The best example of this type may be another artificial mathematical object $i = \sqrt{-1}$. This symbol is no less abstract than the anticommuting variables χ (although the reader is definitely more accustomed to i than to the anticommuting variables). However, suppose that we have to compute the integral

$$I = \int\limits_{-\infty}^{\infty} \frac{\cos mx}{x^2 + a^2} dx \qquad (14.1)$$

and suppose that we were not good students and did not study or we have forgotten the residue theory. Then, evaluation of the integral I would not be simple. This integral contains real variables only and the idea of using artificial objects like i for calculations would not seem natural for a freshman. So, one first has to invest some labor in studying the theory of complex variable functions. This certainly takes time, but with knowledge of the residue theory one can compute the integral I and many others in one line. Because of their usefulness in many applications, fewer people try to avoid complex numbers nowadays.

The situation with the anticommuting variables χ_i is similar, and I believe that, soon, these variable will be as common in physics as complex numbers. In fact, the Grassmann anticommuting variable χ is an extremely natural extension of numbers. Real numbers r are such numbers that $r^2 > 0$. Imaginary numbers s obey the inequality $s^2 < 0$. It is natural to complement these sets by numbers χ such that $\chi^2 = 0$ and we come to the Grassmann variables.

Although the book contains solutions of some important problems, it is absolutely clear that a lot has to be done. Let me list some important unsolved problems that, I hope, can be solved and for which the supersymmetry technique seems to be the most adequate calculational tool.

This book contains the words *disorder* and *chaos* in its title. However, these two different disciplines remain to be unified analytically. Although the equivalence of the Wigner–Dyson random matrix theory and the 0D supermatrix σ-model has been proved (see, e.g., Chapter 6), the equivalence of random matrix theory and problems of quantum chaos is based mainly on numeric and semiclassical considerations. It would be extremely interesting to start from, for example, a chaotic quantum billiard and derive the random matrix theory or the 0D σ-model. I believe that trying to derive the σ-model is more promising. At least with respect to disorder problems, the σ-model has been derived, whereas it is completely unclear how to obtain random matrix theory directly. A derivation of the σ-model from chaotic billiards would help clarify conditions under which a billiard manifests quantum chaos.

The same connection also has to be established between such problems as that of the kicked rotator and the 1D σ-model. Although, again, the one-dimensional σ-model can be derived from the theory of random band matrices, the equivalence of the latter theory and the kicked rotator model is based on (quite plausible) assumptions.

The Anderson metal–insulator transition in three dimensions is still a problem. It is clear that the one-parameter scaling theory based on the renormalization theory in $2 + \epsilon$ dimensions (Chapter 5) leads to a completely different scenario with respect to the one obtained in the effective medium approximation (Chapter 12); the latter approximation is well controlled in a high dimensionality. The question of which of these scenarios wins in three dimensions is open, but I believe that essential features of the effective medium approximation scenario must also be present there.

Very interesting problems remain unsolved in two dimensions. It would be very interesting to describe the transitions between the plateaus in the quantum Hall effect, and the σ-model with the topological term seems to be adequate for this problem. In the absence of the magnetic field or if the field is weak, properties of large samples also are not clear. The nontrivial saddle-point configurations discussed in Chapter 13 can completely change the renormalization group equations derived in Chapter 5. It is quite possible that one can arrive at a functional renormalization group that is definitely different from the conventional one-parameter scaling picture. Then, the question of the character of the localization in two dimensions becomes open. I would not exclude the possibility of a power law localization, although the exponential localization is the most probable prediction.

The list can be continued, but it is already clear that there are many different problems that can be attacked by using the supersymmetry technique. At the same time, it is difficult to foresee interesting new directions of research. For example, the possibility of calculating correlation functions for 1D models of interacting particles (Chapter 10) by using a 0D supermatrix σ-model was not anticipated. Maybe the supersymmetry method can become useful for studying continuous matrix models popular in field theory, but I do not exclude that something completely new can appear.

So, I finish this book hoping that important new results will be obtained with the super-symmetry technique in the very near future.

Appendix 1

Calculation of the Jacobian

To derive the Jacobian (Berezinian) of the transformation to the variables $\hat{\theta}$, u, v one can again write, as in Section 2.5 and Section 6.1, the elementary length, which is just str $(dQ)^2$ in the case considered. Using Eq. (6.47) we have

$$dQ = dU\,Q_0\bar{U} + U\,dQ_0\bar{U} + U\,Q_0d\bar{U} \tag{1.1}$$

Rewriting the elementary length with the help of Eqs. (6.47) and the relation $d\bar{U} = -\bar{U}dU\bar{U}$ one can reduce this quantity to a more explicit form

$$\text{str}\,(dQ)^2 = \text{str}\,(dQ_0)^2 - 2\,\text{str}\,(\delta u)^2 - 2\,\text{str}\,(\delta v)^2$$
$$+ 2\,\text{str}\left(2\delta u\sin\hat{\theta}\,\delta v\sin\hat{\theta} + (\delta u\cos\hat{\theta})^2 + (\delta v\cos\hat{\theta})^2\right) \tag{1.2}$$

where $\delta u = \bar{u}du = -\delta u$, $\delta v = \bar{v}dv = -\delta\bar{v}$. Let us rewrite Eq. (1.2) explicitly separating the anticommuting Grassmann variables δu_{12}, δu_{21}, δv_{12}, δv_{21} from the conventional commuting variables δu_{11}, δu_{22}, δv_{11}, δv_{22}

$$\text{str}\,(dQ)^2 = \text{str}\,(dQ_0)^2 + \left[\text{str}\,(dQ)^2\right]_1 + \left[\text{str}\,(dQ)^2\right]_2 \tag{1.3}$$

where

$$\text{str}\,(dQ_0)^2 = 2\,\text{str}\left(d\hat{\theta}\right)^2 \tag{1.4}$$

$$\left[\text{str}\,(dQ)^2\right]_1 = 4\,\text{tr}[2\delta u_{12}\sin\hat{\theta}_{22}\delta v_{21}\sin\hat{\theta}_{11}$$
$$- \delta u_{12}\delta u_{21} + \delta u_{12}\cos\hat{\theta}_{22}\delta u_{21}\cos\hat{\theta}_{11} - \delta v_{12}\delta v_{21}$$
$$+ \delta v_{12}\cos\hat{\theta}_{22}\delta v_{21}\cos\hat{\theta}_{11}] \tag{1.5}$$

$$\left[\text{str}\,(dQ)^2\right]_2 = 2\,\text{tr}[-(\delta u_{11})^2 + \left(\delta u_{11}\cos\hat{\theta}_{11}\right)^2$$
$$- (\delta v_{11})^2 + \left(\delta v_{11}\cos\hat{\theta}_{11}\right)^2 + 2\delta u_{11}\sin\hat{\theta}_{11}\delta v_{11}\sin\hat{\theta}_{11}$$
$$+ (\delta u_{22})^2 - \left(\delta u_{22}\cos\hat{\theta}_{22}\right)^2 + (\delta v_{22})^2 - \left(\delta v_{22}\cos\hat{\theta}_{22}\right)^2$$
$$- 2\delta u_{22}\sin\hat{\theta}_{22}\delta v_{22}\sin\hat{\theta}_{22}] \tag{1.6}$$

The subdivision of str $(dQ)^2$ into three parts, Eqs. (1.4–1.6), is convenient because the first part, Eq. (1.4), contains only the elements of the matrix $\hat{\theta}$; the second part, Eq. (1.5), contains only the anticommuting elements of δu_{12}, δu_{21}, δv_{12}, δv_{21}, and the third part, Eq. (1.6), is the contribution due to commuting elements of δu_{11}, δu_{22}, δv_{11}, δv_{22}. The total Jacobian (Berezinian) is equal to the product of the Jacobians corresponding to these parts. Using the parametrization for Q suggested in Section 6.3 one can rewrite the quadratic form Eq. (1.3) through the independent variables η, κ, M, ϕ, and χ.

Let us first present the calculation for Model II (unitary ensemble). For this model using Eq. (6.44) we have

$$\text{str} \, (d Q_0)^2 = 4 \left((d\theta)^2 + (d\theta_1)^2 \right) \tag{1.7}$$

Using the simple equality

$$\delta u = \bar{u}_2 \delta u_1 u_2 + \delta u_2 \tag{1.8}$$

one obtains $(\delta u_{12})_{11} = 2d\eta_1 z$, $(\delta u_{12})_{22} = -2d\eta_1^* z^*$, $(\delta u_{21})_{11} = -2d\eta_1^* z^*$, $(\delta u_{21})_{22} = -2d\eta_1 z$, where $z = \exp(i(\chi - \phi))$. Similar relations hold for the matrices $(\delta\kappa)_{12}$ and $(\delta\kappa)_{21}$ and one can write Eq. (1.5) as

$$\left[\text{str} \, (dQ)^2 \right]_1 = 32[(d\eta_1 d\eta_1^* - d\kappa_1 d\kappa_1^*)(1 - \cos\theta \cosh\theta_1) \\ + \sin\theta \sinh\theta_1 (z d\eta_1 d\kappa_1^* + z^* d\kappa_1 d\eta_1^*)] \tag{1.9}$$

For the commuting elements of the matrices δu and δv one has

$$\delta u_{11} = (d\phi - 2i(\eta_1 d\eta_1^* - d\eta_1 \eta_1^*))i\tau_3$$
$$\delta u_{22} = (d\chi - 2i(\eta_1 d\eta_1^* - d\eta_1 \eta_1^*))i\tau_3 \tag{1.10}$$
$$\delta v_{11} = \delta v_{22} = -2(\kappa_1 d\kappa_1^* - d\kappa_1 \kappa_1^*)\tau_3$$

Substituting Eqs. (1.10) into Eq. (1.6) we have

$$\left[\text{str} \, (dQ)^2 \right]_2 = 4 \left((d\tilde{\phi})^2 \sin^2\theta + (d\tilde{\chi})^2 \sinh^2\theta_1 \right) \tag{1.11}$$

where $d\tilde{\phi} - d\phi = d\tilde{\chi} - d\chi = -2i(\eta_1 d\eta_1^* - d\eta_1 \eta_1^* - \kappa_1 d\kappa_1^* + d\kappa_1 \kappa_1^*)$.

Reconstructing the Jacobians $J_1^{(II)}$ and $J_2^{(II)}$ from the quadratic forms, Eqs. (1.7–1.11) one comes to Eqs. (6.67) (the determinant corresponding to the quadratic form of the Grassmann differentials, Eq. (1.9), goes to the denominator in $J_1^{(II)}$). At first glance, one should write the product $d\tilde{\phi} d\tilde{\chi}$ instead of $d\phi d\chi$. However, these products being multiplied by $d R_1^{(II)}$ are equal to each other. Let us remember that from the beginning integration over complex numbers was weighted by π, Eq. (2.74), and that is why one has π^2 in the denominator in $J_2^{(II)}$, Eq. (6.67).

Calculation of the Berezinians for Models I and III, although as straightforward as for Model II, is more lengthy. In fact, Models I and III are very similar, as one can see from the formulae of Section 6.3, and so, it is enough to present the derivation for Model I.

Eq. (1.4) can be written for the Model I as

$$\text{str} \, (d Q_0)^2 = 4 \left[(d\theta)^2 + (d\theta_1)^2 + (d\theta_2)^2 \right] \tag{1.12}$$

In order to rewrite Eq. (1.5) explicitly one has to calculate the matrices δu_{12}, δu_{21}, δv_{12}, δv_{21}, $\sin\hat{\theta}_{22}$, $\cos\hat{\theta}_{22}$, using Eqs. (6.43, 6.57, 6.57–6.59). The matrices $\cos\theta_{11}$, $\sin\theta_{11}$ are proportional to the unit matrix. After simple manipulations we have

$$\cos\hat{\theta}_{22} = \begin{pmatrix} \cosh\theta_1 \cosh\theta_2 & \sinh\theta_1 \sinh\theta_2 \\ \sinh\theta_1 \sinh\theta_2 & \cosh\theta_1 \cosh\theta_2 \end{pmatrix}, \tag{1.13}$$

$$\sin\hat{\theta}_{22} = i \begin{pmatrix} \sinh\theta_1\cosh\theta_2 & \cosh\theta_1\sinh\theta_2 \\ \cosh\theta_1\sinh\theta_2 & \sinh\theta_1\cosh\theta_2 \end{pmatrix}, \tag{1.14}$$

$$(\delta u)_{12} = 2\bar{F}_1[d\eta - 2d\eta\,(\bar{\eta}\eta + \eta\bar{\eta}) \\ - 2\eta\bar{\eta}d\eta\bar{\eta}\eta - 12\eta\bar{\eta}\eta d\bar{\eta}\eta]F_2 = 2\bar{F}_1 d\tilde{\eta}F_2 \tag{1.15}$$

where

$$d\tilde{\eta} = d\eta \begin{pmatrix} 1 - 8\eta_1\eta_1^*\eta_2\eta_2^* & -4\eta_1^*\eta_2 \\ -4\eta_2^*\eta_1 & 1 - 8\eta_1\eta_1^*\eta_2\eta_2^* \end{pmatrix} \tag{1.16}$$

The same equation can be written for the matrix δv_{12}. The matrices δu_{21} and δv_{21} can be found easily by using the relations $\overline{(\delta u)} = -\delta u$, $\overline{(\delta v)} = -\delta v$. The determinant of the transformation from the differentials $d\eta$ to $d\tilde{\eta}$ is equal to unity, and therefore one can write the quadratic form in terms of the differentials $d\tilde{\eta}$. Moreover, it is convenient to use the differentials

$$d\eta' = \bar{F}_1 d\tilde{\eta} F_2 \tag{1.17}$$

The determinant of this transformation is again unity. The corresponding transformation should be done for the differentials $d\kappa$ also, and finally the quadratic form, Eq. (1.5), can be represented as a scalar product

$$\left[\mathrm{str}\,(dQ)^2\right]_1 = 32\,(d\vec{\rho},\,Ad\vec{\rho}^*) \tag{1.18}$$

where $d\vec{\rho} = (\ d\eta_1'\ \ d\eta_2'\ \ d\kappa_1'\ \ d\kappa_2'\ )$ and the matrix A is

$$A = \begin{pmatrix} a & b \\ b & -a \end{pmatrix} \tag{1.19}$$

with the 2×2 blocks a and b

$$a = \begin{pmatrix} 1 - \cosh\theta_1\cosh\theta_2 & -\sinh\theta_1\sinh\theta_2 \\ -\sinh\theta_1\sinh\theta_2 & 1 - \cosh\theta_1\cosh\theta_2 \end{pmatrix}\cos\theta,$$

$$b = \begin{pmatrix} \sinh\theta_1\cosh\theta_2 & \cosh\theta_1\sinh\theta_2 \\ \cosh\theta_1\sinh\theta_2 & \sinh\theta_1\cosh\theta_2 \end{pmatrix}\sin\theta$$

Calculating $\det A$ one comes to $J_1^{(\mathrm{I})}$, Eq. (6.66).

To write the expression $\left[\mathrm{str}\,(dQ)^2\right]_2$, Eq. (1.6), one can use the relations

$$\delta F_1 = -2i\,(1 - iM)^{-1}\,dM\,(1 + iM)^{-1},$$

$$\delta F_2 = id\phi\tau_3, \qquad \delta\Phi = id\chi\tau_3, \qquad \sin\hat{\theta}_{11} = \sin\theta \tag{1.20}$$

Substituting Eqs. (1.20) into Eqs. (6.51, 1.6) one can reduce $\left[\mathrm{str}\,(dQ)^2\right]_2$ to the form

$$\left[\mathrm{str}\,(dQ)^2\right]_2 = \left[\mathrm{str}\,(dQ)^2\right]_2' + \left[\mathrm{str}\,(dQ)^2\right]_2'' \tag{1.21}$$

where

$$\left[\text{str}\,(d\,Q)^2\right]_2' = \frac{16\sin^2\theta\left((dm)^2 + |dm_1|^2\right)}{\left(1 + m^2 + |m_1|^2\right)^2} \tag{1.22}$$

$$\left[\text{str}\,(d\,Q)^2\right]_2'' = (d\vec{\phi},\; B d\vec{\phi}) \tag{1.23}$$

In Eq. (1.23), $d\vec{\phi} = \begin{pmatrix} d\phi & d\chi \end{pmatrix}$ and the matrix B in the scalar product is equal to

$$B = \begin{pmatrix} \sinh^2\theta_1 + \sinh^2\theta_2 & \sinh^2\theta_1 - \sinh^2\theta_2 \\ \sinh^2\theta_1 - \sinh^2\theta_2 & \sinh^2\theta_1 + \sinh^2\theta_2 \end{pmatrix} \tag{1.24}$$

Calculating det B and using Eqs. (1.12, 1.24) we obtain $J_2^{(1)}$ in Eq. (6.66).

Appendix 2

Magnetic field parametrization

It is not difficult to check that, with parametrization, Eqs. (8.45–8.51), all symmetries of supermatrix Q, and in particular Eqs. (4.38, 4.48) are satisfied. The inequalities imposed on the variables $\theta_{c,d}$ and $\theta_{1c,d}$ are chosen in such a way that double counting is prevented. To write these inequalities one need not worry about the Grassmann variables. The most economical way to obtain them is to compare formulae for the magnetic field parametrization and the standard one by writing the "skeletons" of the supermatrix Q consisting of conventional variables. The 4×4 blocks Q^{12} already contain all independent variables such that it suffices to compare the blocks Q^{12} given by Eqs. (8.45–8.50) and by Eqs. (6.42, 6.43, 6.50–6.52). Equating the respective 2×2 boson–boson blocks yields

$$\sinh{(\theta_{1d})} \cosh{(\theta_{1c})} \exp{(i\chi_d)} = \sinh{(\theta_1)} \cosh{(\theta_2)} \exp{(i\,(\phi - \chi))}$$

$$\cosh{(\theta_{1d})} \sinh{(\theta_{1c})} \exp{(i\phi_c)} = \cosh{(\theta_1)} \sinh{(\theta_2)} \exp{(i\,(\phi + \chi))} \tag{2.1}$$

with $0 < \theta_1, \theta_2 < \infty$ and $0 < \phi, \chi < 2\pi$. From Eq. (2.1), the following correspondence can be derived

$$\theta_{1d} \leftrightarrow \theta_1, \qquad \theta_{1c} \leftrightarrow \theta_2, \qquad \chi_d \leftrightarrow \phi - \chi, \qquad \chi_c = \chi + \phi \tag{2.2}$$

and one comes to the inequalities for θ_{1c}, θ_{1d}, χ_c, χ_d written in Eqs. (8.50, 8.51).

Comparing the fermion–fermion blocks we have

$$\sin\theta_d \cos\theta_c \leftrightarrow \sqrt{t}\sin\theta, \qquad \sin\theta_c \leftrightarrow \sin\theta \sqrt{1 - t} \tag{2.3}$$

and

$$\phi_d \leftrightarrow \alpha, \ \phi_c \leftrightarrow \beta \tag{2.4}$$

where $0 < \theta < \pi$ and the matrix $\mathcal{F}_1$ in Eq. (6.52) is reparametrized as

$$\mathcal{F}_1 = \begin{pmatrix} \sqrt{t}\exp{(i\alpha)} & -\sqrt{1-t}\exp{(i\beta)} \\ \sqrt{1-t}\exp{(-i\beta)} & \sqrt{t}\exp{(-i\alpha)} \end{pmatrix} \tag{2.5}$$

with $0 < t < 1$ and $0 < \alpha, \beta < 2\pi$. Supposing that $0 < \phi_c, \phi_d < 2\pi$ one comes, using Eq. (2.4), to the inequalities $\sin\theta_d > 0$, $\cos\theta_c > 0$, $\sin\theta_c > 0$, and hence to the first two inequalities in Eq. (8.50). It is worth emphasizing that Eqs. (2.2–2.4) are correspondence equations that can be used to find the limits of integration but are not equalities. This means that one may not use these equations when changing from standard to magnetic field parametrization. Exact equations relating these parametrizations to each other contain all anticommuting variables and are much more complicated.

In order to write the Berezinian for the parametrization, Eqs. (8.45–8.51), one can start again, as in the previous appendix, writing the square of the elementary length in the superspace of the matrix elements, $\mathrm{str}\,(d\,Q)^2$. Using Eq. (8.49) we have

$$\mathrm{str}\,(d\,Q)^2 = \mathrm{str}\,[\delta V,\ \Lambda]^2 \tag{2.6}$$

with $\delta V = \bar{V}\,dV$. The fact that the elementary length is expressed in terms of the commutator of δV with Λ in Eq. (2.6) means that the blocks $(\delta V)^{12}$ and $(\delta V)^{21}$ can be considered independent differentials. The final goal is to express them in terms of the differentials of the independent variables $\theta_{c,d},\ \theta_{1c,d},\ \phi_{c,d},\ \chi_{c,d},\ \eta_{c,d},\ \eta^*_{c,d},\ \kappa_{c,d}$ and $\kappa^*_{c,d}$, from Eqs. (8.46–8.51). It is more convenient to perform the evaluation of the Berezinian in several steps, first calculating the Jacobian of transformation of the variables contained in supermatrix $(\delta V)^\perp$ to variables contained in supermatrices $(\delta V_c)^\perp$ and $(\delta V_d)^\perp$, where the superscript $\perp$ denotes the part of the supermatrices anticommuting with Λ. Using Eq. (8.49) one can write

$$\delta V = \delta V_c + \bar{V}_c \delta V_d V_c \tag{2.7}$$

The elements in the supermatrix $(\delta V)_\perp$ anticommuting with τ_3 correspond to the cooperon degrees of freedom, and those of $(\delta V)_\parallel$ to diffuson ones. Using the block structure of the supermatrices V_d and V_c we see that

$$(\delta V)^\perp_\parallel = \bar{V}^\parallel_c (\delta V_d)^\perp V^\parallel_c + \bar{V}^\perp_c (\delta V_d)^\perp V^\perp_c \tag{2.8}$$

so that $(\delta V_c)^\perp$ does not enter this equation and the corresponding determinant vanishes

$$\frac{(\delta V)^\perp_\parallel}{(\delta V_c)^\perp} = 0 \tag{2.9}$$

This gives the possibility of considering the Berezinian of the transformations from $(\delta V)^\perp_\parallel$ to $(\delta V_d)^\perp$ and from $(\delta V)^\perp_\perp$ to $(\delta V_c)^\perp$ independently. As concerns the Berezinian of the latter transformation we can conclude immediately, using Eq. (2.7), that it is equal to unity

$$\frac{(\delta V)^\perp_\perp}{(\delta V_c)^\perp} = 1 \tag{2.10}$$

To calculate the Berezinian of the transformation to the variables $(\delta V_d)^\perp$ let us write this matrix in the form

$$(\delta V_d)^\perp = i\begin{pmatrix} 0 & B \\ \bar{B} & 0 \end{pmatrix} \tag{2.11}$$

with

$$B = \begin{pmatrix} a & i\sigma \\ \rho^+ & ib \end{pmatrix}, \qquad \bar{B} = \begin{pmatrix} a^+ & \rho \\ i\sigma^+ & ib^+ \end{pmatrix}$$

and

$$a = \begin{pmatrix} a & 0 \\ 0 & a^* \end{pmatrix}, \qquad b = \begin{pmatrix} b & 0 \\ 0 & b^* \end{pmatrix}, \qquad \sigma, \rho = \begin{pmatrix} \sigma, \rho & 0 \\ 0 & -\sigma^*, -\rho^* \end{pmatrix}$$

Eqs. (2.11) provide the necessary symmetry of the supermatrix $(\delta V_d)^{\perp}$ that follows from the relation $\delta V = -\delta \bar{V}$. First, let us make the transformation

$$
\begin{pmatrix} 0 & B \\ \bar{B} & 0 \end{pmatrix} \rightarrow \begin{pmatrix} u_c & 0 \\ 0 & v_c \end{pmatrix} \begin{pmatrix} 0 & B \\ \bar{B} & 0 \end{pmatrix} \begin{pmatrix} \bar{u}_c & 0 \\ 0 & \bar{v}_c \end{pmatrix} \tag{2.12}
$$

which does not change the structure of the matrix $(\delta V_d)^{\perp}$ but removes the matrices u_c and v_c from Eq. (8.48) for V_c. Then, using Eq. (2.8), denoting by Q_{c0} the central part of the supermatrix Q_c, Eq. (8.46), we can write

$$
\begin{aligned}
\text{str} \left[(\delta V)_{\parallel}, \Lambda \right]^2 &= -2 \, \text{str} \left((\delta V)_{\parallel}^{\perp} \right)^2 = \text{str} \left[(\delta V_d)^{\perp}, Q_c \right]^2 \\
&\rightarrow 2 \, \text{str} [C_0 \sin \hat{\theta}_c \bar{B} C_0 \sin \hat{\theta}_c \bar{B} + C_0^T \sin \hat{\theta}_c B C_0^T \sin \hat{\theta}_c B \\
&\quad + 2 \cos \hat{\theta}_c B \cos \hat{\theta}_c \bar{B} + 2 B \bar{B}]
\end{aligned} \tag{2.13}
$$

Writing the symbol $\rightarrow$ in Eq. (2.13) implies that, in order to pass from the first to the second line, one should make the transformation, Eq. (2.12). The quadratic form, Eq. (2.13), should be rewritten explicitly through the commuting and anticommuting variables, and these two parts decouple. The part corresponding to the commuting variables a and b takes the form

$$
16 \left(|a|^2 \cos^2 \theta_c + |b|^2 \right) \tag{2.14}
$$

and the second part is written as

$$
8 \left[(\sigma^* \sigma + \rho \rho^*) (1 + \cosh \theta_{1c} \cos \theta_c) + (\sigma \rho + \sigma^* \rho^*) \sin \theta_c \sinh \theta_{1c} \right] \tag{2.15}
$$

From Eqs. (2.14, 2.15), the Berezinian J_{cd} of the transformation from the variables $(\delta V)_{\parallel}^{\perp}$ to the variables $(\delta V_d)^{\perp}$ can easily be extracted and written in the form

$$
J_{cd} = 4 \left(\frac{\cos \theta_c}{\cosh \theta_{1c} + \cos \theta_c} \right)^2 \tag{2.16}
$$

Now, the only thing that remains to be done is to write the Berezinians of the transformations from the variables $(\delta V_c)^{\perp}$ and $(\delta V_d)^{\perp}$ to the variables $\theta_{c,d}, \theta_{1c,d}, \phi_{c,d}, \chi_{c,d}, \eta_{c,d}, \eta_{c,d}^*, \kappa_{c,d}$, and $\kappa_{c,d}^*$. This is a simple task because both Berezinians J_c and J_d coincide with the Berezinian of the corresponding transformation in the unitary ensemble (cf. Eq. (6.67)). As a result, the total Berezinian J reads as

$$
J = J_c J_d J_{cd} \tag{2.17}
$$

and we come to Eq. (8.52).

Appendix 3

Density–density correlation function at $k = 0$

In this appendix a direct computation of the integral I_d, Eq. (7.46), is presented to check Eq. (7.48), which is a consequence of particle conservation and must hold for an arbitrary magnetic field. Of course, it would not be reasonable to try to obtain simple relations by making complicated calculations. The main purpose of this appendix is to give one more example of calculations with the magnetic field parametrization introduced in Chapter 8.

Using the magnetic field parametrization, Eq. (8.45–8.51), one has to overcome the difficulties due to the singularities at $\lambda_{1c} = \lambda_c$ and $\lambda_{1d} = \lambda_d$ in the Jacobian Eq. (8.52). The procedure of avoiding the singularities has been explained in Section 8.3. Following this procedure one can rewrite the integral I_d, Eq. (7.46), as

$$I_d(\omega) = (I_d(\omega))_{\text{unit}} + \lim_{X_0 \to \infty} \int \text{str}\left[k(1 - \tau_3) Q^{12} k(1 - \tau_3) Q^{21}\right]$$

$$\times \left(\exp\left(-F_1[Q, X]\right) - \exp\left(-F_1[Q, X_0]\right)\right) dQ \tag{3.1}$$

where $(I_d(\omega))_{\text{unit}}$ is the integral, Eq. (7.46), calculated in the limit $X_0 \to \infty$ (unitary ensemble) and the function $F_1[Q, X]$ is determined by Eq. (8.42). First, let us calculate the integral $(I_d(\omega))_{\text{unit}}$. Using the parametrization for the supermatrices Q for the unitary ensemble, Eqs. (6.42, 6.44, 6.46, 6.51, 6.53, 6.60); the Jacobian Eq. (6.67), the relation

$$\int \bar{u}_d k(1 - \tau_3) u_d d\eta_d d\eta_d^* = 8(1 - \tau_3) \tag{3.2}$$

and the corresponding relation for integration over κ_d, κ_d^* one has

$$(I_d(\omega))_{\text{unit}} = 16 \int\limits_1^\infty \int\limits_{-1}^1 \frac{\lambda_1 + \lambda}{\lambda_1 - \lambda} \exp\left(i(x + i\delta)(\lambda_1 - \lambda)\right) d\lambda_1 d\lambda \tag{3.3}$$

The integral in Eq. (3.3) can be most conveniently computed by using the variables

$$t = \lambda_1 - \lambda, \qquad s = \lambda_1 + \lambda \tag{3.4}$$

Integrating first over the variable s and then over the variable t we obtain

$$(I_d(\omega))_{\text{unit}} = \frac{32}{-ix} \equiv \frac{32\Delta}{-i\pi\omega} \tag{3.5}$$

In fact, one can anticipate general arguments that this is already the final result for $I_d(\omega)$ because a magnetic field cannot influence the particle conservation. However, our task now is to make an exercise on calculation of integrals over supermatrix Q and therefore let us proceed further.

Following the procedure of Section 8.3 one should integrate with a more general free energy function $\tilde{F}[Q, X, Y]$, Eq. (8.57). The integral in the second term in Eq. (3.1) is to be subdivided into integrals with $Z_1(Q)$ and $Z_2(Q)$, Eqs. (8.58). The integral I_2 with $Z_2(Q)$ reads

$$I_2 = \int \text{str} \left[k(1 - \tau_3) Q^{12} k(1 - \tau_3) Q^{21} \right] Z_2(Q) \, dQ \tag{3.6}$$

In the limit $Y_0 \to \infty$ all diffuson degrees of freedom are "frozen," and one has to integrate over the cooperon degrees. Then in Eq. (3.6) one should replace supermatrix Q by Q_c, Eq. (8.46), and integrate over Q_c using the cooperon part of the Jacobian in Eq. (8.52). Integrating first over the anticommuting variables one obtains

$$\int \text{str} \left[(\bar{u}_c k(1 - \tau_3) u_c) C_0 \sin \hat{\theta}_c (\bar{v}_c k(1 - \tau_3) v_c) C_0^T \sin \hat{\theta}_c \right] d\eta_c d\eta_c^* d\kappa_c^* d\kappa_c$$

$$= 64 \, \text{str} \left[(1 - \tau_3) C_0 \sin \hat{\theta}_c (1 - \tau_3) C_0^T \sin \hat{\theta}_c \right] \tag{3.7}$$

$$= 64 \, \text{str} \left[(1 - \tau_3) \sin \hat{\theta}_c (1 + \tau_3) \sin \hat{\theta}_c \right] = 0$$

Computation of the integral I_1

$$I_1 = \int \text{str} \left[k(1 - \tau_3) Q^{12} k(1 - \tau_3) Q^{21} \right] Z_1(Q) \exp(-F_0[Q]) \, dQ \tag{3.8}$$

is considerably more lengthy because one has to integrate over both the diffuson and cooperon degrees of freedom. Elements of the supermatrices u_d, v_d enter only the term in the square brackets in the integrand in Eq. (3.8), and therefore it is convenient to integrate over them first. Using the parametrization, Eqs. (8.45–8.48), and Eq. (3.2) for the integral of the square brackets over the supermatrices u_d, v_d one has

$$\int \text{str} \left[\bar{u}_d k(1 - \tau_3) u_d \left(V_{d0} Q_c \bar{V}_{d0} \right)^{12} \bar{v}_d k(1 - \tau_3) v_d \left(V_{d0} Q_c \bar{V}_{d0} \right)^{21} \right]$$

$$\times d\eta_d d\eta_d^* d\kappa_d^* d\kappa_d = 64 \, \text{str} \left[(1 - \tau_3) \left(V_{d0} Q_c \bar{V}_{d0} \right)^{12} (1 - \tau_3) \left(V_{d0} Q_c \bar{V}_{d0} \right)^{21} \right]$$

$$= 16 \, \text{str} \left[(1 - \tau_3)(1 + Q_{d0}) Q_c (1 - \tau_3)(1 - Q_{d0}) Q_c \right] \tag{3.9}$$

where $Q_{d0} = \bar{V}_{d0} \Lambda V_{d0}$, and V_{d0} is the central part of Eq. (8.47). Substituting the result of the integration over the diffuson anticommuting variables, Eq. (3.9), into Eq. (3.8) one should then integrate over the cooperon anticommuting variables. They arise not only from the supermatrices Q_c in Eq. (3.9) but also from the function $\exp(-F_0[Q])$. Using Eq. (8.53) one can see that the latter function can be represented as a sum of three terms: The first term does not contain the anticommuting variables, the second term is proportional to $\eta_c \eta_c^* - \kappa_c \kappa_c^*$, and the third term to $\eta_c \eta_c^* \kappa_c \kappa_c^*$.

Integrating the third term over the anticommuting variables one may replace all Q_c in the preexponential by Q_{c0}, and the result of the integration is equal to

$$-64 x^2 (\lambda_{1c} - \lambda_c)^2 (\lambda_{1d} - \lambda_d)^2 \left(\lambda_c^2 \left(1 - \lambda_d^2 \right) + \lambda_{1c}^2 \left(\lambda_{1d}^2 - 1 \right) \right) \tag{3.10}$$

(In Eq. (3.10) only the preexponential is written.) The result of integration with the first term of the expansion of $\exp(-F_0[Q])$ is obtained by integrating the last line of Eq. (3.9) over the cooperon anticommuting variables

$$32 \int \mathrm{str} \left[(1 - \tau_3) \sin \hat{\theta}_d \bar{u}_c \cos \hat{\theta}_c u_c \, (1 - \tau_3) \sin \hat{\theta}_d \bar{v}_c \cos \hat{\theta}_c v_c \right]$$

$$\times \, d\eta_c d\eta_c^* d\kappa_c^* d\kappa_c = 128 \, (\lambda_{1c} - \lambda_c)^2 \left(\lambda_{1d}^2 - \lambda_d^2 \right) \tag{3.11}$$

When calculating the integral in the first line of Eq. (3.11) it is convenient to represent the matrix $\cos \hat{\theta}_c$ as

$$\cos \hat{\theta} = \frac{1}{2} (\lambda_{1c} + \lambda_c) - \frac{k}{2} (\lambda_{1c} - \lambda_c)$$

and then use Eq. (3.2), which also holds for the cooperon variables.

Calculation of the integral with the second term of the expansion of $\exp(-F_0[Q])$ is lengthy although still straightforward. When performing the integration it is convenient to use the following formula:

$$u_c \cos \hat{\theta}_c \bar{u}_c = \cos \hat{\theta}_c + 4\eta_c \bar{\eta}_c (\lambda_{1c} - \lambda_c) + 2 (\lambda_{1c} - \lambda_c) \begin{pmatrix} 0 & \eta_c \\ \bar{\eta}_c & 0 \end{pmatrix}$$

The corresponding preexponential after this integration reads

$$- 128ix \, (\lambda_{1c} - \lambda_c)^2 \, (\lambda_{1d} - \lambda_d) \left[(\lambda_{1c} - \lambda_c) (\lambda_{1d}\lambda_d + 1) - 2 \left(\lambda_{1c}\lambda_{1d}^2 - \lambda_c\lambda_d^2 \right) \right] \tag{3.12}$$

Picking up all the terms, Eqs. (3.10–3.12), and writing the Jacobian and the exponentials one obtains that the second term in Eq. (3.1) is proportional to $\tilde{I}$

$$\tilde{I} = \int \left[\frac{2(\lambda_{1d} + \lambda_d)}{\lambda_{1d} - \lambda_d} - \frac{2ix \left((\lambda_{1c} - \lambda_c)(\lambda_{1d}\lambda_d + 1) - 2 \left(\lambda_{1c}\lambda_{1d}^2 - \lambda_c\lambda_d^2 \right) \right)}{\lambda_{1d} - \lambda_d} \right.$$

$$\left. - x^2 \left(\lambda_{1c}^2 \left(\lambda_{1d}^2 - 1 \right) - \lambda_c^2 \left(\lambda_d^2 - 1 \right) \right) \right] \exp \left(ix \, (\lambda_{1d}\lambda_{1c} - \lambda_d\lambda_c) \right)$$

$$\times \, (\lambda_{1c} - \lambda_c)^{-2} \exp \left(X^2 \left(\lambda^2 - \lambda_1^2 \right) \right) d\lambda_{1d} d\lambda_d d\lambda_{1c} d\lambda_c \tag{3.13}$$

We see that the integral, Eq. (3.13), is still rather nontrivial, and, in order to see that finally it turns to zero, one should integrate over λ_{1d} and λ_d. Integration of the third term in Eq. (3.13) over these variables is simple and one obtains

$$\frac{4}{x} \exp(ix\lambda_{1c}) \left(\frac{\sin \lambda_c x}{\lambda_c} - \frac{\cos \lambda_c x}{i\lambda_{1c}} \right) \tag{3.14}$$

(the functions of λ_{1c} and λ_c from the third line in Eq. (3.13) are not written in Eq. (3.14)). Integration of the first two terms in the integrand in Eq. (3.13) can be performed by using the parametrization, Eq. (3.4). Integrating first over the variable s one comes to the following integrals over t:

$$- 2 \int_0^2 \left[\exp \left(ix \, (t\lambda_{1c} + \lambda_{1c} - \lambda_c) \right) + \exp \left(ix \, (t\lambda_c + \lambda_{1c} - \lambda_c) \right) \right] dt \tag{3.15}$$

$$- 2 \int_2^\infty \left[\exp \left(ix \, (t\lambda_{1c} + \lambda_{1c} - \lambda_c) \right) + \exp \left(ix \, (t\lambda_{1c} + \lambda_c - \lambda_{1c}) \right) \right] dt$$

The integration over t is simple and gives the following result:

$$-\frac{4}{x} \exp{(ix\lambda_{1c})} \left[\frac{\sin \lambda_c x}{\lambda_c} - \frac{\cos \lambda_c x}{i\lambda_{1c}} \right] \tag{3.16}$$

Comparing Eqs. (3.14) and (3.16) we see that they cancel each other, and, thus, the second integral in Eq. (3.1) is equal to zero. This proves Eq. (7.46).

Appendix 4

Effective medium approximation as a saddle point

A4.1 Effective Lagrangian

The mean field theory can be obtained by seeking the minimum of the Landau expansion for the order parameter (cf. Section 12.1). This is a well-known fact that allows one to study fluctuations in a very convenient manner. Can one write such an effective Lagrangian that its extremum would give the effective minimum approximation? This might be the first step in studying corrections due to fluctuations, determining the upper critical dimensionality, and writing renormalization group equations near this dimensionality, which is the conventional scheme in theory of phase transitions.

It turns out that such a Lagrangian can be written in an extended space consisting of the real one and of the space of the matrix elements of the supermatrix Q (Efetov (1990)). To construct the Lagrangian, suppose that the model specified by Eq. (12.33) is defined on a d-dimensional hypercubic lattice and that only nearest neighbors interact in this equation. Before deriving the Lagrangian it is convenient to obtain several useful relations.

One can introduce the operator $\hat{N}$, which acts as follows:

$$\hat{N} \Phi (Q) = \int N (Q, \, Q') \, \Phi (Q') \, dQ' \tag{4.1}$$

where the kernel $N (Q, \, Q')$ is given by Eq. (11.17). Let $\Phi_{\mathcal{E}}$ denote an eigenfunction of $\hat{N}$, that is,

$$\hat{N} \Phi_{\mathcal{E}} (Q) = \mathcal{E} \Phi_{\mathcal{E}} (Q) \tag{4.2}$$

The inverse operator $\hat{N}^{-1}$ is defined in the usual way

$$\int N^{-1} (Q, \, Q'') \, N (Q'', \, Q') \, dQ'' = \delta (Q - Q') \tag{4.3}$$

The δ-function entering Eq. (4.3) is defined in Eqs. (11.20, 11.21). The operator $\hat{N}^{-1}$ has the eigenvalues $\mathcal{E}^{-1}$ and the eigenfunctions $\Phi_{\mathcal{E}} (Q)$.

Using the function $N^{-1} (Q, \, Q')$ one can construct a quadratic form R

$$R = B \int \bar{x} (Q) \, N^{-1} (Q, \, Q') \, x (Q') \, dQ \, dQ' \tag{4.4}$$

where B is a number.

In Eq. (4.4), $x (Q)$ and $\bar{x} (Q)$ are functions that depend on elements of supermatrices Q. An arbitrary function $x (Q) (\bar{x} (Q))$ is represented by the expansion in a series in the eigenfunctions $\Phi_\mathcal{E}(Q)$,

$$x(Q) = \sum_\mathcal{E} x_\mathcal{E} \Phi_\mathcal{E}(Q), \qquad \bar{x}(Q) = \sum_\mathcal{E} \Phi_\mathcal{E}^*(Q) x_\mathcal{E}^* \tag{4.5}$$

It is assumed that the products $x_\mathcal{E} \Phi_\mathcal{E}(Q)$ and $\Phi_\mathcal{E}^*(Q) x_\mathcal{E}^*$ are commuting variables, although $x_\mathcal{E}, x_\mathcal{E}^*, \Phi_\mathcal{E}$ and $\Phi_\mathcal{E}^*$ separately may be both commuting and anticommuting ones. Because of this convention the functions $x(Q)$ and $\bar{x}(Q)$ are commuting variables.

Using the quadratic form, Eq. (4.4), and the expansions, Eq. (4.5), one can construct Gaussian integrals

$$\frac{\int x(Q)\,\bar{x}(Q') \exp(-R)\, Dx(Q)\, D\bar{x}(Q)}{\int \exp(-R)\, Dx(Q)\, D\bar{x}(Q)} = N(Q,\,Q') \tag{4.6}$$

Eq. (4.6) enables us to rewrite the partition function

$$\int \exp(-F[Q])\, DQ$$

and the correlation function $Y^{00}(\mathbf{r},\,\omega)$, Eq. (12.34), in an unusual form that can, however, be useful for studying the Anderson metal–insulator transition. In order to derive this representation let me present one more formula for an effective Green's function $g_{rr'}^{\alpha\beta}(Q,\,Q')$. This function can be introduced as follows:

$$g_{\mathbf{rr'}}^{\alpha\beta}(Q,\,Q') = \int g_{\mathbf{k}}^{\alpha\beta}(Q,\,Q') \exp\left[i\mathbf{k}(\mathbf{r}-\mathbf{r}')\right] \frac{d^d\mathbf{k}}{(2\pi)^d},$$

$$g_{\mathbf{k}}^{\alpha\beta}(Q,\,Q') = \frac{\int y_{\mathbf{k}}^\alpha(Q)\,\bar{y}_{\mathbf{k}}(Q') \exp(-\mathcal{L}_0[y])\, Dx\, D\bar{x}}{\int \exp(-\mathcal{L}_0[y])\, Dx\, D\bar{x}} \tag{4.7}$$

where

$$\mathcal{L}_0[y] = \sum_{\alpha=1}^d \int \exp(-ik_\alpha)\,\bar{y}_{\mathbf{k}}^\alpha(Q)\,N^{-1}(Q,\,Q')\,y_{\mathbf{k}}^\alpha(Q')\,dQ\,dQ' \frac{d^d\mathbf{k}}{(2\pi)^d} \tag{4.8}$$

In Eqs. (4.7, 4.8), the variables $y_{\mathbf{k}}^\alpha$ are related to $x_{\mathbf{k}}^\alpha$ as follows:

$$y_{\mathbf{k}}^\alpha(Q) = \text{sign}(\cos k_\alpha) \cdot x_{\mathbf{k}}^\alpha, \quad \text{where } \text{sign}\,t = \begin{cases} 1, & t > 0 \\ -1, & t < 0 \end{cases} \tag{4.9}$$

The use of the additional factor $\text{sign}(\cos k_\alpha)$ in Eq. (4.9) provides the convergence of the integrals in Eq. (4.7). The variables $\alpha,\,\beta$ in Eqs. (4.7–4.9) stand for the directions of the d-dimensional cubic lattice. (It is assumed that the distance between neighboring sites is equal to unity.)

Using Eq. (4.6) for Gaussian integrals one can easily obtain the following equality:

$$g_{\mathbf{rr'}}^{\alpha\beta}(Q,\,Q') = N(Q,\,Q')\,\delta_{\alpha\beta}\delta(\mathbf{r}+\mathbf{e}_\alpha - \mathbf{r}') \tag{4.10}$$

where $\mathbf{e}_\alpha$ is the unit vector directed along the α-axis.

Now, let us introduce the effective Lagrangian $\mathcal{L}$,

$$\mathcal{L}[y] = \mathcal{L}_0[y] + \mathcal{L}_{\text{int}}[y] \tag{4.11}$$

where $\mathcal{L}_0[y]$ is defined by Eq. (4.8) and $\mathcal{L}_{int}[y]$ is equal to

$$\mathcal{L}_{int}[y] = -\sum_r \int Z_0(Q) \prod_\alpha \bar{y}_\mathbf{r}^\alpha(Q) y_\mathbf{r}^\alpha(Q) dQ \qquad (4.12)$$

where the function $Z_0(Q)$ is defined in Eq. (11.17).

The Lagrangian $\mathcal{L}$, Eqs. (4.11, 4.12), is a formal construction and I do not try to attribute any physical sense to it.

Having defined the Lagrangian $\mathcal{L}$, Eqs. (4.11, 4.12), one can write the basic formulae of the proposed method. For the partition function one writes the following identity:

$$\int \exp(-F[Q]) DQ = A^{-1} \int \exp(-\mathcal{L}[y]) Dx(Q) \qquad (4.13)$$

where

$$A = \int \exp(-\mathcal{L}_0[y]) Dx$$

In order to prove Eq. (4.13) one should expand the exponential on the right-hand side of this equation, representing it in the form of a series in power of $\mathcal{L}_{int}$ and calculate Gaussian integrals over $x(Q)$. As usual, by such an expansion one can represent each term of the series by a graph. We already have the expression, Eq. (4.10), for the bare Green function, which can be represented by a segment connecting two neighboring sites. This segment has only one direction. We have the functions $Z_0(Q)$ as vertices. One should integrate over Q at each site. Only one line can enter each site along an axis and only one line can go out of the site because only one direction of segments along each axis is possible. In what follows it is assumed that cyclic boundary conditions are used.

Using the rules formulated one can see that only one graph in the expansion is not equal to zero. This graph consists of N vertices where N is the number of sites on the lattice, and only one vertex is at each site. Using Eqs. (11.17, 4.10) it is not difficult to realize that this graph gives the exact partition function of the system. Of course, the partition function of the supermatrix model is exactly equal to unity but the identity Eq. (4.13) holds for models with arbitrary objects Q.

The main quantity that has to be calculated in the supermatrix model involved is the density–density correlation function $Y^{00}(\mathbf{r}, \omega)$, Eq. (12.34). Using the same arguments presented for the derivation of Eq. (4.13) one can obtain

$$Y^{00}(\mathbf{r}, \omega) = -2\pi^2 v^2 q^2 \frac{\int M_{33}^{12}(0; y) M_{33}^{21}(r; y) \exp(-\mathcal{L}[y]) Dx}{\int \exp(-\mathcal{L}[y]) Dx} \qquad (4.14)$$

where

$$M_{kl}^{ij}(r, y) = \int Q_{kl}^{ij} Z_0(Q) \prod_\alpha \bar{y}_\mathbf{r}^\alpha(Q) y_\mathbf{r}^\alpha(Q) dQ,$$

$$y_\mathbf{r}^\alpha(Q) = \int y_\mathbf{k}^\alpha(Q) \exp(i\mathbf{kr}) \frac{d^d \mathbf{k}}{(2\pi)^d}$$

Eq. (4.14) is the basis of the Lagrangian formulation. It is relevant to emphasize that Eq. (4.14) was obtained from Eq. (12.34) without making any approximations and is exact. At first glance, it is much more complicated than Eq. (12.34). However, this representation is very convenient for studying the metal–insulator transition. This transition can be obtained in the saddle-point approximation. Eq. (4.14) describes a field theoretical model in an extended superspace (real space + space of supermatrices Q).

For one-dimensional systems the Lagrangian $\mathcal{L}$ is quadratic and hence the functional integral in Eq. (4.14) can be calculated exactly. Proceeding in this way one comes to the transfer-matrix equations studied in Chapter 11. The theory in this case becomes "trivial" and no phase transition can be obtained.

A4.2 Saddle point

Let us now try to evaluate the functional integrals over x by using the saddle-point method. Calculating the functional derivative of $\mathcal{L}$, Eqs. (4.8, 4.11, 4.12), over y one obtains the following equation:

$$y_{\mathbf{r}-\mathbf{e}_\alpha}^\alpha (Q) = \int N(Q, Q') Z_0(Q') \left(\prod_{p\neq\alpha} \bar{y}_{\mathbf{r}}^\beta (Q') y_{\mathbf{r}}^\beta (Q') \right) y_{\mathbf{r}}^\alpha (Q') dQ' \qquad (4.15)$$

An analogous equation can be obtained for $\bar{y}$.

Let us seek solutions of Eq. (4.15) that do not depend on coordinates and α

$$y_{\mathbf{r}}^\alpha (Q) = \bar{y}_{\mathbf{r}}^\alpha (Q) = \Psi(Q) \qquad (4.16)$$

Substituting Eq. (4.16) into Eq. (4.15) for $\Psi(Q)$ we obtain

$$\Psi(Q) = \int N(Q, Q') Z_0(Q') \Psi^m(Q') dQ', \qquad m = 2d - 1 \qquad (4.17)$$

Eq. (4.17) coincides exactly with Eq. (12.47), which is the main equation in the effective medium scheme. We see that the effective medium approximation can be obtained from the extremum conditions of the Lagrangian, Eq. (4.8, 4.11, 4.12). Corrections to this solution can be considered as fluctuations and studied by methods of phase transition theory.

The formulation of the metal–insulator transition problem in terms of a field theory with the Lagrangian specified by Eq. (4.8, 4.11, 4.12) appears to be very similar to the description of phase transitions where the free energy functional in the Ginzburg–Landau form is used. Usually the conditions of the minimum of the free energy functional give equations for the order parameter. Using this analogy one may identify the solution $\Psi(Q)$ (or, for example $-\ln \Psi(Q)$) of Eq. (4.17) with an order parameter of the problem under consideration. However, this order parameter is a nontrivial function below the transition. This can be similar to the situation existing in spin glasses (Parisi (1979)), where the order parameter is also a function depending on an additional variable.

The density–density correlation function as well as other correlators can also be computed by using Eq. (4.14) and the saddle-point method. In the zero-order approximation one might replace all $y_{\mathbf{r}}^\alpha (Q)$ in Eq. (4.14) by $\Psi(Q)$. However, such a substitution gives zero because the integrals of M_{kl}^{ij} yield zero after integration over the anticommuting elements of the supermatrices Q (the function $\Psi(Q)$ depends only on the variables $\hat{\theta}$ and the anticommuting elements can only come from Q_{kl}^{ij}). In order to get a nonzero contribution one should represent the variable $x(Q)$ as

$$x_{\mathbf{k}}(Q) = \Psi(Q) \delta(\mathbf{k}) + \tilde{x}_{\mathbf{k}}(Q) \qquad (4.18)$$

and make an expansion in terms of $\tilde{x}_{\mathbf{k}}(Q)$. The Lagrangian $\mathcal{L}$, Eqs. (4.8, 4.11, 4.12), is still quadratic with respect to $\tilde{x}(Q)$ in one dimension and an exact solution is possible. For $d > 1$ the Lagrangian also has terms of higher order. These higher-order terms can be considered as perturbations describing fluctuations near the saddle point. Investigation of these fluctuations is not simple and has not been done yet. Here, only the part of the Lagrangian quadratic in $\tilde{x}(Q)$ is taken into account. The behavior of the density–density correlation function Y^{00} is most interesting at large distances $r \gg 1$. In this limit only small momenta are important. For small k, the variables $y_{\mathbf{k}}(Q)$ coincide with the variables $x_{\mathbf{k}}(Q)$. Besides, in the preexponential in Eq. (4.14) one may keep only the second power of $\tilde{x}(Q)$; all other $y(Q)$ are replaced by $\Psi(Q)$. As a result of this expansion one obtains

$$\mathcal{L}[x] = \mathcal{L}^{(0)} + \mathcal{L}^{(2)}[\tilde{x}],$$

$$\mathcal{L}^{(0)} = N[(m+1)\int \Psi(Q) N^{-1}(Q, Q') \Psi(Q') dQ dQ'$$

$$- \int Z_0(Q) \Psi^{m+1}(Q) dQ], \qquad (4.19)$$

$$\mathcal{L}^{(2)}[\tilde{x}] = \int \left\{ \sum_\alpha \int \exp(-i\mathbf{k}\mathbf{e}_\alpha) \bar{\tilde{x}}_{\mathbf{k}}^\alpha(Q) N^{-1}(Q, Q') \tilde{x}_{\mathbf{k}}^\alpha(Q') dQ dQ' \right.$$

$$- \int Z(Q) \left(\frac{1}{2} \sum_{\alpha \neq \beta} \left[\tilde{x}_{\mathbf{k}}^\alpha(Q) \tilde{x}_{-\mathbf{k}}^\beta(Q) + \bar{\tilde{x}}_{\mathbf{k}}^\alpha(Q) \bar{\tilde{x}}_{-\mathbf{k}}^\beta \right] \right.$$

$$\left. + \sum_{\alpha\beta} \bar{\tilde{x}}_{\mathbf{k}}^\alpha(Q) \tilde{x}_{\mathbf{k}}^\beta(Q) \right) dQ \right\} \frac{d^d\mathbf{k}}{(2\pi)^d} \qquad (4.20)$$

where $Z(Q)$ is given by Eq. (12.48).

The zero-order term $\mathcal{L}^{(0)}$ is not interesting and does not contribute to the physical quantities. In the second term the variables $\tilde{x}$ are written instead of y because only small k are important. Of course, $\exp(-i\mathbf{k}\mathbf{e}_\alpha)$ must also be expanded when making explicit computations.

For the function M_{kl}^{ij} from Eq. (4.14) one gets in the same approximation

$$M_{kl}^{ij}(r, x) = \sum_\alpha \int Q_{33}^{12} \Psi^m(Q) Z_0(Q) \left[\bar{\tilde{x}}_r^\alpha(Q) + \tilde{x}_r^\alpha(Q) \right] dQ \qquad (4.21)$$

Now the calculation of the density–density correlation function is straightforward. One has to calculate Gaussian integrals only. In order to carry out these calculations explicitly, it is convenient to introduce eigenvalues $\mathcal{E}$ and eigenfunctions $\phi_\mathcal{E}$ according to the following integral equation

$$\mathcal{E}\phi_\mathcal{E}(Q) = \int N(Q, Q') Z^{1/2}(Q) Z^{1/2}(Q') \phi_\mathcal{E}(Q') dQ' \qquad (4.22)$$

Expanding the function $x_{\mathbf{k}}(Q)$ in a series in the functions $\phi_\mathcal{E}(Q)$ one can diagonalize the quadratic form, Eq. (4.20), and compute the Gaussian integrals. Keeping only the lowest orders in k for the density–density correlation function $Y^{00}(\mathbf{k}, \omega)$ in the momentum representation one obtains

$$Y^{00}(\mathbf{k}, \omega) = 2\pi^2 v^2 q V \left(1 + \frac{1}{m} \right) \sum_\mathcal{E} B_\mathcal{E} \left(1 - m\mathcal{E} + \frac{k^2}{2(m-1)} \right)^{-1} \qquad (4.23)$$

where the coefficients $B_\mathcal{E}$ are determined by Eq. (12.90). In Eq. (4.23), only the contribution of states with $\mathcal{E}$ close to m^{-1} is written. When calculating the correlation function Y^{00} in the coordinate representation at large distances these are the states that give the main contribution. In the limit $m \gg 1$, Eq. (4.23) agrees with the corresponding Eq. (12.89). Eq. (4.23) also works in one dimension.

It is interesting to note that a continuum σ-model can be derived starting from the Lagrangian $\mathcal{L}$, Eqs. (4.8, 4.11, 4.12). One should notice first that if a function $\Psi_0(Q)$ is a solution of Eq. (4.17) at $\beta = 0$, then the function $\Psi_0\left(V Q \bar{V}\right)$ for an arbitrary V such that $V\bar{V} = 1$ is also a solution. In order to take slow fluctuations into account one should assume that Ψ_0 does not depend on coordinates but $V(\mathbf{r})$ varies slowly in space. Replacing $y(Q)$ in Eqs. (4.8, 4.11, 4.12) by $\Psi_0\left(V Q \bar{V}\right)$ and expanding in $\nabla V(\mathbf{r})$ and β, one arrives at the continuum σ-model with the free energy F

$$F = \int \mathrm{str}\left[D\left(\nabla \tilde{Q}\right)^2 + \frac{\beta}{4}\Lambda \tilde{Q}\right] d\mathbf{r} \qquad (4.24)$$

where $\tilde{Q} = V\Lambda\bar{V}$. The diffusion coefficient in Eq. (4.24) depends on the parameter α of the bare lattice σ-model and coincides with the value given by the effective medium approximation, Eq. (12.132).

Thus, we see that calculating within the effective medium approximation is equivalent to finding the saddle point of the Lagrangian $\mathcal{L}$, determined by Eqs. ((4.8, 4.11, 4.12), and integrating around this point in the Gaussian approximation.

It is important to note that the preceding procedure may be useful for studying phase transitions in noncompact models only. For any compact model the zero harmonics with respect to the additional matrix coordinates is only important, which means that one returns to the real space. This effect is analogous to the vanishing of the quantum effects near the transition at any finite temperature in the model of a granulated superconductor considered at the end of Section 12.1.

In principle, one can study fluctuations near the saddle-point solution $\Psi(Q)$. Proceeding in this way it seems to be possible to clarify the question about the upper critical dimensionality and to find out how the saddle-point (effective medium approximation) results change. Still, it has not been done because of technical difficulties related to rather large sizes of supermatrices Q. The procedure proposed in this Appendix may help to clarify the critical behavior near the Anderson transition.

References

Abou-Chacra, R., Anderson, P. W., and Thouless, D. J. (1973) *J. Phys. C* **6**, 1734.

Abou-Chacra, R., and Thouless, D. J. (1974) *J. Phys. C* **7**, 65.

Abrahams, E., Anderson, P. W., Licciardello, D. C., and Ramakrishnan, T.V. (1979) *Phys. Rev. Lett.* **42**, 673.

Abrahams, E., Anderson, P. W., and Ramakrishnan, T. V. (1980) *Phil. Mag.* **42**, 827.

Abrahams, E., and Ramakrishnan, T. V. (1980) *J. Non-Cryst. Sol.* **35**, 15.

Abrikosov, A. A. (1988) *Fundamentals of the Theory of Metals*, North Holland, Amsterdam.

Abrikosov, A. A., Gorkov, L. P., and Dzyaloshinskii, I. E. (1963) *Methods of Quantum Field Theory in Statistical Physics*, Prentice Hall, New York.

Abrikosov, A. A., and Ryzhkin, I.A. (1978) *Adv. Phys.* **27**, 147.

Affleck, I. (1984) *J. Phys. C* **17**, 2323.

Aharonov, Y., and Bohm, D. (1959) *Phys. Rev.* **115**, 485.

Ahmed, M. T., Kaiser, A. B., Roth, S., and Migahed, M. D. (1992) *J. Phys. D* **25**, 79.

Akkermans, E. (1991) *Europhys. Lett.* **15**, 709.

Akkermans, E., and Montambaux, G. (1992) *Phys. Rev. Lett.* **68**, 642.

Alhassid, Y., and Attias, H. (1995) *Phys. Rev. Lett.* **74**, 4635.

Altland, A., and Fuchs, D. (1995) *Phys. Rev. Lett.* **74**, 4269.

Altland, A., Iida, S., and Efetov, K. B. (1993) *J. Phys. A* **26**, 3545.

Altland, A., Iida, S., Müller-Groeling, A., and Weidenmüller, H. A. (1992) *Ann. Phys. NY* **219**, 148.

—(1993) *Europhys. Lett.* **20**, 155.

Altshuler, B. L. (1985) *Pis'ma Zh. Eksp. Teor. Fiz.* **51**, 530 (*Sov. Phys. JETP Lett.* **41**, 648).

Altshuler, B. L., and Aronov, A. G. (1981) *Pis'ma Zh. Eksp. Teor. Fiz.* **33**, 315 (*Sov. Phys. JETP Lett.* **33**, 499).

—(1985) in *Electron–Electron Interactions in Disordered Systems*, ed. A. L. Efros and M. Pollak, Elsevier, New York, p. 1.

Altshuler, B. L., Aronov, A. G., Khmelnitskii, D. E., and Larkin, A. I. (1983) in *Quantum Theory of Solids*, ed. I. M. Lifshits, MIR Publishers, Moscow.

Altshuler, B. L., Aronov, A. G., and Spivak, B. Z. (1981) *Pis'ma Zh. Eksp. Teor. Fiz.* **33**, 101 (*Sov. Phys. JETP* **33**, 1981).

Altshuler, B. L., Gefen, Y., and Imry, Y. (1991) *Phys. Rev. Lett.* **66**, 88.

Altshuler, B. L., and Ioffe, L. B. (1992) *Phys. Rev. Lett.* **69**, 2979.

Altshuler, B. L., and Khmelnitskii, D. E. (1985) *Pis'ma Zh. Eksp. Teor. Fiz.* **52**, 291 (*Sov. Phys. JETP Lett.* **42**, 359, 1986).

Altshuler, B. L., Khmelnitskii, D. E., Larkin, A. I., and Lee, P. A. (1980) *Phys. Rev. B* **20**, 5142.

Altshuler, B. L., Kravtsov, V. E., and Lerner, I. V. (1985) *Pis'ma Zh. Eksp. Teor. Fiz.* **43**, 342 (*Sov. Phys. JETP Lett.* **43**, 441, 1986).

—(1986) *Zh. Eksp. Teor. Fiz.* **91**, 2276 (*Sov. Phys. JETP* **64**, 1352).

—(1989) *Phys. Lett. A* **134**, 488.

—(1991) in *Mesoscopic Phenomena in Solids*, ed. B. L. Altshuler, P. A. Lee, and R. A. Webb, North Holland, New York, p. 449.

Altshuler, B. L., Lee, P., and Webb, R. A., eds. (1991) *Mesoscopic Phenomena in Solids*, North Holland, New York.

Altshuler, B. L., and Prigodin, V. N. (1989) *Zh. Eksp. Teor. Fiz.* **95**, 348 (*Sov. Phys. JETP* **68**, 198).

Altshuler, B. L., and Shklovskii, B. I. (1986) *Zh. Eksp. Teor. Fiz.* **91**, 220 (*Sov. Phys. JETP* **64**, 127).

Ambegaokar, V., and Eckern, U. (1990) *Phys. Rev. Lett.* **65**, 381.

Amit, D. J. (1978) *Field Theory, the Renormalization Group, and Critical Phenomena*, McGraw-Hill, New York.

Anderson, P. W. (1958). *Phys. Rev.* **109**, 1492.

Anderson, P. W., Abrahams, E., and Ramakrishnan, T. V. (1979). *Phys. Rev. Lett.* **43**, 718.

Anderson, P. W., Thouless, D. J., Abrahams, E., and Fisher, D. S. (1980) *Phys. Rev. B* **22**, 3519.

Andersson, M. et al. (1992) in *Proceedings of the International Conference on Science and Technology of Synthetic Metals*, Göteborg, Sweden.

Ando, T., and Aoki, H. (1981) *Sol. St. Comm.* **38**, 1079.

Ando. T., and Fukuyama, H. (1988) *Anderson Localization, Proceedings of the International Symposium, Tokyo, Japan*, Springer-Verlag, New York.

Ando, T., and Uemura, Y. (1974) *J. Phys. Soc. Jpn.* **36**, 359.

Andreev, A. V., and Altshuler, B. L. (1995) *Phys. Rev. Lett.* **75**, 902.

Aoki, H. (1983) *J. Phys. C* **16**, L205.

—(1986) *Phys. Rev. B* **33**, 7310.

—(1987) *Rep. Prog. Phys.* **50**, 655.

Apenko, S. M., and Lozovik, Y. E. (1985) *Zh. Eksp. Teor. Fiz.* **89**, 573 (*Sov. Phys. JETP* **62**, 328).

Argaman, N. (1995) *Phys. Rev. Lett.* **75**, 2750.

Argaman, N., Imry, Y., and Smilansky, U. (1993) *Phys. Rev. B* **47**, 4440.

Aronov, A. G., Mirlin, A. D., and Wölfle, P. (1994) *Phys. Rev. B* **49**, 16609.

Aronov, A. G., and Sharvin, Y. V. (1987) *Rev. Mod. Phys.* **59**, 755.

Azbel, M.A. (1984) *Localization, Interaction and Transport Phenomena*, ed. B. Kramer, G. Bergmann, and Y. Bruynsraede, Springer-Verlag, New York, p. 162.

Baranger, H. U., and Mello, P. A. (1994) *Phys. Rev. Lett.* **73**, 142.

Baranger, H. U., and Stone, A. D. (1989) *Phys. Rev. B* **40**, 8169.

Bardeen, J., Cooper, L. N., and Schrieffer, J. R. (1957) *Phys. Rev.* **108**, 1175.

Beenakker, C. W. J. (1991) *Phys. Rev. B* **44**, 1646.

—(1993) *Phys. Rev. Lett.* **70**, 4126.

—(1994) *Phys. Rev. B* **50**, 15170.

Beenakker, C. W. J., and van Houten, H. (1991) in *Solid State Physics*, Vol. 44, ed. by H. Ehrenreich and D. Turnbull, Academic, New York, pp. 1–228.

Beenakker, C. W. J., and Rejaei, B. (1993) *Phys. Rev. Lett.* **71**, 3689.

—(1994) *Phys. Rev. B* **49**, 7499.

Belavin, A. A., and Polyakov, A. M. (1975) *Pis'ma Zh. Eksp. Teor. Fiz.* **22**, 503 (*Sov. Phys. JETP Lett.* **22**, 245).

Belitz, D., and Götze, W. (1983) *Phys. Rev. B* **28**, 5445.

Belitz, D., and Kirkpatrick, T. R. (1994) *Rev. Mod. Phys.* **66**, 261.

Benedict, K. A. (1987) *Nucl. Phys. B* **280 [FS18]**, 549.

Benedict, K. A., and Chalker, J. T., ed. (1991) *Localization*, in *Proceedings of the International Conference on Localization, London*, Institute of Physics, Conference Series.

Benzi, R., Paladin, G., Parisi, G., and Vulpiani, A. (1984) *J. Phys. A.: Math. Gen.* **17**, 3521.

—(1985) *J. Phys. A.: Math. Gen.* **18**, 2157.

Berezin, F. A. (1961). *Dokl. Akad. Nauk SSR*, **137**, 31.

—(1965) *Metod vtorichnogo kvantovaniya "Nauka" (The Method of Second Quantization)* (English translation published by Academic Press, New York).

—(1967) *Math. Zam.*, **1**, 3.

—(1979) *Yadernaya Fiz.*, **29**, 1670 (*Nucl. Phys.*).

—(1983) *Vvedenie v algebru i analiz s antikommutiruyushchimi peremennymi, "Izdatelstvo Moskovskogo Universiteta."*

—(1987) *Introduction to Superanalysis*, MPAM, Vol. 9, D. Reidel, Dordrecht.

Berezinsky, V. L. (1970) *Zh. Eksp. Teor. Fiz.* **59**, 907 (*Sov. Phys. JETP*, **32**, 493).

— (1973) *Zh. Eksp. Teor. Fiz.* **65**, 1251 (*Sov. Phys. JETP* **38**, 620, 1974).

Berezinsky, V. L., and Gorkov, L. P. (1979) *Zh. Eksp. Teor. Fiz.* **77**, 249 (*Sov. Phys. JETP* **50**, 1209).

Bergmann, G. (1982) *Phys. Rev.B* **25**, 2937.

—(1984) *Phys. Rep.* **101**, 1.

Berry, M. V. (1985) *Proc. R. Soc. Lond. A* **400**, 229.

Berry, M. J., Baskey, J. H., Westervelt, R. M., and Gossard, A. C. (1994a) *Phys. Rev. B* **50**, 8857.

—(1994b) Berry, M. J., Katine, J. A., Marcus, C. M., Westervelt, R. M., and Gossard, A.C. *Surf. Sci.* **305**, 480.

Binder, K., and Young, A. P. (1986) *Rev. Mod. Phys.* **58**, 801.

Blanter, Y., and Mirlin, A. D. (1995) preprint, cond-mat/9509097.
Bloch, F. (1965) *Phys. Rev.* **137**, A787.
—(1968) *Phys. Rev.* **166**, 415.
—(1970) *Phys. Rev. B* **2**, 109.
Blümel, R., and Smilansky, U. (1988) *Phys. Rev. Lett.* **60**, 477.
—(1989) *Physica D* **36**, 111.
—(1990) *Phys. Rev. Lett.* **64**, 241.
Bogdanovich, S., Dai, P., Sarachik, M. P., and Dobrosavljevic, V. (1995) *Phys. Rev. Lett.* **74**, 2543.
Bogoliubov, N. N., and Shirkov, D. V. (1976) *Vvedenie v Teoriyu Kvantovannykh Polei* (*Introduction to Theory of Quantized Fields*), Moscow, Nauka.
Bohigas, O. (1991) in *Chaos and Quantum Physics*, ed. M.-J. Gianonni, A. Voros and J. Zinn-Justin, Elsevier, Amsterdam.
Bohigas, O., and Gianonni, M.-J. (1984) in *Mathematical and Computational Methods in Nuclear Physics*, ed. J. S. Dehesa, J. M. G. Gomez, and A. Polls, *Lecture Notes in Physics*, Vol. 209, Springer-Verlag, Heidelberg.
Bouchiat, H., and Montambaux, G. (1989) *J. Phys. (Paris)* **50**, 2695.
Bouchiat, H., Montambaux, G., and Sigeti, D. (1991) *Phys. Rev. B* **44**, 1682.
Braun, D., and Montambaux, G. (1994) *Phys. Rev. B* **50**, 7776.
Brezin, E., Gross, D. J., and Itzykson, C. (1984) *Nucl. Phys. B* **235 [FS11]**, 24.
Brezin, E., Hikami, S., and Zinn-Justin, J. (1980) *Nucl. Phys. B* **165**, 528.
Brezin, E., Itzykson, C., Parisi, G., and Zuber, J. (1978) *Comm. Math. Phys.* **59**, 35.
Brezin, E., and Zinn-Justin, J. (1976) *Phys. Rev. B* **14**, 3110.
Brezin, E., Zinn-Justin, J., and Le Guillou, J. C. (1976) *Phys. Rev. D* **14**, 2615.
Broderix, K., Heldt, N., and Leschke, H. (1991) *J. Phys. A* **24**, L825.
Brody, T. A., Flores, J., French, J. B., Mello, P. A., Pandey, A., and Wong, S. S. M. (1981) *Rev. Mod. Phys.* **53**, 385.
Brom, H. B., Fritschij, F. C., van der Putten, D., Hannemann, F. A., de Jongh, L. J., and Schmid, G. (1994) A ^{195}Pt-NMR study of mesoscopic fluctuations in monodisperse Pt-clusters, preprint.
Brouwer, P. W. (1995) *Phys. Rev. B* **51**, 16878.
Brouwer, P. W., and Beenakker, C. W. J. (1994) *Phys. Rev. B* **50**, 11263.
Büttiker, M. (1985) *Phys. Rev. B* **32**, 1846.
—(1986) *Phys. Rev. Lett.* **57**, 1761.
—(1988) *Phys. Rev. B* **38**, 9376.
—(1992) in *Semiconductors and Semimetals*, ed. M. Reed, Academic Press, Boston.
Büttiker. M., Imry, Y., and Landauer, R. (1983) *Phys. Lett.* **96A**, 365.
Byers, N., and Yang, C. N. (1961) *Phys. Rev. Lett.* **7**, 46.
Calogero, F. (1969) *J. Math. Phys.* **10**, 2191; **10**, 2197.
Casati, G., ed. (1985) *Chaotic Behavior in Quantum Systems*, Plenum Press, New York.
Casati, G., Chirikov, B. V., Ford, J., and Izrailev, F. M. (1979) *Lect. Notes in Phys.* **93**, 334.
Casati, G., Chirikov, B. V., Guarneri, I., and Shepelyanski, D. L. (1987) *Phys. Rep.* **154**, 77.
Casati, G., Guarneri, F., Izrailev, F. M, Fishman, S., and Molinari, L. (1992) *J. Phys. C* **4**, 149.
Casati, G., Molinary, L., and Izrailev, F. M. (1990) *Phys. Rev. Lett.* **64**, 1851.
Castellani, C., and Peliti, L. (1986) *J. Phys. A* **19**, L429.
Castilla, G. E., and Chakravarty, S. (1993) *Phys. Rev. Lett.* **71**, 384.
Chakravarty, S., and Schmid, A. (1986) *Phys. Rep.* **140**, 193.
Chalker, J. T., and Coddington, P. D. (1988) *J. Phys. C* **21**, 2665.
Chan, I. H., Clarke, R. M., Marcus, C. M., Campman, K., and Gossard, A. C. (1995) *Phys. Rev. Lett.* **74**, 3876.
Chandrasekhar, S. (1943) *Rev. Mod. Phys.* **15**, 1.
Chandrasekhar, V., Webb, R. A., Brady, M. J., Ketchen, M. B., Gallagher, W. J., and Kleinsasser, A. (1991) *Phys. Rev. Lett.* **67**, 3578.
Chang, A. M., Baranger, H. U., Pfeiffer, L. N., and West, K.W. (1994) *Phys. Rev. Lett.* **73**, 2111.
Chang, A. M., Baranger, H. U., Pfeifer, L. N., West, K. W., and Chang, T. Y. (1996) *Phys. Rev. Lett.* **76**, 1695.
Chapon, C. M., Gillet, M. F., and Henry, C. R. (1989) *Z. Phys. D* **12**, 145.
Cheung, H. F., Gefen, Y., Riedel, E. K., and Shih, W. H. (1988a) *Phys. Rev. B* **37**, 6050.
Cheung, H. F., Gefen, Y., and Riedel, E. K. (1988b) *IBM J. Res. Develop.* **32**, 359.
Cheung, H. F., Riedel, E. K., and Gefen, Y. (1989) *Phys. Rev. Lett.* **62**, 587.
Clarke, R. M., Chan, I. H., Marcus, C. M., Duruöz, Harris, Jr., Campman, K., and Gossard, A. C. (1995) *Phys. Rev. B* **52**, 2656.

Collet, P., Eckmann, J.-P. (1980) *Iterated Maps of the Interval as Dynamical Systems*, Birkhäuser, Boston.

Cornfeld, I. P., Fomin, S. V., and Sinai, Ya.G. (1982) *Ergodic Theory*, New York, Springer-Verlag.

Corwin, L., Neemann, J., and Sternberg, S. (1975) *Rev. Mod. Phys.* **47**, 573.

David, F. (1985) *Nucl. Phys. B* **257**, 45; **257**, 543.

De Gennes, P. G. (1966) *Superconductivity of Metals and Alloys*, W. A. Benjamin, New York, Amsterdam.

Denton, R., Mühlschlegel, B., and Scalapino, D. J. (1973) *Phys. Rev. B* **7**, 3589.

Deutscher, G., Imry, Y., and Gunter, L. (1974) *Phys. Rev. B* **10**, 4598.

Devaty, R. P., and Sievers, A. J. (1980) *Phys. Rev. B* **22**, 2123.

DeWitt, B. (1984) *Supermanifolds*, Cambridge University Press, Cambridge.

Di Francesco, P., Ginsparg, P., and Zinn-Justin, J. (1995) *Phys. Rep.* **254**, 1.

Dittrich, T., Doron, E., and Smilansky, U. (1994) *J. Phys. A.: Math. Gen.* **27**, 79.

Doron, E., Smilansky, U., and Frenkel, A. (1990) *Phys. Rev. Lett.* **65**, 3072.

Dorokhov, O. N. (1982) *Pis'ma Zh. Eksp. Teor. Fiz.* **36**, 259 (*Sov. Phys. JETP Lett.* **36**, 318).

—(1983) *Zh. Eksp. Theor. Fiz.* **85**, 1040 (*Sov. Phys. JETP* **58**, 606).

Dorozhkin, S. I., Shashkin, A. A., Kravchenko, G. V., Dolgopolov, V. T., Haug, R. J., von Klitzing, K., and Ploog, K. (1993) *Zh. Eksp. Teor. Fiz.* **57**, 55 (*JETP Lett.* **57**, 58).

Du, R. R., Stormer, H. L., Tsui, D. C., Pfeiffer, L. N., and West, K. W. (1993) *Phys. Rev. Lett.* **70**, 2944.

—(1994) *Sol. State. Comm.* **90**, 71.

Dyson, F. J. (1962a) *J. Math. Phys.* **3**, 140, 157, 166.

—(1962b) *J. Math. Phys.* **3**, 1199.

Edwards, S. F. (1958) *Phil. Mag.* **3**, 33, 1020.

Edwards, S. F., and Anderson, P. W. (1975) *J. Phys. F* **5**, 965.

Efetov, K. B. (1980) *Zh. Eksp. Teor. Fiz.* **78**, 2017 (*Sov. Phys. JETP*) **51**, 1015.

—(1982a) *Zh. Eksp. Teor. Phys.* **82**, 872 (*Sov. Phys. JETP* **55**, 514).

—(1982b) *Zh. Eksp. Teor. Phys.* **83**, 833 (*Sov. Phys. JETP* **56**, 467); *J. Phys. C* **15**, L909.

—(1983) *Adv. Phys.* **32**, 53.

—(1984) *Pis'ma Zh. Eksp. Teor. Fiz.* **40**, 17 (*Sov. Phys. JETP Lett.* **40**, 738); *Zh. Eksp. Teor. Fiz.* **88**, 1032 (*Sov. Phys. JETP* **51**, 606).

—(1987a) *Zh. Eksp. Teor. Fiz.* **92**, 360 (*Sov. Phys. JETP* **65**, 360).

—(1987b) *Zh. Eksp. Teor. Fiz.* **93**, 1125 (*Sov. Phys. JETP* **66**, 634).

—(1988) *Zh. Eksp. Teor. Fiz.* **94**, 357 (*Sov. Phys. JETP* **67**, 199).

—(1990) *Applications of Statistical and Field Theory Methods to Condensed Matter*, ed. D. Baeriswyl, A. R. Bishop, and J. Carmelo, Plenum Press, New York and London.

—(1990a) *Physica A* **167**, 119.

—(1991) *Phys. Rev. Lett.* **66**, 2794.

—(1995) *Phys. Rev. Lett.* **74**, 2299.

—(1996) *Phys. Rev. Lett.* (forthcoming).

Efetov, K. B., and Iida, S. (1993) *Phys. Rev. B* **47**, 15794.

Efetov, K. B., and Larkin, A. I. (1975) *Zh. Eksp. Teor. Fiz.* **69**, 764 (*Sov. Phys. JETP* **42**, 390).

Efetov, K. B., and Larkin, A. I. (1983) *Zh. Eksp. Teor. Fiz.* **85**, 764 (*Sov. Phys. JETP* **58**, 444).

Efetov, K. B., Larkin, A. I., and Khmelnitskii, D. E. (1980) *Zh. Eksp. Teor. Fiz.* **79**, 1120 (*Sov. Phys. JETP* **52**, 568).

Efetov, K. B., and Marikhin, V. G. (1989) *Phys. Rev. B* **40**, 12126.

Efetov, K. B., and Prigodin, V. N. (1993) *Phys. Rev. Lett.* **70**, 1315; *Mod. Phys. Lett. B* **7**, 981.

Efetov, K. B., and Viehweger, O. (1992) *Phys. Rev. B* **45**, 11546.

Eliashberg, G. M. (1961) *Zh. Eksp. Teor. Fiz.* **41**, 1241 (*Sov. Phys. JETP* **14**, 886, 1962).

Entin-Wohlman, O., Gefen, Y., Meir, Y., and Oreg, Y. (1992) *Phys. Rev. B* **45**, 11890.

Evangelou, S. N., and Economou, E. N. (1990) *Phys. Lett. A* **151**, 345.

Faas, M., Simons, B. D., Zotos, Z., and Altshuler, B. L. (1993) *Phys. Rev. B* **48**, 5439.

Falko, V. I. (1995) *Phys. Rev. B* **51**, 5227.

Falko, V. I., and Efetov, K. B. (1994) *Phys. Rev. B* **50**, 11267.

—(1995a) *Europhys. Lett.* **32**, 627.

—(1995b) *Phys. Rev. B* **52**, 17413.

Fastenrath, U., Janßen, M., and Pook, W. (1992) *Physica A* **191**, 401.

Feder, J. (1988) *Fractals*, Plenum, New York.

Feingold, M., Leitner, D. M., and Wilkinson, M. (1991) *Phys. Rev. Lett.* **66**, 986.

Feynman, R. P. (1948) *Rev. Mod. Phys.* **20**, 367.

Feynman, R. P., and Hibbs, A. R. (1965) *Quantum Mechanics and Path Integrals*, McGraw-Hill, New York.

Finkelshtein, A. M. (1983) *Zh. Eksp. Teor. Fiz.* **84**, 168 (*Sov. Phys. JETP* **57**, 97).

—(1984) *Z. Phys. B* **56**, 189.

Fisher, D. S., and Lee, P. A. (1981) *Phys. Rev. B* **23**, 6581.

Fishman, S., Grempel, D. R., and Prange, R. E. (1982) *Phys. Rev. Lett.* **49**, 509.

Folk, J. A., Patel, S. R., Godijn, S. F., Huibers, A. G., Cronenwett, S. M., Marcus, C. M., Campman, K., and Gossard, A. C. (1996), *Phys. Rev. Lett.* **76**, 1699.

Foote, M. C., and Anderson, A. C. (1987) *Rev. Sci. Inst.* **58**, 130.

Frahm, K. (1995) *Phys. Rev. Lett.* **74**, 4706.

Frölich, H. (1937) *Physica* **4**, 406.

Fu, Y., and Anderson, P. W. (1986) *J. Phys. A* **19**, 1605.

Fyodorov, Y, V., and Mirlin, A. D. (1991) *Phys. Rev. Lett.* **67**, 2405.

—(1991a) *Phys. Rev. Lett.* **67**, 2049.

—(1992) *Phys. Rev. Lett.* **69**, 1093.

—(1994) *JETP Lett.* **60**, 790.

Fyodorov, Y. V., and Sommers, H.-J. (1995) *Phys. Rev. E* **51**, R2719.

Gaspard, P., Rice, S. A., Mikeska, H. J., and Nakamura, K. (1990) *Phys. Rev. A* **42**, 4015.

Gefen, Y., Imry, Y., and Azbel, M. Y. (1984) *Phys. Rev. Lett.* **52**, 109.

Girvin, S. M., and Johnson, M. (1980) *Phys. Rev. B* **22**, 2671.

Gogolin, A. A. (1976) *Zh. Eksp. Teor. Fiz.* **71**, 1912 (*Sov. Phys. JETP* **44**, 1003).

—(1982) *Phys. Rep.* **66**, 1.

Gogolin, A. A., Melnikov, V. I., and Rashba, E. I. (1975) *Zh. Eksp. Teor. Fiz.* **69**, 327 (*Sov. Phys. JETP* **42**, 168).

Goldberg, J., Smilansky, U., Berry, M. V., Schweizer, W., Wunner, G., and Zeller, G. (1991) *Nonlinearity* **4**, 1.

Goldman, V., Shayegan, M., and Tsui, D. (1988) *Phys. Rev. Lett.* **61**, 881.

Goldsheid, I. Y., Molchanov, S. A., and Pastur, L. A. (1977) *Funct. Anal. Appl.* **11**, 1.

Gol'fand, Y. A., and Lichtmann, E. P. (1972) *Problemy teoreticheskoi fiziki* (collected articles in memory of I. E. Tamm), Nauka, Moscow, p. 37.

Gorkov, L. P., Dorokhov, O. N., and Prigara, F. V. (1983) *Zh. Eksp. Teor. Fiz.* **84**, 1440 (*Sov. Phys. JETP* **54**, 838).

Gorkov, L. P., and Eliashberg, G. M. (1965) *Zh. Eksp. Teor. Fiz.* **48**, 1407 (*Sov. Phys. JETP* **21**, 940).

Gorkov, L. P., Larkin, A. I., and Khmelnitskii, D. E. (1979) *Pis'ma Zh. Eksp. Teor. Fiz.* **30**, 248 (*Sov. Phys. JETP Lett.* **30**, 228).

Götze, W. (1981) *The Conductor–Nonconductor Transition in Strongly Disordered Three-Dimensional Systems*, in *Recent Developments in Condensed Matter Physics*, ed. J. T. Devreese, Vol. 1, Plenum Press, New York; *Phil. Mag.* **43**, 219.

—(1985) *Localization, Interaction and Transport Phenomena*, ed. B. Kramer, G. Bergmann, and Y. Bruynseraede, Springer-Verlag, New York, p. 62.

Gradshteyn, I. S., and Ryzhik, I. M. (1966) *Tables of Integrals, Series and Products*, Academic Press, New York.

Gross, D. J., and Klebanov, I. R. (1991) *Nucl. Phys. B* **352**, 671.

Gross, D. J., Piran, T., and Weinberg, S., eds. (1991) *Two Dimensional Quantum Gravity and Random Surfaces*, Proceedings of the Jerusalem Winter School for Theoretical Physics 1990, World Scientific, Singapore.

Grussbach, H., and Schreiber, M. (1992) *Physica A* **191**, 394.

—(1993) *Phys. Rev. B* **48**, 6650.

—(1995) *Phys. Rev. B* **51**, 663.

Gunter, L., and Imry, Y. (1969) *Sol. State. Comm.* **7**, 1391.

Gutzwiller, M. C. (1991) *Chaos in Classical and Quantum Mechanics*, Springer-Verlag, New York.

Haake, F. (1992) *Quantum Signatures of Chaos*, Springer, Berlin.

Hackenbroich, G., and Weidenmüller, H. A. (1995) *Phys. Rev. Lett.* **74**, 4118.

Haldane, F. D. M. (1981) *Phys. Rev. Lett.* **47**, 1840; *J. Phys. C* **14**, 2585.

Halperin, B. I. (1982) *Phys. Rev. B* **25**, 2185.

Halperin, W. P. (1986) *Rev. Mod. Phys.* **58**, 533.

Halperin, B. I., Lee, P. A., and Read, N. (1993) *Phys. Rev. B* **47**, 7312.

Halsey, T. C., Jensen, M. H., Kadanoff, L. P., Procaccia, I., and Shraiman, B. I. (1986) *Phys. Rev. A* **33**, 1141.

Hartmann, U., and Davis, E. D. (1989) *Comput. Phys. Comm.* **54**, 353.

Hauser, W., and Feshbach, H. (1952) *Phys. Rev.* **87**, 366.

Heeger, A. J., Kivelson, S., Schrieffer, J. R., and Su, W. P. (1988) *Rev. Mod. Phys.* **60**, 781.

Helgason, S. (1984) *Groups and Geometric Analysis*, Academic Press, Orlando, Fl.

Henneke, M., Kramer, B., and Ohtsuki, T. (1994) *Europhys. Lett.* **27**, 389.

Hentschel, H. G. E., and Procaccia, I. (1983) *Physica D* **8**, 435.

Hikami, S. (1981) *Phys. Rev. B* **24**, 2671.

—(1983) *Nucl. Phys. B* **215**, 555.

—(1990) *Lecture Notes of NATO ASI on Interference Phenomena in Mesoscopic Systems*, **254**, ed. B. Kramer, Plenum, New York, p. 429.

—(1992) *Prog. Theor. Phys. Suppl.* **107**, 213.

Hikami, S., Larkin, A. I., and Nagaoka, Y. (1980) *Prog. Theor. Phys.* **63**, 707.

t'Hooft, G. (1974) *Nucl. Phys. B* **72**, 461.

Hopkins, F., Burns, M. J., Rimberg, A. J., and Westervelt, R. M. (1989) *Phys. Rev. B* **39**, 12708.

Hornung, M., Ruzzi, A., Schlager, H. G., Stupp, H., and Löhneysen, H. (1994) *Europhys. Lett.* **28**, 43.

Houghton, A., Jevicki, A., Kenway, R. D., and Pruisken, A.M.M. (1980) *Phys. Rev. Lett.* **45**, 394.

Huang, K. (1987) *Statistical Machanics*, Wiley, New York.

Huckestein, B. (1990) *Physica A* **167**, 175.

—(1992) *Europhys. Lett.* **20**, 451.

Huckestein, B., and Kramer, B. (1989) *Sol. State. Comm.* **71**, 445.

—(1990) *Phys. Rev. Lett.* **64**, 1437.

Huo, Y., and Bhatt, R. N. (1992) *Phys. Rev. Lett.* **68**, 1375.

Iida, S., Weidenmüller, H. A., and Zuk, J. A. (1990) *Ann. Phys.* **200**, 219.

Imry, Y. (1993) *Phys. Rev. Lett.* **71**, 1868.

Iordansky, S. V. (1982) *Sol. State. Comm.* **48**, 1.

Ishiguro, T., Kaneko, H., Nogami, Y., Ishimoto, H., Nishiyama, H., Tsukamoto, J., Takahashi, A., Yamaura, M., Hagiwara, T. and Sato, K. (1992) *Phys. Rev. Lett.* **69**, 660.

Isihara, A. (1971) *Statistical Physics*, Academic Press, New York and London.

Itzykson, C., and Drouffe, J. M. (1989) *Statistical Field Theory*, Vol. 1, Cambridge University Press, Cambridge.

Izrailev, F. M. (1990) *Phys. Rep.* **196**, 299.

Jalabert, R. A., Baranger, H. U., and Stone, A. D. (1990) *Phys. Rev. Lett.* **65**, 2442.

Jalabert, R. A., Pichard, J.-L., and Beenakker, C. W. J. (1994) *Europhys. Lett.* **27**, 255.

Jalabert, R. A., Stone, A. D., and Alhassid, Y. (1992) *Phys. Rev. Lett.* **68**, 3468.

Jain, D. K. (1989) *Phys. Rev. Lett.* **63**, 199; *Phys. Rev. B* **40**, 8079.

—(1990) *Phys. Rev. B* **41**, 7653.

Janßen, M., Viehweger, O., Fastenrath, U., and Hajdu, J. (1994) *Introduction to the Theory of the Integer Quantum Hall Effect*, VCH, Weinheim.

Javadi, H. H. S., Chakraborty, A., Li, C., Theophilou, N., Swanson, D. B., MacDiarmid, A. G., and Epstein, A. J. (1991) *Phys. Rev. B* **43**, 2183.

Kamenev, A., and Braun, D. (1994) *J. Phys. (Paris)* **4**, 1049.

Kane, C. L., Serota, R. A., and Lee, P. A. (1988) *Phys. Rev. B* **37**, 6701.

Kang, W., Stormer, H. L., Pfeiffer, L. N., Baldwin, K. W., West, K. W. (1993) *Phys.Rev. Lett.* **71**, 3850.

Kanter, I., and Sompolinsky, H. (1987) *Phys. Rev. Lett.* **58**, 164.

Karawarabayashi, T., and Ohtsuki, T. (1995) *Phys. Rev. B* **51**, 10897.

Kastner, M. A. (1992) *Rev. Mod. Phys.* **64**, 849.

Katsumoto, S. (1991) in *Localization, Proceedings of the International Conference on Localization, London*, ed. K. A. Benedict and J. T. Chalker, p. 17. Institute of Physics, Conference Series, Bristol.

Kazakov, V. A. (1985) *Phys. Lett. B* **150**, 282.

Keller, M. W., Millo, O., Mittal, A., Prober, D. E., and Sacks, R. N. (1994) *Surf. Sci.* **305**, 501.

Kettemann, S., and Efetov, K. B. (1995) *Phys. Rev. Lett.* **74**, 2547.

Khmelnitskii, D. E. (1983) *Pis'ma Zh. Eksp. Teor. Fiz.* **38**, 454 (*Sov. Phys. JETP Lett.* **38**, 552).

Khveshchenko, D. V., and Meshkov, S. V. (1993) *Phys. Rev. B* **47**, 12051.

Kirk, W. P., and Reed, M. A. (1992) *Nanostructures and Mesoscopic Systems*, Academic Press, Boston.

Kirkpatrick, S. (1973) *Rev. Mod. Phys.* **45**, 574.

von Klitzing, K., Dorda, G., and Pepper, M. (1980) *Phys. Rev. Lett.* **45**, 494.

Kobayashi, S., Takahashi, T., and Sasaki, W. (1971) *J. Phys. Soc. Jpn.* **31**, 1442.

—(1972) *J. Phys. Soc. Jpn.* **32**, 1234.

Kopietz, P., and Efetov, K. B. (1992) *Phys. Rev. B* **46**, 1429.

Korepin, V. E., Izergin, G., and Bogoliubov, N. M. (1992) *Quantum Inverse Scattering Method, Correlation Functions and Algebraic Bethe Ansatz*, Cambrige University Press, Cambridge.

Kostant, B. (1977). *Lecture Notes in Mathematics*, **570**, Springer-Verlag, New York, p. 177.

Kosterlitz, J. M. (1974) *J. Phys. C* **7**, 1046.

Kosterlitz, J. M., and Thouless, D. J. (1973) *J. Phys. C* **6**, 1181.

Kramer, B., Bergmann, G., and Bruynseraede, Y. (1985) *Localization, Interaction, and Transport Phenomena, Proceedings of the International Conference, Braunschweig, Germany*, Springer-Verlag, New York.

Kramer, B., and MacKinnon, A. (1993) *Rep. Prog. Phys.* **56**, 1469.

Kramer, B., Ono, Y. and Ohtsuki, T. (1988) *Surf. Sci.* **196**, 127.

Kramers, H. A. (1930) *Proc. Acad. Sci. Amsterdam*, **33**, 959.

Kravchenko, S. V., Furneaux, J. E., and Pudalov, V. M. (1994) *Phys. Rev. B* **49**, 2250.

Kravtsov, V. E., Lerner, I. V., and Yudson, V. I. (1988) *Zh. Eksp. Teor. Fiz.* **94**, 255 (*Sov. Phys. JETP* **68**, 1441).

—(1989) *Phys. Lett.* **A134**, 245.

Kravtsov, V. E. and Mirlin, A. D. (1994) *JETP Lett.* **60**, 657.

Kravtsov, V. E., and Zirnbauer, M. (1992) *Phys. Rev. B* **46**, 4332.

Kubo, R. (1962) *J. Phys. Soc. Jpn.* **17**, 975.

—(1969) in *Polarization, Matiere et Rayonnement*, livre jubile en l'honneur du Professeur A. Kastler, Presses Universitaires de France, Paris, p. 325.

Kudrolli, A., Kidambi, V., and Sridhar, S. (1995) *Phys. Rev. Lett.* **75**, 822.

Kunz, H., and Suillard, B. (1980) *Commun. Math. Phys.* **78**, 201.

—(1983) *J. Phys. Lett. (Paris)* **44**, L411.

Kunz, H., and Zumbach, G. (1989) *J. Phys. A* **22**, L1043.

Kurchatov, I. V. (1933) *JETP* **3**, 181.

Landau, L. D., and Lifschitz, E. M. (1959) *Quantum Mechanics*, Pergamon, London.

—(1968) *Statistical Physics*, Chap. 13, Pergamon, Oxford.

Landauer, R. (1957) *IBM J. Res. Develop.* **1**, 233.

—(1970) *Phil. Mag.* **21**, 863.

—(1985) *Localization, Interaction and Transport Phenomena,* ed. B. Kramer, G. Bergmann, and Y. Bruynseraede, Springer, New York, p. 38.

Lane, A. M., and Thomas, R. G. (1958) *Rev. Mod. Phys.* **30**, 257.

Langer, J. S. (1967) *Ann. Phys.* **41**, 108.

Langer, J. S. and Neal, T. (1966) *Phys. Rev. Lett.* **16**, 984.

Langreth, D. C., and Abrahams, E. (1981) *Phys. Rev. B* **24**, 2978.

Larkin, A. I., and Khmelnitskii, D. E. (1969) *Zh. Eksp. Teor. Fiz.* **56**, 2087 (*Sov. Phys. JETP* **29**, 1123).

Laughlin, R. B. (1981) *Phys. Rev. B* **23**, 5632.

Lee, P. A. and Fisher, D. S. (1981) *Phys. Rev. Lett.* **47**, 882

Lee, P. A., and Ramakrishnan, T. V. (1985) *Rev. Mod. Phys.* **57**, 287.

Lee, P. A., and Stone, A. D. (1985) *Phys. Rev. Lett.* **55**, 1622.

Lee, P. A., Stone, A. D., and Fukuyama, H. (1987) *Phys. Rev. B* **35**, 1039.

Leites, D. A. (1975) *Usp. Mat. Nauk* **30**, 3, 156.

Lerner, I. V. (1988) *Phys. Lett. A* **133**, 253.

Lerner, I. V., and Imry, Y. (1995) *Europhys. Lett.* **29**, 49.

Lerner, I. V., and Wegner, F. (1990) *Z. Phys. B* **81**, 95.

Levine, H., Libby, S. B., and Pruisken, A. M. M. (1983) *Phys. Rev. Lett.* **51**, 1915.

—(1984) *Nucl. Phys. B* **240 [SF12]**, 30; 49; 71.

Lévy, L. P., Dolan, G., Dunsmuir, J., and Bouchiat, H. (1990) *Phys. Rev. Lett.* **64**, 2074.

Lichtenberg, A. J., and Lieberman, M. A. (1983) *Regular and Stochastic Motion*, Springer-Verlag, New York.

Lifshitz, I. M., Gredescul, S., and Pastur, L. A. (1988) *Introduction to the Theory of Disordered Systems*, Wiley, New York.

Lucini, C., Bishop, D. J., Kastner, M. A., and Melngailis, J. (1985) *Phys. Rev. Lett.* **55**, 2987.

Lurui, S., and Kazarinov, R. F. (1983) *Phys. Rev. B* **37**, 1386.

Ma, S. K. (1976) *Modern Theory of Critical Phenomena*, Benjamin, New York.

MacDiarmid, A. G. et al. (1992) in *Proceedings of the International Conference on Science and Technology of Synthetic Metals*, Göteborg, Sweden.

Macedo, A. M. S., and Chalker, J. T. (1992) *Phys. Rev. B* **46**, 14985.

MacKinnon, A., and Kramer, B. (1981) *Phys. Rev. Lett.* **47**, 1546.

—(1983) in *The Application of High Magnetic Fields in Semiconductor Physics*, Vol. 177, *Lecture Notes in Physics*, ed. G. Landwehr, Springer, Berlin, 74.

Mailly, D., Chapelier, C., and Benoit, A. (1993) *Phys. Rev. Lett.* **70**, 2020.

Mandelbrot, B. B. (1983) *The Fractal Geometry of Nature*, W. H. Freeman, San Francisco.

Marcus, C. M., Rimberg, A. J., Westervelt, R. M., Hopkins, P. F., and Gossard, A.C. (1992) *Phys. Rev. Lett.* **69**, 506.

—(1993) *Phys. Rev. B* **48**, 2460.

Marichev, O. I. (1983) *Handbook of Integral Transforms of Higher Transcendental Functions*, Ellis Horwood, New York, pp. 136–142.

Martin, I. L. (1959) *Proc. R. Soc. A* **251**, 536.

Mathur, H., and Stone, A. D. (1991) *Phys. Rev. B* **44**, 10957.

Mc Kane, A. J. (1980) *Phys. Lett. A* **76**, 33.

Mehta, M. L. (1991) *Random Matrices*, Academic Press, San Diego.

Mehta, M. L., and Dyson, F. J.(1963) *J. Math. Phys.* **4**, 713.

Mehta, M. L., and Pandey, A. (1983) *J. Phys. A* **16**, 2655.

Meir, Y., Wingreen, N., and Lee, P. A. (1991) *Phys. Rev. Lett.* **66**, 3048.

Mello, P. A. (1988) *Phys. Rev. Lett.* **60**, 1089.

Mello, P. A., Pereyra, P., and Kumar, N. (1988) *Ann. Phys. N.Y.* **181**, 290.

Mello, P. A., and Stone, A. D. (1991) *Phys. Rev. B* **44**, 3559.

Mezard, M., and Parisi, G. (1985) *J. Phys. (Paris)* **47**, 1285.

—(1987) *Europhys. Lett.* **3**, 1067.

Miek, B. (1990) *Europhys. Lett.* **13**, 453.

Milnikov, G. V., and Sokolov, I. M. (1988) *Pis'ma Zh. Eksp. Teor. Fiz.* **48**, 494 (*JETP Lett.* **48**, 536).

Mirlin, A. D., and Fyodorov, Y. V. (1991) *Nucl. Phys. B* **366**, 507.

—(1991a) *J. Phys. A* **24**, 2273.

—(1993) *J. Phys. A* **26**, L551.

—(1994) *Europhys. Lett.* **25**, 669; *J. Phys. (Paris)* **4**, 655.

Mirlin, A. D., Müller-Groeling, A., and Zirnbauer, M. R. (1994) *Ann. Phys.* **236**, 325.

Molchanov, S. (1981) *Comm. Math. Phys.* **78**, 429.

Montambaux, G., Bouchiat, H., Sigeti, D., and Friesner, R. (1990) *Phys. Rev. B* **42**, 7647.

Montgomery, H. L. (1973) in *Proceedings of the Symposium on Pure Mathematics*, American Mathematical Society, Providence, RI, p. 181.

—(1974) in *Proceedings of the International Congress of Mathematicians*, Vancouver, B.C., p. 379.

Moser, J. (1975) *Adv. Math.* **16**, 197.

Mott, N. F. (1972) *Phil. Mag.* **26**, 1015.

Mott, N. F., and Davis, E. A. (1971) *Electron Processes in Non-Crystalline Materials*, Clarendon Press, Oxford.

Mott, N. F., and Twose, W. D. (1961) *Adv. Phys.* **10**, 107.

Mucciolo, E. R., Prigodin, V. N., and Altshuler, B. L. (1995) *Phys. Rev. B* **51**, 1714.

Muzykantskii, B. A., and Khmelnitskii, D. E. (1995) *Phys. Rev. B* **51**, 5480.

Nagaev, E. L. (1992) *Phys. Rep.* **222**, 199.

Nagaoka, Y. (1985) in *Anderson Localization, Prog. Theor. Phys.* **84**, 1.

Nakamura, K., and Lakshmanan, M. (1986) *Phys. Rev. Lett.* **57**, 1661.

Narayan, O., and Shastry, B. S. (1993) *Phys. Rev. Lett.* **71**, 2106.

Nelson, D. R., and Pelcovits, R. A. (1977) *Phys. Rev. B* **16**, 2191.

Nishioka, H., Verbaarschot, J. J. M., Weidenmüller, H., and Yoshida, S. (1986) *Ann. Phys.* **67**, 172.

Nogami, Y., Kaneko, H., Ito, H., Ishiguro, T., Sasaki, T., Toyota, N., Takahashi, A., and Tsukamoto, J. (1991) *Phys. Rev. B* **43**, 11829.

Odlyzko, A. M. (1987) *Math. Comp.* **48**, 273

—(1989) *AT&T Bell Lab Report*.

Oh, S., Zyuzin, A. Y., and Serota, R. A. (1991) *Phys. Rev. B* **44**, 8858.

Ohtsuki, T., Kramer, B., and Ono, Y. (1993) *J. Phys. Soc. Jpn.* **62**, 224.

Ono, Y. (1982) in *Anderson Localization*, ed. Y. Nagaoka, H. Fukuyama, Springer-Verlag, Berlin.

Ono, Y., Ohtsuki, T., and Kramer, B. (1989) *J. Phys. Soc. Jpn.* **58**, 1705.

von Oppen, F. (1994) *Phys. Rev. Lett.* **73**, 798.

von Oppen, F., and Riedel, E. K. (1991) *Phys. Rev. Lett.* **66**, 84.

Ott, E. (1981) *Rev. Mod. Phys.* **53**, 655.

Pakhomov, V. F. (1974) *Mat. Zam.* **16**, 1.
Paladin, G., and Vulpiani, A. (1987) *Phys. Rep.* **156**, 147.
Pandey, A., and Mehta, M. L. (1983) *Comm. Math. Phys.* **87**, 449.
Parisi, G. (1979) *Phys. Rev. Lett.* **43**, 1754.
Parisi, G., and Sourlas, N. (1979) *Phys. Rev. Lett.* **43**, 744.
—(1981) *J. Phys. (Paris)* **41**, L403.
Pastur, L. A., and Figotin, A. L. (1978) *Theor. Math. Fiz.* **35**, 193.
Patashinskii, A. Z., and Pokrovskii, V. L. (1979) *Fluctuation Theory of Phase Transitions*, Pergamon Press, Oxford.
Pechukas, P. (1983) *Phys. Rev. Lett* **51**, 943.
Pichard, J.-L., Sanquer, M., Slevin, K., and Debray, P. (1990) *Phys. Rev. Lett.* **65**, 1812.
Pines, D., and Nozières, P. (1992) *The Theory of Quantum Liquids*, Vol. 1, Addison-Wesley, Reading, MA.
Pluhar, Z., Weidenmüller, H. A., and Zuk, J. A. (1994) *Phys. Rev. Lett.* **73**, 2115.
Polyakov, A. M. (1975) *Phys. Lett. B* **59**, 79.
Pook, W., and Janßen, M. (1991) *Z. Phys. B* **82**, 295.
Porter, C. E. (1965) *Statistical Theories of Spectra: Fluctuations*, Academic Press, New York.
Prange, R. E., and Girvin, S., eds. (1990) *The Quantum Hall Effect*, Springer-Verlag, New York.
Prigodin, V. N. (1995) *Phys. Rev. Lett.* **74**, 1566.
Prigodin, V. N., and Efetov, K. B. (1993) *Phys. Rev. Lett.* **70**, 2932.
Prigodin, V. N., Efetov, K. B., and Iida, S. (1993) *Phys. Rev. Lett.* **71**, 1230.
—(1995) *Phys. Rev. B* **51**, 17223.
Prigodin, V. N., Taniguchi, N., Kudrolli, A., Kidambi, V., and Sridhar, S. (1995) *Phys. Rev. Lett.* **75**, 2392.
Pruisken, A. M. M. (1984) *Nucl. Phys. B* **235**, 277.
—(1985) in *Localization, Interaction, and Transport Phenomena*, ed. B. Kramer, G. Bergmann, and Y. Bruynseraede, Springer-Verlag, New York.
—(1987) *Nucl. Phys. B* **285**, 719; *Nucl. Phys. B* **290**, 61.
Rice, M. J., Schneider, W. R., and Strässler, S. (1973) *Phys. Rev. B* **8**, 474.
Riemann, B. (1876) *Gesammelte Werke*, Teubner, Leipzig (reprinted by Dover, New York, 1973).
Rittenberg, V., and Scheunert, M. (1978) *J. Math. Phys.* **19**, 709.
Rodgers, G. J. and Bray, A. J. (1988) *Phys. Rev. B.* **37**, 3557.
Saher, D., Haake, F., and Gaspard, P. (1991) *Phys. Rev. A* **44**, 7841.
Salomaa, M. M., and Volovik, G. E. (1987) *Rev. Mod. Phys.* **59**, 533.
Schäfer, L., and Wegner, F. (1980) *Z. Phys. B* **38**, 113.
Schick, M. (1968) *Phys. Rev.* **166**, 401.
Schmid, A. (1991) *Phys. Rev. Lett.* **66**, 80.
Schreiber, M. (1985) *Phys. Rev. B* **31**, 6146.
—(1990) in *Localization*, ed. K. A. Benedict and J. T. Chalker, p. 65, Institute of Physics, Conference Series, Bristol.
Schreiber, M., and Grussbach, H. (1991) *Phys. Rev. Lett.* **67**, 607.
Seligman, T. H., and Nishioka, H., eds. (1986) *Quantum Chaos and Statistical Nuclear Physics, Lecture Notes in Physics*, Vol. 263, Springer-Verlag, Berlin.
Seligman, T. H., Verbaarschot, J. J. M., and Zirnbauer, M. R. (1984) *Phys. Rev. Lett.* **53**, 215.
Serota, R. A. (1992) *Sol. State Comm.* **84**, 843.
—(1994) *Mod. Phys. Lett. B* **8**, 1243.
Serota, R. A., Feng, S., Kane, C., and Lee, P. A. (1987) *Phys. Rev. B* **36**, 5031.
Serota, R. A., and Zyuzin, A. Y. (1993) *Phys. Rev. B* **47**, 6399.
Shapiro, B. (1986) *Phys. Rev. B* **34**, 4394.
—(1987) *Philos. Mag. B* **56**, 1031.
Sharvin, Y. V. (1984) *Physica B* **126**, 288.
Sharvin, D. Y., and Sharvin, Y. V. (1981) *Pis'ma Zh. Eksp. Teor. Phys.* **34**, 285 (*Sov. Phys. JETP Lett.* **34**, 281).
Simons, B. D., and Altshuler, B. L. (1993) *Phys. Rev. Lett.* **70**, 4063; *Phys. Rev. B* **48**, 5422.
Simons, B. D., Hashimoto, A., Courtney, M., Kleppner, D., and Altshuler, B. L. (1993) *Phys. Rev. Lett.* **71**, 2899.
Simons, B. D., Lee, P. A., and Altshuler, B. L. (1993) *Phys. Rev. Lett.* **70**, 4122.
—(1993a) *Nucl. Phys. B* **409** [FS], 487.
—(1994) *Phys. Rev. Lett.* **72**, 64.
Sivan, U., and Imry, Y. (1987) *Phys. Rev. B* **35**, 6074.
Sivan, U., Imry, Y., and Aronov, A. (1994) *Europhys. Lett.* **28**, 115.

Sivan, U., Milliken, F. P., Milkove, K., Rishton, S., Lee, Y., Hong, J. M., Boegli, V., Kern, D., and DeFranza, M. (1994) *Europhys. Lett.* **25**, 605.

Skyrme, T. (1961) *Proc. R. Soc. Lond.* **262**, 237.

Slichter, C. P. (1980) *Principles of Magnetic Resonance*, Springer Series in Solid State Sciences, Vol. 1, Springer-Verlag, Berlin.

Smilansky, U. (1991) in *Chaos and Quantum Physics*; see Bohigas (1991).

Smrcka, L., and Streda, P. (1977) *J. Phys. C* **10**, 2153.

Sommers, H. J., and Iida, S. (1994) *Phys. Rev. E* **49**, R2513.

Soukoulis, C. M., and Economou, E. N. (1984) *Phys. Rev. Lett.* **52**, 565.

Sridhar, S. (1991) *Phys. Rev. Lett.* **67**, 785.

Srivastava, V. (1990) *Phys. Rev. B* **41**, 5667.

Staveren van, M. P. J., Brom, H. B., and de Jongh, L. J. (1991) *Phys. Rep.* **208**, 1.

Stein, J., and Stöckmann, H.-J. (1992) *Phys. Rev. Lett.* **68**, 2867.

Stöckmann, H.-J., and Stein, J. (1990) *Phys. Rev. Lett.* **64**, 2215.

Stone, M., ed. (1992) *Quantum Hall Effect*, World Scientific, Singapore.

Stone, A. D., Mello, P. A., Muttalib, K. A., and Pichard, J.-L. (1991) in *Mesoscopic Phenomena in Solids*, ed. B. L. Altshuler, P. A. Lee, and R. A. Webb, North Holland, Amsterdam.

Stone, A. D., and Szafer, A. (1988) *IBM J. Res. Develop.* **32**, 384.

Strässler, S., Rice, M. J., and Wyder, P. (1972) *Phys. Rev. B* **6**, 2575.

Sutherland, B. (1971) *J. Math. Phys.* **12**, 246; **12**, 251; *Phys. Rev. A* **4**, 2019.

—(1972) *Phys. Rev. A* **5**, 1372.

Szafer, A. and Altshuler, B. L. (1993) *Phys. Rev. Lett.* **70**, 587.

Taniguchi, N., Andreev, A., and Altshuler, B. L. (1994) *Europhys. Lett.* **29**, 515.

Taniguchi, N., Hashimoto, A., Simons, B. D., and Altshuler, B. L. (1994) *Europhys. Lett.* **27**, 335.

Taylor, K. T., ed. (1987) *Atomic Spectra and Collisions in External Fields, Physics of Atoms and Molecules*, Plenum Press, Oxford.

Thaha, M., Blümel, R., and Smilansky, U. (1993) *Phys. Rev. E* **48**, 1993.

Thomas, G. A. (1986) *Localization and Interactions in Disordered and Doped Semiconductors*, ed. D. M. Finlays, SUSSP, Edinburgh, p. 172.

Thouless, D. J. (1974) *Phys. Rep.* **13**, 93.

—(1975) *J. Phys. C* **8**, 1803.

—(1977) *Phys. Rev. Lett.* **39**, 1167.

Trugman, S. (1983) *Phys. Rev. B* **27**, 7539.

Tsui, C. D., Stormer, H. L., and Gossard, A. C. (1982) *Phys. Rev. Lett.* **48**, 1559.

Uhlenbeck, G. E., and Ornstein, L. S. (1930) *Phys. Rev.* **36**, 823.

Umbach, C. P., Washburn, S., Laibowitz, R. B., and Webb, R. A. (1984) *Phys. Rev. B* **30**, 4048.

Vaks, V. G., Larkin, A. I., and Pikin, S. A. (1966) *Zh. Eksp. Teor. Fiz.* **51**, 361 (*Sov. Phys. JETP* **24**, 240).

Verbaarschot, J. J. M. (1988) *Nucl. Phys. B* **300**, 263.

Verbaarschot, J. J. M., Weidenmüller, H. A., and Zirnbauer M. R. (1984) *Phys. Rev. Lett.* **52**, 1597.

—(1985) *Phys. Rep.* **129**, 367.

Verbaarschot, J. J. M., and Zirnbauer, M. R. (1984) *Ann. Phys. N.Y.* **158**, 78.

—(1985) *J. Phys. A* **17**, 1093.

Viana, L., and Bray, A. Y. (1985) *J. Phys. C* **18**, 3037.

Viehweger, O., and Efetov, K. B. (1990) *J. Phys. Cond. Mat.* **2**, 7049.

—(1991) *J. Phys. Cond. Mat.* **3**, 1695; *Phys. Rev. B* **44**, 1168.

Vilenkin, N. J. (1986) *Special Functions and the Theory of Group Representations*, AMS Translation, Providence.

Volkov, D. V,. and Akulov, V. P. (1973) *Phys. Lett. B* **46**, 109.

Volkov, D. V. and Soroka, V. A. (1973) *Pis'ma Zh. Eksp. Teor. Fiz.* **18**, 529 (*Sov. Phys. JETP Lett.* **18**, 312).

Vollhardt, D. and Wölfle, P. (1980) *Phys. Rev. Lett.* **45**, 842; *Phys. Rev. B* **22**, 4666.

—(1992) *Electronic Phase Transitions*, ed. W. Hanke and Y. V. Kopaev, North Holland, Amsterdam, p. 1.

Wang, M. C., and Uhlenbeck, G. E. (1945) *Rev. Mod. Phys.* **17**, 323.

Washburn, S., and Webb, R. A. (1986) *Adv. Phys.* **35**, 375.

Weaver, R. L. (1989) *J. Acoust. Soc. Am.* **85**, 1005.

Webb, R. A., Washburn, S., Umbach, C. P., and Laibowitz, R. B. (1985) *Phys. Rev. Lett.* **54**, 2696.

Wegner, F. (1976) *Z. Phys. B* **25**, 327.

—(1979) *Z. Phys. B* **35**, 207.

—(1980) *Z. Phys. B* **36**, 209.

—(1982) *Z. Phys. B* **44**, 9.

—(1983) *Z. Phys. B* **51**, 279.

—(1985) in *Localization, Interaction and Transport Phenomena*, ed. B. Kramer, G. Bergmann, and Y. Bruynseraede, Springer-Verlag, New York, p. 99.

—(1987) *Nucl. Phys. B* **280**, 193.

—(1989) *Nucl. Phys. B* **316**, 663.

—(1990) *Z. Phys. B* **78**, 33.

Weidenmüller, H. A. (1984) *Ann. Phys. N.Y.* **158**, 120.

—(1987) *Nucl. Phys. B* **290**, 87.

Weidenmüller, H. A., and Zirnbauer, M. R. (1988) *Nucl. Phys. B* **305 [FS25]**, 339.

Weller, W., Prigodin, V. N., Firsov, Y. A. (1982) *Phys. Stat. Sol.* **110**, 143.

Wess, G., and Zumino, B. (1974) *Nucl. Phys. B* **70**, 39.

Wigner, E. P. (1951) *Ann. Math.* **53**, 36.

—(1958) *Ann. Math.* **67**, 325.

Willet, R. L., Paalanen, M. A., Ruel, R. R., West, K. W., Pfeifer, L. N., and Bishop, D. J. (1990) *Phys. Rev. Lett.* **65**, 112.

Willet, R. L., Ruel, R. R., West, K. W., and Pfeiffer, L. N. (1993) *Phys. Rev. Lett.* **71**, 3846.

Wilson, K. G., and Kogut, J. B. (1974) *Phys. Rep.* **12**, 77.

Wölfle, P., and Vollhardt, D. (1987) *The Superfluid Phases of* ^3He, Taylor and Francis, London.

Wong, K. Y., and Sherrington, D. (1987) *J. Phys. A* **20**, L785.

Yee, P., and Knight, W. D. (1975) *Phys. Rev.* **11**, 3261.

Yoon, C.-O., Park, Y. W., Akagi, K., and Shirikawa, H. (1992) in *Proceedings of the International Conference on Science and Technology of Synthetic Metals*, Göteborg, Sweden.

Yukawa, T. (1985) *Phys. Rev. Lett.* **54**, 1883.

Zakrzewski, J., and Delande, D. (1993) *Phys. Rev. E* **47**, 1650.

Zhang, S.-C., Kivelson, S., and Lee, D.-H. (1992) *Phys. Rev. Lett.* **69**, 1252.

Ziegler, K. (1982) *Z. Phys. B* **48**, 293.

—(1983) *Phys. Lett. A* **99**, 19.

Zinn-Justin, J. (1993) *Quantum Field Theory and Critical Phenomena*, Clarendon Press, Oxford.

Zirnbauer, M. R. (1986) *Nucl. Phys. B* **265 [FS15]**, 375; *Phys. Rev. B* **34**, 6394.

—(1991) *Comm. Math. Phys.* **141**, 503.

—(1992) *Phys. Rev. Lett.* **69**, 1584.

—(1993) *Nucl. Phys.* A**560**, 95.

—(1994) *Ann. Phys.* **3**, 513.

Author index

Subject index